ADVANCES IN ASTROFUNDAMENTAL PHYSICS

THE SCIENCE AND CULTURE SERIES – ASTROPHYSICS
Series Editor: A. Zichichi

International School of Astrophysics
"D Chalonge"

ADVANCES IN ASTROFUNDAMENTAL PHYSICS

Erice, Italy 1994

edited by

N. Sanchez
Observatoire de Paris, Demirm

A. Zichichi
Ettore Majorana Centre for Scientific Culture

World Scientific
Singapore • New Jersey • London • Hong Kong

Published by

World Scientific Publishing Co Pte Ltd
5 Toh Tuck Link, Singapore 596224
USA office: 27 Warren Street, Suite 401-402, Hackensack, NJ 07601
UK office: 57 Shelton Street, Covent Garden, London WC2H 9HE

British Library Cataloguing-in-Publication Data
A catalogue record for this book is available from the British Library.

International School of Astrophysics "D. Chalonge"
ADVANCES IN ASTROFUNDAMENTAL PHYSICS

ISBN-13 978-981-02-2379-3
ISBN-10 981-02-2379-X

CONTENTS

IV. GRAVITY, COSMOLOGY AND STRINGS

PREFACE

The main purpose of the Courses of the International School of Astrophysics "D. Chalonge" -in the framework of the Galileo Galilei Celebrations- is to provide an up dated understanding of the progress and current problems in the interplay between fundamental physics, astrophysics and cosmology.

In the last years, the cross-section between these fields has been increasing, both at the theoretical and experimental levels: particle physics experiments, astronomical observations, space satellite data. Such interplay fruitfully influenced research activity setting up *Astrofundamental physics*.

Topics are covered with both theory and observations, and allowing for different approaches and for different lines of research. Topics included in this volume are: dark matter, neutrinos in astrophysics, gamma-ray astrophysics, and a section devoted to gravity, cosmology and strings. Each of these topics has matter enough to constitute a Course by its own. The inter-relation between these topics is important and a source of problems at the frontiers of present knowledge and experimental limits. Latest available data are constraining theory and models in these topics. These Lecture Notes provide an occasion to review achievements, to confront theory and models with observations and among themselves, to exchange information on the latest developments and to discuss future prospects.

The Course brings together experimentalists and theoreticians, -physicists, astrophysicists and astronomers- from a variety of backgrounds. The Course is addressed to students at post-doctoral level, senior scientists and to advanced graduate students as well.

The School is named after the pioneering french astrophysicist Daniel Chalonge (1895-1977) for his work in experimental and theoretical astrophysics. Precursor and creator in France of stellar spectroscopy and precision spectrophotometry, he worked on the conception and construction of new instruments like the hydrogen tube and the microphotometer named after him. His main contributions include : study of the continuum spectrum of atomic hydrogen in stars, the basis of a spectral stellar classification based on two and three new parameters (position, size and gradient flux of the Balmer discontinuity), and accurate measurements of the atmosphere ozone layer. He was a personality of his time, one of the founders of the Institut d'

Astrophysique de Paris. He also worked at the Observatoire de Paris, at the Observatoire d'Haute Provence, and at the Jungfraujoch Scientific Station in Switzerland, where particle physics experiments-such as the Manchester-Cern collaboration-were performed.

A Museum dedicated to Daniel Chalonge was inaugurated during the first Course of this School in Setember 1991. The "Daniel Chalonge Museum, Erice the City of Science" is located at the San Domenico Institute of the Ettore Majorana Centre. It exhibits a permanent collection of pictures, documents and instruments retracing the work of D. Chalonge and the institutions where he worked.

The School is sponsored by the Italian Ministry of Education, the Italian Ministry of University and Scientific Research, the Sicilian Regional Government, the European Physical Society, the French Ministry of Foreign Affairs, the French Ministry of Higher Education and Research. The Course is part of the Galileo Galilei Celebrations taking place in Europe.

We wish to express our grateful thanks to all the lecturers who did so much to make this Course successful, and to both participants and lecturers for contributing so much to the outstanding discussions and to create such a stimulating atmosphere at the Course.

We specially thank the reception secretaries, Mrs. Nicole Letourneur, Sylvie Gordon, Nathalie Grabar, and the scientific secretaries of the Course: Luis Bettencourt, Manuela Campanelli, Antonio Campos, Miquel Dorca, Richard Easther, Maria Cristina Falvella, Massimo Giovannini, Da-Shin Lee and Carlos O. Lousto, for their efficient assistance all along the Course.

N. Sanchez
Director of the Course

I. DARK MATTER

BARYONIC DARK MATTER CONSTRAINTS AND OBSERVATIONAL PROSPECTS

Bernard Carr
Astronomy Unit, Queen Mary & Westfield College, London E1 4NS

ABSTRACT

This lecture complements and updates the lecture which I gave on baryonic dark matter during the First Course on "Current Topics in Astrofundamental Physics" in 1991. Topics covered in my earlier lecture are therefore mentioned only cursorily, although figures and tables are updated where appropriate. Topics emphasized here include: the evidence for baryonic dark matter, lensing and dynamical constraints on baryonic candidates, and the brown dwarf scenario. The 1991 article contains a more detailed discussion of the Population III scenario, the infrared background constraints on the black hole scenario, and the cooling flow scenario. The material of both lectures is covered more comprehensively in my recent article in Annual Reviews of Astronomy and Astrophysics, although some material here postdates that review.

1. Evidence for Dark Matter

Evidence for dark matter has been claimed in four different contexts: (i) there may be local dark matter in the Galactic disk with a mass comparable to that in visible form ($M_{dark}{\sim}M_{vis}$); (ii) there may be dark matter in the halo of our own and other galaxies with a mass which depends upon the (uncertain) halo radius R_h and is of order $M_{dark}{\sim}10M_{vis}(R_h/100\text{kpc})$; (iii) there may be dark mass associated with clusters of galaxies ($M_{dark}{\sim}10M_{vis}$), which may or may not derive from the original galactic halos; (iv) if one believes the inflationary scenario, there may also be smoothly distributed background dark matter, required in order that the total cosmological density have the critical value ($M_{dark}{\sim}100M_{vis}$). Baryonic dark matter may not be implicated in all of these, so one must make a clear distinction between them.

2.1 Local Dark Matter

Measurements of the stellar velocity and density distribution perpendicular to the Galactic disk provide an estimate of the total disk density. This turns out to be about 0.1 $M_{\odot}pc^{-3}$ and it has long been suspected that this exceeds the density in visible stars. *The possibility of disk dark matter is very important in the present context because - of all the dark matter problems - this is the one most likely to have a baryonic solution.* Unfortunately, the evidence is very controversial. Bahcall (1984a,b,c) used counts of F dwarfs and K giants to conclude that the density of unseen material must be at least 50% that of the visible material. He also concluded that the disk dark matter must have an exponential scale height of less than 700 pc, so that it must itself be confined to a disk. However, Bahcall assumed a particular model and Bienayme et al. (1987), using a different model, found a best-fit dark matter density of only 0.01 $M_{\odot}pc^{-3}$ and even this could be removed if the halo was slightly flattened. Knapp (1988) came to the same conclusion by studying the velocity dispersion and scale height of molecular hydrogen.

Further doubt was cast in a series of papers by Kuijken & Gilmore (1989a,b,c), who used the full distribution function for the velocities and distances of K dwarfs rather than assuming a particular model. They argued that no local dark matter is required if there is a thick disk. Although Gould (1990) used a maximum likelihood analysis to conclude that Kuijken & Gilmore's data were not inconsistent with the Bahcall et al. claim, Kuijken & Gilmore (1991) disagreed with this. More recently Bahcall et al. (1992a) have concluded from another analysis of K giants that the no-disc-dark-matter hypothesis is only consistent with the data at the 14% level and their best-fit model has a dark density of 0.15 $M_{\odot}pc^{-3}$, which corresponds to 53% more dark matter than visible matter. For present purposes the existence of disk dark matter will be regarded as an open question.

1.2 Spiral Galaxies.

The best evidence for dark matter in galaxies comes from the rotation curves of spirals, the dependence of the rotation speed V upon galactocentric distance R being a measure of the density profile $\rho(R)$. An important feature of our own and many other spiral galaxies is that the rotation speed, after an initial rise, remains approximately constant with increasing R (Rubin et al. 1980). This implies that the mass within radius R increases like R, which is faster than the increase of visible mass. Although the dark matter does not dominate within the optical galaxy (at least for bright galaxies), neutral hydrogen observations suggest that V continues to remain constant well beyond the visible stars (Sancisi & van Albada 1987). *In considering the baryonic contribution to galactic halos, the crucial issue is how far the halos extend.* For our galaxy the minimum halo radius consistent with rotation curve measurements, the local escape speed and the kinematics of globular clusters and satellite galaxies is 35 kpc but the dynamics of the Magellanic Stream and the Local Group of galaxies may require a halo radius of 70kpc (Fich & Tremaine 1991). We will see later that these values are marginally consistent with a baryonic halo. However, Zaritsky et al. (1993) argue from observations of satellite systems that spirals typically have 200 kpc halos and this would not be.

Another relevant issue concerns the *roundness* of galactic halos. If galactic halos are baryonic, one would expect their formation to involve dissipation, in which case they should be flatter than in the non-baryonic case: N-body experiments show that dissipationless collapse does gives some flattening but the resultant halos are rarely flatter than E6 (Frenk et al. 1988, Dubinski & Carlberg 1991). Thus evidence for halos flatter than this would be evidence for baryonic dark matter. Polar ring galaxies probably provide the best probe of halo shape and these do seem to indicate flattening (Whitmore et al. 1987), sometimes (eg. for NGC 4650A) as high as E6 (Sackett & Sparke 1990), but it is not clear that it is ever more than this. Note that the model discussed in Combes' lecture requires the halo dark matter to be distributed in a disk.

1.3 Elliptical Galaxies.

The mass distribution in ellipticals can be probed by measuring the velocity dispersion of the stars and globular clusters. Unfortunately, the velocities do not determine the density profile uniquely and this method gives no evidence for dark matter within the *central* regions of ellipticals (de Zeeuw 1992), although the dynamics of globular clusters does provides evidence for dark matter around M87 (Huchra & Brodie 1987, Mould et al.1990). The best information therefore comes from X-ray observations of hot gas. These do, in fact, provide evidence for dark matter and, in many cases, one finds the same $M \sim R$ law which characterizes spirals (Forman et al. 1985, Sarazin 1986). The evidence for dark matter in ellipticals is not as good as it is for spirals but it is still impressive. Giant ellipticals are sometimes the focus of cooling flows and this suggests that at least some of the dark matter in ellipticals may be baryonic (Fabian 1994).

1.4 Dwarf Galaxies

Some of the dwarf irregulars are extremely gas-rich, which means that their HI rotation curves can be traced to many optical scale lengths. Many of them seem to have much higher dark mass fractions than bright spirals, with their dark halos dominating even within the optical regions. Particularly striking examples are DDO 154 (Carignan & Freeman 1988), for which the dark-to-luminous mass ratio exceeds 10 at the last measured point of the rotation curve, GR 8 (Carignan et al. 1990) and DDO 170 (Lake et al.1990). Dwarf spheroidals also seem to have dark halos (Lin & Faber 1983, Aaronson 1983, Aaronson & Olszewski 1987). This claim is based on measurements of velocity dispersions and tidal radii for the six dwarf spheroidals within the Local Group. *The presence of dark matter in dwarf galaxies is crucial in the present context because it requires that halos consist either of baryonic or cold non-baryonic dark matter.* Lake (1990) has argued that the observations are more consistent with the first possibility: for if the formation of the halos was dissipationless, their central densities imply that they must have formed at a redshift exceeding 30, whereas they should form at a redshift of 10 in the Cold Dark Matter scenario.

1.5 Groups and Clusters of Galaxies

Galaxies are clumped on various scales (as members of binaries, small groups and rich clusters) and velocity dispersion measurements indicate that the dynamical mass exceeds the visible mass on all these scales. Binaries can only be studied statistically (because one does not know the orbital inclination in any individual case), so the data is less clear-cut here, but there is compelling evidence for dark mass in clusters of galaxies. This is confirmed by X-ray data on the gas temperature (which provide an independent measure of the gravitational potential). In rich clusters, the dark mass dominates by at least a factor of 10 and the recent discovery of hot gas in two *small* groups of galaxies (HG92 and HCG62) by ROSAT shows that there are comparable amounts of dark matter there (Mulchaey et al. 1993, Ponman and Bertram1993). Further evidence for dark matter in clusters comes from the lensing of background galaxies. The galaxies are distorted into arclets by the cluster potential and the properties of these arclets can be used to infer the dark matter distribution (Tyson et al. 1990).

In assessing whether the dark mass in groups and clusters can be baryonic, it is important to determine whether it is the same as the halo dark matter. Although the cluster dark mass cannot all be associated with individual galaxies *now* - else dynamical friction would result in the most massive galaxies being dragged into the cluster centre (White 1976) - it may still have derived from the galaxies *originally*. Indeed in the hierarchical clustering picture one would expect the galaxies inside a cluster to be stripped of their individual halos to form a collective halo (White & Rees 1978). However, this would only suffice to explain all the cluster dark matter if the original galactic halos were larger than about 200 kpc and, in this case, we will see that they could probably not be purely baryonic.

1.6 Background Dark Matter.

None of the forms of matter discussed above can have the critical density required for the Universe to recollapse: $\rho_{crit} = 3H_0^2/8\pi G = 2x10^{-29}h^{-2}$ g cm^{-3} where $h = H_0/(100$ km s^{-1}Mpc$^{-1})$. However, according to the currently popular inflation theory (Guth 1981), in which the Universe undergoes an exponential expansion phase at some early time, the total density should have almost exactly the critical value (Ω=1). [See, however, Ellis (1988) and Ellis et al. (1991) for a different point of view.] This would have two possible implications: either there is another dark component or galaxy formation is biased (Kaiser 1984, Dekel & Rees 1987) in the sense that galaxies form preferentially in just a small fraction of the volume of the Universe. The second possibility avoids a proliferation of dark matter species but some people now invoke this anyway (Taylor & Rowan-Robinson 1992).

In either case, one would expect the mass-to-light ratio to increase as one goes to larger scales and there is some indication of this from dynamical studies. One can probe the density on scales above 10 Mpc, for example, by analysing large-scale streaming motions (Dressler et al. 1987, Bertschinger & Dekel 1989) or by determining the dipole moment of the IRAS sources (Rowan-Robinson et al. 1990). In all these analyses, the inferred density depends on the bias parameter b (dynamical effects depending on the product $\Omega^{0.6}b^{-1}$). More sophisticated analyses are needed to determine Ω and b separately (Peacock & Dodds 1993, Nusser & Dekel 1993). It is interesting that the IRAS dipole suggests a critical density if the IRAS sources are unbiased (b=1). For a recent review of the evidence for and against Ω=1, see Coles & Ellis (1994).

2. Baryonic versus Non-Baryonic Dark Matter

The main argument for both baryonic and non-baryonic dark matter is associated with Big Bang nucleosynthesis. This is because the success of the standard picture in explaining the primordial light element abundances [viz. $X(^4He)\approx 0.24$, $X(^2D)\sim X(^3He)\sim 10^{-5}$, $X(^7Li)\sim 10^{-10}$] only applies if the baryon density parameter Ω_b is strongly constrained. Walker et al. (1991) find that it must lie in the range

$$0.010h^{-2} < \Omega_b < 0.015h^{-2} \qquad (1)$$

the upper and lower limits coming from the upper bounds on ^{4}He and ^{2}D+^{3}He, respectively. The upper limit implies that Ω_b is well below 1, which suggests that no baryonic candidate could provide the critical density required in the inflationary scenario. This conclusion also seems to apply if one invokes inhomogeneous nucleosynthesis since one requires $\Omega_b < 0.09h^{-2}$ even in this case (Mathews et al. 1993). The standard scenario therefore assumes that the total density parameter is 1, with only the fraction given by eqn (1) being baryonic.

On the other hand, the value of Ω_b allowed by eqn (1) almost certainly exceeds the density of visible baryons Ω_V. A careful inventory by Persic & Salucci (1992) shows that the contributions to Ω_v are 0.0007 in spirals, 0.0015 in ellipticals and spheroidals, $0.00035h^{-1.5}$ in hot gas within an Abell radius for rich clusters, and $0.00026h^{-1.5}$ in hot gas out to a virialization radius in groups and poor clusters. This gives a total of $(2.2+0.6h^{-1.5})\times 10^{-3}$, so eqn (1) implies the fraction of baryons in dark form must be in the range 80% to 95% for $0.5 < h < 1$. It therefore seems that one needs both non-baryonic and baryonic dark matter.

Various provisos should be stressed at this point. Firstly, the Persic-Salucci estimate does not include any contribution from dwarf galaxies (Bristow & Phillipps 1993), low surface brightness galaxies, or neutral gas in the form of Lyman-α clouds. Although the density parameter associated with "damped" clouds is probably no more than $0.003h^{-2}$ (Lanzetta et al. 1991), comparable to the density in galaxies and therefore consistent with the idea that these are protogalactic discs, the density associated with undamped systems is unknown and could be much larger. Secondly, the discrepancy between Ω_b and Ω_v could be resolved if the missing baryons were in a hot intergalactic medium. In this case the temperature would need to be finely tuned since the Gunn-Peterson test requires $\Omega(HI)<10^{-8}h^{-1}$(Sargent & Steidel 1990) and the COBE limit on the Compton distortion of the microwave background ($y<3\times10^{-5}$) requires $\Omega(HII)<0.1(T/10^8K)^{-1}h^{-1}$ (Mather et al. 1994), but there is still a temperature range between 10^4K and 10^7K in which one could have $\Omega_{IGM}\sim\Omega_b$. Note that recent measurements of the deuterium abundance in quasar absorption systems appear to give a value around 10^{-4} and this is an order of magnitude larger than is usually assumed. In this case, the upper limit in eqn (1) is reduced to $0.005h^{-2}$, which is only marginally larger than the Persic-Salucci estimate of Ω_v. However, the evidence for this now looks weak. The issue is discussed in more detail in Hogan's lecture.

Given that there seems to be a need for *some* baryonic dark matter, which of the dark matter problems mentioned in §1 could this explain? It would certainly suffice to explain the alleged dark matter in the Galactic disk: for even if all disks have the 60% dark component envisaged for our Galaxy by Bahcall et al. (1992), this only corresponds to $\Omega_v\approx0.001$ - well below the value required by eqn (1). The more interesting question is whether the baryonic density could suffice to explain the dark matter in galactic halos. If our galaxy is typical, the density associated with galactic halos would be $\Omega_h\approx0.01h^{-1}(R_h/35\text{kpc})$ where R_h is the halo radius. Thus eqn (1) implies that *all* the dark matter in our halo could be baryonic only for $R_h < 50h^{-1}$kpc. We saw in §1.2 that the minimum size for our halo is 70 kpc, which would just be compatible with this. If it is larger, the baryonic fraction could be at most $(R_h/50h^{-1}\text{kpc})^{-1}$. The cluster dark matter has a density $\Omega_c\approx0.1$ and eqn (1) implies that this cannot be purely baryonic unless one invokes inhomogeneous nucleosynthesis. The various values of Ω required by these different arguments are summarized in Figure (1).

If the missing baryons are inside galactic halos, they cannot be in the form of hot gas else they would generate too many X-rays. This suggests the possibility - the main focus of the rest of this lecture - that the dark baryons have been processed into stellar remnants and the term "Massive Compact Halo Object" or "MACHO" has been coined in this context. This contrasts with the possibility that the halo dark matter is non-baryonic and the term "Weakly Interacting Massive Particle" or "WIMP" is then used. Recently, Pfenniger et al. (1994) have argued that halos could be baryonic but in the form of a disk of cold molecular gas clouds. The term "Disk Aggregated Molecular Ensemble" or "DAME" might be used for such objects. This model would explain why the surface density ratio of dark matter and HI gas is constant outside the optical disk (Carignan et al. 1990) and is discussed in detail in Combes' lecture.

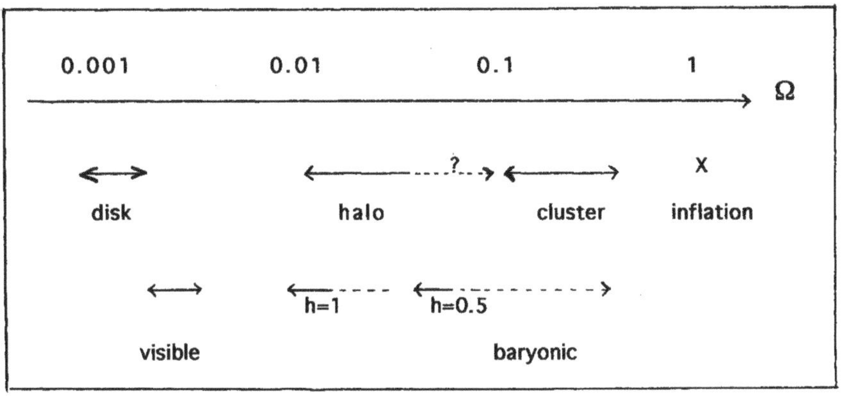

Figure (1). This compares the values of Ω associated with the various dark matter problems to the baryonic density required by cosmological nucleosynthesis (with the inhomogeneous case shown dotted) and the density in visible form. The halo density depends on the typical halo radius and this must exceed 200 kpc to reach the cluster density.

2.4 Variants of the Baryonic Dark Matter Scenario

One may consider several variants of the BDM scenario, depending on how strongly one wishes to retain the three standard cosmological assumptions: (1) homogeneous primordial nucleosynthesis, (2) inflation and (3) non-baryonic dark matter.

* In what we may term the *standard* BDM scenario, one retains all three assumptions, which requires $\Omega_b \approx 0.01h^{-2}$ and $\Omega=1$. In this case, we conclude that 80% to 95% of baryons are dark but MACHOs alone can provide galactic halos only if $R_h<50h^{-1}$kpc. The cluster and critical density dark matter must be WIMPs and, if they are cold, the halo dark matter is expected to be a mixture of MACHOs and WIMPs.

* In the *modified* BDM scenario, one still invokes non-baryonic dark matter but one drops either assumption (1) or (2). This is motivated by the fact that recent X-ray data suggest that the ratio of visible baryon mass (i.e. the mass in the form of stars and hot gas) to total mass in clusters is anomalously high compared to the mean cosmic ratio

implied by eqn (1) (White et al. 1993). Unless one invokes a cosmological constant, this suggests either that the cosmological density is well below the critical value or that the baryon density is higher than implied by the homogeneous nucleosynthesis scenario.

* In the *maximal* BDM scenario, one assumes that the Universe has a critical density of baryons ($\Omega_B=\Omega=1$), thereby relinquishing the need for non-baryonic dark matter without giving up inflation (Fowler 1990). The scenario is inconsistent with homogeneous nucleosynthesis unless one invokes unrealistically low values of H_o. When the upper limit on Ω_b was $0.06h^{-2}$ s (Yang et al. 1984), it was possible to have $\Omega_b=1$ by invoking the only moderately extreme value $H_o=25$ (Shanks 1985) but the new upper limit would require $H_o=10$, which is probably absurd.

* In the *baryon-dominated* scenario, one only assumes the existence of the dark matter for which there is direct dynamical evidence and attributes this solely to baryons ($\Omega_b=\Omega\approx0.1$). In this case, one has to give up all three standard assumptions. In order to explain the observed light element abundances, one then has to invoke some exotic astrophysical process, such as the spallation of primordial helium by high energy photons from accreting black holes (Gnedin & Ostriker 1992). The viability of this scenario also depends on whether the isocurvature baryon-dominated model (Peebles 1987) is compatible with the CMB anisotropy constraints.

These variants of the BDM scenario are summarized in Table (1). Most of the emphasis in this review will be on the standard BDM scenario but, in assessing which baryonic candidates are viable, it is important to bear in mind the more radical proposals.

Table (1). This indicates which standard assumptions are dropped in each BDM scenario. HCN = Homogeneous Cosmological Nucleosynthesis; NBDM = Non-Baryonic Dark Matter.

SCENARIO	HCNS	INFLATION	NBDM
Standard BDM	√	√	√
Modified BDM	X	√	√
	√	X	√
Maximal BDM	X	√	X
Baryon-dominated	X	X	X

3. Population III Stars

There is some confusion in the literature because the term "Population III" has been used in two distinct contexts. It has been applied to describe: (1) the stars which generate the first metals; and (2) the stars hypothesized to provide the dark matter in galactic halos. Stars of type (1) must exist since heavy elements can only be generated through stellar nucleosynthesis. However, the most natural assumption is that these are merely the ones at the high mass end of the Population II mass spectrum, since in this case they would generate the first metals because they evolved fastest. Henceforth I will use the term "Population III" specifically in sense (2). Although there are no observations which unambiguously *demand* that most of the baryons were processed through Population III stars, there are theoretical reasons for anticipating their formation. This is because the existence of galaxies and clusters of galaxies implies that there must have been density fluctuations in the early Universe and, in many scenarios, these fluctuations would also give rise to a population of pregalactic clouds. The precise way in which this occurs depends on the nature of the fluctuations and the nature of the dominant dark matter, as reviewed in my 1991 lecture.

3.1 Expected Mass of Population III Stars

The question arises of what happens to these pregalactic clouds. They could face various possible fates. They might just turn into ordinary stars and form objects like globular clusters. On the other hand, the conditions of star formation could have been very different at early times and several alternatives have been suggested.

* Some people argue that the first stars could have been much *smaller* than at present. Fairly general arguments suggest that the minimum fragment mass could be as low as 0.007 $M_\odot$ (Low & Lynden-Bell 1976, Rees 1976) and it is possible that conditions at early epochs - such as the enhanced formation of molecular hydrogen (Palla et al. 1983, Yoshii & Saio 1986) - could allow the formation of even smaller objects. One might also invoke the prevalence of high pressure pregalactic cooling flows (Ashman & Carr 1988, Thomas & Fabian 1990, Ashman & Carr 1991), analagous to the cluster flows observed at the present epoch (Fabian 1994) but on a smaller scale.

* Other people argue that the first stars could have been much larger than at present. For example, the fragment mass could be increased before metals formed because cooling would be less efficient (Silk 1977). There is also observational evidence that the IMF may become shallower as metallicity decreases (Terlevich 1985), thereby increasing the fraction of high mass stars. Another possibility is that

the characteristic fragment mass could be increased by the effects of the microwave background (Kashlinsky & Rees 1983) or by the absence of substructure in the first bound clouds (Tohline 1980).

* One could also get a mixture of small and large stars. For example, Cayrel (1987) has proposed that one gets the formation of massive exploding stars in the core of the cloud, followed by the formation of low mass stars where the gas swept up by the explosions encounters infalling gas. Kashlinsky & Rees (1983) have proposed a scheme in which angular momentum effects lead to a disk of small stars around a central very massive star. Salpeter & Wasserman (1993) have a scenario in which one gets clusters of neutron stars and asteroids.

* In the baryon-dominated isocurvature scenario, with highly non-linear fluctuations on small scales, the collapse of the first overdense clouds depends on the effects of radiation diffusion and trapping. Hogan (1993) finds that sufficiently dense clouds collapse very early into black holes with a mass of at least $1M_{\odot}$, while clouds below this critical density delay their collapse until after recombination and may produce neutron star or brown dwarf remnants. One of the attractions of this idea is that it allows a baryon density parameter higher than indicated by eqn (1) because the nucleosynthetic products in the high density regions are locked up in the remnants, leaving the the products from the low density regions outside.

* It is possible that the first clouds collapse directly to supermassive black holes (Gnedin & Ostriker 1992). Usually clouds will be tidally spun up by their neighbours as they become gravitationally bound and the associated centrifugal effects then prevent direct collapse. However, just after recombination, Compton drag could prevent this tidal spin-up, especially if the gas becomes ionized or contains dust (Loeb 1993, Umemura et al. 1993). Even if rotation is important, one could still get a supermassive disc which slowly shrinks to form a black hole due to angular momentum transport by viscous effects (Loeb & Rasio 1993). One might even end up with a supermassive binary.

This discussion indicates that, while there is clearly considerable uncertainty as to the fate of the first bound clouds, they could well fragment into stars which are larger or smaller than the ones forming today. As we will see in the next section, one certainly needs this if they are to produce a lot of dark matter. We note that there is no necessity for the Population III stars to form *before* galaxies, just as long as some change in the conditions of star formation makes the IMF different from what it is today. It is possible that the Population III clouds just remain in purely gaseous form and become Lyman-α clouds (Rees 1986), in which case the formation of the dark-matter-producing stars would need to be postponed until the epoch of galaxy formation. Nevertheless, there is at least the possibility that the Population III stars were pregalactic and this would have various attractions. For example, it would permit the Universe to be reionized at high redshifts, thereby hiding small-scale anisotropies in the microwave background (Gouda & Sugiyama 1992), it might explain why the intergalactic medium is ionized back to redshifts of at least 5 (Schneider et al. 1991), and it might account for the substantial heavy element abundances in intracluster gas at low redshifts (Hatsukade 1989) and in intergalactic clouds at redshifts above 3 (Steidel & Sargent 1988).

We note that the epoch of Population III formation will be very important for the relative distribution of baryonic and non-baryonic dark matter, especially if the non-baryonic dark matter is "cold" so that it can cluster in galactic halos. In this case, if the Population III stars form *before* galaxies, one might expect their remnants to be distributed throughout the Universe (White & Rees 1978) with the ratio of the baryonic and non-baryonic densities being the same everywhere and of order 10. If they form at the same time as galaxy formation, perhaps in the first phase of protogalactic collapse, one would expect the remnants to be confined to halos and clusters. In this case, their contribution to the halo density could be larger since the baryons would probably dissipate and become more concentrated. Angular momentum considerations suggest that the local baryon fraction must be increased by at least a factor of 10 (Fall & Efstathiou 1981). If the WIMPs are hot and cannot cluster in halos, halos would consist exclusively of MACHOs.

3.2 The Evolution of Population III Stars

In order to calculate the constraints on Population III stars, we need to know how stars in the different mass ranges evolve. This is summarized in Table (2), which also gives the mass fraction ϕ which ends up in dark relicts.

* *Low Mass Objects (LMOs).* Stars in the range 0.08-0.8 $M_\odot$ are still on the main-sequence but objects in the range 0.001-0.08 $M_\odot$ would never burn hydrogen. Such brown dwarfs (BDs) represent a balance between gravity and degeneracy pressure. Those above 0.01 $M_\odot$ could still burn deuterium; Shu et al. (1987) have argued that this may represent a lower limit to a BD's mass but this conclusion is not definite. Objects below 0.001$M_\odot$ are held together by intermolecular rather than gravitational forces (i.e. they have atomic density) and may be described as snowballs.

* *Massive Objects (MOs).* Stars in the range 0.8-4 $M_\odot$ would leave white dwarf remnants, while those between 8 $M_\odot$ and some mass M_{BH} would leave neutron star remnants. In either case, the remnants would eventually cool and become dark. (Stars in the mass range 4-8 $M_\odot$ could

be disrupted entirely during their carbon-burning stage but this is not certain.) Stars larger than some mass M_{BH} could evolve to black holes; estimates of M_{BH} range from $25M_\odot$ (Maeder 1993) to $50M_\odot$ (Schild & Maeder 1985). Since white dwarfs and neutron stars ars the most abundant remnants of stars forming at the present epoch, some people favour them as dark matter candidates (Ryu et al. 1990, Silk 1993).

* *Very Massive Objects (VMOs).* Stars in the mass range above 100 $M_\odot$ would experience the pair-instability during their oxygen-burning phase (Fowler & Hoyle 1964). This would lead to disruption below some mass M_C but complete collapse above it (Woosley & Weaver 1982, Ober et al. 1983, Bond et al. 1984). VMO black holes may therefore be more plausible dark matter candidates than ordinary stellar black holes. In the absence of rotation, $M_C=200M_\odot$ but it could be as high as $2x10^4M_\odot$ if rotation were maximal (Glatzel et al. 1985). Note that stars with an initial mass above $100M_\odot$, are radiation-dominated and therefore unstable to pulsations during hydrogen-burning. These pulsations lead to considerable mass loss but are unlikely to be completely disruptive.

**Supermassive Objects (SMOs).* Metal-free stars larger than $10^5M_\odot$ would collapse directly to black holes on a timescale $10^4(M/10^5M_\odot)^{-1}y$ due to general relativistic instabilities before any nuclear burning (Fowler 1966). They would therefore have no nucleosynthetic consequences, although they could explode in some mass range above $10^5M_\odot$ if they had non-zero metallicity (Fuller et al. 1986). SMOs would also generate very little radiation, emitting only 10^{-11} of their rest-mass energy in photons. The existence of SMOs is rather less speculative than that of VMOs since supermassive black holes are thought to reside in some galactic nuclei and to power quasars.

3.3 Constraints on Population III Stars

Figure (2) summarizes the constraints on $\Omega_*(M)$, the density of stars of mass M in units of the critical density, which can be placed on Population III stars and their remnants by considering their dynamical effects, their generation of background light and their nucleosynthetic consequences. This is an updated version of Figure (6) from Carr et al. (1984). The constraints on LMOs come mainly from microlensing and infrared searches; the former only become important at masses too low to appear in Figure (2) (see §4.5). The main constraints on MOs come from considering their nucleosynthetic consequences; the enrichment limits shown in Figure (2) are somewhat different from those in Carr et al. (1984), mainly because it now seems possible that 4-8 $M_\odot$ stars do not explode. This would permit a small mass range in which white dwarfs could provide halos. The constraints on SMOs come mainly from dynamical considerations and these are discussed in detail in §5.

Table (2). This shows the fate and type of remnant for stars of different mass

Mass ($M_\odot$)	Fate	Remnant	Name
	Collapse from GR instability before H-burning	Black hole (ϕ=1)	SMO
10^5			
	Collapse due to pair-instability during O-burning	Black hole (ϕ=0.5-1)	VMO
200			
	Explode due to pair-instability during O-burning	None (ϕ=0)	
100			
	Core collapse + envelope ejection after nuclear burning	Black hole (ϕ<0.2)	MO
20-50			
	Core collapse + envelope ejection after nuclear burning	Neutron star (ϕ<0.2)	
8			
	May explode due to degenerate carbon ignition	None (ϕ=0)	
4			
	Stable nuclear burning until collapse at carbon/helium stage	White dwarfs (ϕ=0.2-0.5)	
0.8			
	Still undergoing H-burning but very dim if small	M-dwarfs (ϕ=1)	LMO
0.08			
	Never ignite hydrogen degeneracy-supported	Brown dwarf (ϕ=1)	
0.001			
	Solid objects evaporate if too small	Snowball (ϕ=1)	

The main constraint on VMOs comes from background light limits. As discussed in my 1991 lecture, these are now stronger than indicated by Carr et al. (1984) because of the COBE results. If the radiation from VMOs is affected only by cosmological redshift, then it would presently be in the near-IR. In this case, the DIRBE constraints imply that the VMOs could only have the density required to explain galactic halos if they burn sufficiently early ($z>200$) for their light to be redshifted beyond 10μ where it would be hidden by interstellar dust emission. However, in many circumstances, one would expect the VMO light to be reprocessed into the submillimetre band by pregalactic dust (Bond et al. 1986, 1991), in which case the strong constraints on the spectral distortion of the microwave background imposed by FIRAS exclude almost all scenarios. Indeed Wright et al. (1994) claim that only one of the models considered by Bond et al. (1991) still survives.

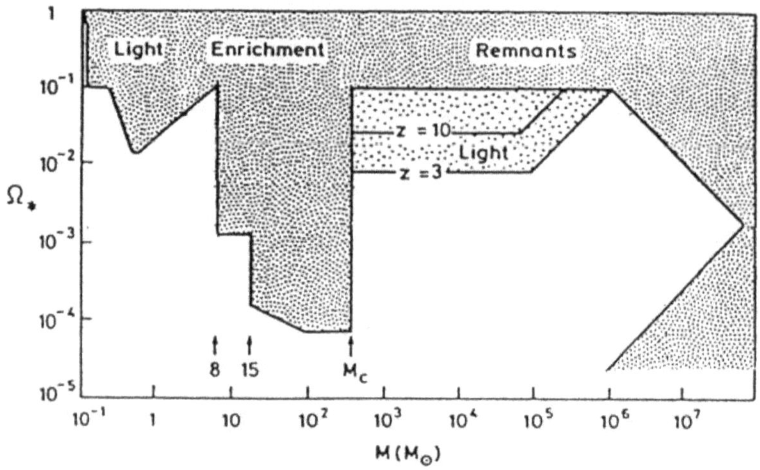

Figure (2). This summarizes the various constraints on the density of Population III stars of mass M. All the limits except the "light" one assume that the objects are inside galactic halos (assumed to have density Ω_h=1). The light limit is only interesting for stars which burn at z<30 unless the light is reprocessed by dust. The nucleosynthesis limits for M>$10^2 M_\odot$ assume that the fraction of mass lost during the hydrogen and helium burning phase is not too large.

4. Gravitational Lensing Effects

One of the most useful signatures of baryonic dark matter candidates is undoubtedly their gravitational lensing effects. Indeed it is remarkable that lensing could permit their detection over the entire mass range $10^{-7}M_\odot$ to $10^{12}M_\odot$. All sorts of astronomical objects can serve as lenses (Blandford & Narayan 1992) but the crucial advantage of Population III objects is that they are compact and spherically symmetric and this makes their effects very clean. To search for them, one requires sources which are numerous, small, bright and with predictable intrinsic variations. The most useful sources to date have been quasars, galaxies, radio jets, gamma-ray bursts and stars; all of these are discussed below. Other possibilities include radio sources, supernovae and pulsars. There are two distinct lensing effects and these probe different but nearly overlapping mass ranges: *macrolensing* (the multiple-imaging of a source) can be used to search for objects larger than $10^4M_\odot$, while *microlensing* (modifications to the intensity of a source) can be used for objects smaller than this. The current constraints on the density Ω_C of compact objects in various mass ranges are brought together in Figure (3).

4.1 Macrolensing by Compact Objects

If one has a population of compact objects with mass M and density parameter Ω_C, then the probability of one of them image-doubling a source at redshift $z \approx 1$ and the separation between the images are given by

$$P \approx (0.1\text{-}0.2)\,\Omega_C, \qquad \theta \approx 6\times10^{-6}(M/M_{\odot})^{1/2}\,h^{1/2}\ \text{arcsec} \qquad (2)$$

(Press & Gunn 1973). One can therefore use upper limits on the frequency of macrolensing for different image separations to constrain Ω_C as a function of M. The usual approach is to derive the "detection volume", defined as the volume between the source and observer within which the lens would need to lie in order to produce an observable effect (Nemiroff 1989, Kassiola et al. 1991). Limits are then obtained by adding the detection volume for each source and comparing this to the volume per source expected for a given Ω_C.

There have been several optical and radio surveys to search for multiply-imaged quasars (Hewitt et al. 1989, Bahcall et al. 1992b). In particular, Hewitt (1986) used VLA observations to infer $\Omega_C(10^{11}\text{-}10^{13}M_{\odot})<0.4$, Nemiroff (1991) used optical QSO data from Crampton et al. (1989) to infer $\Omega_C(M>10^{9.9}M_{\odot})<1$ and $\Omega_C(M>10^{10.3}M_{\odot})<0.25$, and Surdej et al. (1993) used data on 469 highly luminous quasars (including Hubble Space Telescope observations) to infer $\Omega_C(10^{10}\text{-}10^{12}M_{\odot})<0.02$. To probe smaller scales, one must use high resolution radio sources: Kassiola et al. (1991) have used lack of lensing in 40 VLBI objects to infer $\Omega_C(10^{7}\text{-}10^{9}M_{\odot})<0.4$, while a study by Patnaik et al. (1992) of 200 flat spectrum radio sources may lead to a limit $\Omega_C(10^{6}\text{-}10^{9}M_{\odot})<0.01$ (Henstock et al. 1993).

Future observations could strengthen these constraints considerably: speckle interferometry could push $\Omega_C(10^{8}\text{-}10^{10}M_{\odot})$ down to 0.01, while VLBA could push $\Omega_C(10^{5}\text{-}10^{8}M_{\odot})$ down to 0.001 (Surdej et al. 1993). These two limits are shown as broken lines in Figure (3). Another interesting possibility is to search for lensing distortions in radio jets (Kronberg et al. 1991); this would permit the detection of objects with mass around $10^{6}M_{\odot}$ since the Einstein radius for such objects is of order milliarcsecs and therefore comparable to the characteristic jet scale. Of course, jets may be intrinsically kinky but Wambsganss & Paczynski (1992) have pointed out that this poses no problem if one uses VLBI and VLBA maps of the jets in image-doubled quasars since only one of the images would then be kinked. Their numerical simulations show that the effects of supermassive black holes would be numerous and obvious.

4.2 Microlensing in Macrolensed Quasars

Even if a lens is too small to produce resolvable multiple images of a source, it may still induce detectable intensity variations. In particular, one can look for microlensing in quasars which are already macrolensed. This possibility arises because, if a galaxy is suitably positioned to image-double a quasar, then there is also a high probability that an individual halo object will traverse the line of sight of one of the images (Gott 1981) and this will give intensity fluctuations in one but not both images. The effect would be observable for objects bigger than $10^{-4}M_\odot$ but the timescale of the fluctuations, being of order $40(M/M_\odot)^{1/2}$y, would only show up over a reasonable time for $M<0.1M_\odot$.

There is already evidence of this effect for the quasar 2237+0305 (Irwin et al. 1989). This has four images at a redshift of 1.7 and the lens is a galaxy at redshift 0.04. The brightest image brightened by 0.5 magnitudes from September 1987 to August 1988 and then dimmed by 0.15 magnitudes by September 1988. There was no variation in the other images, even though the difference in light-travel time is only hours. The observed timescale for the variation indicates a mass in the range $0.001M_\odot$ to $0.1M_\odot$, although Wambsganss et al. (1990) argue that it might be as high as $0.5M_\odot$, the mass where a standard IMF gives the dominant contribution. (The variable image is almost exactly aligned with the centre of the lensing galaxy, where the density should be dominated by ordinary stars.) Analysis of more extensive data (Corrigan et al. 1991) has strengthened the evidence for microlensing with a mass below $0.1M_\odot$ (Webster et al. 1991).

4.3 The Effect of Microlensing on Quasar Luminosity

Evidence for the microlensing of unmacrolensed quasars may come from studying their luminosity variations (Refsdal & Starbell 1991, Lewis et al. 1993) and there may already be cases of this. In particular, Nottale (1986) claims that lensing by low mass objects may explain some optically violently variable quasars. For example, the quasar 0846+51 brightened by 4 magnitudes in a month and then dimmed by 1 magnitude in a few days. The fact that its line of sight is only 12 arcsec from a galaxy suggests that the variation may result from microlensing by one of the halo objects, in which case the mass of the halo object must be in the range 10^{-4} to $10^{-2}M_\odot$.

More dramatic but no less controversial evidence for the effect of microlensing on quasar luminosity comes from Hawkins (1993), who has been monitoring 300 quasars in the redshift range 1-3 over the last 17 years using a wide field Schmidt camera. He finds quasi-sinusoidal variations of amplitude 0.5m on a timescale 5 y and he attributes this to lenses with mass $\sim 10^{-3}M_\odot$. The crucial point is that the timescale decreases with increasing z, which is the opposite to what one would expect for intrinsic variations (and these would be on a shorter timescale anyway). The timescale also increases with the luminosity of the quasar. He tries to explain this by noting that the luminosity should increase with the size of the accretion disk but this only works if the disk is larger than the Einstein radius of the lens (about 0.01 pc) and this is questionable. Another worrying feature of Hawkins' claim (cf. Schneider 1993) is that he requires the density of the lenses to be close to critical (so that the sources are being transited continuously). In this case, Big Bang nucleosynthesis constraints require the lenses to be non-baryonic, so he is forced to invoke primordial black holes.

4.4 Line-to-continuum Effects for Quasars

In some circumstances, only part of the quasar may be microlensed. In particular, the line and continuum fluxes may be affected differently because they may come from regions which act as extended and pointlike sources respectively. (For a lens at a cosmological distance the Einstein radius is $0.05(M/M_\odot)^{1/2}h$ pc, whereas the size of the optical continuum and line regions are of order 10^{-4} pc and 0.1-1 pc respectively.) This can be used to probe *individual* sources. For example, the variations in the line-to-continuum ratio for different images of the same macrolensed quasar can be used to constrain the mass of the objects in the lensing galaxy. Evidence for such an effect may already exist in the case of the double quasar 2016+112, where variations in the intensity ratios for the different images suggest that the lensing objects have a mass in the range $3\times10^4 M_\odot$ to $3\times10^7 M_\odot$ (Subramanian & Chitre 1987).

The line-continuum effect can also show up in *statistical* studies of many quasars and there is one particularly important effect in this context. One would expect the characteristic equivalent width of quasar emission lines to decrease as one goes to higher redshift because there would be an increasing probability of having an intervening lens. Indeed a third of quasars should have equivalent widths smaller by 2-3 at only a moderate redshift if $\Omega_C=1$. This idea was first studied by Canizares (1982). More recently Dalcanton et al. (1994) have compared the equivalent widths for a high and low redshift sample comprising 835 Einstein Medium Source Survey quasars and 92 Steidel-Sargent absorption systems and find no difference. They infer the following limits:

$$\Omega_C(0.001\text{-}60M_\odot)<0.2,\ \Omega_C(60\text{-}300M_\odot)<1,\ \Omega_C(0.01\text{-}20M_\odot)<0.1 \qquad (3)$$

The mass limits come from the fact that the amplification of even the continuum region would be unimportant for $M<0.001M_\odot$, while the amplification of the broad-line regions would be important (cancelling the effect) for $M>20M_\odot$ if $\Omega_C=0.1$ or $M>60M_\odot$ if $\Omega_C=0.2$ or $M>300M_\odot$ if $\Omega_C=1$. These limits are indicated in Figure (3). This compares with the earlier Canizares (1982) constraint of $\Omega_C(0.01\text{-}10^5 M_\odot)<1$; his upper mass limit was larger because the broadline region was then thought to be bigger. Note that eqn (3) is incompatible with Hawkins' claim that $\Omega_C(10^{-3} M_\odot)\sim 1$, although one would only need to reduce Ω_C or M slightly.

4.5 Microlensing of Stars by Halo Objects in our own Galaxy

Attempts to detect microlensing by objects in our own halo by looking for intensity variations in stars in the Magellanic Clouds and the Galactic Bulge have now been underway for several years and may already have met with success. In this case, the timescale for the variation is $P=0.2(M/M_\odot)^{1/2}$y, so one can seek lenses over the mass range 10^{-8} -10^2 $M_\odot$, but the probability of an individual star being lensed is only $\tau\sim 10^{-6}$, so one has to look at many stars for a long time (Paczynski 1986). The likely event rate is $\Gamma \sim N\,\tau\,P^{-1} \sim (M/M_\odot)^{-1/2} yr^{-1}$ where $N\sim 10^6$ is the number of stars. Thus small masses give frequent short-duration events (eg. 0.01 $M_\odot$ events would last a week and occur a few times a year) and are best sought with CCDs, while large masses give rare long-duration events (eg. 10 $M_\odot$ events would last a year and occur every few years) and are best sought with photographic plates. The key feature of these microlensing events is that the light-curve is time-symmetric and achromatic and this may allow them to be distinguished from intrinsic stellar variations (Griest 1991).

Three groups are involved and each of these now claims to have detected lensing events (Alcock et al. 1993, Auborg et al. 1993, Udalski et al. 1993). The American group (MACHO) has used a dedicated telescope at Mount Stromlo to study 10^7 stars in red and blue light in the LMC, the SMC and the Galactic bulge. They currently have 4 LMC events (with durations of order a month) and around 45 bulge events. The French group (EROS) has been studying stars in the LMC and their approach is two-pronged: they are seeking 1-100 day events (corresponding to 10^{-4}-$1 M_\odot$ lenses) with digitized red and blue Schmidt plates obtained with the ESO telescope in Chile and 1 hour to 3 day events (corresponding to 10^{-7}-$10^{-3} M_\odot$) with CCDs taken at the Observatoire de Haute Provence. The CCD searches have given no results, which implies a limit $\Omega_C(10^{-7}\text{-}10^{-3} M_\odot) < 0.1$, but analysis of 3×10^6 stars on the Schmidt plates yields two events, each with duration of about two months. The Polish collaboration (OGLE) are using the Las Campanas telescope in Chile to look at 7×10^5 stars in the Galactic bulge. They have claimed 11 events but they only have data in one colour. The rough values of M and Ω_C for the LMC events are indicated in Figure (3) but there is considerable uncertainty in both these values. Note that the number of bulge events is anomalously high and this means that the LMC events cannot definitely be associated with halo objects.

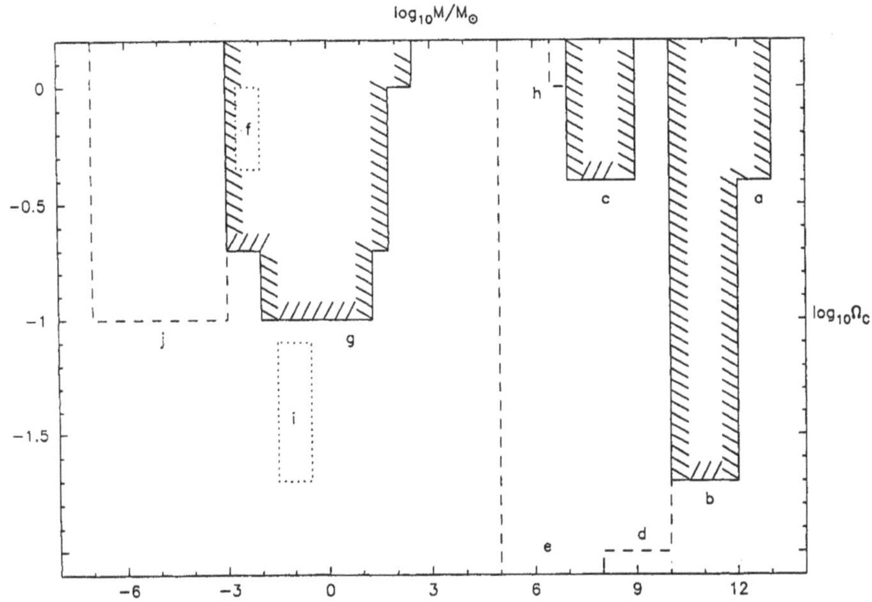

Figure (3). This shows the macrolensing and microlensing constraints on the density parameter for compact objects of mass M. Current limits are shown by shaded lines and potential ones by broken lines. (a) is the VLA limit of Hewitt (1986); (b) is the optical and HST quasar limit of Surdej et al. (1993); (c) is the VLBI limit of Kassiola et al. (1991); (d) and (e) are the potential speckle interferometry and VLBA limits; (f) corresponds to the quasar variations claimed by Hawkins (1993); (g) is the Dalton et al. (1994) quasar line-continuum limit; (h) is the gamma-ray burst limit of Nemiroff et al. (1993), assuming these are at a cosmolgical redshift; (i) corresponds to the MACHO and EROS microlensing results; (j) is the limit associated with the null EROS CCD results.

5. Dynamical Constraints

A variety of constraints can be placed on the mass of any dark compact objects in the disk and halo of our own Galaxy by considering their dynamical effects. The constraints are usually calculated on the assumption that the objects are black holes but most of them also apply for dark clusters of smaller objects. There are also constaints for dark objects in clusters of galaxies or in the intergalactic medium, though these are weaker. The limits are summarized as upper limits on the density parameter $\Omega_B(M)$ for black holes of mass M in Figure (4), where the disk, halo and cluster dark matter are assumed to have densities of 0.001, 0.1 and 0.2, respectively.

5.1 Disk Heating by Halo Holes.

As halo objects traverse the Galactic disk, they will impart energy to the stars there. This will lead to a gradual puffing up of the disk, with older stars being heated more than younger ones. Indeed Lacey & Ostriker (1985) have argued that black holes of around $10^6 M_\odot$ could provide the best mechanism for generating the observed amount of puffing. In particular, this explains why the velocity dispersion of the disk stars σ scales with age as $t^{1/2}$ and why the relative velocity dispersions in the radial, azimuthal and vertical directions. In order to normalize the $\sigma(t)$ relationship correctly, the number density of the holes n must satisfy $nM^2 \approx 3\times10^4 M_\odot^2 pc^{-3}$ and combining this with the local halo density $\rho_h = nM \approx 0.01 M_\odot pc^{-3}$ gives $M = 2\times10^6 M_\odot$.

This argument is no longer compelling because more recent measurements give smaller velocity dispersions for older stars, so that σ may no longer rise as fast as $t^{1/2}$ (Stromgren 1987, Gomez et al. 1990). Heating by a combination of spiral density waves and giant molecular clouds may now give a better fit to the data (Lacey 1991). Nevertheless, one can still use the Lacey-Ostriker argument to place an upper limit on the density in halo objects of mass M (Carr et al. 1984):

$$\Omega_B < \Omega_h \min\,[1,(M/M_{heat})^{-1}]\,, \quad M_{heat} = 3\times10^6 (t_g/10^{10}y)^{-1} M_\odot \qquad (4)$$

where t_g is the age of the Galaxy. Otherwise the disk would be more puffed up than observed. This limit is shown in Figure (4), along with the line which corresponds to having at least one black hole of mass M within the Galaxy.

Although the dependence is not shown explicitly in eqn (4), M_{heat} also scales as σ^2 and ρ_h^{-1}, so by applying the disk-heating argument to galaxies with higher dark matter density or lower stellar velocity dispersion or smaller age, one can obtain stronger constraints. For the gas-rich dwarf galaxy DDO154 (which has σ=17 km s^{-1}, an age of at least 1.5 Gyr and a central dark matter density of $0.009 M_\odot pc^{-3}$), Rix & Lake (1993) find $M < 7\times10^5 M_\odot$. For the dwarf galaxy GR8 (which has σ=4 km s^{-1}, an age of at least 1 Gyr and a central dark matter density of $0.07 M_\odot pc^{-3}$), they find $M < 6\times10^3 M_\odot$. Of course, unless the black holes form pregalactically, there is no reason for expecting the halo objects to have the same mass in different galaxies, so these limits are not shown in Figure (4).

5.2 The Disruption of Stellar Clusters by Halo Objects

Another type of dynamical effect associated with halo objects would be their influence on bound groups of stars (in particular, globular clusters and loose clusters). Every time a halo object passes near a star cluster, the object's tidal field heats up the cluster and thereby reduces its binding energy. If the object is sufficiently large it will disrupt the cluster in a single fly-by. For smaller objects the disruption will be a gradual, requiring the cumulative effect of many traversals. This process was first discussed by Spitzer (1958) for the case in which the disrupting objects are giant molecular clouds. The analysis was extended by Wielen (1985) to halo black holes with the mass of $2\times10^6 M_\odot$ required in the Lacey-Ostriker scenario and by Sakellariadou (1984) and Carr & Sakellariadou (1994) for halo black holes of general mass. By comparing the expected disruption time for clusters of mass m_c and radius r_c with the cluster lifetime t_L, one finds that the local density of halo holes of mass M must satisfy

$$\rho_B < \max\left[\left(\frac{m_c V}{G M t_L r_c}\right), \left(\frac{m_c}{G t_L^2 r_c^3}\right)^{1/2}, \left(\frac{m_c^{2/3} M^{1/3}}{V t_L r_c^2}\right)\right] \tag{5}$$

where V is the characteristic speed of the holes. The increasing mass regimes correspond to disruption by multiple encounters, single encounters and non-impulsive encounters, respectively. Any lower limit on t_L therefore places an upper limit on ρ_B. The crucial point is that the limit is independent of M in the single-encounter regime, so that the limit bottoms out at density which is of order $(\rho_c/G t_L^2)^{1/2}$. The constraint is therefore uninteresting if this exceeds the *observed* halo density ρ_h. In particular, if the clusters survive for the lifetime of the Galaxy, which is essentially the age of the Universe t_o, the limiting density is just $\sqrt{\rho_c \rho_o}$, where ρ_o is the mean cosmological density.

The strongest limit is associated with globular clusters, for which we take $m_c=10^5 M_\odot$, $r_c=10$pc, $V_c=10$ km s^{-1}and $t_L>10^{10}$y. We also assume that the holes have a speed V=300 km s^{-1}. Rather remarkably, due to the "coincidence" that the halo density is the geometric mean of the cosmological density and the globular cluster density, the upper limit on ρ_B is comparable to the actual halo density and this suggests that halo objects might actually *determine* the characteristics of surviving globular clusters (cf. Fall & Rees 1977). Numerical calculations for the disruption of globular clusters by Moore (1993) confirm the general qualitative features indicated above, with gradual mass loss for small halo objects and sudden disruption for larger ones. However, using data for nine particular globular clusters, he infers an upper limit of $10^3 M_\odot$. This is considerably stronger than the limit implied by eqn (5), presumably because his clusters are more diffuse. Because of the uncertainties, the line corresponding to Moore's result is only shown dotted in Figure (4).

5.3 The Effect of Dynamical Friction on Halo Objects

Another important dynamical effect is that halo objects will tend to lose energy to lighter objects and consequently drift towards the Galactic nucleus. In particular, one can show that halo objects will be dragged into the nucleus by the dynamical friction of the Spheroid stars from within a Galactocentric radius

$$R_{df} = (M/10^6 M_\odot)^{2/3}(t_g/10^{10}y)^{2/3}\ \mathrm{kpc} \qquad (6)$$

and the total mass dragged into the Galactic nucleus therefore exceeds the upper observational limit of $3 \times 10^6 M_\odot$ (Spaenhauer et al. 1992) unless

$$M < 10^4 (a/2\mathrm{kpc})(t_g/10^{10}y)^{-1} M_\odot \qquad (7)$$

where a is the halo core radius. This is certainly stronger than the disk-heating limit and it would seem to preclude the Lacey-Ostriker proposal. However, there is an important caveat in this conclusion (Hut & Rees 1992). Eqn (6) implies that about 10^3 holes of $10^6 M_\odot$ would have drifted into the Galactic nucleus by now, corresponding to one arrival every 10^7y. Once two black holes have reached the nucleus, they will form a binary and this will eventually coalesce due to loss of energy through gravitational radiation. If a third hole arrives before coalescence occurs, then the "slingshot" mechanism could eject one of the holes and the remaining pair might also escape due to the recoil. Hut & Rees estimate that the time for binary coalescence is *shorter* than the interval between infalls, suggesting that slingshot is ineffective. However, limit (7) is clearly not completely firm, so it is only shown dotted in Figure (4).

5.4 Is the Halo Made of Dark Clusters?

We have seen that both the cluster disruption and dynamical friction constraints may be incompatible with the Lacey & Ostriker proposal that $2 \times 10^6 M_\odot$ halo black holes generate the observed disk-heating. There is also the problem that supermassive halo black holes might generate too much radiation through accretion as they traverse the disk (Ipser & Price 1977). To circumvent these objections, Carr & Lacey (1987) have proposed that the disk heaters are $2 \times 10^6 M_\odot$ *clusters* of smaller objects rather than single black holes. The accretion luminosity is then reduced by a factor of order the number of objects per cluster and the dynamical friction problem is avoided provided the clusters are disrupted by collisions before they are dragged into the Galactic nucleus by dynamical friction.

One can generalize this idea to a more general cluster scenario (Wasserman & Salpeter 1993, Kerins & Carr 1994, Moore & Silk 1994). If we assume that the clusters all have the same mass M_c and radius R_c, then they will be disrupted by collisions within the Galactocentric radius (6) at which dynamical friction operates providing

$$R_C > 1.4(a/2kpc)^2(t_g/10^{10}y)^{-1}pc \qquad (8)$$

If this condition is not satisfied, then Mc must be less than the value indicated by eqn (7). In order to avoid the clusters evaporating as a result of 2-body relaxation, one also requires another lower limit on R_C. An *upper* limit on R_C comes from requiring that the clusters do not disrupt at our own Galactocentric radius. Note that together these limits eqns require the cluster components to be smaller than $10^2M_\odot$ and this probably excludes their being VMO black holes.

5.5 Constraints on Dark Objects Outside Halos

Dynamical constraints on dark objects in the Galactic disc are generally stronger than the halo limits. In particular, Bahcall et al. (1985) have argued that the disk dark matter could not comprise objects larger than $2M_\odot$ else they would disrupt the wide binaries observed by Latham et al. (1984). This an important constraint because, if correct, it rules out disk dark matter comprising stellar black holes. However, the Bahcall et al. conclusion has been disputed by Wasserman & Weinberg (1987) on the grounds that there is no sharp cut-off in the distribution of binary separations above 0.1pc. The limit is therefore weakened (somewhat arbitrarily) to $10M_\odot$ in Figure (4).

Dynamical constraints on dark objects in clusters of galaxies are weaker than the halo limits. For example, one does not get an interesting constraint by applying eqn (5) to the disruption of cluster galaxies by cluster black holes because the upper limit on ρ_B exceeds the cluster density. However, one does get an interesting constraint from upper limits on the fraction of galaxies with unexplained tidal *distortions.* Van den Bergh (1969) applied this argument to the Virgo cluster and inferred that black holes binding the cluster could not be bigger than $10^9M_\odot$. If we assume that Virgo is typical, we obtain the limit indicated in Figure (4). We also show the limit corresponding to there being at least one black hole of mass M within the cluster.

The dynamical constraints on intergalactic black holes are much weaker. The most interesting one comes from the fact that, if there were a population of huge intergalactic black holes, each galaxy would have a peculiar velocity due to its gravitational interaction with the nearest one (Carr 1978). If the holes were smoothly distributed and had a number density n, one would expect every galaxy to have a peculiar velocity of order $GMn^{2/3}t_g$. Since the CMB dipole anisotropy shows that the peculiar velocity of our own Galaxy is only 600 km s^{-1}, one infers a limit $\Omega_B < (M/10^{16}M_\odot)^{-1/2}$ and this is also shown in Figure (4). The limit on the bottom right corresponds to the requirement that there be at least one object of mass M within the current particle horizon.

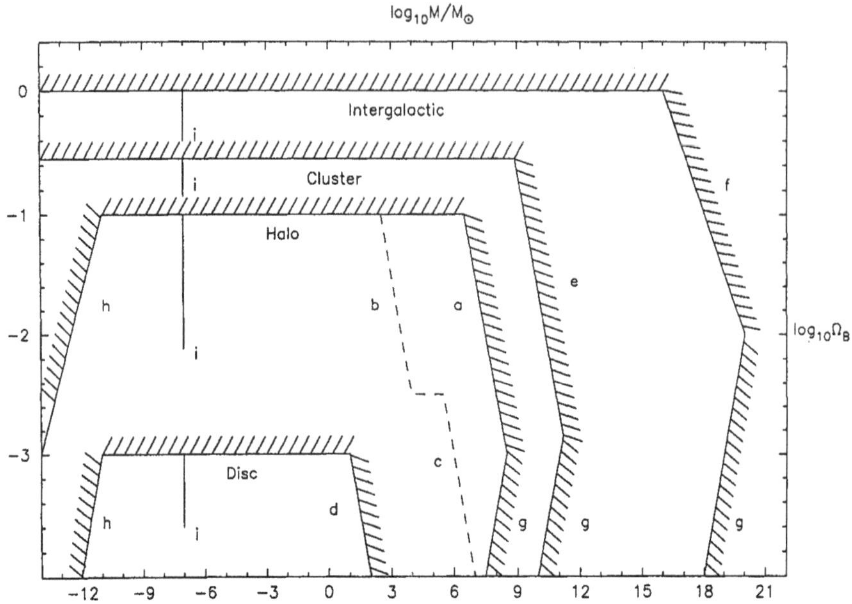

Figure (4). This summarizes the dynamical constraints on the density parameter Ω_B for black holes of mass M located in the Galactic disk, the Galactic halo, clusters of galaxies and the intergalactic medium. The total dark matter density in these cases is taken to be 0.001, 0.1, 0.2 and 1, respectively. The limits come from: (a) disk heating; (b) globular cluster disruption; (c) dynamical friction; (d) binary disruption; (e) galaxy distortions; (f) galaxy peculiar velocities; (g) one black hole per disk/halo/cluster/Universe; (h) comet observations. Limits (b) and (c) are shown by broken lines because they are not so secure. The evaporation limit (i) is non-dynamical and corresponds to the fact snowballs smaller than $10^{-8}M_\odot$ evaporate within the age of the Universe (De Rujula et al. 1992).

6. Low Mass Objects

In this section we will focus specifically on the LMO scenario. There are several reasons why LMOs currently seem to be the most plausible option. Firstly, there may be direct evidence from cluster cooling flows that baryons can turn into low mass stars with high efficiency even at the present epoch. Secondly, recent data on the stellar IMF in our own galaxy suggests there may be a higher fraction of LMOs when the metallicity is low. Thirdly, as we saw in §4, microlensing data may already indicate that there is dark matter in the form LMOs.

6.1 M-dwarfs versus Brown Dwarfs

In determining how small LMOs would need to be to provide the disk or halo dark matter, important information comes from red and infrared observations. From searches for sources in our own halo Richstone et al. (1992) and Bahcall & Soneira (1984) find that stars in these mass ranges can only have a small fraction of the halo density. Even stronger constraints come from Gilmore & Hewett (1983), who find that the local number density of stars in the mass range 0.08-$0.1M_\odot$ can be at most 0.01 pc^{-3} . This is a hundred times too small to explain the local dark matter problem and ten times too small to explain the halo problem.

A similar conclusion is indicated by infrared observations of other spiral galaxies. For example, the K-band mass-to-light ratio exceeds 50 for NGC 4565 (Boughn et al. 1981), 100 for M87 (Boughn & Saulson 1983), 64 for NGC 5907 (Skrutskie et al. 1985) and 140 for NGC 100 (Casali & James 1994). Since the mass-to-light ratio is less than 60 for stars bigger than $0.08M_\odot$, the lower limit for hydrogen-burning (D'Antona & Mazzitelli 1985), this suggests that any hydrogen-burning stars are excluded. These observations suggest that the halo dark matter must be in the form of brown dwarfs. On the other hand, Sackett et al. (1994) have recently claimed to detect a faint luminous halo around NGC 5907 and brown dwarfs would be too dim to account for this.

6.2 Evidence from Population I and Population II

The evidence for stars in the brown dwarf mass range (Simons & Becklin 1992, Steele et al. 1993) is controversial but this merely reflects the fact that they are hard to find (Stevenson 1991) and it would be very surprising if the IMF happened to cut off just above $0.08M_\odot$. Although it is difficult to observe BDs themselves, one can study the IMF of stars in the mass range above the hydrogen-burning limit and infer whether its extrapolation would permit a lot of BDs. If one assumes that the IMF has the power-law form

$$dN/dm \sim m^{-x} \quad \text{for} \quad m_{min} < m < m_{max} \tag{9}$$

(at least over some mass range), then most of the mass is in the smallest stars for $x>2$ and the largest ones for $x<2$. Determining the value of x in the LMO range is difficult: partly because obtaining the luminosity function is hard and partly because there are large uncertainties in the mass-luminosity relation as one approaches the hydrogen-burning limit.

Let us first consider the possibility that the disk dark matter (if it exists) is in the form of BDs. Early studies of the luminosity function for nearby stars (Reid & Gilmore 1982, Gilmore & Reid 1983, Gilmore et al. 1985) suggested that the IMF is too shallow for BDs to have an intersting density. These results were initially contradicted by the results of Hawkins (1985) and Hawkins & Bessell (1988), who went to somewhat fainter magnitudes and claimed that the observations were consistent with an IMF which steepened enough to put the all dark mass in BDs. However, the data of Tinney et al. (1992, 1993) makes it quite clear that the IMF flattens off below 0.2 $M_\odot$ and, unless it rises again below $0.08M_\odot$, the contribution to the local dark matter must be small (Tinney 1993). This is also consistent with the results of Kroupa et al. (1993) who find $x=2.7$ for $m>1M_\odot$, $x=2.2$ for $0.5<m<1M_\odot$ and $0.7<x<1.8$ for $0.08M_\odot<m<0.5M_\odot$. This suggests that stars of 0.5 $M_\odot$ should dominate the disk density. BDs may dominate the *number* density but, unless the value of x changes below $0.08M_\odot$, they can only contain 1% of the mass.

The situation is less clear-cut when one considers Population II stars. Richer et al. (1991) claim that metal poor Globular clusters have $x=3.6$ below $0.5M_\odot$ down to at least $0.14M_\odot$, while Richer & Falman (1992) claim that stars in the Galactic Spheroid have $x=4.5\pm1.2$ in the same mass range. This *does* allow the possibility that most of the mass is in the smallest objects; indeed BDs could explain all the halo dark matter if the IMF extended down to $M_{min}\sim0.01M_\odot$. However, Richer & Falman also point out that the rotation curve of the Galaxy requires that the total spheroid mass cannot exceed $7\times10^{10}M_\odot$ and this implies that the IMF cannot extend below $0.05M_\odot$. It is therefore unlikely that Population II stars themselves could explain the halo dark matter. The main point of these results is that they lend support to the suggestion that low metallicity enhances the fraction of mass in low mass objects.

It should be stressed that there is an important difference between attributing the disk and the halo dark matter to BDs. If the disk dark matter comprises BDs, one would expect them to represent the low mass tail of the Population I IMF since all disc stars presumably form at the same time. However, there may be no connection between the dark halo stars and Population II stars because they probably form at a different time and place. One should thus be wary of attempts to exclude halo BDs on the grounds that Population II stars have a particular IMF, as do Hegyi & Olive (1983, 1986, 1989).

6.3 Infrared Searches for Brown Dwarfs

Even though BDs do not burn hydrogen, they still generate some luminosity in the infrared. They radiate first by gravitational contraction (for about 10^7 yr) and then by degenerate cooling. If the disk or halo dark matter is in the form of BDs, it is therefore important to consider whether they can be detected via this infrared emission. Current constraints on BDs are rather weak (Beichmann et al. 1990, Nelson et al. 1993) but the prospects of detection will be much better with impending space satellites. The problem has been addressed in various contexts by several authors. Karimabadi & Blitz (1984) have calculated the expected intensity from BDs with a discrete IMF comprising an $\Omega=1$ cosmological background. Adams & Walker (1990) have discussed the possibility of detecting the collective emission of the brown dwarfs in our own Galactic halo for both a discrete and power law IMF. Daly & McLaughlin (1992) have considered the prospects of detecting the emission of individual halo brown dwarfs of a given mass and age in the Solar vicinity, as well as the collective emission of brown dwarfs in other galaxy halos. Kerins & Carr (1994) have considered the possibility that the BDs are assembled into dark clusters and also discuss how infrared observations at different wavelengths could be used to probe the mass spectrum of the brown dwarfs.

As an illustration of the feasibility of detecting radiation from BDs, let us consider the prospects of detecting the nearest one in our halo. If the BDs all have the same mass m, then the local halo density (ρ_0=0.01$M_\odot pc^{-3}$) implies that the expected distance to the nearest one is $0.55(m/0.01M_\odot)^{1/3}$pc. The expected spectra are shown in Figure (5) and compared to the sensitivities of IRAS and ISO. This assumes the temperature and luminosity of Stevenson (1986) where the BD age and opacity are taken to be 10^{10}yr and 0.01cm^2g^{-1} (corrresponding to electron-scattering). This shows that IRAS gives no useful constraints (it is too weak by a factor of 2 even for the optimal mass of 0.07$M_\odot$) but the ISOCAM instrument on ISO could detect 0.08$M_\odot$ BDs in a few hours, 0.04$M_\odot$ BDs in a few days and 0.01$M_\odot$ BDs in a few months. Note that disk BDs would be younger, locally more numerous and more opaque than halo BDs, increasing the peak flux by 6 and decreasing the peak wavelength by 0.6. Indeed IRAS already implies that BDs with a discrete IMF could provide the disk dark matter only for m< 0.01$M_\odot$.

One might expect the BDs to be easier to detect if they are in clusters. This is because, although the distance to the nearest source is increased by a factor $(M_c/m)^{1/3}$, the luminosity is increased by (M_c/m), giving an increase in flux of $(M_c/m)^{1/3}$. Indeed Rix & Lake (1993) have already used this to exclude the cluster scenario. However, this assumes that the clusters are point sources and the dynamical constraints discussed in §5.4 then imply that the clusters will always be extended sources. In fact, the IRAS extended source sensitivity at 12μ is too low to permit the detection of clusters. The ISO extended source sensitivity (ESS) at 6.75μ will suffice but the time required to find them is very sensitive to their mass and radius. Note that the halo clusters will cover the sky if they are large enough, in which case detecting the clusters is equivalent to detecting the halo background. ISO would take several months to detect the Galactic background, even in the optimal case with m=0.08 $M_\odot$ (Kerins & Carr 1994).

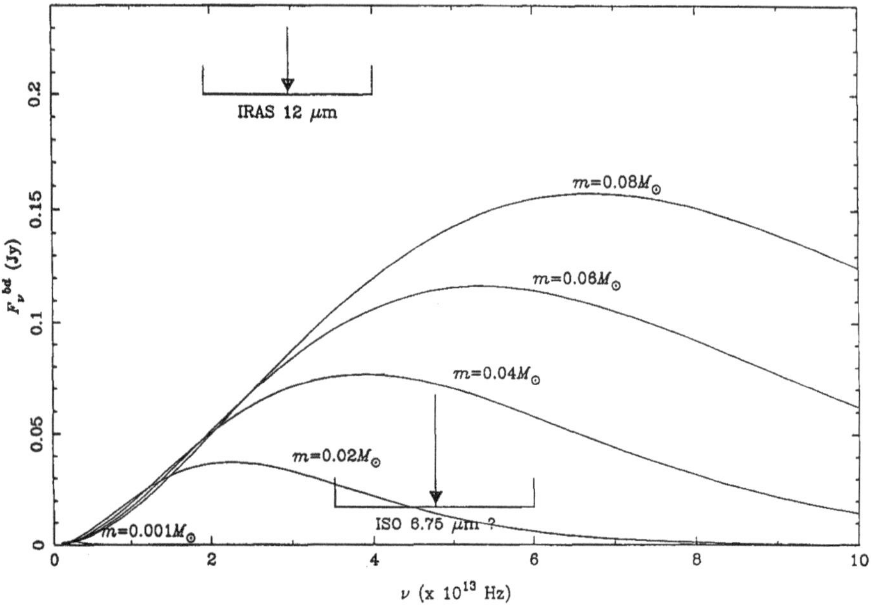

Figure (5). This shows the expected flux from the nearest halo BD for various values of the BD mass. The IRAS point source sensitivity at 12μ is shown and this a factor of two above the predicted flux even in the optimal case. The expected 3σ ISO 6.75μ sensitivity is also shown, assuming an observation time of 10 days and a 100s integration time.

7. Conclusions and Future Prospects

The various constraints on the form of baryonic dark matter discussed in this review are brought together in Table (2), the shaded regions being excluded by either dynamical, nucleosynthetic, lensing or light constraints. The dotted regions may also be excluded but this is less certain. Table (2) is different from the equivalent figure in my 1991 lecture only in that the dotted region is extended to include halo white dwarfs and VMOs. Whether the "cluster" dotted region in Table (2) is excluded depends on whether one believes that the primordial nucleosynthesis constraint permits the cluster dark matter to be baryonic. This is possible only if one invokes inhomogeneous cosmological nucleosynthesis but this possibility should still be taken seriously. On the other hand, it seems clear that even inhomogeneous nucleosynthesis will not permit baryons to have the critical density. Indeed a critical baryon density is excluded in most mass ranges anway. The prospects for the "maximal BDM" scenario therefore seem bleak.

The prime message of Table (2) is that one could not expect any single candidate to explain all four dark matter problems. On the other hand, the table does constrain the possible solutions:

* The *local dark matter* (if it exists) could be white dwarfs or brown dwarfs but observations of the Population I IMF gives no reason for expecting this; presumably it could not be inos since these are non-dissipative and so would not settle into a disk.

* The *halo dark matter* could be brown dwarfs or white dwarfs or VMO black holes; all of these possibilities require a departure from the standard Population II IMF but the first probably requires the least radical departure since observations may already indicate a preponderance of low mass stars at small metallicity. The white dwarf option - although in a sense the most conservative - requires cutting the IMF off at both ends.

* The *cluster dark matter* may be partly baryonic, especially if galactic halos are baryonic, but we have seen that it could only be dominated by baryonic dark matter if one invokes inhomogeneous cosmological nucleosynthesis.

* The *background dark matter* (if it exists) would have to be inos; if the inos are cold, one would expect both the halo and cluster dark matter to be a *mixture* of WIMPs and MACHOs.

Table (2): Constraints on baryonic dark matter candidates, shaded regions being excluded by the limits discussed in the text and the dotted regions being improbable. SMO, VMO and BH refer to the black hole remnants of Supermassive Objects, Very Massive Objects, and ordinary stars respectively; WD=white dwarf; MD=M-dwarf; BD=brown dwarf.

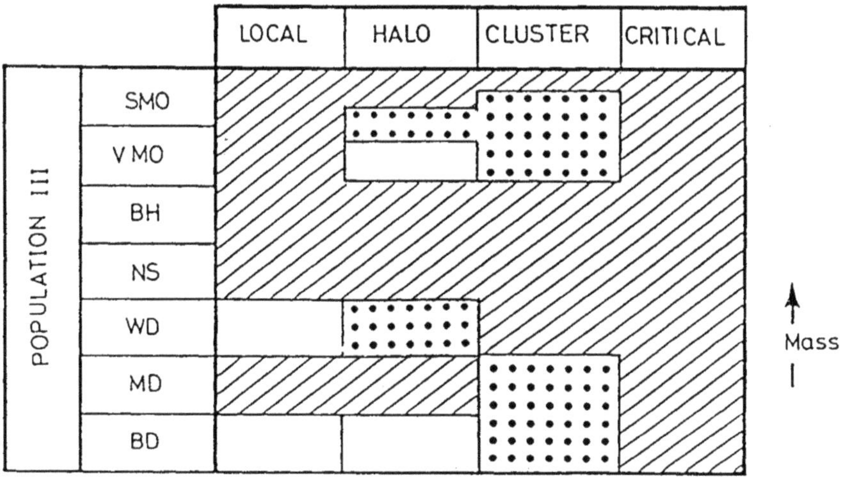

Some of the most exciting developments in this field have come from gravitational lensing studies and we can expect a proliferation of these efforts in the next few years. Macrolensing searches are placing ever more stringent limits of the number of *high* mass baryonic objects and microlensing searches may have already provided evidence for *low* mass ones. It is surely significant that microlensing evidence from both quasars and stars all points towards lens masses in the range $0.001\text{-}0.1M_\odot$. Admittedly, the masses indicated by the microlensing results are marginally too high; the most likely values are all in the M-dwarf range, whereas light constraints require the halo objects to be smaller than $0.08M_\odot$. Nevertheless, there is a fairly broad probability distribution for the lensing masses, depending on the velocity and spatial distribution of the halo objects (Kerins 1994).

It should be stressed that, even if the lensing results are due to MACHOs, this does not preclude WIMPs from providing some or even most of the dark halo. This is because the microlensing searches only probe the part of the halo at Galactocentric radii from 10-20 kpc, whereas the halo itself could extend much further than this. There could therefore be plenty of WIMPs further out, especially if the dark baryons are preferentially concentrated as a result of dissipation. Even within the 10-20 kpc region, the number of microlensing events observed merely suggests that the fraction of halo mass in MACHOs must exceed 10% (Gates & Turner 1993). It is therefore important that WIMP seachers should not be too discouraged by the success of the MACHO searchers. It still seems likely that we need both MACHOs and WIMPs.

As discussed in Mather's lecture, another source of exciting developments has been the COBE satellite. We have seen that the FIRAS constraints on the microwave background spectral distortions and the DIRBE measurements of the infrared background density already severely restrict any scenario in which the dark matter is in the remnants of massive stars. Indeed the only hope for these scenarios may be that the radiation remains in the near-IR (unmodified by dust) where it may be hidden by interplanetary dust emission. The FIRAS constraints on the Compton y-parameter may also exclude the supermassive accreting black hole scenario.

Looking further to the future, two more develoments will have an important impact. If the halo dark matter is in brown dwarfs, then we have seen that the next generation of infrared space satellites will either detect these or push their mass down below $0.01M_\odot$. In this case, we have stressed the importance of knowing whether the brown dwarfs are clustered since this determines whether one is seeking discrete or extended sources. If the halo dark matter is in black holes, then the next generation of gravitational wave detectors (either ground-based or space-based interferometers) will have an excellent chance of detecting the associated gravitational radiation.

REFERENCES.

Aaronson, M. (1983), Ap.J.Lett., 266, L11.
Aaronson,M. & Olszewski,E. (1987), in *Dark Matter in the Universe*, ed. J.Kormendy & G.R.Knapp, p 153 (Dordrecht, Reidel).
Adams, F.C. & Walker, T.P. (1990). Ap.J., 359, 57.
Alcock, C. et al. (1993). Nature, 365, 621.
Ashman, K.M. & Carr, B.J. (1988). MNRAS, 234, 219.
Ashman, K.M. & Carr, B.J. (1991). MNRAS, 249, 13.
Auborg, E. et al. (1993). Nature, 365, 623.
Bahcall, J.N. (1984 a,b,c). Ap.J., 276, 156; 276,169; 287, 926.
Bahcall, J.N. & Soneira, R.M. (1984). Ap.J.Supp., 55, 67.
Bahcall, J.N., Hut, P. & Tremaine, S. (1985). Ap.J., 290, 15.
Bahcall,J.N., Flynn,C. & Gould,A. (1992a). Ap.J., 389, 234.
Bahcall, J.N. et al. (1992b). Ap.J., 387, 56.
Begelman, M.C., Blandford,R.D. & Rees, M.J. (1980). Nature, 287, 307.
Beichmann, C.A. et al. (1990) Astron.J. 99, 1569.
Bertschinger, E. & Dekel, A. (1989). Ap.J.Lett., 336, L5.
Bienayme,O., Robin,A.C. & Creze,M. (1987). Astron.Astrophys. 180, 94.
Blanchard, A., Valls-Gabaud,D. & Mamon, G.A. (1992). Astron.Astrophys., 264, 365.
Blandford, R.D. & Narayan, R. (1992). Ann.Rev.Astron.Astrophys., 30,311.
Bond,J.R.,Arnett, W.D. and Carr,B.J. (1984). Ap.J., 280, 825.
Bond,J.R., Carr,B.J. & Hogan,C.J. (1986). Ap.J., 306, 428.
Bond,J.R., Carr,B.J. & Hogan,C.J. (1991). Ap.J., 367, 420.
Boughn,S.P., Saulson,P.R. & Seldner,M. (1981). Ap.J.Lett., 280, L15.
Boughn,S.P. & Saulson,P.R. (1983). Ap.J.Lett., 265, L55.
Bristow, P. & Phillipps, S. (1994). MNRAS, 267, 13.
Canizares, C.R. (1982). Ap.J., 263, 508.
Carignan, C. & Freeman, K.C. (1986). Ap.J.Lett., 332, L33.
Carignan,C., Beaulieu,S. & Freeman,K.C. (1990). Astron.J.,99,178.
Carr, B.J. (1978). Comm.Astrophys.,7, 161.
Carr, B.J. (1992). In *1st Course in Current Topics in Astrofundemantal Physics*, ed. N.Sanchez & A.Zichichi, p 135 (World Scientific)
Carr, B.J. (1994). Ann.Rev.Astron.Astrophys., 32, 531.
Carr, B.J., Bond, J.R. & Arnett, W.D. (1984). Ap.J., 277, 445.
Carr, B.J. & Lacey, C.G. (1987). Ap.J., 316, 23.
Carr,B.J. & Sakellariadou, M. (1994). Preprint.
Casali,M. & James, P. (1994). MNRAS, in press.
Cayrel,R. (1987). Astr.Astrophys., 168, 81.
Coles, P. & Ellis. G.F.R. (1994). Nature, 340, 609.
Corrigan, R. et al. (1991). Astron.J., 102, 34.
Crampton, D. et al. (1989). Astron.J., 98, 1188.
Crawford, M. & Schramm, D.N. (1982). Nature, 298, 538.
Dalcanton, J., Canizares,C., Granados,A., Steidel, C.C. & Stocke, J.T. (1994). Ap.J., in press.

Daly, R.A. & McLaughlin, G.C. (1992). Ap.J., 390, 423.
D'Antona, F. & Mazzitelli, I. (1985). Ap.J., 296, 502.
Dekel, A. & Rees, M.J. (1987). Nature, 326, 455.
De Rujula,A., Jetzer, Ph. & Masso, E. (1992). Astron.Astrophys., 254, 99.
de Zeeuw, P.T. (1992). in *Morphological and Physical Classification of Galaxies*, ed. G. Busarello, M.Capacciolli & G.Longo, p ? (Dordrecht, Kluwer).
Dressler, A., Faber, S.M., Burstein, D., Davies, R.L., Lynden-Bell, D., Terlevich, R.L. & Wegner, G. (1987). Ap.J.Lett., 313, L37.
Dubinski, J. & Carlberg, R.G. (1991). Ap.J., 378, 496.
Ellis, G.F.R. (1988). Class.Quant.Grav., 5, 891.
Ellis, G.F.R., Lythe, D.H., & Mijic, M.B. (1991). Phys.Let.B., 271, 52.
Fabian, A.C. (1994). Ann.Rev. Astr.Astrophys., 32, 277.
Fall,S.M. & Rees,M.J. (1977). MNRAS, 81, 37P.
Fall,S.M. & Efstathiou, G. (1981), MNRAS, 193, 189.
Fich,M. & Tremaine, S. (1991). Ann.Rev.Astron.Astrophys., 29, 409.
Forman, W., Jones, C. & Tucker, W. (1985). Ap.J., 293, 102.
Fowler, W. (1966). Ap.J., 144, 180.
Fowler, W. (1990). In *Baryonic Dark Matter*, ed. Lynden-Bell, D. & Gilmore, G., p 257 (Kluwer).
Fowler, W. & Hoyle, F. (1964). Ap.J.Supp., 9, 201.
Frenk, C.S., White, S.D.M., Davis, M. & Efstathiou,G. (1988). Ap.J., 327, 507.
Fuller, G.M., Woosley, S.E. & Weaver, T.A. (1986). 307, 675.
Gates,E. & Tuner, M.S. (1994). Phys.Rev.Lett., 72, 2520.
Gilmore, G. & Hewett, P. (1983). Nature, 306, 669.
Gilmore, G. & Reid, N. (1983). MNRAS, 202, 1025.
Gilmore.G., Reid,N. & Hewett, P. (1985). MNRAS, 213, 257.
Glatzel,W., El Eid,M.F. & Fricke,K.J. (1985) .Astr.Astrophys., 149, 419.
Gnedin, N.Yu. & Ostriker, J.P. (1992). Ap.J., 400, 1.
Gott, J.R. (1981). Ap.J., 243, 140.
Gouda,N. & Sugiyama,N. (1992). Ap.J.Lett., 395, L59.
Gould, A. (1990). MNRAS, 244, 25.
Griest, K. (1991). Ap.J., 366, 412.
Guth, A.H. (1981). Phys.Rev.D., 23, 347.
Hatsukade, I. (1989). PhD thesis (Miyazaki University).
Hawkins, M.R.S. (1985). Phil.Trans.R.Astr.Soc., 320, 553.
Hawkins, M.R.S. & Bessell, M.S. (1988). MNRAS, 234, 177.
Hawkins, M.R.S. (1993). Nature, 366, 242.
Hegyi, D.J. & Olive, K.A. (1983). Phys.Lett.B., 126, 28.
Hegyi, D.J. & Olive, K.A. (1986). Astrophys.J., 303, 56.
Hegyi, D.J. & Olive, K.A. (1989). Astrophys.J., 346, 648.
Henstock, D.R. et al. (1993). In *Gravitational Lenses in the Universe.*, ed. J. Surdej et al. (Univ. Liege).
Hewett, P.C.. et al. (1989). Ap.J.Lett., 346, L61.
Hewitt, J.N. (1986). PhD thesis (Massachussets Institute of Technology).

Hewitt, J.N. et al. (1989). In *Gravitational Lenses*, ed. J.M.Moran, J.N.Hewitt & K.Y.Loh, p 147 (Berlin: Springer-Verlag).
Hogan, C.J. (1993). Ap.J.Lett., 415, L63.
Huchra,J. & Brodie,J. (1987). Astron.J., 93, 779.
Hut,P. & Rees, M.J. (1992). MNRAS, 259, 27P.
Ipser, J.R. & Price R.H. (1977). Ap.J., 216, 578.
Irwin, M.J. et al. (1989). Astron.J. , 98, 1989.
Kaiser, N. (1984). Ap.J.Lett., 284, L9.
Karimabadi, H. & Blitz, L. (1984). Ap.J., 283, 169.
Kashlinsky, A. & Rees, M.J. (1983). MNRAS, 205, 955.
Kassiola,A., Kovner,I. & Blandford,R.D. (1991). Ap.J., 381, 6.
Kayser,R., Refsdal,S. & Stabell, R. (1986). Astron.Astrophys., 166, 36.
Kerins, E. (1994). MNRAS, in press.
Kerins, E. & Carr,B.J. (1994). MNRAS, 266, 775.
Knapp, G.R. (1988), in *The Mass of the Galaxy*, ed. M.Fich, p 35 (CITA Workshop)
Kovner, I. (1991). Ap.J., 376, 70.
Kronberg, P.P. et al. (1991). Ap.J.Lett., 367, L1.
Kroupa,P., Tout, C.A. & Gilmore,G. (1993). MNRAS, 262, 545.
Kuijken, K. & Gilmore.G. (1989). MNRAS, 239,571; 239, 605; 239,651.
Kuijken, K. & Gilmore.G. (1991). Ap.J.Lett., 367, L9.
Lacey,C.G. (1991). In *Dynamics of Disk Galaxies*, ed. B.Sundelius, p 257.
Lacey, C.G. & Ostriker, J.P. (1985). Ap. J., 299, 633.
Lake, G. (1990). Ap.J.Lett., 356, L43.
Lake, G., Schommer, R.A. & van Gorkom,J.H. (1990). Astron.J., 99, 547.
Lanzetta, K.M. et al. (1991) Ap.J.Sup., 77, 1.
Latham, D.W., Tonry,J., Bahcall, J.N., Soneira,R.M. & Schechter,P.S. (1984). Ap.J.Lett., 281, L41.
Lewis, G.F. et al. (1993). MNRAS, 261, 647.
Lin, D.N.C. & Faber, S.M. (1983). Ap.J.Lett., 266, L21.
Loeb, A. (1993). Ap.J., 403, 542.
Loeb, A. & Rasio, F. (1993). Preprint.
Low, C. & Lynden-Bell,D. (1976). MNRAS, 176, 367.
Lynden-Bell, D. & Gilmore, G. (1990), *Baryonic Dark Matter*, Kluwer.
Maeder, A. (1992) Astron. Astrophys., 264, 105.
Malaney, P. & Mathews, G. (1993). Phys.Rep., 229, 147.
Mather, J.C. et al. (1994). Ap.J., 420, 439.
Mathews,G.J., Schramm,D.N. & Meyer, B.S. (1993), Ap.J., 404, 476.
McGaugh, S. (1994). Nature, 367, 538.
Mateo,M., Olszewski,E., Welch,D.L., Fischer,P. & Kunkel,W. (1991) Astron.J., 102, 914.
Moore, B. (1993). Ap.J.Lett., 413, L93.
Moore, B. & Silk, J. (1994). Preprint.
Mould,J.R.,Oke,J.B., de Zeeuw,P.T. & Nemec,J.M. (1990),Astron.J.,245,454.
Mulchaey, J.S., Davis, D.S., Mushotzky, R.F. & Burstein, D. (1993) Ap.J., 404, L9.

Nelson, L.A., Rappaport, S.A. & Joss, P.C. (1993) Ap.J., 404, 723.
Nemiroff, R.J. (1989). Ap.J., 341, 579.
Nemiroff, R.J. (1991). Phys.Rev.Lett., 66, 538.
Nemiroff, R.J. et al. (1993). Ap.J., 414, 36.
Nottale, L. (1986). Astron.Astrophys. 157, 383.
Nusser, A. & Dekel, A. (1993). Ap.J., 405, 437.
Ober, W.W., El Eid, M.F. & Fricke, K.L. (1983).Astr.Astrophys.,119, 61.
Paczynski, B. (1986). Ap.J., 304, 1; 308, L43. (?)
Palla, F., Salpeter, E.E. & Stahler, S.W. (1983). Ap.J., 271, 632.
Patnaik,A.R., Browne, I.W.A., Wilkinson,P.A. & Wrobel,J.M. (1992). MNRAS, 254, 655.
Peacock, J.A. & Dodds, S.J. (1994). MNRAS., 267, 1020.
Peebles, P.J.E. (1987). Ap.J.Lett., 315, L73.
Persic, M. & Salucci,P. ,1992. MNRAS, 258, 14P.
Pfenniger,D., Combes,F. & Martinet,L.(1994). Astron.Astrophys., 285,79.
Ponman, T.J. & Bertram, D. (1993), Nature, 363, 51.
Press, W.H. & Gunn, J.E. (1973). Ap.J., 185, 397.
Rees, M.J. (1976). MNRAS, 176, 483.
Rees,M.J. (1986). MNRAS, 218, 25P.
Refsdal, S. & Stabell, R. (1991). Astron.Astrophys., 250, 62.
Reid, N. & Gilmore, G. (1982). MNRAS, 201, 73.
Richer, H.B. & Fahlman, G.G. (1992). Nature, 358, 383.
Richer, H.B. et al. (1991). Ap.J., 381, 147.
Richstone, D. et al. (1992). Ap.J. 388, 354.
Rix, H. & Lake, G. (1993). Ap.J., 417, 1.
Rowan-Robinson, M. et al. (1990). MNRAS, 247, 1.
Rubin, V.C., Ford, W.K. & Thonnard, N. (1980). Ap.J., 238, 471.
Ryu, D., Olive, K.A. & Silk, J. (1990). Ap.J., 353, 81.
Sackett,P.D. & Sparke, L.S. (1990). Ap.J., 361, 408.
Sakellariadou, M. (1984). MSc thesis (Cambridge).
Salpeter, E.E. & Wasserman,I.M. (1993). Astron.Soc.Pac., 36, 345.
Sackett, P.D., Morrison, H.L., Harding, P. & Boroson, T.A. (1994). Nature, 370, 441.
Sancisi, R. & van Albada, T.S. (1987). In *Dark Matter in the Universe*, ed. J.Kormendy & G.Knapp , p 67 (Reidel, Dordrecht).
Sarazin, C. (1986). Rev.Mod.Phys., 58, 1.
Sargent. W. L.W. & Steidel, C.C. (1990). In *Baryonic Dark Matter*, ed. Lynden-Bell, D. & Gilmore, G., p 223 (Kluwer).
Schild, H. & Maeder, A. (1985). Astron.Astrophys.Lett., 143, L7.
Schneider, D.P., Schmidt,M. & Gunn, J.E. (1991). Astron.J., 102, 837.
Schneider, P. (1993). Astron. Astrophys., 279,1.
Shanks, T. (1985). Vistas in Astron., 28, 595
Shu,F., Adams, F. & Lizano, S. (1987). Ann.Rev.Astron.Astrophys., 25, 23.
Silk, J. (1977). Ap.J., 211, 638.
Silk, J. (1991). Science, 251, 317.
Silk, J. (1993). Phys.Rep., 227, 143.

Simons,D.A. & Becklin,E.E. (1992). Ap.J., 390, 431.
Skrutskie, M.F., Shure, M. & Beckwith, S. (1985). Ap.J. 299, 303.
Speanhauer, A., Jones,B.F. & Whitford, A.E. (1992). Astron.J., 103, 297.
Spitzer, L. (1958). Ap.J., 127, 17.
Steele, I.A., Jameson, R.F. & Hambly, N.C. (1993). MNRAS, 263, 647.
Steidel,C.C. & Sargent,W.L.W. (1988). Ap.J.Lett., 333, L5.
Stevenson, D.J. (1986). In *Astrophysics of Brown Dwarfs*, ed. M.C. Kafatos et al., p 66 (Cambridge University Press).
Stevenson, D.J. (1991). Ann.Rev.Astron.Astrophys., 29, 163.
Stromgren, B. (1987). In *The Galaxy*, ed. G.Gilmore & R.Carswell, p 229 (Reidel)
Subramanian, K. & Chitre, S.M. (1987). Ap.J., 313, 13.
Surdej, J. et al. (1993). Astron.J., 105, 2064.
Taylor, A.N. & Rowan-Robinson,M. (1992). Nature, 359, 396.
Terlevich,R.J. (1985). In *Star Forming Dwarf Galaxies*, p 395, eds. D.Kunth,T.X.Thuan & J.Tran Thanh Van, Editions Frontieres (Paris).
Thomas, P. & Fabian, A.C. (1990), MNRAS, 246, 156.
Tinney, C.G. (1993). Ap.J., 414, 379.
Tinney, C.G., Mould, J.R. & Reid, I.N. (1992). Ap.J., 396, 173,
Tinney, C.G., Reid, I.N. & Mould, J.R. (1993). Ap.J., 414, 254.
Tohline, J.E. (1980). Ap.J. 239, 417.
Tyson, J.A., Valdes, F. & Wenk, R.A. (1990). Ap.J.Lett., 349, L1.
Umemura,M., Loeb. A. & Turner, E.L. (1993), Ap.J.Lett., 419, 459.
Udalski. A. et al. (1993). Acta Astron., 43, 289.
Van den Bergh, S. (1969). Nature, 224, 891.
Walker, T., Steigman,G, Schramm,D.N., Olive, K.A. & Kang, H.S. (1991), Ap.J., 376, 51.
Wambsganss, J. & Paczynski, B. (1992). Ap.J.Lett., 397, L1.
Wambsganss, J., Paczynski,B. & Schneider,P. (1990). Ap.J.Lett., 358,L33.
Wasserman, I. & Weinberg, M.D. (1987). Ap.J., 312, 390.
Wasserman, I. & Salpeter,E.E. (1993). Preprint.
Webster, R.L., Hewett, P.C., Harding, M.E. & Wegner, G.A. (1988). Nature, 336, 358.
Webster, R.L. et al. (1991). Astron.J., 102, 1939.
White, S.D.M. (1976). MNRAS, 174, 19.
White, S.D.M. & Rees, M.J. (1978). MNRAS, 183, 341.
White, S.D.M., Navarro,J.F., Evrard, A.E. & Frenk, C.S. (1993). Nature, 366, 429.
Whitmore,B.C.,McElroy,D.B. &Schweizer,F. (1987). Ap.J., 314, 439.
Wielen, R. (1985). In *Dynamics of Star Clusters*, ed. J.Goodwin & P.Hut, p 449 (Reidel: Dordrecht)
Woosley, S.E. & Weaver, T.A. (1982). In *Supernovae: A Survey of Current Research*, ed. M.J.Rees & R.J.Stoneham, p 79 (Reidel: Dordrecht)
Wright, E.L. et al. (1994). Ap.J., 420, 450.
Yoshii, Y. & Saio, H. (1986). Ap.J., 301, 587.
Zaritsky,D., Smith,R., Frenk,C. & White,S.D.M. (1993), Ap.J., 405, 464.

Primordial Deuterium Abundance and Cosmic Baryon Density

CRAIG J. HOGAN
Astronomy and Physics Departments, University of Washington
Seattle, WA 98195, USA

ABSTRACT

The comparison of cosmic abundances of the light elements with the density of baryonic stars and gas in the universe today provides a critical test of big bang theory and a powerful probe of the nature of dark matter. A new technique allows determination of cosmic deuterium abundances in quasar absorption clouds at large redshift, allowing a new test of big bang homogeneity in diverse, very distant systems. The first results of these studies are summarized, along with their implications. The quasar data are confronted with the apparently contradictory story from the helium-3 abundances measured in our Galaxy. The density of baryonic stars and gas in the universe today is reviewed and compared with the big bang prediction.

1. Primordial Deuterium and Galactic ^{3}He

People never seem to tire of telling or hearing the triumphant story of the cosmic light element abundances— for example, about how the observed predominance of hydrogen, and the helium mass fraction of roughly 25%, confirm the basic tenets of the big bang theory back to a time of one second, or about how the precisely determined ^4He mass fraction in low-metal galaxies, 0.228±0.005, constrains the cosmic baryon density to be less than $\Omega_b = 0.015h^{-2}$. Here however I will disappoint the avid listeners by focussing attention on the more ambiguous story of the primordial deuterium abundance.

The cosmic baryon-to-photon ratio η is the one astrophysical parameter of big bang nucleosynthesis theory. Everything else follows from pure physics, including predictions for the abundances of light elements. Since we know the photon density today, and since the entropy has changed rather little since nucleosynthesis (as indicated by the perfect Planck spectrum of the background radiation), η determines the physical density of baryons today. In units of the cosmic critical density, the density of baryons is

$$\Omega_b = 3.73 \times 10^{-3} h^{-2} \eta_{10}$$

where $\eta_{10} = 10^{10}\eta$. The theory predicts[1,2] a mass fraction of ^4He

$$Y_p = 0.228 + 0.01\eta_{10}$$

and a primordial number ratio of deuterium to hydrogen

$$(D/H)_p = 4.6 \times 10^{-4} \eta_{10}^{-5/3} = 2.5 \times 10^{-4} \left(\frac{\Omega_b h^2}{0.005}\right)^{-5/3} = 4.5 \times 10^{-5} \left(\frac{\Omega_b h^2}{0.015}\right)^{-5/3},$$

where the last two "high" and "low" values are chosen to illuminate the arguments below. The helium is the best verification of the big bang theory, but the deuterium is the best measure of baryon density, since it is fairly sensitive to η.

The trick is to measure the primordial abundance of this fragile element. Since deuterium is almost impossible to make outside of the big bang, almost all cosmic deuterium is a relic of primordial nucleosynthesis, so it is easy to get a lower limit on $(D/H)_p$. But since the big bang, successive generations of stars have destroyed most of it by burning to heavier elements, which makes any upper limit less straightforward. Indirect constraints on $(D/H)_p$ have been deduced from D abundances in the reprocessed material in the Galactic interstellar medium and from the local abundances of the principal immediate product of D burning, ^3He. Using models of galactic chemical evolution (a weak spot in the argument), the solar abundance of ^3He is used to give an upper limit on the primordial $(D/H)_p$, and thereby a lower limit on the η from standard big bang nucleosynthesis theory[1,2]. This lower limit is higher than observed density of baryons, and therefore provides an argument for abundant baryonic dark matter. If this lower limit is relaxed, the range of allowed η includes lower values where the predicted ^4He abundance lies more comfortably close to observations, where the bound on the number of particle species is relaxed, and where there is almost no baryonic dark matter.

Advances in instrumentation and telescope aperture, especially the Keck telescope and the HIRES spectrograph, have recently enabled observations of high-redshift quasars at high spectral resolution, of the order of 10km sec^{-1}. This has enabled detailed study of the Lyman series absorption lines of hydrogen and its deuterium counterpart (shifted by one part in 3700 to the blue) in high-column-density foreground absorbers[3,4]. The first such observations revealed absorption consistent with a high abundance of deuterium in one very metal-poor absorbing cloud in one quasar (Q0014+813), around $D/H \approx 2 \times 10^{-4}$ by number, which is about a factor of five higher than previous guesses for the primordial D/H based on observations of local ^3He abundances, and an order of magnitude greater than the interstellar value[5]. As of this writing, the abundance in one other, slightly more enriched cloud has been measured[6] closer to the ISM value. One Ly series absorption line system can always be dismissed as a chance hydrogen interloper masquerading as deuterium[7], while a low D/H can be attributed to destruction, so we require more systems for a clearer picture to emerge of the true primordial value. The technique will in time be applied to many other quasars and absorbers, establishing critical tests we have never had before of big bang theory: abundances at cosmological distances, in different environments and at different cosmic epochs, and in pristine material which has undergone relatively little chemical evolution (see figure 2). If more than one high D/H value is found, it will lead to a major revision of thinking about this isotope. While not conclusive, the first data have raised at least the possibility that the primordial D/H might be rather high, and the global baryon density rather low, motivating a reexamination of the interpretation of cosmic abundances and the data on the density of baryons.

The measure of $^3He/H$ most widely used for estimating $(D/H)_p$ comes from measurements of the solar wind, both from direct exposure experiments and from meteorites, of $^3He/^4He$[8], which are used to infer that the abundances of the presolar nebula by number were $D/H = 2.6 \pm 1.0 \times 10^{-5}$, $^3He/H = 1.5 \pm 0.3 \times 10^{-5}$, $(D + {}^3He)/H = 4.1 \pm 1.0 \times 10^{-5}$. These are taken as Galactic or solar-circle averages for the purpose of defining constraints on primordial abundances. A skein of theory then connects the presolar abundance with the primordial one. The main reason why the new Keck D/H observations come as a surprise was that they give a D/H a factor of five larger than ^3He abundances in the solar system.

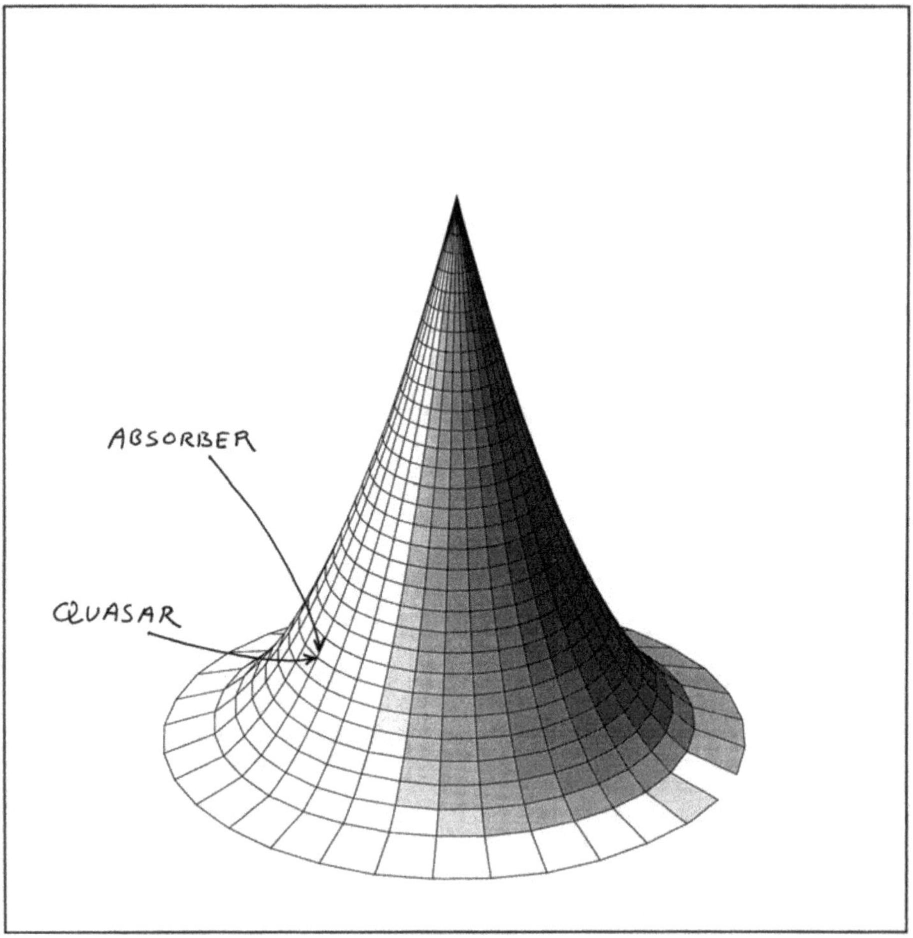

Figure 1– Comoving lightcone in an Einstein-deSitter model, showing space-time location of Q0014+83 and the "Chaffee cloud" where Songaila et al. estimated D/H from Lyman series absorption. Vertical position corresponds to cosmic scale factor; the bottom of the cone is $t = 0$, the top is the present epoch. Horizontal position is the present-day proper distance; the apex is our position, and the classical particle horizon is the bottom of the cone. The great distance of the cloud tests the cosmological principle applied to primordial abundances, the first measurement probing the early past world lines of distant points— the distant interior of our past light cone. The high redshift of the absorber provides a direct probe of pristine material from the big bang; this cloud has metal abundances estimated at less than 10^{-3} of the solar value.

In the standard picture the bulk of primordial D in the Galaxy is burned to ^3He in protostellar collapse. Galactic chemical evolution models[9,10,11] show that D/H can be reduced in this way to its present interstellar value ($\approx 1.5 \times 10^{-5}$, ref. 5) from any plausible initial value. However, in the low mass stars which now dominate the chemical recycling of the interstellar medium (ISM), the bulk of the material is assumed to be never heated to the higher temperature required to burn the ^3He, so the bulk of the primordial D reappears in the ISM as ^3He when the envelopes are ejected. For the galaxy as a whole, the sum $(D+^3He)/H$ therefore only increases with time, so that even the solar ^3He abundance can be used to set constraints on $(D/H)_p$. This is why the evolution of ^3He is critical.[12]

The only other useful measure of cosmic $^3He/H$ comes from radio emission maps of highly ionized HII regions in the Galaxy[13,14]. The column density of $^3He^+$ is estimated from the brightness in the 8.665 GHz hyperfine transition line, and the column (squared) density of H^+ or $^4He^+$ is estimated from radio recombination lines. Balser et al. use this data and a simple model of the gas distribution to obtain reliable estimates of $^3He/H$ in 7 Galactic HII regions, and "preliminary" abundances and limits in 7 more. Two of the most reliable ones are W43 and W49, with low values $^3He/H = 1.13 \pm 0.1 \times 10^{-5}$ and $^3He/H = 0.68 \pm 0.15 \times 10^{-5}$ respectively. There appears to be a real range of values, with W3 for example measured at $^3He/H = 4.22 \pm 0.08 \times 10^{-5}$, and some are consistent with still higher values. There may be a trend with galactocentric radius in the sense that lower values tend to lie within the solar orbit and higher values outside it.

Note that these results do not mesh with the standard interpretation of the solar ^3He; quite aside from the Keck D/H observations, empirical evidence in the Galaxy suggest that stellar populations on average actually get rid of ^3He.

- The Solar System value $(D+\ ^3He)/H$ is greater than the interstellar one; if $(D+\ ^3He)/H$ were steadily increasing, it ought to be less, because of the elapsed time since the formation of the solar system.
- The ISM shows large variations in $^3He/H$, which argues that one ought not take any one point, such as the solar system, as an average of the Galactic abundance, and that simple uniform-mixing models are unlikely to accurately model the abundances at any given point, such as the solar system.
- The gradient with Galactic radius goes the wrong way; if stars are creating ^3He on average, it ought to be highly enriched towards the Galactic center, like other heavier elements are.
- If we adopt instead the lowest $^3He/H$ value in ISM (W49) as the primordial one, to be consistent with the idea that $^3He/H$ cannot decrease, thereby assuming that the additional ^3He found at other sites is Galactic in origin as required in the standard picture, *then* SSBN requires a large $\Omega_b h^2 = 0.075$, in which case it also predicts an excessive ^4He abundance $Y_p \approx 0.26$. The observed value is $Y_p = 0.228 \pm 0.005$ [15,16], which is marginally inconsistent even with the SBBN prediction for solar $^3He/H$, $Y_p = 0.242$.

A destruction mechanism is therefore desirable both for improving the consistency of big bang theory and for interpreting the Galactic $^3He/H$ data. It is not clear whether such a mechanism operates in the Galaxy. One recently proposed mechanism[12] is based on mixing envelope material in low mass stars down to high temperature after they reach the giant branch, so that the ^3He is destroyed before the material is ejected. This process, originally postulated to explain the observed change in C and N isotope abundances[17,18] as low mass stars ascend the giant branch, would also destroy ^3He. It remains to be seen how important it is for the population as a whole, but the possibility of such effects motivates caution in using highly processed material for estimating $(D/H)_p$.

2. Cosmic Baryon Bookkeeping

It is interesting to compare the density of baryons inferred from either SBBN argument with the density of baryons and dark matter found in the universe.

Let us review a number of different measures of global densities, summarized in figure 2. Each column shows both estimated statistical errors in the method and the variation with the (still uncertain) Hubble constant h, where $h = H_0/100\text{km sec}^{-1}\ \text{Mpc}^{-1}$.

The first column shows the contribution of baryons to the mean density, estimated from standard Big Bang nucleosynthesis (SBBN). The current canonical range[1,2, 9] is shown, $\Omega_b h^2 = 0.010 - 0.015$, which leads to the best concordance with a low value of $(D/H)_p$. Most reliable is the upper limit of this range, which appears firmly fixed by a variety of abundances. Indeed, it represents a 3σ departure from the best value[15] for the primordial ^4He abundance, $Y_P = 0.228 \pm 0.005$. Some have argued[19] that one should instead fit the best value of Y_P, requiring $\Omega_b h^2 \approx 0.005$, the lower indicated range. This estimate of $\Omega_b h^2$ agrees with the recent possible detection[3] of high deuterium.

A direct lower limit on gas density is imposed by quasar absorption line statistics, shown in the second column. A large sample of quasars provides an accurate census of all neutral hydrogen in the universe at $z < 4$ through Lyα absorption along their sightlines. At high redshift, the bulk of the HI is in high column density damped Lyman α (DLyα) absorbers, with column density in the range $N(HI) \approx 10^{20-22}\text{cm}^{-2}$. These contribute[20] an integrated density $\Omega_{\text{DLy}\alpha} h = 2.9 \pm 0.6 \times 10^{-3}$, which should be taken as a lower limit on the total density of such absorbers[21].

The third column shows an example of traditional baryonic bookkeeping[22]: estimate the mass-to-light ratio of a population, then use the mean cosmic luminosity density (here, in V band and solar units) to find a contribution to the mean mass density. We use spiral galaxies, as they dominate the luminosity density. Our Galactic disk out to 700pc height has $(M/L)_V = 5\,\text{M}_\odot/\text{L}_\odot$, with most of the mass contained in stars. If all spiral disks have the same mass-to-light ratio, we can use an estimate[22] of the luminosity density ($j_0 = 1.7 \pm 0.6 \times 10^8\,\text{L}_\odot/Mpc^3$ in V) to get the integrated density of all the material in spiral galaxy disks. (Note that the errors shown are just those from the estimate of j_0). Similarly there are several mass measurements of the Galaxy halo mass from local group satellite galaxy orbits[23] and from local group timing, which yield masses of at least $1 \times 10^{12}\,\text{M}_\odot$. With a Galactic luminosity[22] of $1.4 \times 10^{10}\,\text{L}_\odot$ this corresponds to an overall lower limit of $(M/L)_V = 71\,\text{M}_\odot/\text{L}_\odot$ for the Galaxy. If we assume that all spirals have a similar amount of halo material per disk light, we obtain the estimate shown for spiral halos, $\Omega_{\text{halos}} h > 4.4 \pm 1.5 \times 10^{-2}$.

A variant of this argument is shown in column four. Persic and Salucci[24] have integrated the luminosity functions of spiral and elliptical galaxies separately, with their M/L estimated directly from dynamics, allowing h to be eliminated. They estimate $\Omega_b = 1.5 \times 10^{-3}$ from ellipticals and $\Omega_b = 0.7 \times 10^{-3}$ from spirals, which is systematically lower, and probably more accurate, than the previous argument. I have estimated errors to again be at the 30% level. It is interesting that the total contribution from gas in groups and clusters is comparable to these, $\Omega_b = 1.5 \times 10^{-3} h_{50}^{-1.3}$, in spite of the fact that rich clusters contain only about 1% of galaxies, and that the clusters out to the Abell radius were assembled from about 1% of the total comoving volume, but there ought to be large errors attached to this estimate.

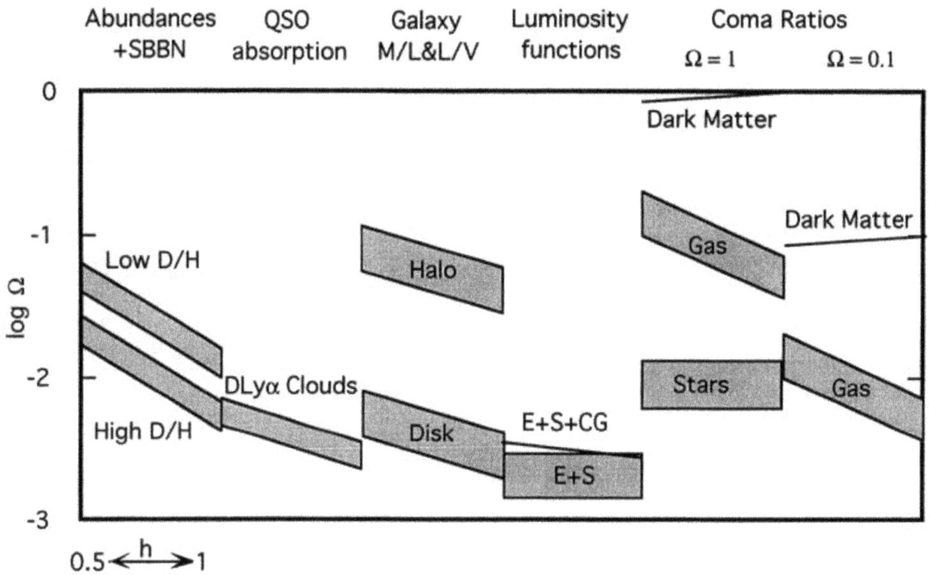

Figure 2– Estimated contributions of various components to the global density, in units of the cosmic critical density. Each column shows a vertical range from estimated internal errors, as well as a variation across each column due to the range of possible values of the Hubble constant, $0.5 < h < 1$. Column 1 shows the range for Ω_b allowed by SBBN, both for the canonical limits derived from solar system ^3He abundance, and for the lower value estimated from the recent possible detection of high primordial D/H in a single QSO absorber and the best measured value of primordial ^4He abundance Y_p. Column 2 shows the contribution of neutral hydrogen in quasar DLyα absorbers at $z \approx 3.5$; although Wolfe's errors are shown, this should be taken as a lower limit for the HI density. Column 3 shows estimates of the global density of spiral galaxy disk stars and halos, obtained from local (Galactic) M/L estimates combined with the mean cosmic luminosity density. The lower band represents $M/L = 5\,\mathrm{M}_\odot/\mathrm{L}_\odot$, representative of the local disk material out to about 700 pc, and the upper band (Ω_{halos}) represents $M/L = 71\,\mathrm{M}_\odot/\mathrm{L}_\odot$, corresponding to a Galaxy halo mass of 1×10^{12} solar masses. Column 4 shows a similar estimate but based on integration of luminosity functions and dynamically estimated M/L; the band represents Persic and Salucci's estimate for spiral and elliptical galaxies, with errors added by me, and the upper line represents their estimate including cluster gas. Columns 5 and 6 show global densities estimated from the ratio of components in the Coma cluster. Column 5 shows the dark matter, gas and star components where the sum is assumed to have $\Omega = 1$; Column 6 shows the same ratios, but where the total density is fixed to be 0.1.

Column five is based on just the Coma cluster, where we have the most uniformly reliable data[25]: it shows cosmic densities based on the estimated mass for Coma ($M = 1.1 \pm 0.18 \times 10^{15}h^{-1}\,M_\odot, M_{gas} = 5.45 \pm 0.98 \times 10^{13}h^{-5/2}\,M_\odot, M_{stars} = 1.0 \pm 0.2 \times 10^{13}h^{-1}\,M_\odot$), assuming that Coma is representative of their cosmic ratios and that $\Omega_{PDM} = 1$. The high baryon density is in apparent conflict with SBBN. Ironically, the high M/L of galaxy clusters, regarded since Zwicky as the strongest evidence for plentiful cosmic dark matter, is now apparently due to low L (i.e., the bulk of the baryons being in gas rather than stars), and not to high M (i.e., having a more representative sample of the high cosmic dark matter density.) The well-studied central region of Coma has about the same mix of visible baryons and dynamical dark matter as the Milky Way. This is shown in column five, which again shows the empirical ratios of mass to gas and stars in the Coma cluster, only now assuming instead that $\Omega = 0.1$ for the dark matter.

Where then are the baryons? Common thinking is that most of the baryons reside in a photoionized IGM— everywhere outside of clusters, there is the same large amount of gas per galaxy as there is inside, but it is not seen because it is not hot and dense enough to emit X-rays. But it remains difficult to reconcile the large number of cluster baryons with the small Ω_b required by even the highest SBBN limits. Galaxy halos contain about the same amount of material as the total baryonic density for canonical (low D/H) SBBN; a popular option is to make the halos out of compact objects (MACHOs; see Carr's contribution in this volume). But making halos out of MACHOs prevents us from using these baryons in the IGM; baryons would have to be converted into MACHOs in galaxies, and gas in clusters, so the formation of the MACHOs would need to be at a low redshift, accessible to observation. The lower value of $\Omega_b h^2 = 0.005$ does not provide even enough baryons to make galaxy halos. There is no easy way to reconcile the large baryon abundance of clusters with this very low baryon density; a cluster like Coma would need to gather baryons from a volume twenty times bigger than the volume from which it gets galaxies.

One route to reconciliation is the introduction of inhomogeneities in the baryon distribution within the early big bang, on either small or large scales. Small scale inhomogeneities, which can arise naturally from QCD or electroweak phase transitions, create zones of neutron-rich nucleosynthesis[26, 27, 28, 29], which naturally creates extra deuterium[26] compared to a homogeneous model; these models however do not appear capable of exceeding the usual SBBN upper limit[29]. There could also be more baryons if some of them gravitationally collapse into a compact form at or before recombination, due to large-scale inhomogeneities [30,31]. Some noise distributions naturally produce approximately correct abundance distributions in the diffuse baryons that do not collapse, so in principle the total density of baryonic matter could be greater than classical nucleosynthesis bounds, although they cannot easily solve the Coma cluster problem, which necessarily involves diffuse gas. Such ideas can be constrained or eliminated by a variety of gravitational microlens probes of the nature of dark matter[32,33,34,35,36], and can also be tested directly, by measuring abundances in different locations (eg, different quasar absorbers.) It is important to emphasize that the overall concordance between the different light element abundances remains good, which indicates that the basic SBBN picture is a good approximation— and that there is no reason to think that this situation will change.

This work was supported by NASA grant NAGW-2523 and NSF grant AST 9320045 at the University of Washington.

5. References

1. Walker, T. P., Steigman, G., Schramm, D. N., Olive, K. A., Kang, H. S. *Astrophys. J.*, **376**, 51 (1991).
2. Smith, M. S., Kawano, L. H. & Malaney, R. A. *Astrophys. J. Suppl.*, **85**, 219 (1993).
3. Songaila, A., Cowie, L. L., Hogan, C. J., & Rugers, M. *Nature*, submitted (1994)
4. Carswell, R. F., Rauch, M., Weymann, R. J., Cooke, A. J., and Webb, J. K., 1994, MNRAS, 268, L1-L4
5. Linsky, J. L., et al., 1992, ApJ, 402, 694
6. Tytler, D., and Fan, X., private communication
7. Steigman, G., 1994, MNRAS, 269, L53
8. Geiss, J. 1993, in Origin and Evolution of the Elements, ed. N. Prantzos, E. Vangioni-Flam and M. Cassé (Cambridge: Cambridge Univ. Press), p. 89
9. Steigman, G. & Tosi, M. *Astrophys. J.*, **401**, 150 (1992).
10. Vangioni-Flam, E., Olive, K. A., and Prantzos, N., 1994, ApJ, 427,618
11. Galli, D., Palla, F., Ferrini, F., Penco, U., 1994, ApJ in press, Arcetri preprint 20/94
12. Hogan, C. J., 1994, *Astrophys. J.*, submitted
13. Balser, D. S., Bania, T. M., Brockway, C. J., Rood, R. T., and Wilson, T. L., 1994, ApJ, in press
14. Wilson, T. L., and Rood, R. T., 1994, Ann. Rev. A. Ap., 32, in press
15. Pagel, B. E. J., Simonson, E. A., Terlevich, R. J., & Edmunds, M. G., *Mon. Not. Roy. astr. Soc.*, *255*, 325-345 (1992)
16. Skillman, E. D., and Kennicutt, R. C., 1993, ApJ, 411, 655
17. Charbonnel, C. 1994, AA, 282,811
18. Brown, J. A., Wallerstein, G. W., and Oke, J.B. 1990, AJ , 100, 1561
19. Vangioni-Flam, E. & Audouze, J. *Astron. Astrophys.*, **193**, 81 (1988).
20. Wolfe, A. M. *Ann. N. Y. Acad. Sci.*, **688**, 281 - 296 (1993).
21. Fall, S. M. & Pei, Y. C., *Astrophys. J.*, **402**, 479-492 (1993)
22. Binney, J. & Tremaine, S., *Galactic Dynamics*, Princeton University Press, (1987)
23. Zaritsky, D., Olszewski, E. W., Schommer, R. A., Peterson, R. C., and Aaronson, M. *Astrophys. J.*, **345**, 759-769 (1989)
24. Persic, M. and Salucci, P., 1992, MNRAS, 258, 14P
25. White, S. D. M., Navarro, J. F., Evrard, A. E., & Frenk, C. S., *Nature*,**366**, 429-433 (1993)
26. Applegate, J. H., and Hogan, C. J., 1985, Phys. Rev. D 30, 3037
27. Hogan, C. J., 1991, Ann. N.Y. Acad. Sci., 1991, 647, 76
28. Malaney, R. A., and Mathews, G. J., 1993, Phys Rep 229, 145
29. Jedamzik, K., Fuller, G., and Mathews, G., 1994, ApJ, in press
30. Hogan, C. J. *Astrophys. J.*, **415**, L63 - L66 (1993)
31. Gnedin, N. Y., Ostriker, J. P., and Rees, M. J., 1994, ApJ, in press
32. Alcock, C., et al., *Nature*, **365**, 621-623 (1993).
33. Aubourg, E., et al., *Nature*, **365**, 623-625 (1993)
34. Dalcanton, J. J., Canizares, C. R., Granados, A., Steidel, C. C., and Stocke, J. T., *Astrophys. J.*, 424, 550 (1994).
35. Hawkins, M. R. S., *Nature*, 366, 242 (1993).
36. Crotts, A. P., *Astrophys. J.*, **399**, L43-L46, (1992)

COLD MOLECULAR GAS AS DARK MATTER

Françoise COMBES
DEMIRM, Observatoire de Paris
61 Av. de l'Observatoire, F-75 014 Paris, France

and

Daniel PFENNIGER
Geneva Observatory
University of Geneva, CH-1290 Sauverny, Switzerland

ABSTRACT

The visible baryonic mass in galaxies contributes to only a fraction Ω_b = 0.005 of the critical density of the Universe. From the virial dynamics and rotation curves, it can be derived that galaxies are embedded in extended haloes of dark matter, that contribute to $\Omega = 0.05$, while the big-bang nucleosynthesis constrains the total baryonic mass between $0.01 < \Omega_b < 0.06$. From this, we can deduce that there must exist some dark baryonic matter, and that the dark matter around galaxies could be only baryonic. This has motivated the search of brown dwarfs in the galactic halo (MACHO-EROS-OGLE experiments), with the preliminary result that compact objects can not make all the dark matter in the Galactic halo.

Could these dark baryons be in fact in the form of cold molecular gas? This hypothesis has not yet been seriously considered, although it is suggested by numerous observational facts: the existence of more visible gas in interacting galaxies, the constant ratio between surface densities of HI gas and dark matter in external parts of spiral galaxies, the rate of gas consumption by star formation in galaxies, the numerous absorption systems gas clouds in quasar spectra, etc...

Also, the evolution of galaxies along the Hubble sequence implies that a significant fraction of dark matter is transformed into stars in a Hubble time: this is only possible with baryonic dark matter; furthermore, it has to be in a sufficiently dilute form to remain available for star formation, which excludes compact objects like brown dwarfs, white dwarfs, and black holes.

Inspired by the observational properties of cold interstellar gas a model is developed where the cold molecular gas has a fractal structure, in isothermal interaction with the cosmic background radiation. The gas never reaches an asymptotic equilibrium: the Jeans instability tends to fragment hierarchically the medium, and the gravothermal instability e-vaporates growing singularities; the medium is collisional and the collisions between fragments lead to coalescence or disruption, with the thermal me-diation of the 3 K background. The smallest scales of the fractal are "clumpuscules" of the order of Jupiter masses, with sizes of about 30 AU, at the transition from isothermal to adiabatic regime.

The model can also explain the existence of large masses of hot gas in galaxy clusters, the decrease of the total to luminous mass ratio in rich clusters and the dark matter concentration towards the cluster center, observed in X-rays and confirmed by gravitational lenses.

1. Introduction

The nature of most of the mass in the Universe is still unknown, although a multitude of candidates, more or less exotic, have been proposed to account for the preponderant dark matter. If we follow the Big Bang scenario, the primordial nucleosynthesis predicts that from the census of all the baryons most of the baryonic dark matter remains to be discovered. In fact, the dark matter required around galaxies to account for the observed rotation curves could be only baryonic.

The observation that our Universe is homogeneous and isotropic at the largest scale, and nearly flat today, and the reluctance to consider arbitrarily small numbers and parameter fine-tuning, have led theoreticians to contemplate the hypothesis of inflation, and adopt the critical value for the density $\Omega = 1$. The hypothesis provides also a physical basis for a scale-invariant spectrum of primordial density fluctuations, which is compatible with the COBE observations. However a Universe containing only baryons, with a smaller value of Ω, compatible with primordial nucleosynthesis, is not precluded. Moreover, dynamical observations are not compatible with $\Omega = 1$, unless we accept that galaxies and galaxy clusters are a biased mass tracer.

Most baryonic candidates considered for dark matter are massive compact objects, like black holes or brown dwarfs. Recently, three groups have surveyed millions of stars towards the Large Magellanic Cloud or the Galactic Center, and detected a dozen of gravitational microlensing events, proving the existence of massive compact objects (MACHOs) between 0.1 and 1 $M_\odot$ in large numbers (Alcock et al 1993, Aubourg et al. 1993, Udalski et al. 1994). It is not yet clear whether these objects represent a large fraction of dark matter, or only the extrapolation of the stellar mass function towards stars of small mass. The highest number of events have occurred towards the Galactic Center, which suggests that we are probing the anisotropic distribution of stars in the Milky Way bar (Kiraga & Paczinsky 1994). It is interesting to note that the non-detection of microlensing of very small duration, already rules out that small objects of much less than a Jupiter mass, be the main component of the halo dark matter.

Another candidate for the baryonic dark matter around the Galaxy is cold molecular gas, distributed in a fractal structure. Homogeneous gas filling the intergalactic volume can be ruled out by the Gunn-Peterson test, but condensed fragments of cold gas, with a very low surface filling factor, would be extremely difficult to detect either in emission or absorption. Such a fractal structure is already observed in the cold ($5 - 100\,\mathrm{K}$) interstellar medium, so above the $3\,\mathrm{K}$ background, when it is slightly heated by stars and by-products like cosmic rays.

We present here the astrophysical motivations for such a model, and describe the physics of the fractal gas.

2. Rotation Curves of Spiral and Irregular Galaxies

The best evidence until now of the existence of dark matter is provided by the HI gas rotation curves of spiral galaxies. Already, the optical rotation curves obtained with ionized gas (Hα) were found to be surprisingly flat (e.g. Rubin et al. 1980), but they do not extend enough outside the optical disk to unambiguously require the presence of dark matter. On the contrary, the radio rotation curves obtained up to 3–4 optical radii with HI at 21 cm, are maintained flat so far out that the presence of a dark component is unavoidable (e.g. Bosma 1981, van Albada et al. 1985).

2.1 Dark Matter Fraction and Morphological Type

More recently, it has been realised that the fraction of dark matter depends on the morphological type: it is increasing along the Hubble sequence, from Sa to Sd (also called "early" and "late" galaxies respectively), and maximizes for dwarf irregulars. The Hubble sequence was until now characterized by five main parameters monotonously varying from Sa to Sd: 1) the decreasing bulge-to-disk luminosity ratio, indicating that the mass concentration decreases along the sequence; 2) the increasing gas fraction; 3) the decreasing total mass; 4) the decreasing fraction of heavy elements, processed in stars; and 5) the increasing pitch angle, i.e., the spiral arms being more wound for early-type galaxies.

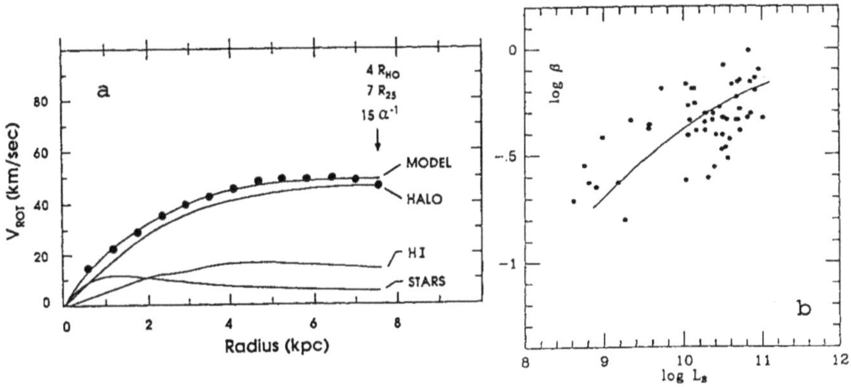

Fig. 1: a) Rotation curve and mass model for DDO154, from Carignan & Freeman (1988). In this dwarf galaxy, the HI mass dominates the stellar mass, and the dark halo is even more massive. b) The fractional visible mass β at R_{opt} as a function of luminosity L_B, from Persic et al (1993).

From enhanced sensitivity HI observations in gas-poor early-type galaxies, falling rotation curves have been discovered (Casertano & van Gorkom 1991, Salucci & Frenk 1989). On the other side of the sequence, dwarf irregulars are observed with rising rotation curves (e.g. Carignan & Freeman 1988). These objects contain a large fraction of gas, sometimes the mass in HI gas is much larger than the visible mass in stars, yet the missing mass is even larger (Fig. 1a).

This means that early-type galaxies *have proportionally less dark matter*, and dwarfs spirals *have proportionally more dark matter*. Since the Hubble sequence is also a mass (or luminosity) sequence, it emerges that the dark-to-luminous mass decreases with total visible mass, as shown in Fig. 1b (Persic & Salucci 1993, Persic et al 1993).

Another observational feature is the high resemblance of rotation curves shapes among all galaxy types. If the rotation velocity continues to increase with radius in dwarf galaxies, and begin to fall in massive galaxies, it might be only a question of size of the visible galaxy. The rotation curve of a dwarf corresponds to the inner part of the rotation profile of a luminous early-type galaxy. When calibrating with two main parameters, the maximum of the rotation curve, and the radius where this maximum is reached, it is possible to build a universal rotation profile (Persic & Salucci 1991). The main direct features of rotation curves are therefore only function of their total luminosity.

On the intermediate scale, results on binary galaxies and satellites of galaxies, have suggested the trend towards a constant mass around spiral galaxies, of about 10^{12} $M_\odot$ at 100kpc, with an uncertainty of $30-50\%$ (e.g. Ostriker 1985). The large range of variation of the rotational velocity at small scale, due to various stages of mass concentrations, would then be washed out at large distances. This could be reconciled with the observed rotation curves, if the rotation profiles of massive galaxies are falling, while those of dwarf galaxies are rising outside the measured domain.

These new results should be re-enforced with larger statistics. It is of prime importance, for instance, to find several galaxies with no evidence at all of dark matter. This will provide strong constraints on non-baryonic dark matter models, and even rule out modified Newtonian gravity hypotheses.

2.2 Maximum Disk

How well do we know the radial distribution and exact amount of dark matter in spiral galaxies? How much mass is associated with visible stars?

The M/L mass-to-light ratio for the stellar population depends on age, metallicity, and the overall colors. Tinsley (1981) has computed the variations of M/L with (B-V) color, from 1 to 10 for red to blue galaxies. She deduced already at that time that since the total (dynamical) mass to luminosity ratio was not varying as much, the fraction of dark matter was higher in blue galaxies, i.e. in the late-types of the Hubble sequence. The possibility to vary M/L for stars, according to their age and color, makes the maximum disk hypothesis tenable for most galaxies: i.e., it is possible to explain the rotation curve in the inner parts of the galaxy by maximizing the contribution of the visible matter in fitting M/L. No dark matter is then required within the optical disk. Recently, Buchhorn et al. (1994) have succeeded to fit the observed Hα rotation curves of 97% galaxies in a sample of about 500: they compute the gravitational potential of the galaxy from the red image (less perturbed by extinction), assuming a constant M/L as a function of radius. The derived rotation curve fits surprisingly well the kinematical Hα data;

it remains quite flat, except for small-scale wiggles, tracing the streaming motions associated with spiral arms. Of course, the choice of a lower M/L for the stars is still possible, within the uncertainties; however the precise fits of these wiggles is a strong argument in favor of the maximum disk, since any spherically distributed dark matter does not produce wiggles so tightly related to the spiral arms.

When the rotation curve is flat, the radial distribution of dark matter corresponds to an isothermal component, the density varying as r^{-2} in the extended parts of the disk, and the projected surface density as r^{-1}. In such a mass distribution the inner mass *diverges* as $\sim r$. In some cases, however, not only isothermal distributions, but also r^{-3} or r^{-4} asymptotic laws for the density are possible (Lake & Feinswog 1989).

The main argument against the maximum disk hypothesis rises within non-baryonic dark matter scenarios. It is indeed difficult to imagine a scenario producing a hollow non-baryonic dark matter component, with a central hole filled in by the baryonic matter, if the dark and luminous matter components are physically different and independent (one is spherical, the other one flattened). However, this argument is no longer valid if both types of matter are two forms belonging to the same physical component, and simply one form converts progressively into the other one at some radius.

2.3 Conspiracy

The flatness of most rotation curves has been considered a puzzle, since it implies a conspiracy between two supposedly independent dynamical components: the visible disk and the spherical dark halo. We have seen already that this flatness is no longer universal, rotation curves having tendencies to rise for late-types and to fall for early-type galaxies. But yet, most intermediate types have a flat curve.

When the curve is flat already within the optical disk, the visible matter alone can account for it (e.g. Freeman 1993). The problem is to explain the continuity of the curve in the outer parts, dominated by dark matter. Several scenarios have been proposed, arguing that dark matter is dragged by baryonic infall (Blumenthal et al. 1986). But in dark matter dominated dwarfs the predicted core radius of the total mass is too short with respect to the observations (Moore 1994).

The conspiracy is of course no longer a problem if the dark and luminous matter are two aspects of the same dynamical component, one phase being progressively transformed in the other, during galaxy evolution.

3. Hubble Sequence

3.1 Characteristics of the Sequence

The classification of galaxies, as presented by Hubble (1926) was mainly morphological. Hubble introduced the terms "early" and "late", but as a convenience in his sequence going from elliptical to spirals and irregulars. He has not claimed at this epoch that there could be any temporal evolution from one type to another, and what was the direction of evolution. However, Sandage (1961) in the Hubble Atlas, suggests already that, if there exists an evolution, it should go in the reverse sense, i.e. from Sc to Sa. He notices that Irr, Sc and SBcs are galaxies remarkable from their amount of young stars: dust, gas and OB stars are there very prominent, while star formation appears to have ceased in the tightly wound Sa spirals. He proposed that a galaxy must begin its life as an Sc; when the bright stars die, it become an Sa; and when all arms are gone and no stars are being formed, the system becomes an S0.

What are the main physical parameters characterizing the sequence?

1. – The bulge-to-disk luminosity ratio increases from late to early types spiral galaxies. According to the M/L ratios of both components, this means that the mass concentration increases. If there is an evolution, it can only be towards increasing mass concentration, therefore from Sc to Sa.

2. – The percentage of gas mass decreases, from Sc to Sa. This is mainly the fact of atomic hydrogen, molecular hydrogen being more homogeneously distributed (Young & Scoville 1991).

3. – The total (visible) mass increases from Irr and Sc to S0. Again, this points towards an evolution direction.

4. – The percentage of dark matter decreases from Sc to Sa. If the late-type galaxies have been observed in large numbers, and their flat rotation curves are well known, this is due to their huge amount of atomic hydrogen. The rotation curves of early-types are less known, in particular far from their centers, only because of HI sensitivity problems. As presented in section 2.1, there is now observational evidence that the rotation curves of early-types tend to decrease with radius at large distance from the center, and that the dark-to-visible mass ratio decreases along the sequence.

5. – Early-type spirals arms are more wound, and late-types more open. The pitch angle regularly decreases along the sequence. This was one of the main parameter of Hubble's morphological classification.

3.2 Evolution along the Hubble Sequence

The motor of evolution is mass accretion. There exist several pieces of evidence that late-type galaxies are continuously accreting gas along their lives. The first is the star-formation history, derived from star formation rate (SFR) tracers at various epochs: Hα luminosities give the current SFR ($\leq 10^8$ yr), blue luminosities inform about stars formed over the last few 10^9 yr, and the mass of the galaxy measures the integrated SFR over the galaxy life. A number of studies have derived from these indicators that for late-type spirals and irregulars, the SFR has been nearly constant over the galaxies' lifetimes (Kennicutt 1983, Gallagher et al. 1984). This involves a continuous gas resupply, or gas infall as already suggested by Larson et al. (1980), to solve the problem of rapid gas consumption in star formation. Another argument comes from the necessity of spiral structure maintenance (e.g. Toomre 1990). The spiral waves transfer gas from the disk towards the center, or the inner Lindblad resonance. If the dissipative medium disappears, the spiral structure is washed away on a time-scale of a rotation period.

Mass concentration is only possible if there exists a mechanism to evacuate the excess angular momentum of the accreted gas. Viscous torques are by far insufficient to drive the gas inwards in a Hubble time, but gravitational torques provide the solution. Trailing spiral waves and bars transfer efficiently the angular momentum outwards. As soon as the percentage of rotational kinetic energy is too large, the bar instability sets in, and the disk dynamical equilibrium is self-regulated by those instabilities.

The bar instability can reshape the galaxy morphology in a few rotations, as shown by N-body simulations. Not only it concentrates the galaxy disk mass towards the center, but also vertical Lindblad resonances excite instabilities *transverse* to the plane, thickening the disk in the central regions, and forming a boxy or peanut-shaped bulge (Combes et al. 1990). Moreover, the accumulation of gas in the central regions is able to dissolve the bar into a hot spheroid or bulge (Pfenniger & Norman 1990; Hasan & Norman 1990). This indicates that bars can be short-lived objects, if the galaxy mass increases. Several bar instabilities can occur in a galaxy's lifetime, contributing to mass concentration and bulge formation, accompanying the galaxy along the Hubble sequence, from Sd to Sa.

Mass accretion can also occur more violently by swallowing small galaxies. Pfenniger (1991) has shown that bars are destroyed by such merging events, and at the same time big bulges are developed.

In summary, there are several pieces of evidence that galaxies are not eternally confined in their morphological type. On the contrary they may evolve in a short time-scale (less than 10^9 yr) along the sequence, through numerous dynamical instabilities, such as the formation and dissolution of spiral waves and bars (Pfenniger et al. 1994). The non-axisymmetric waves produce gravity torques that drive the mass towards the center: a galaxy then evolves from late to early types (Sd to Sa), increasing its luminous mass, and its mass concentration; at the same time, star-formation consumes the gas and increases the heavy element content. The evolution cannot be the other way around, since mass concentration, star formation, and nucleosynthesis are irreversible processes.

The fact that during the evolution, the luminous mass increases, while the dark mass fraction decreases suggests that the dark matter has been transformed into stars, particularly because disk galaxies like the Galaxy are too fragile structures to afford a substantial mass accretion in or at the periphery of the optical disk (Tóth & Ostriker 1992). This is possible if the dark matter around galaxies is composed mainly of hydrogen and helium gas, not yet condensed in compact objects.

4. Clues from the Gas

4.1 Radial Distribution of Gas and Dark Matter

The atomic gas serves as a tracer of dark matter, since only HI is visible in the outer parts of spiral disks. There is moreover some evidence that the gas and dark matter are intimately related. From the flat rotation curves, the surface density of dark matter $\sigma_{\rm DM}$ varies asymptotically as $1/r$, and the HI surface density $\sigma_{\rm HI}$ is also observed to vary in $1/r$. Bosma (1981) was the first to notice a constant ratio of $\sigma_{\rm DM}/\sigma_{\rm HI}$ as a function of radius in spiral galaxies. The constant ratio has been confirmed by many authors (Sancisi & van Albada 1987, Puche et al. 1990), and varies between 10 and 20 according to the morphological type (Broeils 1992).

4.2 3D Distribution: Polar Rings and Thickness of HI Planes

Albeit many observational efforts, it has not been possible until now to determine with certainty the 3D-shape of the dark matter distribution.

Polar rings, i.e., material orbiting almost perpendicular to the main plane of the galaxy were a good hope to test the 3D distribution of mass, and in particular the shape of dark matter. Much work has been devoted to this problem but with controversial results (Whitmore et al 1987; Sackett & Sparke 1990). The modeling is in fact confronted with many difficulties, the main one being intrinsic to the formation mechanism: gas settling in the polar ring and then forming stars. Due to gas clouds collisions, a polar ring cannot form in a galaxy that has already a gaseous component at the same radii in the equatorial plane. Therefore, the equatorial velocities are not determined at the same radius as the polar ones, and the flattening of the halo cannot be directly determined (Reshetnikov & Combes 1994).

Warps and their problematic longevity have long been an argument in favor of spherical dark halos, since the time-scale of differential precession is then considerably enlarged. However, the warp lines of nodes are not observed to be wound as they should even at radii where the disk dominates the mass, and other explanations involving gas accretion and short life-time are now considered more plausible (Binney 1992).

As well as the HI kinematics in the galaxy plane being used to test the radial distribution of mass, the HI velocity dispersion and height can be used to test the mass distribution perpendicular to the plane. In the isothermal sheet model of a thin plane, the height $h(r)$ of the gaseous plane is $h(r) = \sigma_v^2(r)/\pi G\mu(r)$, where $\sigma_v(r)$ is the z-velocity dispersion, and $\mu(r)$ the total surface density, within the gaseous plane. Since the HI gas z-velocity dispersion has been observed constant in the outer parts of face-on galaxies (Shostak & van der Kruit 1984; Dickey et al. 1990), the HI characteristic height as a function of radius gives the amount of dark matter $\mu(r)$ included within the HI plane. In the Milky Way, the height $h(r)$ is increasing linearly with radius (Merrifield 1992), and its precise value is that expected if all the dark matter was flattened as much as the flaring HI plane. In M31, Brinks & Burton (1984) also discovered a flaring linear with radius.

4.3 Gas Reservoirs around most Spiral Galaxies

It has long been known that intermediate-type spiral galaxies have formed stars at about a constant rate along the Hubble time (Kennicutt 1983). Infall of gas has been invoked to solve the problem of gas consumption by star formation (Larson et al. 1980), and even the maintenance of spiral structure needs the replenishment of the dissipative component (Toomre 1990). Recently Braine & Combes (1993) have found that interacting galaxies appear to possess more visible gas than isolated ones, as if galaxy interactions would suddenly reveals more gas. This could be due to the driving and heating of the gas towards the center by tidal torques, if there exists an outer gas reservoir.

5. A New Candidate for Dark Matter : Cold H_2

All these arguments about galaxy evolution and gas properties led us to propose a new model for dark matter around spiral galaxies (Pfenniger, Combes, & Martinet 1994; Pfenniger & Combes 1994). The idea of diffuse gas has been eliminated since the beginning by the Gunn-Peterson test (Gunn & Peterson 1965): if the Universe was filled by an homogeneous diffuse gas, we should detect it in the spectrum of remote quasars, as an absorption trough starting at the redshift of the quasar. This test could be done for atomic gas (21 cm line), or ionized gas (Lyα) etc...

But the gas could be in a clumpy and fragmented structure as it is observed in nearby molecular clouds.

The obvious problem with gaseous dark matter is then to explain why the mass of gas as usually determined from observations is, although proportional to dark matter, insufficient by a large factor (≈ 20). By critically examining the knowledges we have about interstellar gas, we observe the following: interstellar gas is typically very inhomogeneous and multi-phase, covering decades of density and temperature ranges; in gravitating systems such as galaxies only gas at a temperature higher than the virial temperature ($\approx 10^4$ K in galaxies, $\approx 10^7$ K in clusters) tends to be smooth. Colder gas tends to be highly lumpy, without necessarily forming stars or Jupiters (since it is still there!). This is indicative that gravitation rules the gas state through the Jeans' instability condition. Also observations of the local interstellar gas show that most of the gas mass, as usually measured, is in a cold form and fills only a few percents of the volume, an unusual fluid indeed.

It has been proposed by several authors that the nearby cold interstellar medium is distributed in a fractal structure, over at least 4 orders of magnitude in scale (e.g., Scalo 1985; Falgarone et al. 1992). This medium is well known through its emission of molecular lines, and in particular carbon monoxide, CO. Power-law relations have been established between mass and size, or line-width and size, along several magnitudes ranges. This yields a rough estimate of the fractal dimension of this medium: of the order of 1.8 ± 0.2, so less than a surface.

Now a dimension less than 2 has a crucial significance for observations. Indeed, a mathematical theorem about projection of fractals states that the projection of a fractal on a surface has most of the times the same dimension that the fractal, if its

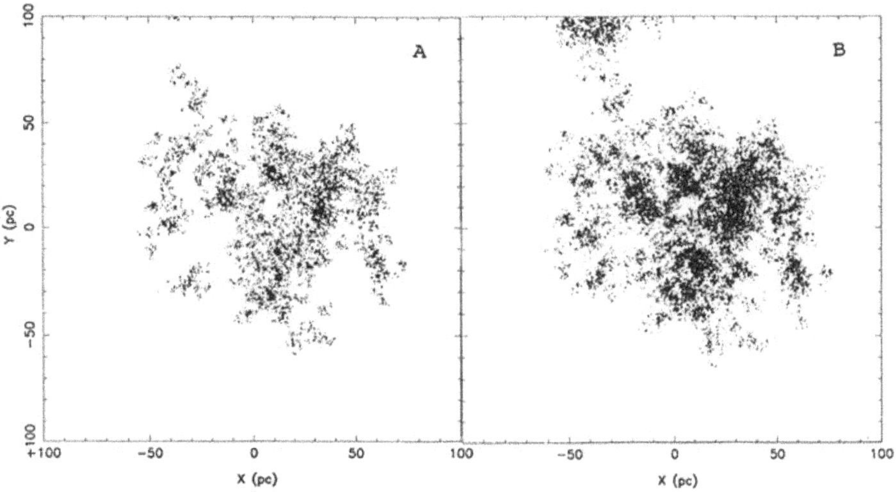

Fig. 2: a) Contours of the projected density of a typical molecular cloud, represented by a fractal structure over 9 levels, containing 4×10^7 clumpuscules, with a fractal dimension of D=1.8. b) same with D=2.2

dimension is less than 2 (Falconer 1990). (Like a line projects generally into a line.) In practice no physical object may keep a fractal behavior over an infinite range. So we conclude that the observational properties (which depend on the projection properties) of a physical object with a fractal dimension less than 2 depend directly on *the smallest scale at which the fractal behavior stops.* For example, the effective sky covering factor of half of the mass of an interstellar cloud may be very small when its dimension is less than 2. Furthermore, this half mass is typically extremely dense in projection. Both factors contribute to underestimate the amount of mass when the instruments are unable to resolve the smallest scale. Computer simulations of fractal clouds have fully confirmed such properties (Pfenniger & Combes 1994), most of the mass may be missed in a fractal object with dimension less than 2 because a tiny fraction of the sky is covered by it at high density (cf Fig. 2).

We extrapolate the observed fractal in the interstellar cold gas down by 2 orders of magnitude in scale, since structures as small as 10 – 20 AU are detected in the interstellar gas (HI VLBI interferometry: Diamond et al. 1989, proper motions when passing in front of quasars: Fiedler et al. 1987, or pulsars: Cognard et al. 1993).

We postulate that large amounts of such molecular gas exist in the outer parts of galaxies, providing enough mass to account for the rotation curves. Since this gas is far away from the stars and stellar formation activity, it is devoid of heating sources, and must be in near thermal equilibrium with the 3 K cosmic background. Isothermal fragmentation occurs in the medium through Jeans instabilities, from small-scale to large-scale (bottom-up fragmentation). The smallest fragments are determined by the transition to the adiabatic regime, i.e. they cannot exchange energy fast enough with the background (here in the millimetric wavelength domain).

We call these smallest fragments, the building blocks of the fractal, the "clumpuscules": they have masses of the order of Jupiters (10^{-3} $M_\odot$), volumic densities of $10^{10}\,\mathrm{H\,cm^{-3}}$, and surface densities of $10^{25}\,\mathrm{H\,cm^{-2}}$. In these conditions, the gas is essentially in molecular form.

So it appears that with today's observational constraints a large range of a priori possible clumps at subparsec scales can hardly be detected. These clumps might be even colder, down to 2.7 K, than those we can measure ($\gtrsim 5$ K). Neglecting the highly clumpy state of cold gas at sub-resolution scale is evaluated to account for the systematic discrepancy between dark matter and HI in the outer gaseous galactic disks (Pfenniger & Combes 1994).

5.1 Physics of Cold Gravitating Gas

Many questions of astrophysics would appear clearer if the following idealized problem would be first understood: *What is the asymptotic state of an ideal gas (small perfectly elastic particles) contained in a box and maintained at a temperature sufficiently low to reach the gravitational (Jeans) instability?* Before introducing more complex physical ingredients, it is important to understand this simple model, because otherwise one might miss important aspects of cosmic gas physics.

In a spherical vessel of radius r the equilibrium state of a hot enough self-gravitating gas is the isothermal sphere (see e.g. Binney & Tremaine 1987 p. 500). However, below a critical temperature ($kT < 0.4GMm/r$), no equilibrium state does exist. Some parts may temporarily collapse and heat indefinitely (the gravothermal catastrophe), but this departure from thermodynamical equilibrium diverges toward a singular state. This indefinitely dense state is equivalent to a finite mass isolated system, which is known to evaporate completely in a finite time. This occurs by increasing the central density and temperature, but also at the same time by decreasing the total collapsing mass (see e.g. Binney & Tremaine 1987 p. 523).

This contradictory behaviors between the beginning and end of a singularity show that in a such Jeans unstable isothermal gas any tendency to collapse is soon prevented by evaporation, but also any tendency to uniformity is prevented by Jeans' instability. The asymptotic state can only be dynamical.

Also, we expect that the average behavior should become scale-free, since no scale between the largest and smallest sizes (given by the box and the particle) is involved in the process. Of course in space there is no wall constraining gas, yet the problem remains because an isothermal sphere requires then an infinite mass. Having a finite mass the whole blob evaporates in about 100 crossing times (see e.g. Binney & Tremaine 1987 p. 525). So the paradox persists, in no way a cold ideal gas in isothermal conditions can find an asymptotic state of rest.

A solution to this paradox is suggested by observations and simulations. As stated above molecular clouds are observed to be fractal over at least 4 decades from 100 to 0.1 pc (Scalo 1990, Falgarone 1992), and shearing sheet experiments of self-gravitating and Jeans unstable layers develop long-range and time-dependent correlations which appear fractal (Toomre & Kalnajs 1991). The asymptotic state is indeed scale-free but dynamical: a fractal state in which clumps exist at every scale, but collide, merge, or disrupt so that the statistical ensemble is in average scale-free.

A fractal state for interstellar gas is very different from the classical picture of fiducial round clouds with a given scale. Not only it has never been possible to show that such clouds are stable over a long time, but also the particular retained size of these clouds has been chosen for purely convenience reasons. In the literature several attempts have been made to "deconstruct" the fiducial interstellar cloud model, yet the model is still alive by lack of alternative pictures.

5.2 Real Cold Gas

In practice no gas is ideal at every scale, and there always exists an upper and lower scale at which the geometrical fractal model breaks. A largest scale in a galaxy is determined by differential rotation ($\sim (d \ln \Omega / dr)^{-1}$), but the smallest scale must be determined by the microphysics. Without invoking complications such as magnetic fields, a simple possibility has been given (Pfenniger & Combes 1994) involving rough opacity properties of ideal gas in a black-body radiation field (Rees 1976): the scale at which there is a transition from the isothermal to the adiabatic regime. In cold conditions near 3 K, for a mix of H_2 and He this transition occurs at solar system sizes, with a corresponding mass of the order of Jupiter. This mass is proportional to $T^{1/4}\mu^{-9/4}$, so is slowly variable with the temperature T, but more strongly dependent on the molecular weight μ. Interestingly, above the H_2 and H ionization temperatures, the critical mass increases by a factor ~ 20, exceeding suddenly the brown dwarf limit of about $0.08\,M_\odot$ by a decrease of μ from about 2.3 to 0.63.

Recently, Ferland et al. (1994) have computed in detail the cooling of uniform gas in cluster conditions taking into account many coupling channels of matter with radiation. They find that indeed cooling may proceed efficiently from millions of K down to 3 K.

In star formation problems it is usually assumed that an adiabatic proto-star cloud collapses in isolation. Yet we have estimated that on the contrary, if the cold gas has a fractal dimension $\approx 1.5 - 2.0$, as suggested by observations of molecular clouds, the collision time is similar or shorter than a crossing time at every scale, so the isolation hypothesis does not hold (Pfenniger & Combes 1994). The clump collisions are then slightly supersonic, which is appropriate to disrupt clumps trying to collapse further in the adiabatic regime. At higher temperature, above the H-ionization, the fractal dimension becomes smaller than 1; then the collisions are subsonic and the collision time increases allowing deeper adiabatic collapses, and eventually star formation.

5.3 Is That Gas Observable around Galaxies?

Most of the mass of the fractal is in a dense enough state to be rapidly in the form of H_2 molecules. Unfortunately the H_2 molecule does not radiate at low excitation, since by symmetry it has no dipole moment (no emission of rotational lines). In molecular clouds, H_2 is traced by the CO emission. This is highly dependent on metallicity, and CO emission is expected to disappear far away from the optical disk, where the gas is deficient in star-synthesised heavy elements. The largest radius where weak CO emission has been found, around and HII region, is 28kpc (de Geus et al 1993). Recently, CO absorption has been found along many lines of sight within the optical disk (cf Lequeux et al 1993, Phillips et al 1994), but it is likely that this gas is diffuse and warm, since its surface filling factor is high. Its excitation tamperature could be very low only because its density is not enough for collisional excitation with H_2 molecules.

High-energy γ rays are also a good tracer of all nucleons, since they are produced in the interaction between cosmic rays and matter. However, the main sources of cosmic rays are supernovae and other stars (particles are accelerated in supernovae shocks), and therefore this tracer is again restricted to the optical disk. In fact, the cosmic-ray distribution is not known, and is presently deduced from the γ ray and gas (HI + H_2) observed distributions. It has been remarked that this indirectly deduced cosmic-ray distribution has a radial scale much larger than what was expected from the supernovae distribution (e.g. Bloemen 1989). This could be due to the extended cold and dark gas distribution that we propose here.

Since the gas is cold, any emission would not be easily detected against the background, and maybe the best hope is to detect it in absorption. Already many absorption lines corresponding to the HI atomic gas are detected in the spectrum of quasars: along some line of sights, up to 100 Lyα absorbing systems have been detected, and represent the well-known Lyα forest. These absorption lines correspond to moderate column densities, between 10^{15} to $10^{20}\,\mathrm{H\,cm^{-2}}$; if they are associated to galaxies they imply an equivalent cross-section, corresponding to a radius of about $R = 200 - 400\,\mathrm{kpc}$, at $z = 2.5$ (Sargent 1988). UV observations with the Hubble Space Telescope have recently revealed that the absorbing systems are still numerous at low redshift (Morris et al. 1991; Bruhweiler et al. 1993). These absorbing systems could be the envelopes of dense and massive molecular clumps.

At the present temperature of the background radiation (2.7 K), it is possible that some of the H_2 freezes. At the average density of the clumpuscules, the phase transition occurs at 3 K. If H_2 ice is present in the shape of snow flakes, they can be efficiently coupled to the background radiation, and radiate themselves like a black-body. This occurs also in the presence of a small amount of dust. The emission of these particles could then be detected by faint fluctuations of the microwave background at small angular scale.

Let us mention also that these clumpuscules are not compact enough to produce micro-lensing for stars in the Large Magellanic Clouds. However, on the line of sight of a remote quasar (at least at 100 Mpc distance), intervening clumpuscules or larger clumps of the fractal structure could produce, by gravitational micro-lensing, the low-frequency part ($\lesssim 10^{-6}\,\mathrm{Hz}$) of the observed variability of these quasi-punctual sources (cf. Hawkins 1993).

6. Galaxy Clusters

The above considerations have some applications to galaxy clusters. Taking the recent observational facts at face: the morphology-density relation shows that nearly all spiral galaxies are transformed into early-type galaxies or completely destroyed

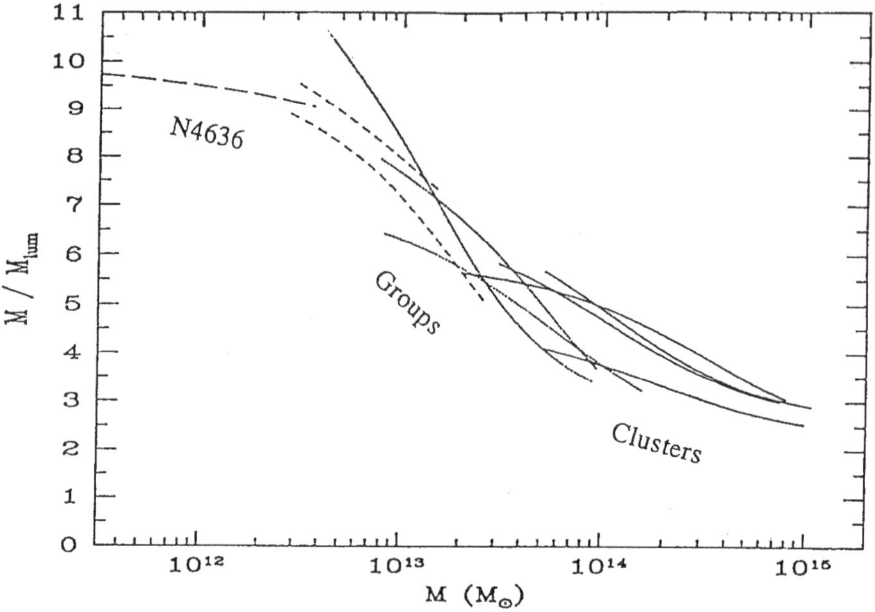

Fig. 3: Ratio of total mass-to-luminous mass in the elliptical galaxy NGC 4636 (long dashed line), the groups NGC 5044 and HC62 (short dashed lines), the cool clusters A262, A359, A2589 (dotted lines), and the hot clusters A1795, A85, A2029, A478 (solid lines), from David et al (1994).

by a single crossing of a cluster. The high hot gas mass fraction ($\sim 30\%$) found in some clusters implies that dark matter in spirals was also at least for such a fraction in form of gas, presumably cold. If cold gas in spiral galaxies is able to escape detection, the same can be expected to happen in clusters. The most natural scale ($\approx 10 - 50\,\mathrm{AU}$) of cold gas at low temperature is hardly accessible with present instruments, even in the solar neighborhood. Since cluster dark matter profiles follow hot gas profiles except near the center, it is therefore likely that a substantial fraction of the cluster dark matter is also in the form of cold fractal gas. The higher fraction of dark matter in cluster centers might well result from the higher cooling rate of the hot gas in denser conditions.

The X-ray emission of the hot gas in galaxy clusters has been explored with enhanced sensitivity by the satellite ROSAT, for 4 years now. More clusters have been detected, and in the brightest clusters, the hot gas has been studied at a much larger distance from the cluster center. Within the hypothesis of hydrodynamical equilibrium, the hot gas is a tracer of the cluster potential, and the total mass has been determined at larger cluster radii. A comparative analysis of groups and clusters have shown that the total mass-to-light ratio decreases with the size of the object (David et al. 1994, Fig. 3). For the largest clusters, which contain also the greatest fraction of hot gas, the M/L ratio begins to decreases, the dark matter becomes luminous. This M/L ratio never rises much above that for spiral galaxies. Previous estimates of larger M/L relied on virial estimates, and on the hypothesis of isotropic motions in the outer parts of clusters, which is not verified since these outer parts are not relaxed.

The radial distribution of dark matter has been determined more accurately from a deeper analysis of X-ray data. The results are consistent with a very small (unresolved) core radius of dark matter: much smaller than that of the hot gas (Gerbal et al. 1992; Briel et al. 1992; Durret et al. 1994). This is confirmed independently by the gravitational lens method for total mass determination, when arcs and mini-arcs are present (Hammer 1991, Wu & Hammer 1993). This more concentrated distribution suggests that the dark matter is baryonic in order to dissipate energy and cool to a lower temperature than the hot gas (Briel et al. 1992).

6.1 Cooling Flows

The hot gas in the center of clusters reaches the density threshold for thermal instability, and its cooling time falls below the Hubble time: it then cools down very quickly (Rees & Ostriker 1977). The rate of cooling has been measured, of the order of $\dot{M} = 10^2\,M_\odot/yr$ (e.g. Fabian 1987). But the cooled gas has never been detected, a puzzle that several scenarios have tried to get round, including star formation with biased initial mass functions. Recently, X-ray absorption has revealed the existence of a large mass of gas in the center of clusters (White et al. 1991; Allen et al. 1993).

Since this gas is not seen in HI (emission, absorption, O'Dea et al. 1994), it could be hidden in the form of molecular clouds (White et al. 1991; Ferland et al. 1994). Several tentatives of CO detection have been made in vain (Grabelsky & Ulmer 1990; Mc Namara & Jaffe 1994; Antonucci & Barvainis 1994; Braine & Dupraz 1994). The low temperature of the molecular gas, and its low filling factor, if it is distributed in small dense clumps, can easily explain these non-detections.

In the hypothesis of cold gas as dark matter, the hot gas comes from the heating of some of the cold gas of the outer parts of galaxies during cluster formation. Since the mass of this cold gas in late-type galaxies is one or two orders of magnitudes than the visible mass, it is easy to explain the domination of the hot gas mass in clusters, between 2 and 8 times the galaxy visible masses (Edge & Stewart 1991). During cooling flows the gas resumes its cold phase, and falls down the cluster potential well, accounting for the small core radii.

6.2 Baryon Fraction of the Total Mass of a Cluster

The new X-ray data allow to make a more accurate mass account within galaxy clusters. From stellar luminosity, hot X-ray gas, the visible baryonic matter is already a fraction $0.15\,h_{50}^{-3/2}$ of the total mass (White et al. 1993).

But standard big-bang nucleosynthesis constrain Ω_b below $0.09\,h_{50}^{-2}$ (Smith et al. 1993). This is hardly compatible with $\Omega = 1$, unless most of the non-baryonic dark matter has been maintained outside of clusters, by some formation mechanism. White et al. (1993) show that this cannot be accounted for by dissipative effects during cluster formation, and conclude:

- either that Ω is much lower than 1,
- or the theory of standard big-bang nucleosynthesis is wrong.

Let us mention that a model of universe, where dark matter is only baryonic, is not ruled out by observations. In X-ray clusters, the visible mass, including the hot gas, is only 10 times lower than the total dynamical mass. The total Ω inferred is of the order of 0.05, still compatible with the standard big-bang nucleosynthesis.

The enhanced sensitivity of ROSAT X-ray data have not revealed the presence of more dark matter on large scales. In fact, the dark matter becomes visible in rich clusters as hot emitting gas. This implies that the fraction of baryonic matter in clusters becomes larger and larger, incompatible with the joint hypothesis of standard big-bang nucleosynthesis (BBN) and $\Omega = 1$. On the other hand, at the galaxy scale, the amount of baryons is still insufficient to meet the lower bound of BBN, so that there must exist some baryonic dark matter. This could exist in the shape of brown dwarfs, as might have been seen by micro-lensing experiments (e.g. Alcock et al 1993, Aubourg et al 1993). Or this baryonic dark matter could be in the form of cold molecular gas in the outer parts of galaxies. This gas is then available for star formation, all along the galaxy life-time, explaining galaxy evolution along the Hubble sequence.

7. Conclusions

We have proposed and supported by a series of physical arguments the existence of $10-20$ times more mass in the form of baryons, more specifically in the form of cold molecular hydrogen. Most of the understanding on the structure and evolution of galaxies, as well as those of galaxy clusters would seem much simpler if a cold gas state would exists which would be hardly detectable with today's instruments.

The recently disclosed fractal structure of the interstellar cold gas at $5-15\,\mathrm{K}$ offers unexplored theoretical ways to refine the understanding and observation interpretation about the interstellar medium. This extremely inhomogeneous "gas" in fact bears little resemblance with common fluids, and probably a word other than "gas" should be used to convey the idea that most of the mass fills only a tiny fraction of the volume, like in a foam. In any case, this inhomogeneous state means that observations may indeed miss most of the cold gas mass by having insufficient dynamical range and spatial resolution.

Physical considerations about the smallest fragmentation clumps in a fractal gas suggest that most of the mass in galaxies would then be in the form of cold ($\sim 3\,\mathrm{K}$) H+He clumpuscules, $\sim 30\,\mathrm{AU}$ in size, with a mass of the order of a Jupiter, so much less dense than Jupiter. The physical state of these smallest structures of the fractal gas has been described. It turns out that the clumpuscules can both hide most of the mass (their average column density is $10^{24}\,\mathrm{H\,cm^{-2}}$) and allow to explain why cold gas does not collapse immediately into stars or Jupiters.

An undetermined but probably small fraction of the mass might be in solid H_2 form. Indeed the clumpuscules have favorable conditions for H_2 freezing. However, the density of nucleation sites and their efficiency is not known; and the energy released by ice formation is too high (≈ 100k per H_2 molecule) to be absorbed for a significant ice fraction. The coincidence that the clumpuscule conditions are just favorable to build H_2 ice at a fraction of degree from today microwave background temperature is astonishing (see also Wickramasinghe & Reddish 1968). An interesting possibility is that H_2 dimers form just around the gas-solid transition, and some continuum and line radiation from these dimers be detected (Shaefer 1994).

This new candidate for dark matter has the advantage of providing a natural origin of the hot gas in clusters, and explaining the observed high baryonic fraction in strong X-ray clusters. It provides also a scenario to interpret the cooling flow phenomenon, and the high dark matter central concentration in clusters.

In the near future we expect that our model might be tested by several types of observations, for example: high resolution observations of nearby molecular clouds, H_2 dimers radiation, absorption line measurements of point sources, detailed mapping of the far-infrared and sub-millimetric emission of the Galaxy, fluctuations in the microwave background at high spatial resolution, gravitational lensing or occultation of distant quasars.

References

Alcock C. et al: 1993, *Nature* **365**, 621

Allen S.W., Fabian A.C., Johnstone R.M., et al.: 1993, *Mon. Not. Roy. Ast. Soc* **262**, 901

Antonucci R., Barvainis R.: 1994, *Astron. J.* **107**, 448

Aubourg E. et al: 1993, *Nature* **365**, 623

Binney J., Tremaine S., 1987, Galactic Dynamics, Princeton Univ. Press, Princeton

Binney J.J.: 1992, *Ann. Rev. Astron. Astrophys.* **30**, 51

Bloemen H.: 1989, *Ann. Rev. Astron. Astrophys.*, **27**, 469

Bosma A.: 1981, *Astron. J.* **86**, 1971

Blumenthal G.R., Faber S.M., Flores R., Primack J.R: 1986, *Astrophys. J.* **301**, 27

Braine J., Combes F.: 1993, *Astron. Astrophys.* **269**, 7

Braine J., Dupraz C.: 1994, *Astron. Astrophys.* **283**, 407

Briel U.G., Henry J.P., Böhringer H.: 1992, *Astron. Astrophys.* **259**, L31

Broeils A.: 1992, "Dark and visible matter in spiral galaxies", PhD Thesis, Rijksuniversiteit Groningen

Bruhweiler F.C., Boggess A., Norman D.J., Grady C.A., Urry M.C., Kondo Y.: 1993, *Astrophys. J.* **409**, 199

Buchhorn M. et al.: 1994, preprint (work presented in Freeman 1993)

Brinks E., Burton W.B.: 1984, *Astron. Astrophys.* **141**, 195

Carignan C., Freeman K.C.: 1988, *Astrophys. J.* **332**, L33

Casertano S., van Gorkom J.H.: 1991, *Astron. J.* **101**, 1231

Cognard I., Bourgois G., Lestrade J-F. et al.: 1993, *Nature* **366**, 320

Combes F., Debbasch F., Friedli D., Pfenniger D.: 1990, *Astron. Astrophys.***233**, 82

David L.P., Forman W., Jones C.: 1994, *Nature* , in press

de Geus E.J., Vogel S.N., Digel S.W., Gruendl R.A.: 1993, *Astrophys. J.* **413**, L97

Diamond P.J., et al.: 1989, *Astrophys. J.* **347**, 302

Dickey J.M., Hanson M.M., Helou G.: 1990, *Astrophys. J.* **352**, 522

Durret F., Gerbal D., Lachièze-Rey M., Lima-Neto G., Sadat R: 1994, *Astron. Astrophys.*, **287**, 733

Edge A.C., Stewart G.C.: 1991, *Mon. Not. Roy. Ast. Soc* **252**, 428

Evrard A.E., 1993, in: Physics of Nearby Galaxies, Nature or Nurture?, T.X. Thuan, C. Balkowski, J.T.T. Van (eds.), Editions Frontières, Gif-sur-Yvette, p. 375

Fabian A.C.: 1987, in "Large Scale Structures in the Universe", Saas-Fee Advanced Course 17, L. Martinet, M. Mayor (eds.), Geneva Observatory, Geneva, p. 1

Falconer K.J.: 1990, Fractal Geometry, Wiley, Chichester

Falgarone E., 1992, in: Astrochemistry of Cosmic Phenomena, P.D. Singh (ed.), Kluwer, Dordrecht, p. 159

Falgarone E., Puget J.-L., Pérault M.: 1992, *Astron. Astrophys.* **257**, 715

Ferland G.J., Fabian A.C., Johnstone R.M.: 1994, *Mon. Not. Roy. Ast. Soc* **266**, 399

Fiedler R.L., Dennison B., Johnston K.J., Hewish A.: 1987, *Nature* **326**, 675

Freeman K.C.: 1993, in "Physics of Nearby Galaxies, Nature or Nurture?" T.X. Thuan, C. Balkowski, J.T.T. Van (eds.), Editions Frontières, Gif-sur-Yvette, p. 201

Gallagher J.S., Hunter D.A., Tutukov A.V.: 1984, *Astrophys. J.* **284**, 544.

Gerbal D., Durret F., Lima-Neto G., Lachièze-Rey M.: 1992, *Astron. Astrophys.* **253**, 77

Grabelsky D.A., Ulmer M.P.: 1990, *Astrophys. J.* **355**, 401

Gunn J.E., Peterson B.A.: 1965, *Astrophys. J.* **142**, 1633

Hammer F.: 1991, *Astrophys. J.* **383**, 66

Hasan H., Norman C.: 1990, *Astrophys. J.* **361**, 69.

Hawkins M.R.S.: 1993, *Nature* **366**, 242

Hubble E.: 1926, *Astrophys. J.* **64**, 321.

Kennicutt R.: 1983, *Astrophys. J.* **272**, 54

Kiraga M., Paczinsky B.: 1994 *Astrophys. J.*, **430**, L101

Lake G., Feinswog L.: 1989, *Astron. J.* **98**, 166

Larson B., Tinsley B.M., Caldwell C.N.: 1980, *Astrophys. J.* **237**, 692

Lequeux J., Allen R.J., Guilloteau S.: 1993, *Astron. Astrophys.* **280**, 23

McNamara B.R., Jaffe W.: 1994, *Astron. Astrophys.* **281**, 673

Merrifield M.R.: 1992, *Astron. J.* **103**, 1552

Moore B.: 1994 *Nature* **370**, 629

Morris S.L., Weymann R.J., Savage B.D., Gilliland R.L.: 1991, *Astrophys. J.* **377**, L21

O'Dea C.P., Baum S.A., Maloney P.M. et al.: 1994, *Astrophys. J.*, in press

Ostriker J.P.: 1985, in "Dark Matter in the Universe", IAU 117, ed. J. Kormendy & G.R. Knapp, Reidel, p. 85

Persic M., Salucci P.: 1991, *Astrophys. J.* **368**, 60

Persic M., Salucci P.: 1993, *Mon. Not. Roy. Ast. Soc* **261**, L21

Persic M., Salucci P., Ashman K.M.: 1993, *Astron. Astrophys.* **279**, 343

Pfenniger D.: 1991, in *Dynamics of Disk Galaxies* ed. B. Sundelius, Göteborg, p. 191

Pfenniger D., Norman C.A.: 1990, *Astrophys. J.* **363**, 391.

Pfenniger D., 1993, in: Physics of Nearby Galaxies, Nature or Nurture?, T.X. Thuan, C. Balkowski, J.T.T. Van (eds.), Editions Frontières, Gif-sur-Yvette, p. 519

Pfenniger D., Combes F., Martinet L.: 1994, *Astron. Astrophys.* **285**, 79

Pfenniger D., Combes F.: 1994, *Astron. Astrophys.* **285**, 94

Phillips J.A. et al : 1994, IAU Symposium S169 "Unsolved problems of the Milky Way", ed. L. Blitz, p.

Puche D., Carignan C., Bosma A.: 1990, *Astron. J.* **100**, 1468

Rees M.J.: 1976, *Mon. Not. Roy. Ast. Soc* **176**, 483

Rees M.J., Ostriker J.: 1977, *Mon. Not. Roy. Ast. Soc* **179**, 541

Reshetnikov V., Combes F.: 1994, *Astron. Astrophys.*, **291**, 57

Rich M., 1993, in: Physics of Nearby Galaxies, Nature or Nurture?, T.X. Thuan, C. Balkowski, J.T.T. Van (eds.), Editions Frontières, Gif-sur-Yvette, p. 153

Rubin V.C., Ford W.K., Thonnard N.: 1980, *Astrophys. J.* **238**, 471

Sackett P.D., Sparke L.S.: 1990, *Astrophys. J.* **361**, 408

Salucci P., Frenk C.S.: 1989, *Mon. Not. Roy. Ast. Soc* **237**, 247

Sancisi R., van Albada T.S.: 1987, in "Dark Matter in the Universe" IAU Symp. 117, J. Kormendy, G.R. Knapp (eds.), Reidel, Dordrecht, p. 67

Sandage A.: 1961, *Hubble Atlas of Galaxies* Carnegie Institution of Washington

Sargent W.L.W.: 1988, in "QSO absorption lines", J. Blades, D. Turnshek, C. Norman (eds.), Cambridge University Press, p. 1

Scalo J.M.: 1985, in "Protostars and Planets II", D.C. Black, M.S. Matthews (eds.), University of Arizona Press, Tucson, p. 201

Scalo J., 1990, in: Physical Processes in Fragmentation and Star Formation, R. Capuzzo- Dolcetta et al. (eds.), Kluwer, Dordrecht, p. 151

Shaefer J.: 1994, *Astron. Astrophys.* **284**, 1015

Shostak G.S., van der Kruit P.C.: 1984, *Astron. Astrophys.* **132**, 20

Smith M.S., Kawano L.H., Malaney R.A.: 1993, *Astrophys. J. Supp.* **85**, 219

Tinsley B.M.: 1981, *Mon. Not. Roy. Ast. Soc* **194**, 63

Toomre A.: 1990, in "Dynamics and Interactions of Galaxies", R. Wielen (ed.), Springer-Verlag, p. 292

Toomre A., Kalnajs A.J.: 1991, in "Dynamics of Disc Galaxies", B. Sundelius (ed.), Göteborg Univ., p. 341

Tóth G., Ostriker J.P., 1992, *Astrophys. J.* **389**, 5

Udalski A., Szymanski M., Kaluzny J. et al (1994) *Astrophys. J.* **426**, L69

van Albada T.S., Bahcall J.N., Begeman K., Sancisi R.: 1985, *Astrophys. J.* **295**, 305

White D.A., Fabian A.C., Johnstone R.M. et al.: 1991, *Mon. Not. Roy. Ast. Soc* **252**, 72

White S.D.M., Navarro J.F., Evrard A.E., Frenk C.S.:1993, *Nature* **366**, 429

Whitmore B.C., McElroy D.B., Schweizer F.: 1987, *Astrophys. J.* **314**, 439

Whitmore B.C., 1993, in: Physics of Nearby Galaxies, Nature or Nurture?, T.X. Thuan, C. Balkowski, J.T.T. Van (eds.), Editions Frontières, Gif- sur-Yvette, p. 351, 425

Wickramasinghe N.C., Reddish V.C.: 1968, *Nature* **218**, 661.

Wu X.P., Hammer F.: 1993, *Mon. Not. Roy. Ast. Soc* **262**, 187

Young J.S., Scoville N.Z.: 1991, *Ann. Rev. Astron. Astrophys.* **29**, 581.

DARK MATTER AND SOLAR NEUTRINOS

MICHEL SPIRO

CEA, DAPNIA/Service de Physique des Particules, CE Saclay, 91191 Gif-sur-Yvette, France

ABSTRACT

This short review discusses the principles of dark matter detection and solar neutrinos. It summarizes the main experimental results obtained up to now. Prospects and new ideas are presented.

1. Introduction

The estimate of the value of Ω, the ratio of the mean energy density in the universe to the critical energy density, is one of the main issue in modern cosmology. We can measure the components of Ω in various ways:

- from luminous matter (stars), Ω_{lum}
- from the dynamical behaviour of stars in spiral galaxies (galactic halos), Ω_{halo}
- from primordial nucleosynthesis (baryons), Ω_{bar}. We know from observations that the contributions from dust or gas to Ω are negligible. The estimates for the values of Ω_{lum}, Ω_{halo} and Ω_{bar} are shown in fig.1 and compared to the magic value $\Omega = 1$ which is the preferred value for aesthetical and theoretical reasons (to avoid fine tuning in initial conditions, and to agree with inflation theories). From all these values, one can draw two main conclusions:

1) All these estimates are below 1. However, the value $\Omega = 1$ is not at all excluded. The allowed range for the total contribution to Ω is $0.001 < \Omega_{tot} < 2$ where the lower limit comes from visible matter and the upper limit comes from the minimum estimate of the age of the universe combined with the measured expansion rate. To reach $\Omega = 1$, it seems unavoidable to invoke intergalactic non baryonic dark matter, such as:

WIMPs (Weakly Interacting Massive Particles: heavy neutrinos ν_h, or lightest supersymmetric particle LSP...)

or Light Neutrinos such as 10 to 100 eV ν_e or ν_μ or ν_τ (a light LSP could also be in principle possible).

2) The comparison of the allowed range for Ω_{lum}, Ω_{halo} and Ω_{bar} suggests that the halos of spiral galaxies, like our own galaxy, could be partly or totally made of MACHOs (Massive Astrophysical Compact Halo Objects) which is almost the only possibility left for baryonic dark matter: these MACHOs could be either aborted stars (brown dwarves, planet like objects), or star remnants (white dwarves, neutron stars, black holes) [1].

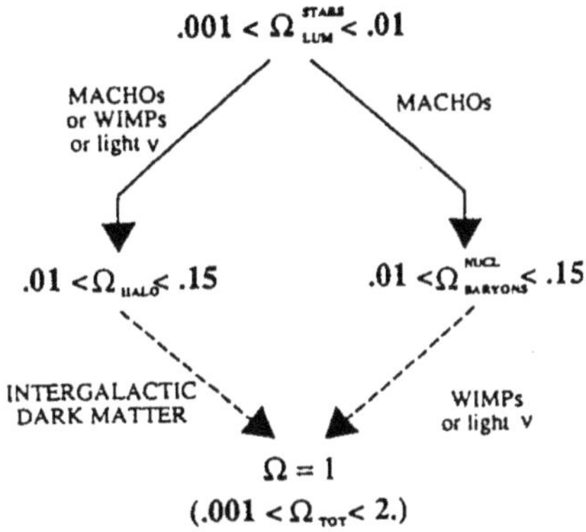

Fig.1 - Present status of Ω determinations

2. LEP + $p\bar{p}$ collider constraints on WIMPs candidates with $\Omega_{WIMP} = 1$

2.1 Heavy Dirac or Majorana Neutrino ν_H

The LEP results on the Z^0 width, combined with the upper limits on the ν_e, ν_μ and ν_τ masses (10 eV, 250 keV and 35 MeV) excludes the possibility of any Dirac or Majorana neutrino in the mass range 35 MeV to 40 GeV. In particular, this excludes (fig.2) the 3 to 7 GeV range, which would naturally give $\Omega_{\nu_H} = 1$ from the relic density which can be reliably estimated for heavy neutrinos in standard cosmology.

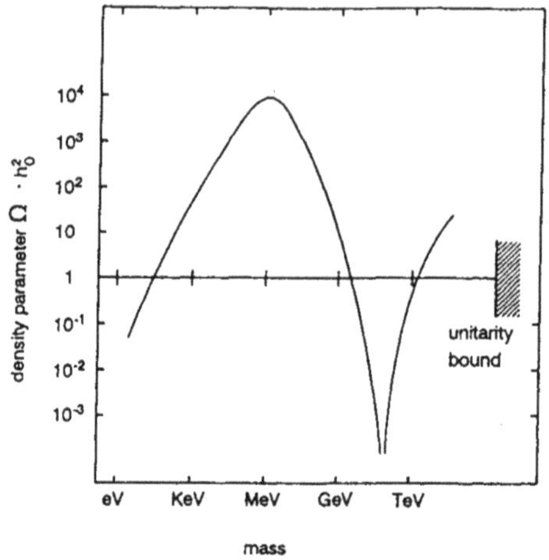

Fig.2 - Ω_ν as a function of the Dirac ν mass

2.2 Lightest Supersymmetric Particle (LSP)

1) Boson

An illustrative candidate of that type is the sneutrino $\tilde{\nu}$. The contribution of a $\tilde{\nu}$ to the width of the Z^0 is half that of a standard neutrino. Again, from the LEP results one can exclude a $\tilde{\nu}$ with mass < 35 GeV at the 3σ level.

2) Fermion

The LSP could naturally be in that case the neutralino χ, a linear combination of the photino, the zino and the two higgsinos which are necessary in the minimal supersymmetric theory.

- Again, the residual cosmological density of neutralinos having survived annihilation can be reliably estimated. The decoupling time is determined by their low-energy annihilation cross section which depends mostly on the neutralino mass and the lightest scalar-fermion mass (fig. 3). From CDF ($p\bar{p}$ experiment), we know already that the sfermion masses are > 100 GeV. Demanding $\Omega_\chi = 1$, this implies that 15 GeV < M_χ < 3 TeV.

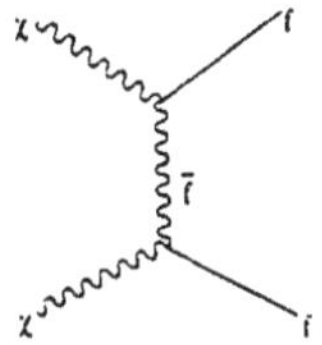

Fig. 3 - Annihilation diagram for neutralinos

- What do we learn from LEP? The Z^0 is not coupled to photino-photino, nor to Zino-Zino but does couple to two higgsinos. This implies that the higgsino masses are higher than 40 GeV although there are some loopholes in the derivation. Furthermore, the wino mass is greater than 40 GeV (again from LEP), which favours also a Zino mass greater than 40 GeV in any natural model. In summary LEP cuts already most of the parameter phase space for neutralinos below 30 GeV. This is bad news for direct detection experiments. As the neutralino masses get higher and higher, the rate of WIMPs scattering decreases.

- The natural scale of supersymmetry is below few TeV, to provide a natural explanation of the hierarchy of the grand unification compared to the electroweak unification. Combined with LEP results, we find surprisingly the same allowed mass range as inferred from cosmological arguments, namely between 30 GeV and few TeV.

3. Is our galactic halo made of MACHO's or WIMP's ?

To answer this question, the idea [2] is to survey for several months, at 10% accuracy or better, the luminosity of a large number of stars in the Large Magellanic Cloud (LMC), see fig. 4, and search for stars that undergo a characteristic brightening due to a MACHO passing near the line of sight and producing a gravitational microlensing event: when a spatially-small massive object (a deflector D which is the MACHO) happens to lie near enough to the line of sight between a light source (a star S from the LMC) and an observer O, then the observer collects more light from the star than he would in the absence of this deflector.

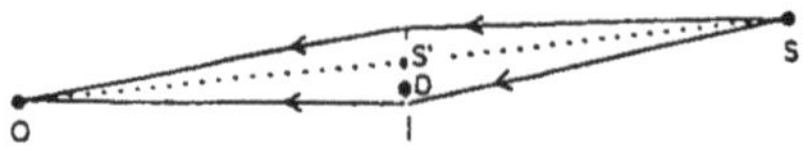

Fig. 4 - Microlensing of a LMC star (S) by a MACHO (D)

Of course, the observer, deflector and source stars are all moving (with typical galactocentric velocities of 200 km/s), so that the brightening of the star is time-dependent. The important characteristics of this light curve are its symmetry, unicity (the brightening occurs only once) and its achromaticity (the brightening is identical whatever the filter used by the observer).

The probability that a star undergoes a brightening due to microlensing with an amplification larger than 1.34 is simply the probability that the line OS crosses the zone of gravitational influence of area πR_0^2 associated to a deflector where $R_0^2 = 4GMd/c^2$ (M mass of the deflector) is the Einstein ring. For various values of deflector mass (MACHO masses), we give in table 1 the mean values of the excitation time (above 1.34 amplification) and the mean number of microlensing events expected for a survey of 10^6 stars in the LMC monitored during 10^7 seconds (4 months) [3].

Two groups , one in the US (Livermore, Berkeley), one in France (Saclay, Orsay, Paris, Marseille) are running experiments based on these principles. The experiments are located in Chile and maybe also in Australia. This survey is of interest for both particle physics and astrophysics. If MACHOs are indeed found, this will give a strong indication that non-baryonic dark matter is not clustered around galaxies, meaning that present searches for WIMPs are likely to be ineffective. On the other hand, if such compact objects are not found, then there would be an added incentive for all kind of WIMPs searches. So far, few candidates have been found, but however somewhat significantly less than expected for a standard spherical halo.

Table I: Microlensing events characteristics

Deflector Mass ($M_\odot$)	Mean R_0 (km)	Mean μlensing time	Nb of μlensing events
> 10	> 3 10^9	> 1 year	0.5
1	1 10^9	3 mths	1.0
10^{-2}	1 10^8	9 days	5
10^{-4}	1 10^7	1 day	50
10^{-6}	1 10^6	2 hrs	500
10^{-8}	1 10^5	12 mins	5000

Table 1 : Microlensing events characteristics

4. Direct and indirect detection of galactic WIMP's.

The hypothesis that dark matter particles are gravitationally trapped in the galaxy leads to the conclusion that, like stars, they should have a local Maxwell velocity distribution with a mean spread of 250 km/s. Then, the mean kinetic energy E_r received by a nucleus of mass M_n (in units of GeV) in an elastic collision with a dark matter particle of mass M_x is:

$$E_r = 2 \text{ keV } M_w M_x^2/(M_x+M_n)^2$$

The energy distribution is roughly exponential . The expected event rate for elastic scattering on a given nucleus, assuming that 0.4 GeV/cm^3 is the local density of the halo (needed to account for the flat rotation curve of stars) depends only on the mass and interaction cross section .

These recoil events can be detected in well shielded deep underground devices such as semiconductor diodes, scintilltors and cryogenic bolometers. Experimental limits on the cross section of wimps for scattering on germanium [4,5,6,7], CaF_2 [8], and NaI [8] have been published. These limits are shown in fig. 5 and 6 , translated in terms of single nucleon effective coupling strength depending on whether we are dealing with pure axial coupling or with coherent N^2 coupling where N is the number of nucleons in the target nucleus.

Wimps can also be seen indirectly by observing their annihilation products [9]. The most sensitive indirect technique uses the fact that Wimps can be trapped in the Sun or the Earth. Wimps with galactic orbits that happen to intersect an astronomical body will be trapped if, while traversing the body, they suffer an elastic collision with a nucleus that leaves the wimp with a velocity below the escape velocity [10,11]. The capture rate is proportional to the elastic cross sections on the nuclei of the astronomical body. A steady state will eventually be reached when the capture rate is balanced by the annihilation rate. Annihilations in the Sun or the Earth will yield a flux of high energy neutrinos either directly or by decay of annihilation products. The muon neutrinos can be observed in underground detector through their interactions in the rock below the detector yeildingupward going muons pointing towards the Sun or the Earth. Presently, the most sensitive limits are those from Kamiokande [12,13] (see fig. 7). The flux limits obtained so far on upward going muons can be interpreted in terms of a limit on the effective elastic cross section (fig.5 and 6). These can be then compared with the limits from direct searches.

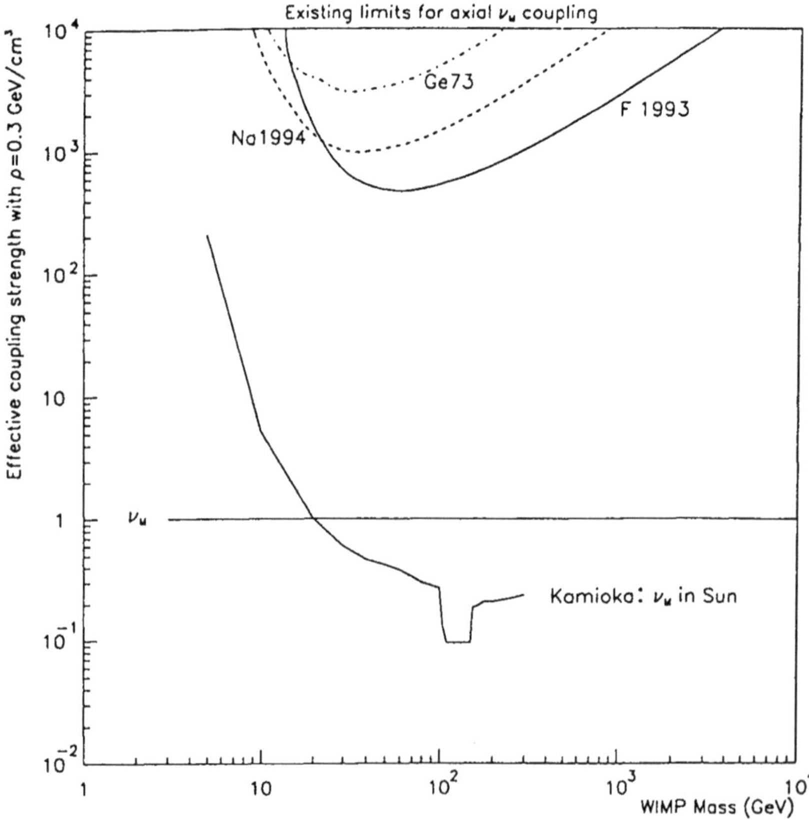

Fig. 5 : Upper limits on the interction strength for particles with spin dependent interactions [14]

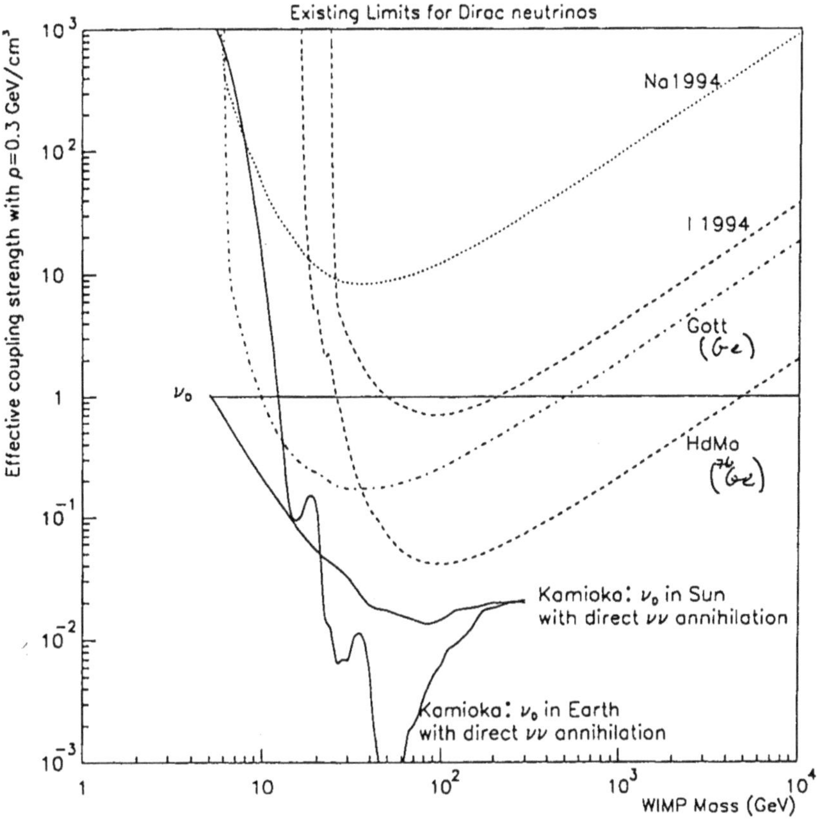

Fig. 6 : Upper limits on the interaction strength for particles with spin-independent cross sections [14].

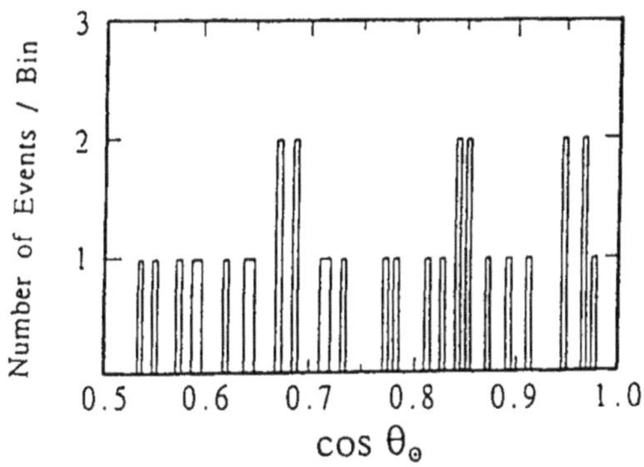

Fig.7 : Search for upward going muons coming from the sun with the Kamiokande detector

Dirac neutrinos with masses from 11 GeV to few TeV are excluded. Combined with LEP results and direct mass determinations of known neutrino species, this closes almost completely the remaining allowed window by cosmology (fig. 2). Neutralinos are expecting to range 2 to 5 orders of magnitude below the present sensitivity of direct detection searches, depending on the exact values of minimal supersymmetry parameters. Indirect detection experiments start to explore the highest part , in terms of cross section, of the neutralino domain.

For the future one can note that the indirect searches are limited by the fixed background from atmospheric neutrinos and can therefore expand their limits only in proportion of the square root of the exposure time or detector area. Direct searches have no known fundamental limit on their background from radioactive impurities in the detector elements. Considerable progress has been made in lowering this background. Promising techniques to further eliminate the backgrounds include simultaneous detection of phonons and ionization in cryogenic germanium detectors, pulse shape analysis in NaI detectors and search for the expected seasonal variation (added velocity due to the movement of the Earth around the Sun) which will be ultimately necessary to confirm any observed signal.

5. The case for light neutrinos.

Although they are not favoured by theories dealing with small scale structure formation (galaxy formation), 30 to 100 eV neutrinos are quite appealing to explain the nature of our halo [15].

Note that if our galactic halo is made of light neutrinos there should be a sharp resonant absorption line in the spectrum of ultrahigh energy intergalactic neutrinos reaching the earth (fig. 8). The detection of such a narrow line would be a proof of the neutrino halo. Its position would provide for the neutrino mass: $2\, M_\nu E_\nu = (M_{Z^0})^2$.

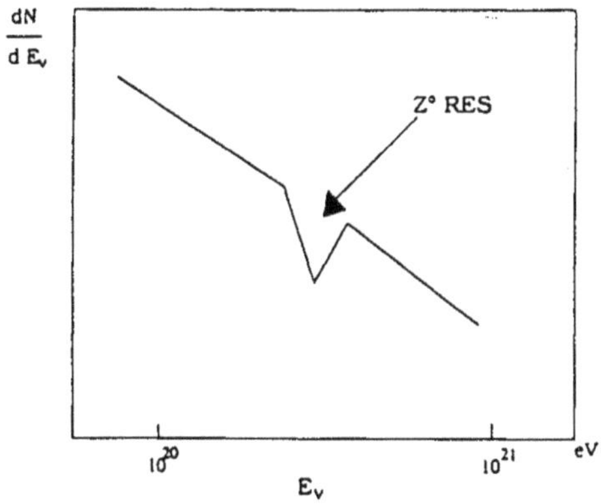

Fig.8 - ν resonant absorption by our galactic halo.

The ν_e is now excluded due to the severe upper limit on the mass (4.5 eV) [16]. However the ν_μ and ν_τ are perfectly viable candidates. Obviously the direct (laboratory or far supernova detection) or indirect (oscillations) measurements of the ν_μ and ν_τ masses are of crucial importance in the context of the dark matter problem. A likely scenario could be that the ν_τ is much heavier (30 eV) than the ν_μ and that the ν_μ is much heavier than theν_e.In that sense the deficit of solar neutrinos could be a hint for neutrino ν_e - ν_μ oscillations and then for neutrino masses. The search for ν_μ–ν_τ oscillations in the range of Δm^2 of eV^2 to few hundred eV^2 (range of cosmological interest) has started at CERN with the CHORUS and NOMAD experiments. Even if the favored range of masses to explain the solar neutrino deficit is 10^{-4} to 10^{-2} eV, this could be the range of mass for the ν_μ with a much lower mass for the ν_e and a much higher one for the ν_τ (30 eV). Neutrino masses could then both solve the dark matter and solar neutrino problems.

6. The solar neutrino problem and neutrino masses.

6.1 Predictions of solar models.

The most firm and solid prediction we have on the solar neutrino flux is based on energy conservation and steady state of the sun. We know that these two well admitted assumptions imply that the total power radiated by the solar surface (the luminosity L^*) should be equal to the thermonuclear power generated by the fusion of hydrogen into helium. For four protons to combine into a ^4He nucleus, two electrons must be involved in the initial state for electric charge conservation, and then two ν_e must be emitted in the final state. The overall reaction is then:

$$4\,p + 2e^- \rightarrow {}^4He + 2\nu_e + 27\ MeV \qquad (1)$$

where 27 MeV is the difference of the masses between the particles involved in the initial state and those involved in the final state (the energy of the neutrinos and the kinetic energy of the nuclei can be neglected in this approximate relation). It is then easy to derive the total flux of neutrinos expected to reach the earth :

$$N_\nu = \frac{2L^*}{27\ MeV\ 4\pi\ d^2} = 6.5\ 10^{10}\ cm^{-2}\ s^{-1} \quad (2)$$

where d is the distance from the Earth to the Sun. Gallium target detectors are so far the most appropriate to measure the total number of neutrinos. This is because :

- of the very low threshold (233 keV) of the capture reaction $\nu_e + {}^{71}Ga \rightarrow {}^{71}Ge + e^-$ which makes gallium target detector sensitive to the bulk of the solar neutrino energy spectrum
- of the high natural abundance of the stable ^{71}Ga isotope (40%)
- of the relatively easy identification of even a few radioactive ^{71}Ge atoms in a large quantity of gallium (30 tons).

However the firm prediction on the total number of solar neutrinos is not enough to compute the capture rate of solar neutrinos on a given target nucleus. To compute the energy spectrum one needs to go through solar modelling and through the exact chain of reactions which combine hydrogen into helium. There are mostly three cycles of reactions : ppI, ppII and ppIII.

- In ppI the two neutrinos are coming from the $pp \rightarrow {}^2H + e^+ + \nu_e$ reaction (2 ν_{pp})
- In ppII one neutrino is a ν_{pp}, the other comes from the decay (through electron capture) of ^7Be (ν_{7Be})
- In ppIII one neutrino is a ν_{pp}, the other comes from the β decay of ^8B (ν_{8B}).

The ν_{pp} spectrum extends from 0 to 450 keV. Only the Gallium experiments are sensitive to those neutrinos. The ν_{7Be} are monoenergetic with a line at 860 keV. Both the gallium and chlorine ($\nu_e + {}^{37}Cl \rightarrow {}^{37}Ar + e^-$, threshold 820 keV) are sensitive to those neutrinos. Finally the ν_{8B} neutrino spectrum extends from 0 to 15 MeV. All the presently running experiments (gallium, chlorine and Kamiokande) are sensitive to them. The Kamiokande experiment is based on the detection of the recoil electron in the elastic scattering of a ν_e with an experimental threshold of about 7 MeV on the energy of the recoil electron. From the Solar Standard Model [17,18] the ν_{pp}, ν_{7Be} and ν_{8B} intensities are computed to be 90%, 8% and 10^{-4} of the total flux. Although the relative intensity of the ν_{8B} neutrino is very small, they contribute significantly to the capture rate, even in the Gallium experiment, due to their high energy. Notice however that the ν_{7Be} and ν_{8B} fluxes are highly sensitive to the ingredients of the SSM. If, for instance, one changes the input parameters, with, as a result, a change in the central temperature T_c prediction, it has been shown that the ν_{8B} flux will vary as T_c^{18}, the ν_{7Be} as T_c^{8} and the ν_{pp} flux only as $T_c^{-1.2}$.

The predictions of the SSMs are shown in Table 2, for the Gallium experiments, in terms of SNU (Solar Neutrino Units). One SNU corresponds to a capture rate of 10^{-36} per second per target nucleus (in this case for Ga).

We see that the Bahcall et al. SSM which is generally considered as giving high SNU values, predicts fluxes only slightly higher than the Turck-Chièze et al. SSM which is generally considered as giving low SNU values. So one might say that the predictions of the SSM for Gallium experiments are rather firm. Notice also, that although the n_{pp} are expected to represent 90% of the total flux of solar neutrinos, their contribution to the capture rate amounts only to 71 SNU out of 127. This is due to their low energy.

	Capture rate (SNU)	
Source	Turck-Chièze	Bahcall
pp	70.6	71.3
pep	2.795	3.07
^{7}Be	30.6	32.9
^{8}B	9.31	12.31
^{13}N	3.87	2.68
^{15}O	6.50	4.28
^{17}F	-	0.04
Total	**124**	**127**

Table 2: Standard Solar Model predictions from Turck-Chièze et al, and Bahcall et al., for the gallium experiments.

6.2 Are the predictions right ?

ν_{8B} flux

The Kamiokande experiment uses a water Cerenkov detector. The basic process is neutrino scattering on electrons which then give detectable Cerenkov light. They measure two quantities, the energy of the recoil electron and its direction. A clear peak can be seen in the direction of the Sun and the excess in that direction is then taken as coming from solar neutrinos (figure 9). However the flux of ν_{8B} they measure since January 1987 [19] is only 0.54 ± 0.08 of the Bahcall et al. SSM, so 0.5 10^{-5} of the total flux.

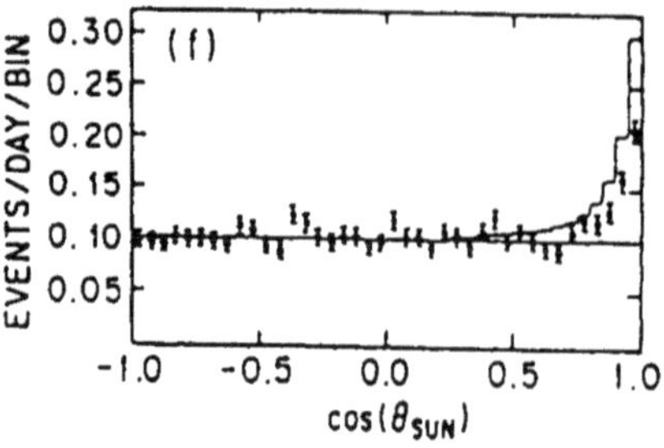

Figure 9 : Counts of the Kamiokande detector plotted against the cosines of the angle of the electron to the Sun's direction ($\cos\theta_{SUN}$, 1557 days of data, electron energies greater than 9.3 MeV for 449 days, 7.5 MeV for 794 days and 7.0 MeV for 314 days).

ν_{7Be} flux

Since 1967 Davis and co-workers have performed a pioneering experiment by extracting ^{37}Ar from a tank of 615 tons of tetrachloroethylene (C_2Cl_4). The ^{37}Ar decays by electron capture. The resulting hole in the K shell can give X rays and Auger electrons with a total energy of 2.8 keV. The counter of 0.5 cm^3 volume is designed to measure this energy. The half life of the decay is 35 days. A typical run consists of a 50 day exposure of the big tank followed by an extraction of the Argon atoms which are then introduced in the small counter. The counting lasts for 260 days. For the period 1970-1984 the data were analysed and give 339 counts of ^{37}Ar. This gives a non-corrected ^{37}Ar counting rate of 5 per run. The data are analysed by a maximum likelihood method assuming a flat background (as a function of time) plus a ^{37}Ar decaying component.

The result [19] is 3.6 ± 0.4 times lower (figure 10) than expected in the Bahcall et al. SSM. This implies, taking into account the fact that the experiment is sensitive to both the ν_{7Be} and ν_{8B} components and taking into account the Kamiokande result (reduction of a factor 2 on the ν_{8B} component) that the ν_{7Be} flux is lower by a factor > 4 than the prediction of Bahcall et al. SSM.

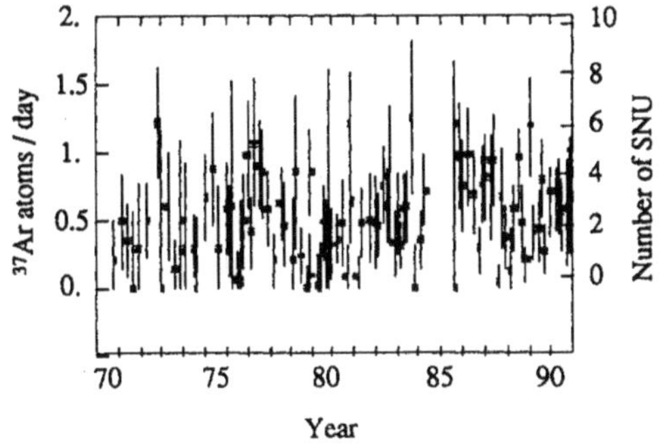

Figure 10 : ^{37}Ar production rate in the Homestake chlorine solar neutrino detector (0.4 atom/d corresponds to 2 SNU).

These deficits are the basis of the solar neutrino problem. The reductions are very hard to reconcile with any modification of the SSM since we expect that any reduction on the ν_{7Be} component should be accompanied by a stronger reduction for the ^{8}B neutrino component [6]. Neutrino masses and mixing could reconcile these reductions with the SSM through ν_e, ν_μ, ν_τ oscillations. However, before invoking new physics in the neutrino sector, the results of the gallium experiments were eagerly awaited. The expectations are much less sensitive to SSM and we can derive absolute lower limits for the capture rate based only on energy conservation and steady state of the Sun.

6.3 Consistent predictions for gallium experiments

Since we know experimentally that the flux of ν_{8B} is reduced by a factor two and that the flux of ν_{7Be} is reduced by a factor greater than 4 compared to SSM predictions, we can deduce that the number of ν_{pp} should be increased to 1.08 ± 0.02 of the SSM to conserve the total number of neutrinos insuring energy conservation. One obtains expectations which range from 80 to 105 SNU. These are not SSM predictions but rather predictions which are consistent with the basic understanding of the Sun (energy conservation and steady state) and with the two experimental results coming from the chlorine and Kamiokande experiments.

6.4 Results of the gallium experiments

Two experiments are now underway, SAGE in Russia which published the first results in January 1991 and GALLEX in Italy, which published their first results in June 1992. The recipes are the same : introduce 1 mg of inactive stable germanium in the 30 tons of gallium, expose the gallium to solar neutrinos in a low background environment, extract by a chemical method the solar neutrinos produced ^{71}Ge atoms together with the inactive germanium, transform into a counting gas (GeH_4), fill a proportional counter and count the decays of ^{71}Ge (11.4 d half life). The main difference is that the SAGE experiment uses metallic liquid gallium target while the GALLEX experiment uses an acidic aqueous gallium chloride solution. This induces important differences in the chemistry.

SAGE

The Soviet-American Gallium Experiment is located in the Baksan Valley in the Caucasus mountains (Russia) under about 4700 meter water equivalent. The expected rate for 30 tons target and 132 SNU is 1.2 ^{71}Ge atom created per day. Taking into account all the efficiencies, one expect only 3 counts per run (a run is 4 week exposure) due to ^{71}Ge K electron capture ($^{71}Ge + e^-_K \rightarrow {}^{71}Ga + \nu$ + X-rays + Auger electrons). Most of the runs in 1990 have preferred values of 0 SNU. Altogether they published in 1991 [20] a preferred value of 20 SNU with upper limits of 55 SNU (68% C.L.) and 79 SNU (90% C.L.). More recently they announced the results they obtained in the last runs when they increased the total mass of gallium from 30 tons to 60 tons . This is shown on figure 11. A signal seems now to emerge. In 1992 at Dallas, the quoted result was 58 ± 20 (stat.) $\pm$ 14 (syst.) SNU . It is now 70 ± 19 (stat.) $\pm$ 10 (syst.) SNU [21] .

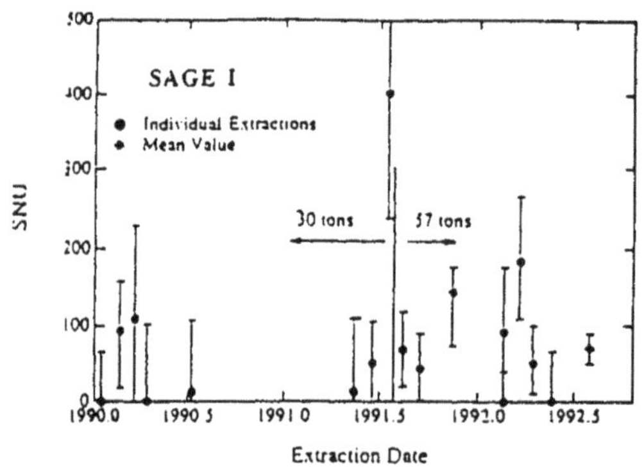

Figure 11 : Results for all runs of the SAGE experiment.. The last point on the right shows the combined result.

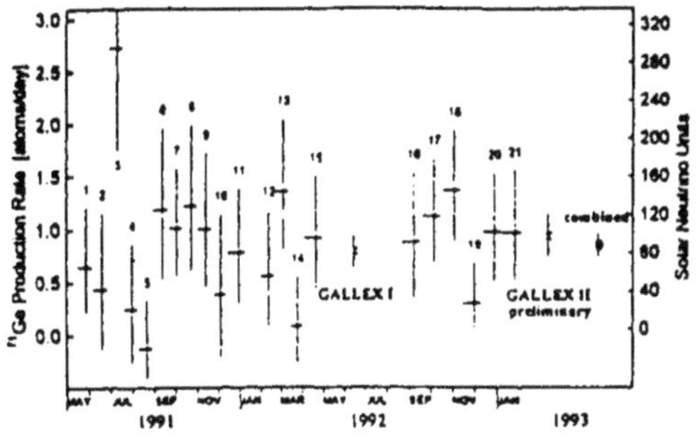

Figure 12 : Final results the first period GALLEX I (before May 1992) and preliminary results for the second period GALLEX II (after August 1992). The combined values for GALLEX I, GALLEX II and GALLEX I + GALLEX II are also shown.

GALLEX

This experiment is located in the Gran Sasso Underground Laboratory in Italy. The 30 tons of Gallium are in the form of a solution of $GaCl_3$ acidified in HCl. The Ge atoms form the volatile compound $GeCl_4$. At the end of 3 week exposures, these molecules are swept out by bubbling a large flow of inert gas (N_2) through the solution. The experiment is sensitive to both K-shell and L-shell electron captures in the decay of ^{71}Ge atoms. Seven counts are then expected after each run, in the K and L regions. The data used in the analysis consist of 21 runs taken from may 1991 to may 1993. They are now published [22], [24]. There is compelling evidence for a signal : the peaks in energy at 1.2 keV and 10 keV for L and K electron capture are seen, the 11.3 half life of ^{71}Ge is well identified over a flat background. Figure 12 shows the results for all runs which have to be compared with the combined result of 83 ± 20 SNU, released in June 1992 and now updated at the level of 79 ± 13 (stat.) ± 5 (syst.) SNU [24]. Furthermore, GALLEX will be calibrated with an artificial ^{51}Cr neutrino source (2 MCi) in 1994.

6.4 Interpretations

The SSM is unable to account for the deficit of solar neutrinos as observed by the chlorine and Kamiokande experiments. However, on the basis of these experiments it is impossible to decide whether these discrepancies come from new physics in the neutrino sector or wrong ingredients in the Solar Standard Models. The gallium experiments are in a much better position to do so. First, the predictions of the SSM are more stable to changes in the ingredients (120 to 140 SNU) and second it is impossible to have predictions below 80 SNU from basic simple principles. Consistent predictions for gallium experiments which agree with these basic principles and with the deficits of solar neutrinos observed by the chlorine and Kamiokande experiments are in the 80-105 SNU range, in agreement with the values measured by GALLEX.

By comparing the deficits of solar neutrinos as observed by the 3 experiments (table 3), the indications which favour neutrino oscillations are the facts that the chlorine experiment has a significantly larger suppression factor than the other experiments and that the Gallium experiments give results near the minimum needed to account for the luminosity of the sun, thus leaving not much room for ^{7}Be neutrinos. This would imply a more severe suppression for ^{7}Be neutrinos than for ^{8}B and pp neutrinos which cannot easily be accommodated by a modification of the SSM.

Experimen	Exp. Results	Turck-Chièze et al.	Bahcall et al.
Chlorine	2.33±.2	6.4	7.2
(%)		36 ± 4	33 ± 3
Kamioka(%)	64 ± 8	54 ± 8	
GALLEX	79 ± 15	123 ± 7	127^{-5}_{+7}
(%)		64 ± 12	62 ± 12

Table 3 : Summary table of solar neutrino experiment results (chlorine, Kamiokande and GALLEX) with the comparison to Turck-Chièze et al. and Bahcall et al. SSMs.

A decrease of the central temperature will produce a suppression factor for ^{8}B neutrinos which is larger than for ^{7}Be or pp neutrinos [23].

On the contrary, oscillations (MSW effect) could reconcile the SSM with all 3 experiments. Figure 13 shows the allowed range for neutrino masses and mixing angles. The preferred solution is for $\Delta m^2 \approx 7.10^{-6}$ eV^2 and $\sin^2 2\theta \approx 6.10^{-3}$.

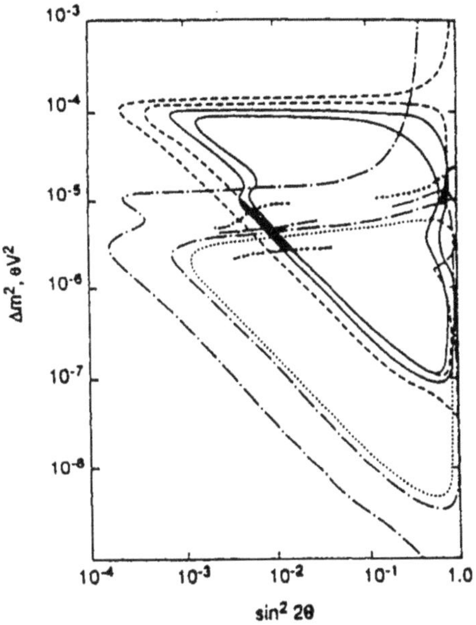

Figure 13 : MSW diagram showing the preferred solution to reconcile experimental neutrino flux measurements and Standard Solar Models.

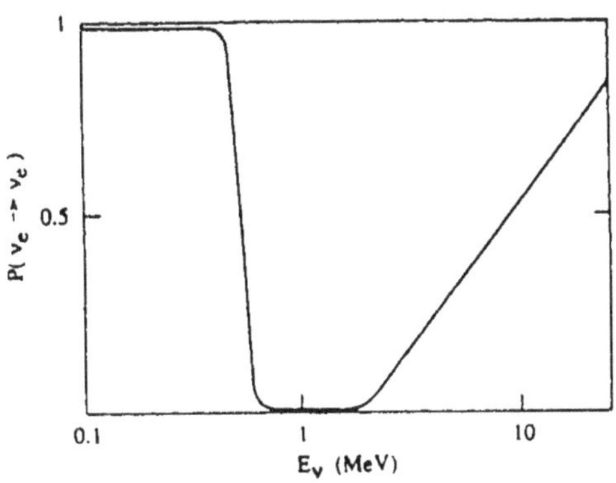

Figure 14 : Suppression factor of neutrino flux as a function of the neutrino energy for the MSW preferred solution.

The suppression factor as a function of the neutrino energy is shown on figure 14 for this solution. It implies a distortion of the ^8B neutrino energy spectrum. Figure 15 shows the recoil electron energy spectrum as observed in Kamiokande [25], normalised to the SSM prediction from Bahcall and Ulrich [26]. The recoil electron energy spectrum is only a smeared reflection of the neutrino energy spectrum.

- A flat suppression (no ν oscillation) gives a c^2 of 16.3 / 13.

- A suppression factor as predicted by the preferred MSW solution gives a c^2 of 18.6 / 13.

It is clear that one has to wait for the Superkamiokande and SNO experiments to establish or reject the small mixing angle MSW solution.

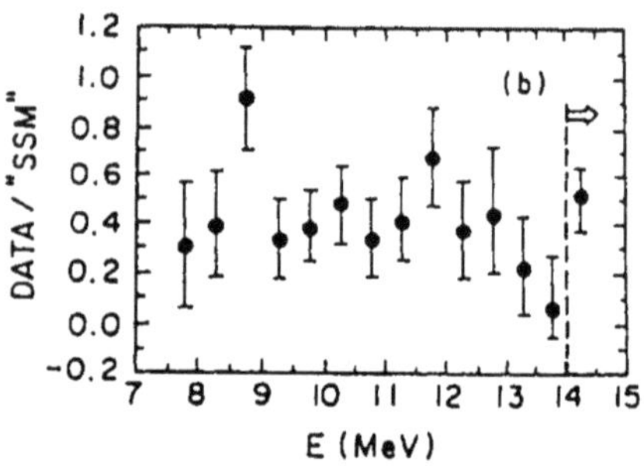

Figure 15 : Recoil electron energy spectrum as observed in Kamiokande, normalised to the SSM prediction from Bahcall and Ulrich [9].

No firm conclusion on neutrino masses can yet be drawn from the present status of solar neutrino experiments and solar modelling. This may not be the case, hopefully, in few years from now when we may expect to have more input to solar models (nuclear cross sections, helioseismology...), better understanding of running experiments (calibrations) and more experiments (SNO, SuperKamiokande...).

7. Conclusion

We are still desperately investigating the dark matter and solar neutrino problems.

Acknowledgements

I thank J. Rich for many helpful discussions, N. Lelièvre for editing the manuscript.

References

1. B. Carr, The Quest For The Cosmological Constant, Moriond 1990, Editions Frontières.

2. B. Paczynski, *Ap. J.* **304** (1986) 1.

3. A. Milsztajn, Proceedings of the Rencontres de Moriond 1990, edited by O. Fackler and J. Tran Thanh Van, Publisher Editions Frontières, p. 481. ; M. Moniez, Orsay preprint, LAL 90-20, 1990 (Moriond Astrophysics, Les Arcs 1990).

4. S. P. Ahlen et al., *Phys. Lett.* **B195** (1987) 603.

5. D. O. Caldwell et al., *Phys. Rev. Lett.* **61** (1988) 510.

6. D. Reusser et al., *Phys. Lett.* **B255** (1991) 143

7. M. Beck et al., TAUP (1993)

8. C. Bacci et al., *Phys. Lett.* **B293** (1992) 460.

9. J. Silk, K. Olive and M. Srednicki, *Phys. Rev. Lett.* **55** (1985) 257.

10. W.H. Press and D.N. Spergel, *Ap. J.* **296** (1985) 679.

11. A. Gould, *Ap. J.* **321** (1987) 571.

12. M. Mori et al., *Phys. Lett.* **B289** (1992) 463.

13. M. Mori et al., *Phys. Rev.* **D48** (1993) 5505.

14. J. Rich and C. Tao, paper in preparation

15. R. Cowsik and J. Mc Clelland, *Phys. Rev. Lett.* **29** (1972) 669 ; J. L. Basdevant, Preprint IAP - July 1984.

16. Lobashev, to be published in the Proc. of the ICHEP Conference, Glasgow, July 1994.

17. J. N. Bahcall, Neutrino Astrophysics, Cambridge University Press, 1989

18. S. Turck-Chieze et al., *Phys. Rep.*, **230**, (1993)

19. A. Suzuki, *KEK preprint 93-96*, August 1993

20. A. I. Abazov et al., *Phys. Rev. Lett.* **67** (1991) 3332

21. V. Gavrin, *Communication at TAUP93*, Gran Sasso, sept. 1993 ; J.N.Abdurashitov et al., *Phys. Lett.* **B328** (1994) 234

22. P. Anselmann et al., *Phys. Lett.* **B285** (1992) 376

23. P. Anselmann et al., *Phys. Lett.* **B285** (1992) 390

24. P. Anselmann et al., *Phys. Lett.* **B327** *(1994) 377*

25. Y. Totsuka, *Proceedings of Texas/PASCOS '92*, p. 344

26. J. N. Bahcall and R.K.Ulrich, *Rev. of Mod. Phys.*,**60** (1988) 297 ; J.N.Bahcall and M.H.Pinsonneault, *Rev. Mod. Phys.* **64** (1992) 885.

II. NEUTRINOS IN ASTROPHYSICS

REVIEW OF NEUTRINO MASSES

DOUGLAS R.O. MORRISON*

ABSTRACT

Neutrinos are the 'Wild West' frontier of physics and astrophysics. New and important results are expected and sometimes found. The role of neutrino masses in astrophysics is explained. A review is given of the search for neutrino masses using various methods — direct, double-beta decay and, in particular, oscillations. The methods and results are critically reviewed and the latest results from Los Alamos, Karmen and reactors are given. Many claims for observed effects that could be ascribed to neutrino masses have subsequently been shown to be invalid (including those for the 17 keV neutrino), and other claims of evidence for neutrino masses — in particular from atmospheric and solar neutrinos — are shown, at present, to be 'not compelling'. Future experiments are indicated, in particular those using accelerator-produced neutrino beams which travel hundreds of kilometres, third-generation detectors, and a fourth-generation detector that can measure neutrino energy accurately. There are many lessons to be drawn for physics, astrophysics and for matters of Scientific Method and responsibility. While the details of solar neutrinos may eventually be forgotten, it is possible that the lessons and history will be remembered.

*Address for correspondence:
CERN
Geneva 23
Switzerland
e-mail: DROM @ VXCERN.CERN CH

1. Introduction — Neutrinos

Neutrinos are the most exciting of particles and they are on the frontier between the known and the unknown. When neutrinos were first suggested by Pauli in 1930, he gave them properties that made them almost impossible to find — neutral, barely interacting, and zero mass. Since then, many surprising properties of neutrinos have been found and even more have been predicted, especially concerning neutrino masses. With so many pioneering experiments, it is not surprising that many apparently significant claims have turned out to be premature, but that only increases the excitement and interest in the field.

Three types of neutrinos have been found: electron, muon and tau. The latest result on the number of types of neutrino found by LEP was quoted by John Ellis at the Texas meeting, as 2.988 ± 0.023. However, it has been realised that neutrinos could have a mass and, even more exciting, if they have different masses, that there is a possibility they could change from one type to another type — called 'neutrino oscillations', because after having changed into another type they can change back again — or go on to the third type. The idea of such strange behaviour is well-established for neutral K-mesons, where the observed K_S and K_L mesons can be considered as mixtures of two 'fundamental' particles, K_1 and K_2. A completely useless and inappropriate piece of information for neutrinos is the mass difference between K_1 and K_2 — useless because neutrinos are completely different. Anyway, this mass difference is 4×10^{-6} eV or, to put it differently, the ratio of the mass difference to the K^0-mass $\delta m/m$ is about 10^{-14} — very small indeed.

With their unusual properties, it is not surprising that neutrino experiments are very difficult and often have low statistics. Thus pioneering neutrino experiments are admirable, but since one frequently does not know what result to expect, very often the pioneering result later needs modification and sometimes even turns out to be wrong. Many claims of measuring a neutrino mass or observing neutrino oscillations have been made. The importance of possibly discovering such a startling new fact makes the search worthwhile, as the consequences for physics and astrophysics would be dramatic.

There are theories relating the ratios of the masses of the three neutrino types to their corresponding charged lepton counterparts — usually linearly or as the square of the charged lepton mass. This also means that if one neutrino has a mass, then the other two should also have mass.

2. Astrophysics

Close relations between neutrino masses and astrophysics are possible.

In the Big Bang hypothesis, accepted by almost all, neutrinos were produced and interacted, but as the Universe expanded they found few particles to interact with and were finally left as a uniform sea, and now have such low energy that their interactions are negligible. Today there are calculated[1] to be 113 relic neutrinos of each of the three types per cm^3. In addition, there are the later neutrinos coming from sources such as type II supernovae, which occur at the rate of about 10 per second in the observable Universe and which each produce about 10^{58} neutrinos of about 10 MeV per explosion. Stars also radiate neutrinos — our Sun gives out about 3% of its power in the form of neutrinos.

If neutrinos have mass, then the sum total of their masses could be a significant part of the mass of the Universe. This leads to an important limit. If the sum of the three neutrino masses were greater than about 30 eV (the exact number varies between 20 and 60 eV), then the Universe would be too massive for current theories of inflation.

More precisely, the parameter used is a ratio, Omega (Ω), which is the ratio of the density, ρ, of the Universe to a critical density, ρ_c. The condition for today's density, ρ_0, to be large enough that the gravitational pull of this large mass would slow the expansion of the Universe, and then lead to contraction and finally the 'big crunch', is that the density be greater than the critical density, $\rho_0 > \rho_c$. This condition means that today's $\Omega_0 > 1$, and the Universe is said to be 'closed'. If, on the other hand, the density is low, $\Omega_0 < 1$, then the Universe expands for ever and is said to be 'open'. If Ω_0 is exactly one, then the Universe will only just manage to expand for ever.

It has been the conventional wisdom that inflation occurred early in the history of the Universe, requiring that Omega have the value of unity. Since the density of luminous (or visible) matter is very small — about 0.005 — and the density of baryonic matter determined from Big Bang nucleosynthesis is also small — about 0.01 to 0.08 — this must mean that over 90% of the matter in the universe is dark matter and that most of it must be non-baryonic matter. These numbers are multiplied by an inverse power of h, which is close to one, as h lies between 0.5 and 0.8, depending on whether the Hubble constant H is 50 or 80 km/sec/Mpc. More exactly, Ω_{lum} is about $0.003h^{-1}$ (based on a 1979 reference[2]), and Ω_{baryon} is between $0.009h^{-2}$ and $0.02h^{-2}$ from Copi et al.[3] (this is wider than earlier estimates). There are a number of candidates for this missing dark matter, with many of them not yet discovered — axions or neutralinos, for example. But neutrinos *are* known to exist which makes them a favoured candidate for all or part of the dark matter.

However neutrinos need not be the sole source of dark matter. If they were, there would be a problem. After matter and anti-matter annihilate, left over is a strong sea of photons (called the cosmic microwave background radiation, CMBR) and relatively few baryons (down by over 10^{-9}). COBE has shown that this sea of photons is homogeneous to 10^{-5} and this means that the sea of baryons must also be smooth. Now the problem is how to create/understand fluctuations in the sea of baryons which would allow stars, galaxies, etc. to be formed in the relatively short time of about a billion years. Neutrinos interact so weakly that they could not do it alone. However it is conceivable that there are other forms of dark matter which could create the desired fluctuations — for example, primordial black holes.

While the limit of the sum of the neutrino masses of about 30 eV seems a good one, another complication begins to emerge. This number was based on the assumption of conventional wisdom that Ω_0 was one. But recent papers have suggested that Omega is about 0.2, (such as Coles and Ellis[4], N. Bahcall[5]). This would reduce the upper limit of about 30 eV to less than 10 eV.

The above is for stable particles; if the massive neutrino is unstable, then it could exist with a mass greater than 30 eV. Thus Simpson proposed that his 17 keV neutrino was unstable and linked it to Supernova 1987A, although this was unjustified[6].

LEP has established that there are no heavy neutral leptons with a mass of below 45 GeV.

3. Direct mass measurements

3.1. Tau Neutrino

The Argus collaboration[7], have measured charged tau leptons decaying into five charged pions with only the expected tau neutrino missing, and have deduced an upper mass limit for the neutrino of 31 MeV with 95% coincidence. The OPAL collaboration[8] used the same reaction at LEP but, having at that time only five events, obtained a higher limit of 74 MeV. However, the ALEPH collaboration[9] used the decay of a tau into five charged pions plus a pi-zero and, based on 24 events, obtained a lower limit of 24 MeV (95% CL).

It may be noted that although there is strong evidence that the tau neutrino exists, it has not actually been directly identified. Two new experiments at CERN, Nomad and Chorus, are now running and hope to prove the existance of the tau neutrino.

3.2. Muon Neutrino

Measurements by a PSI group[10] of the pion decay into a muon has given an upper limit of 160 keV.

3.3. Electron Neutrino

Here the best method is to study the end-point of the electron energy spectrum of tritium decay, which is about 18570 eV. If there were a single neutrino of mass m_ν, then the end-point would be at a value lower by m_ν. However, the experiment is exceedingly difficult, as while there are good statistics well below the end-point, there are very few electrons close to the end-point. An early major attempt by an ITEP group, gave a positive mass of about 22 to 32 eV. Despite the low statistics, this stimulated interest and a number of experiments lowered the upper limit to near to 10 eV, the lowest being by Mainz[11], who used a solenoid retarding spectrometer to obtain < 7.2 eV (though care must be taken, as the value depends on the range of electron energy near the end-point selected). What was most interesting was that the five best experiments (INS Tokyo, Zurich, Los Alamos, Livermore and Mainz) all obtained negative values for the square of their electron neutrino mass. After many checks, it was generally agreed that this was a real effect and showed there was at least one factor missing for understanding the analysis of the experiments. Since the centre of the probability distribution of m_ν^2 is negative, this means that the 95% limit is closer to zero than it would be if the centre of the probability distribution had been at zero. Hence it is perhaps better to quote an upper limit slightly higher than 7.2 eV, to include a possible systematic bias.

Recently, a beautiful new experiment by Belesev et al.[12] from the INR and the Kurchatov Institute, performed at Troitsk, was presented. The rather complicated apparatus is shown in Fig. 1. It is 'an integral electrostatic spectrometer with adiabatic magnetic collimation and a gaseous source of tritium', whose features avoid/reduce many of the problems of previous experiments. The essential point is that the electrons see a 8T magnetic field when they enter the drift space, while afterwards the field is only 1 mT — i.e., some 10^{-4} lower. The resolution is then approximately (18570.5 times 1mT/8T) and finally a resolution of 3.7 eV is obtained, which is superior to previous experiments. The finding of $m_\nu^2 = -18 \pm 6$ eV^2 corresponds to an upper limit of 4.5 eV (95% CL). They claim to understand the negative value: 'The effect of negative m_ν^2 may be explained by a spike-like (in differential spectrum) structure corresponding integrally to about 6×10^{-11} of total probability'. However, the spike does not seem to be fully understood.

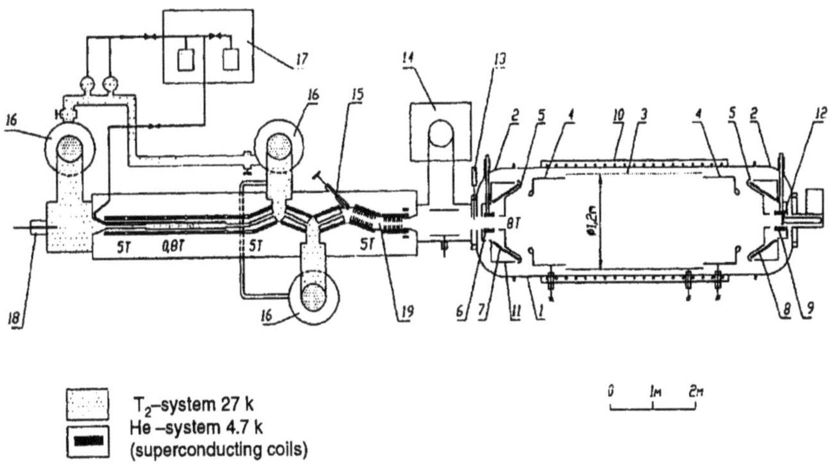

Fig. 1 Experimental set-up of the Troisk apparatus to measure the tritium spectrum: 1,2 — vacuum flask; 3,4 — electrostatic analyser; 5 — grounded electrode; 6, 7, 8, 9 — superconducting solenoids; 10 — warm coil; 11 — liquid N_2 jacket; 13 — fast shutter; 14 — Ti pump; 15 — cold valve; 16 — Hg diffusion pump; 17 — T_2 purification system; 18 — electron gun; 19 — argon pump (Ref. 12).

3.4. Conclusions

The upper limit results for the masses of the three neutrino types are shown in Fig. 2 as a function of the year. It may be seen that there is, on a log scale, continuous improvement.

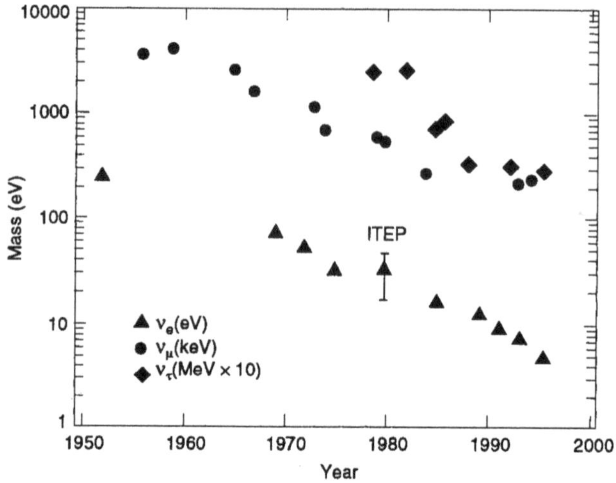

Fig. 2 Variation of the upper limits measured for neutrino masses: ν_e are triangles, ν_μ are circles, and ν_τ are diamonds. For ITEP, a positive value is shown.

The results for the electron neutrino give no indication of a massive neutrino and the upper limit is now down to about 5 eV. However, there are some unexplained factors which probably have nothing to do with neutrino masses. Comparing this value with the astrophysical constraints, this limit is just above the mass expected for the heaviest neutrino, but this neutrino is expected to be the tau and not the electron neutrino according to various theories indicating that the ratios of the neutrino masses are proportional to the corresponding charged lepton $(\text{masses})^1$ or $(\text{masses})^2$. This is discouraging, but perhaps the theories are not optimal.

The limits on the tau and muon neutrinos of 31 000 000 and 160 000 eV, are well above the astrophysical limits (though these limits are for stable neutrinos).

A recent review of this subject has been made by E.W. Otten[13].

4. Double beta decay

Interest in the subject of double-beta decay has been growing steadily[14]. In 1983 there were 20 papers, while in 1993 there were 60 papers. In 1990 there were 30 groups working in this area, while by 1994, there were 40 groups.

Theoretically, there are two types of neutrinos, Dirac and Majorana. With Dirac, the neutrino and its antineutrino are different particles. But with the Majorana particles, they are the same, violating lepton number conservation. The two types can be distinguished either if the neutrino has mass or if the weak interaction has a right-handed component.

Double-beta decay means that an element of atomic number, Z, and atomic mass, A, emits two electrons to give an element with the atomic number Z + 2. Normally (with Dirac neutrinos), the $\beta\beta2\nu$ reaction is:

$$(Z,\ A) \rightarrow (Z+2,\ A) + e^- + e^- + \nu\text{-bar} + \nu\text{-bar} \ .$$

But for Majorana neutrinos, the $\beta\beta0\nu$ reaction is neutrinoless, viz.:

$$(Z,\ A) \rightarrow (Z+2,\ A) + e^- + e^- \ .$$

Calculations for the half-lives of ^{130}Te decays via 2ν or via 0ν give about $10^{22\pm2}$ and $[10^{15\pm2} \times (m^{-2}$ or $\eta^{-2})]$ years resp., where m is the ratio of the neutrino mass to the electron mass and η is the fraction of a right-handed (RH) current. This means that if a 0ν decay rate is measured, the result could be interpreted either as a massive Majorana electron neutrino or as evidence for an RH current.

Over 40 years, some nine examples have been found for Dirac neutrinos in $\beta\beta2\nu$ reactions, and the result has been used to check the complex calculations of the matrix element. It had been hoped that this calculation, plus measurement of the $\beta\beta2\nu$ reaction, could be used as a calibration for the $\beta\beta0\nu$ calculation, but unfortunately it has been shown that this extrapolation is not justified, as the $\beta\beta0\nu$ calculation is too different.

There are three types of experiment: passive or geochemical, radio-chemical, and active.

4.1. Geochemical Experiments

Here, ancient rocks which appear to have been undisturbed are chosen and the amount of daughter (Z + 2) nuclei among the parent Z nuclei is measured — for example, ^{130}Xe in ^{130}Te and also ^{82}Kr among ^{82}Se nuclei. In both these cases, the daughter elements are gases, so that extreme care needs to be taken to ensure that no gas has escaped over such a geologically long period. Pontecorvo suggested that taking the ratio, R, of isotopes, would give a safer value — for example, using ^{130}Te and ^{128}Te, whose half-lives are calculated to be in the ratio of 1.8×10^{-4} if Dirac neutrinos, but lower if Majorana neutrinos.

In 1975, a value of $R = (0.64 \pm 0.2) \times 10^{-5}$ was found, and this gave either a neutrino mass of 40 eV or η of 5×10^{-5}. However, the effect was assumed to have vanished when Kirsten et al.[15] repeated the experiment in 1983 and found an R value of $(1.03 \pm 1.1) \times 10^{-4}$ giving a neutrino mass of less than 5 eV or $\eta < 2 \times 10^{-4}$.

More recently, in 1993, Bernatowski et al.[16] found $R = (3.52 \pm 0.11) \times 10^{-4}$, giving a neutrino mass of less than about 1.1 to 1.5 eV, the lowest limit so far for Majorana neutrinos.

4.2. Radiochemical Experiments

The idea is to store for a few years an element which has a short half-life of about 10 to 100 years, and then measure the amount of accumulated daughter element. Thus Turkevitch et al.[17] used ^{238}U, which decays into ^{238}PU with a half-life of 88 years, and deduced a half-life of $(2.0 \pm 0.6) \times 10^{21}$ years giving a neutrino mass of 10 eV. However, there is concern that deposits from atom bomb tests could affect the results — ^{238}Pu now has an average surface spread of 10^6 atoms per cm^2! and the double-beta signal is only 10^5 atoms.

4.3. Active Experiments

These are counter experiments which measure instantaneous rates of decay of relatively short-lived elements — ^{76}Ge is a favourite element for study. The energy spectrum is measured. For $\beta\beta 2\nu$ decays, the spectrum is zero at the highest energy and has a peak at a little above zero energy, whereas the $\beta\beta 0\nu$ decay has a single line at the maximum energy.

The main results are:

- Caltech–Neuchatel–PSI use ^{136}Xe and find the $m_\nu > 2.8$ eV
- Heidelberg use ^{136}Xe and find the $m_\nu > 1.7$ eV
- NEMO I use ^{100}Mo and find $\tau > 0.9 \times 10^{21}$ years (90% CL), for $\beta\beta 2\nu$ decay
  ^{116}Cd and find $\tau > 10^{19}$ years (99% CL), for $\beta\beta 2\nu$ decay
  ^{116}Cd and find $\tau > 7 \times 10^{21}$ years (90% CL), for $\beta\beta 0\nu$ decay
  ^{106}Cd and find $\tau > 10^{19}$ years (90% CL), for $\beta\beta 2\nu$ decay
- NEMO II and NEMO I use ^{130}Te and find $\tau > 2 \times 10^{21}$ years for $\beta\beta 0\nu$ decay.

The lowest limits have been obtained by the Heidelberg–Moscow collaboration[18] who use enriched ^{76}Ge and find $m_\nu < 1.31$ eV (90% CL).

In the future, bigger enriched sources with less background may be expected, and this should give lower limits, but the progress may be slow because the rate of improvement is (mass x time)$^{1/4}$ if there is background, and (mass x time)$^{1/2}$ without background. A new experiment is Borexino, who hope to reach a limit of 0.3 eV; the ultimate limit may be as low as 0.01 eV.

It will be noted that these limits are almost an order of magnitude lower than the limits obtained from tritium decay measurements, but the difference is that the tritium measurements were for Dirac neutrinos, expected if lepton number is conserved, whereas the double beta decay limits are for Majorana neutrinos.

5. Neutrino oscillations

5.1. Basic Information

In 1957, Pontecorvo[19] suggested that if neutrinos had mass, they could oscillate from one form to another. In 1962, Maki et al.[20], wrote out the formalization for two neutrinos:

$$\nu_e = \nu_1 \cos\theta - \nu_2 \sin\theta$$
$$\nu_\mu = \nu_1 \sin\theta + \nu_2 \cos\theta \ ,$$

where ν_1 and ν_2 are the 'true' neutrinos or mass eigenstates, and the observed ν_e and ν_μ particles are the weak interaction eigenstates. The change from one state to the other proceeds by a mixing matrix as for the two K^0 mesons.

When the third neutrino was discovered, the formalism was extended and the mixing matrix is now 3×3. Experimental results have often been used to derive neutrino masses, but sometimes the input data has later turned out to be unjustified.

Generally, only two neutrinos are considered, so that there are only two parameters and not three. It has become conventional to take these two parameters as the square of the mass difference between the two neutrinos, δm^2, and the mixing angle, θ. It may be noted that the neutrino mass is not directly measured, but is the difference of the masses squared. It could be thought that taking only two neutrinos instead of the correct three would lead to serious misinterpretations, but Shi et al.[21] have shown that the results are not very different (however, it is clearly better to do the job properly and consider all three neutrinos with mass).

Taking the typical case of a beam of muon neutrinos giving some electron neutrinos by oscillations, then:

$$\text{Prob}\,(\nu_\mu \rightarrow \nu_e) = \sin^2 2\theta \sin^2 \Delta$$
$$\text{Prob}\,(\nu_\mu \rightarrow \nu_\mu) = 1 - \sin^2 2\theta \sin^2 \Delta \ ,$$

where $\Delta = 1.27 \times \delta m^2 \times L/E$ and m, L and E are in units of eV, metres, and MeV resp., and where E is the neutrino energy.

The oscillation length is then, $L_0 = 2.5 \times E/\delta m^2$.

The soundness of an experiment is then judged by the L/E value. To search for low masses, wish L to be long and E small. For this, experiments are proposed for sending beams of neutrinos from CERN to Gran Sasso and Nestor, and from Fermilab to Soudan, all distances of about 1000 km, and also to send a lower energy beam the shorter distance from KEK to Kamiokande. The beam energy cannot be too small or it will be too dispersed.

The analysis is not easy as is shown in Fig. 3 of Barger. Here all three neutrinos are considered and because L is fixed it can be seen that very good statistics are needed. For this reason, it is much better not to have only a single detector, but to have at least two so that the result can be checked.

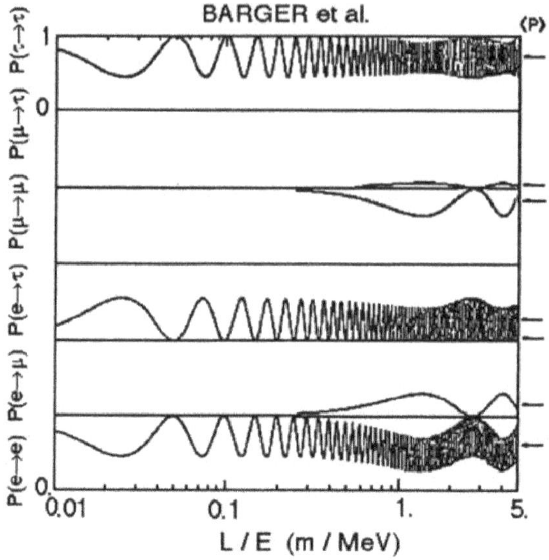

Fig. 3 Example of the probabilities of one neutrino species oscillating into another species as a function of the distance divided by the neutrino energy, L/E, m/MeV.

5.2. Experimental Results.

There have been many experimental results covering various regions of phase space — expressed as regions of the δm^2 versus $\sin^2 2\theta$ plane for each pair of neutrinos — and establishing three plots to be considered for each of the three possible pairs of neutrino types.

Generally, the present results from accelerators extend down to about 0.2 to 1 eV^2. Reactor results extend slightly further down to a few times 10^{-2} eV^2 and there are plans to go down to almost 10^{-3} eV^2. The study of atmospheric neutrinos (that is, neutrinos produced by cosmic ray interactions in the atmosphere) goes even further down, to 10^{-4} eV^2. To go to still lower mass differences, solar neutrinos appear to be the only source: they cover the range from 10^{-4} to 10^{-7} eV^2. Thus most ranges of mass difference seem covered, but the possibility of checking one technique against another is limited. A typical plot given, in this case, by the San Onofre experiment for muon and electron neutrinos, is shown in Fig. 4, where accelerator results are not given as they are at higher mass differences. It will be noted that the range of $\sin^2 2\theta$ values covered is very variable and goes out furthest to 10^{-4}, for solar neutrinos. A new generation of accelerator experiments are starting to explore much lower values of $\sin^2 2\theta$, Nomad and Chorus at CERN, E801 at Fermilab and P803 at BNL. In addition, P860 will go down to about 10^{-1} eV^2 and later down a further factor of three.

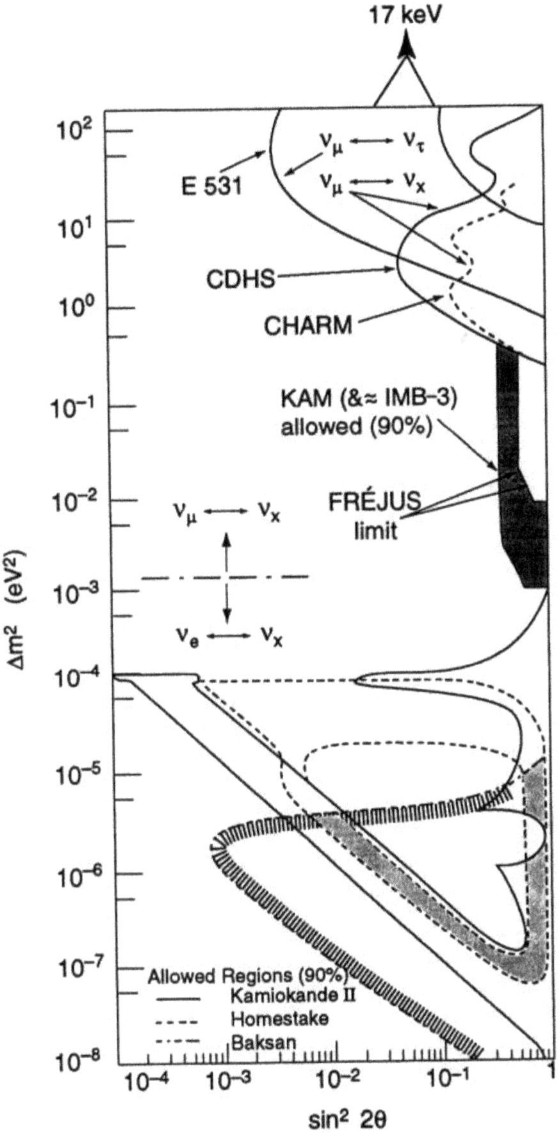

Fig. 4 Example of the plot of the squared neutrino mass difference of two neutrino species, δm^2, versus $\sin^2 2\theta$ where θ is the mixing angle. The allowed regions are shaded, the 90% excluded regions are shown as lines.

Thus it can be seen that the field is very active, with most regions of phase space being covered several times by different experiments and the regions extended to try and answer these important questions.

Until 1994, neither from accelerators nor reactors was there 'compelling evidence' for neutrino oscillations, despite occasional claims which have all faded. Unpublished claims from Los Alamos and from Karmen at RAL are mentioned below in 5.3. Claims made for neutrino oscillations based on atmospheric and solar neutrinos, are discussed in Chapters 7 and 8.

5.3. Recent Experimental Results — Reactors, Los Alamos, Karmen Reactors

At Krasnoyarsk in Russia, important new experimental results have been obtained[22] using reactors with detectors at 57 and 321 metres. They find no evidence for oscillations with δm^2 greater than 7.5×10^{-3} for $\sin^2 2\theta$ of 1.0. At Bugey in France, detectors have been placed 15, 40 and 95 metres from reactors (some 150,000 events counted — a record) and Achkar et al.,[23] have found minimum excluded values for δm^2 and $\sin^2 2\theta$ of 1×10^{-2} eV^2 and 2×10^{-2} resp. This use of multiple detectors is a good feature and can make the results more reliable.

Los Alamos

There have been great discussions recently about results presented in the *New York Times*[24], where it was reported that some evidence had been obtained at Los Alamos for neutrino oscillations. The experiment involves sending a 780 MeV proton beam into a water target to produce mainly anti-muon neutrinos (plus ν_μ and a few ν_e, but no anti-ν_e). A large experimental detector, LSND, with 185 tons of a mixture of baby oil and liquid scintillator, and with 1220 photomultipliers, searches for the production of anti-ν_e using the reaction

$$\text{anti-}\nu_e + p \rightarrow e^+ + n\ .$$

The positron and neutron are then detected, the neutron giving the characteristic gamma of 2.2 MeV from stopping the neutron with a proton, to yield a deuteron and the gamma.

First results based on one-and-a-half months running (1800 hours) gave a graph of the positron energy shown in Fig. 5, where an unexpected peak of about eight events over a background of about one event was observed. Very properly, the group said that these results should be considered preliminary, as the statistics are low and the apparatus has only just begun to be operated.

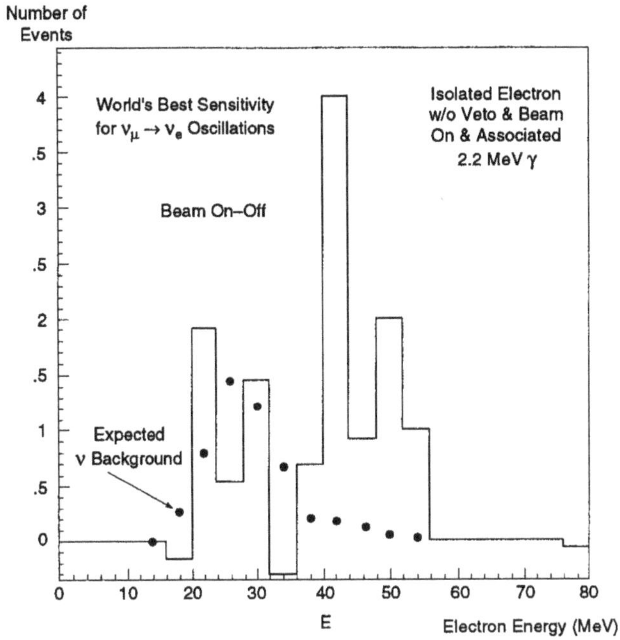

Fig. 5 Number of events as a function of the electron energy for the Los Alamos experiment. The expected background is shown as a series of dots. This was for the first series of measurements and was presented at the Rochester conference in Glasgow, July 1994.

However, a second run three times as long (5000 hours), was said[24,25] to give similar effects, but with much higher statistics, so that 80 events were observed where only 40 were expected. It was also reported that this anomaly corresponded to a neutrino mass difference of 2.4 eV and that both neutrinos must have masses of between 0.5 and 5 eV — just the mass range that would provide much or all of the missing dark matter — and this greatly increased astrophysical interest. It was noted that these mass values were inconsistent with claims based on experiments with solar neutrinos and claims based on atmospheric neutrinos, and that 30 different backgrounds had been studied. Unfortunately, no scientific paper is available to allow comment. It is hoped to have further running later in 1995, after which the LAMPF accelerator will no longer be available. A proposal has been made to transfer the detector to Fermilab and perform an experiment there with the much higher, 8 GeV energy beam.

However, this LSND result is inconsistent with two other published results, though the statistical significances are not large. One is the KARMEN experiment at the Rutherford-Appleton Lab, a well-established experiment which has been running for four years. The other is a BNL experiment which has also been running for some time.

No conclusion can be drawn until a scientific paper has been studied and questions answered. However theoretical papers which assume the result is true are already appearing.

(Note added in proof: At the March Moriond meeting, more results were shown, including a plot very similar to Fig. 5 but with three times the statistics — the peak near 40 MeV is statistically significant. No explanation of the disagreement with previous experiments is offered. No values of δm^2 and $\sin^2 2\theta$ are given apart from those reported in the *New York Times*).

KARMEN

It has been reported in the *New Scientist* magazine[26] and elsewhere that the KARMEN experiment has found evidence of some 100 counts occurring about 3 to 4 microseconds after the beam. If these were caused by particles decaying in flight, they would have a mass about 30 MeV. The nature of such a particle is unknown. It might be, for example, an axion or a heavy neutrino, or even a Wimp. Again, comments and judgement must await study of the scientific paper being prepared.

(Note added in proof: At the March Moriond meeting, observation was claimed of 83 ± 28 events centred at 3.6 ± 0.25 μs with a FWHM of 1.3 ± 0.5 μs, but the plot of the number of events versus time suggests a signal-to-noise ratio of only about 0.3 to 0.4).

5.4. Future — Reactors, Long Baseline Proposals

In view of the importance of the possibility of finding neutrino oscillations, proposals have been made at most major labs with proton beams. Excellent news is that they often propose using several detectors for the same beam line. Below are tabulated the possible average neutrino energies in GeV, the maximum distance in km to a detector, and the lowest value of δm^2 in eV^2 attainable and possible number of detectors and distances.

Laboratory	$\langle E_\nu \rangle$	δm^2	Maximum distance	No. of Detectors (distance)
BNL	1	0.04	25	3
KEK/Kamiokande	1.4	0.006	250	3 (1/2, 25, 250)
CERN /Gran Sasso	7 – 20	few $\times 10^{-3}$	731	1 or 2 (1, 731)
CERN/NESTOR	7 – 20	few $\times 10^{-3}$	1630	1 (1630)
Fermilab/Soudan	10 – 20	0.003	730	2 (1/2, 730)

In general, these values of δm^2 are for $\sin^2 2\theta = 1.0$. The $\sin^2 2\theta$ limits, being of the order of a few times 10^{-2}, are generally not as low as for the new generation of experiments.

It is very important to stress that these are proposals and, except possibly in the case of Kamiokande, which may be able to start in 1988, the numbers are not definitive. They should be treated only as indications.

Previously, the detectors for reactor experiments were near the surface, but to reduce the background from cosmic rays, a second generation of experiments are being prepared where the detector is deep underground or inside a mountain. The three experiments are at San Onofre (USA), Chooz (France) and Perry (USA, at the IMB site). The three detector masses and distances from the reactor, are 12 000, 4800, and 1000 tons, and 680, 1025, and 12 900 metres, respectively. The expected limits (90% CL) on δm^2 are 2×10^{-3}, 1×10^{-3}, and 1×10^{-3} eV^2 resp. The $\sin^2 2\theta$ limit would be about 10^{-1} or possibly better.

It will thus be seen that for these important questions there is a vigorous research programme proposed which covers larger areas of phase space by using new generations of detectors and longer distances.

6. The 17 keV neutrino case history

From 1985 until 1993, the question of the existence of a new neutrino of 17 keV mass was hotly debated — until almost everyone agreed it did not exist. This has become a case history of how Science works and can be considered as an example of the ultimate triumph of the scientific method. Its history can be followed from Fig. 6 where the number of papers (experimentally in favour, null, and theoretical) published every two months is given[6].

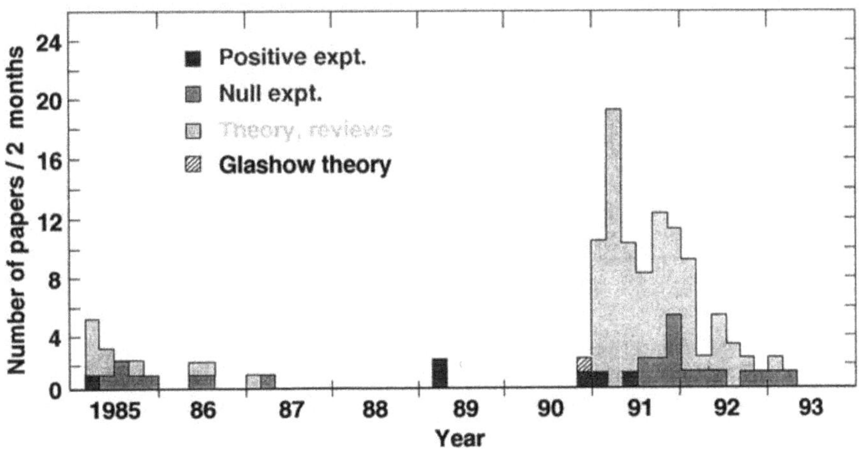

Fig. 6. Plot of the number of papers published every two months during the eight-year period on the 17 keV neutrino, from Ref. 6.

In 1985, John Simpson reported that at Guelph he had observed a change of shape in the spectrum of low-energy electrons from tritium decay and that this could be interpreted as a neutrino of 17 keV mass with a 3% mixing. A few theory papers started appearing, but soon after, four experiments reported data against the existence of the 17 keV particle. Simpson, however, persisted in criticizing the null results and saying that (a) they were mainly using a magnetic field and these fields were not fully understood and, (b) two of them were poorly analysed and that in fact they had evidence in favour of a 17 keV neutrino with mixing of a few percent! Despite this, the general opinion was that the subject was dead, though it stayed in people's memories. Even when Simpson and Hine on 1 April 1989 published two more papers from Guelph showing new positive evidence, there was no reaction.

The situation changed end 1990/early 1991, when new experimental results from Oxford and Berkeley were presented giving evidence for a 17 keV neutrino with a mixing of about 1% — the value Simpson had corrected his first result to. Also, Glashow gave strong theoretical support and then some other groups claimed to find this 17 keV particle. It was claimed to have been observed in decays of ^{3}H, ^{14}C, ^{35}S and ^{63}Ni. What was remarkable was that six experiments all found 17 keV, not 16 or 18, but only 17 keV. Furthermore, Simpson claimed that in supernova 1987A, there was a gap in time of the neutrinos detected and, assuming this was due to the 17 keV neutrino, that it was unstable and had a mean life of (0.6 to 1.6) $\times 10^4$ seconds. Overall, most impressive, so that some believed and many thought it should be taken seriously. As shown in Fig. 6, many theoretical papers were written, often involving astrophysics and using the 17 keV neutrino to explain the claimed solar neutrino deficit.

The natural scientific response was to design new second-generation experiments of higher quality which would answer Simpson's objections to earlier null experiments. This took some time but after rather more than one year, a number of excellent experiments were designed, built and run. They gave a null result and a 17 keV neutrino with 0.8% mixing was excluded by a large number of standard deviations.

Now in the history of wrong results in Science[27], it usually happens that the original group who claimed to have observed an effect reject the null experiments and continue to believe, even though the vast majority of the scientific community reject the claim. For example, after N-rays were exposed by Wood, Blondlot continued to believe and even published a book two years later; Fleishmann and Pons continue with cold fusion and, with a small band of True Believers, organize international meetings. However, in the case of 17 keV neutrinos, something different happened: the experimenters at Oxford and Berkeley concluded that the balance of scientifc evidence was against them and they decided to check their experiments and do more tests. In both cases they were able to find the errors in their experiments and said so, retracting their results. The Oxford paper[28] is clearly entitled 'The Origin of the Spurious 17 keV Neutrino Signal observed in ^{35}S Beta-decay'. This is the manner in which the scientific method is supposed to work and this case should be considered one of its triumphs[6].

7. CCFRW double-vertex events

In 1992, the powerful CCFRW experiment at Fermilab said that they had observed an anomaly in their neutrino experiment. In their long detector they sometimes observed two vertices of events in the same time bin, and when they calculated the expected rate of occurrence of these double vertex events it was significantly less than actually observed. They hypothesised that this could be evidence for a new neutrino emitted in the first event and interacting further downstream. That such a major collaboration had found such an important new result caused great excitement and interest.

The $B^3C^2FHI^3JM^2ORST$ Collaboration[29] were taking photos with the 15 foot-bubble chamber just downstream of the the CCFRW experiment. An important part of the double vertex calculation was how often neutrons and K^0's were produced at the first vertex and what was their energy spectrum. This information about K^0's from both charged and neutral current events is readily obtainable in a bubble chamber as the actual K^0 decays can be observed and measured. No double vertex effect could be observed, but the statistical significance was low.

After presentations by the two collaborations, the CCFRW withdrew their claim, which in fact had not been published officially.

This is a further illustration of the difficulty of neutrino experiments.

8. Atmospheric neutrinos

When cosmic rays enter the atmosphere, they interact. The hadronic products die out fairly quickly but the muons and neutrinos can penetrate into the Earth's crust and can be studied by deep underground detectors. The ratio of muon neutrinos to electron neutrinos is expected to be close to the value of two because of the decay chain

$$\pi \to \mu + \nu_\mu \quad \text{and} \quad \mu \to e + \nu_e + \nu_\mu \ .$$

Detailed Monte Carlo calculations have been made of the ratio, R,

$$R = (\nu_\mu/\nu_e)_{data}/(\nu_\mu/\nu_e)_{MC} \, .$$

The values of R found by five experiments are:

– NUSEX[30]	$0.99^{+0.38}_{-0.25}$
– FREJUS[31]	0.87 ± 0.13
– SOUDAN II[32]	0.64 ± 0.19
– IMB[33]	$0.54 \pm 0.05 \pm 0.07$
– KAMIOKANDE[34]	$0.60^{+0.06}_{-0.05} \pm 0.05$.

Thus two experiments found no effect and three found a low value of R. However, at the Neutrino '94 conference, the *rapporteur* said that, looking at the errors, it may be seen that overall there is evidence for a low value of R.

There was no way of knowing whether there was an excess of muon neutrinos or a deficit of electron neutrinos.

At the Rochester conference at Glasgow, the Kamiokande collaboration[35] presented new results. Previously they had only taken events that were fully contained in their detector; now they added partly-contained events with one or more tracks leaving it. These two samples were different, the former having an average energy of 0.8 GeV and all events with less than 1.33 GeV energy, and the latter sample an average energy of 7 GeV and all having energy above 1.33 GeV. For this new, multi-GeV sample, they found:

$$R = 0.57^{+0.08}_{-0.07} \pm 0.07 \ .$$

It was considered that this was a confirmation of an atmospheric neutrino effect.

An important new contribution was that Kamiokande[35] looked at the zenith angular distribution and found a deficit for upward-going neutrinos — those that had travelled through the earth. This was consistent with neutrino oscillations, and the possible areas of phase space for ν_μ to ν_e and for ν_μ to ν_τ oscillations are shown in Fig. 4. However, it should be noted that some other results shown in the figure are in disagreement with the Kamiokande claim (also Ref. 24).

Later, Olga Ryazhskaya[36] wrote a paper suggesting an alternative explanation. Some of the cosmic muons must interact near the Kamiokande detector and should produce neutrons. She suggested that some of these neutrons traversed the veto counter designed for charged particles. The neutrons interacted in the detector and produced neutral pions, decaying to photons and giving electrons, recorded by the detector and interpreted as electron neutrinos. She calculated the rates and found that the anomaly could be explained by this chain.

Fortunately, there is a very easy way of checking whether Olga's explanation is correct or not by using the procedure employed by Gargamelle for establishing that neutral current events and not only neutron interactions were being observed. Neutrinos have such a low cross-section that most traverse the detector without interacting, hence the distribution of the vertices of events is constant with distance from the wall. On the other hand, neutrons interact strongly and hence most will interact near the wall. Thus a plot of the distribution of the frequency of occurrence of event vertices as a function of the distance from the nearest wall for the electron neutrino candidates and for muon neutrino candidates will show which is correct. If there are no neutrons, only neutrinos, then the two distributions should be flat. If there are neutrons, there will be a maximum near the wall for the electron neutrino candidates.

To determine if there is any evidence for New Physics, an experimental and a theoretical value are compared. But the comparison only makes sense if the errors are known for both. It is difficult to evaluate the theoretical errors — there are statistical (from experimental data fed in) and systematic errors (from lack of knowledge of the theory). The latter are very hard to 'guesstimate' since one rarely knows what one does not know! One technique is to compare different theories, but this suffers from the requirement that the theories be derived independently, which is certainly not the case, as the various theorists meet at conferences and communicate and compare so that the variations tend to be reduced, and thus a systematic theoretical error derived from such a method should be considered only a lower limit (see also critique by Don Perkins[37]).

9. Solar neutrino deficit?

The conventional wisdom for some 20 years has been that there is a discrepancy between the predicted flux of neutrinos from the Sun and the measurements in a Chlorine experiment. But in recent years, with new experiments and checks on the calculations and errors, there have been doubts. A brief summary is given here — a more complete recent review is given in Ref. 38.

The calculations are based on the solar evolutionary model, sometimes called the standard solar model, or SSM. The evolution of the Sun, from its beginning some 4.5 Gyrs ago to the present, is followed. The boundary conditions are the Sun's age and the present-day mass, radius, and luminosity, plus the elemental abundance at the time when the star began. In addition, there are a tremendous amount of other pieces of input data and many assumptions. Despite this, the model generally works well, though the errors have been seriously underestimated by some theoreticians and this is important when a comparison is being made with experimental results that have their own errors.

The evolutionary model has changed with time and some problems still remain, such as the ^{7}Li abundance being a factor of a hundred too low. For low energy neutrinos coming from the dominant pp reaction, the calculations are reliable and are stabilised by the need to have agreement with the Sun's luminosity. For high energy neutrinos the situation is different, as they come essentially from the decay of Boron-8. This is a small channel, only about one per thousand of the total, so that if it were seriously in error, it would affect the calculation of the luminosity. And, in fact, there are serious problems with just this channel, both because of the reaction ^{7}Be(p,γ)^{8}B, which produces the ^{8}B, and the channel ^{3}He(α,γ)^{7}Be, which produces the ^{7}Be that gives the ^{8}B. The Chlorine and Kamiokande experiments are dominated by ^{8}B neutrinos. This has caused Turck-Chieze and Lopez[39] to give errors of 22% and 25% to these two experiments, resp.

A great deal of the pioneering work on solar modelling was done by Bahcall and collaborators, but the errors they quote for ^{8}B neutrino flux are excessively small — 11 to 14%. Recently, Dar and Shaviv[40] have recalculated fluxes based on a new and extensive model calculation of Kovetz and Shaviv[41], using the latest input data, and found lower rates than with older calculations. They quote errors on the calculated fluxes for the Chlorine and Kamiokande experiments of 29% and 30%, resp.

The reaction $^7Be(p,\gamma)^8B$ has been measured down to 117 keV, but one problem is to extrapolate the cross-section down to the energy where the reactions occur in the Sun, which is about 1 to 20 keV. To remove barrier penetration effects, the astrophysical S-factor S_{17} is used. There are uncertainties in the extrapolation, as ^8B is a proton-rich nucleus and the proton wave function is not well-established[42] at the very large radii (about 40 fermis) involved. The second problem is that this reaction has been very poorly measured and the results are highly dispersed, varying by over 100%. Recently, a new measurement by Motobayashi et al.[43] has indicated that the cross-section should be lowered by about 25%. The seriousness of the $^7Be(p,\gamma)^8B$ problem is shown in Fig. 7, where the various estimates of S_{17} based on experiment are shown as a function of time, plus some recent values used by Bahcall et al. and Dar and Shaviv. See Ref. 38 for a fuller discussion.

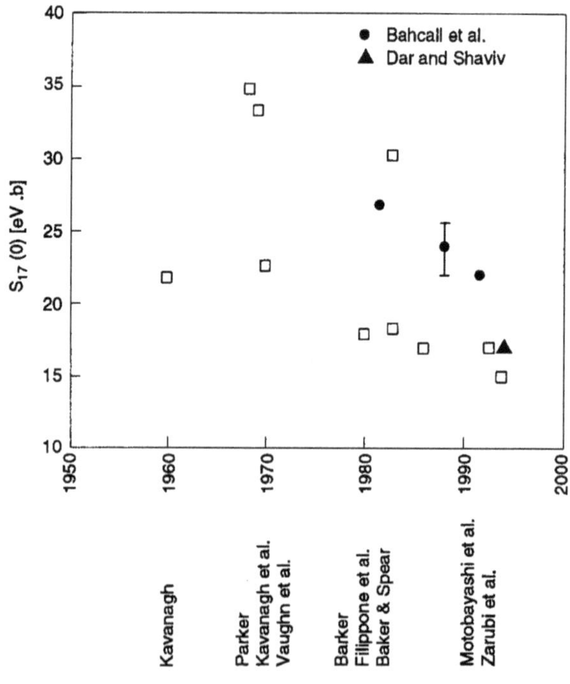

Fig. 7. Plot of the measured values of the astrophysical S-factor, S_{17}, for the reaction $^7Be(p,\gamma)^8B$ as a function of the year. The value taken in the theoretical models of Bahcall et al. are shown as circles and that chosen by Dar and Shaviv is shown as a triangle (Ref. 39).

Four experiments are now measuring the flux of solar neutrinos. Three are radiochemical, where an element absorbs a neutrino and emits an electron to give an element which is radioactive. This new element is then extracted and the radioactivity measured. Thus in the Chlorine experiment, ^{37}Cl gives ^{37}Ar, while in the SAGE and GALLEX experiments, ^{71}Ga yields ^{71}Ge. These experiments are extremely difficult, because although the chlorine experiment starts with some 10^{30} atoms, only some 10^{1} atoms are extracted. A major new work by the GALLEX collaboration[44] was to irradiate their gallium with a ^{51}Cr source of neutrinos and to show for these low energy (< 1 MeV) neutrinos that the counting rate, ± 12%, was in good agreement with that predicted. The fourth experiment is Kamiokande, which observes neutrino scattering on electrons in water by observing Cherenkov light from the electrons. This only works for high-energy neutrinos above 7 MeV which are all from the difficult decay of ^{8}B.

The Kamiokande experimental group are strong and well-funded and give a result[34] which they present as:

$$(\text{Data})/(\text{SSM}) = 0.48 \pm 0.07 \ ,$$

which gives the impression of there being a significant (more than seven standard deviations) solar deficit. However, they have omitted to include the theoretical error, which can be taken as 25% [39]. This reduces the significance to less than two standard deviations.

It is better to take the basic Kamiokande result[34] for the flux of:

$$(2.7 \pm 0.2 \pm 0.3) \times 10^{6}\ \text{cm}^{2}/\text{s}$$

and compare this with the prediction of Ref. 39, reduced by 25% for the new ^{7}Be(p,γ)^{8}B reaction rate, giving:

$$(3.3 \pm 0.8) \times 10^{6}\ \text{cm}^{2}/\text{s} \ ,$$

while Dar and Shaviv[40] predict 2.77×10^{6} cm^{2}/s, making the Kamiokande result in good agreement with the latest model calculations.

The Chlorine experiment is the other one that measures dominantly high energy neutrinos. They started measuring in 1967 but excluded the first three years' data when they found a better trigger (^{37}Ar decays give a fast rise time, while cosmic ray background gives a slow rise time). Their experiment should be divided into two periods, before and after the 18-month shut-down caused by pump failures in 1984–86. The justification for this division comes from their claim, based essentially on the earlier period, that their rates fluctuated strongly with the inverse of the sunspot cycle: they claimed this was significant at the five standard deviation level. These fluctuations were extremely large. Thus in 1972 they claimed that the rate was less than one SNU, whereas the theory of Bahcall et al.[45] was about 8 SNU, and, later, one year's data gave 0.4 ± 0.2 SNU, or some seven standard deviations less than their average value (which was 2.05 SNU but is now 2.55 ± 0.25 SNU) as well as peak values (at the sunspot minima) of over 4 SNU. However, since 1987, Kamiokande has been in operation and has observed no variation of flux with time, and is thus in strong disagreement with the earlier Chlorine results. And GALLEX, too, has observed no variation with time. Since the shutdown, the Chlorine experiment itself has observed no significant variation with time, and says that combining the two sets of data, the effect is now only two standard deviations and no longer significant.

There are several ways of interpreting these results:

a) if one believes that all experimental results have to be believed (and many do) despite the overwhelming historical evidence that pioneering neutrino experiments sometimes need modification), then one has to believe that the neutrino flux varied wildly up to 1985 with the inverse of the sunspot cycle and has then been stabilized since 1986;

b) there were some problems with the pioneering Chlorine experiment, which in its earlier period gave occasional low values.

Hypothesis b) seems more likely and will be adopted.

The data taken after the shutdown gives a rate of 3 SNU, which is only slightly less than the Kamiokande value and hence can be considered in agreement with the latest model calculations.

The GALLEX collaboration are also a strong well-funded group who work very carefully and make many checks and calibrations — for example, Ref. 44. They[46] find a rate of 79 ± 10 ± 6 SNU, which, after their new calibration, was given[47] as 64 ± 20)% of model calculations. This means that it is 36 ± 20% lower than the model — that is, less than a two standard deviation effect (their spokesman, Till Kirsten, commented[47]: 'What's a few standard deviations between friends?').

The SAGE collaboration started with a surprising result from their 1990 runs, finding a very low rate which was finally published[48] as 20 SNU but has since been revised[49] to $(40^{+31}_{-38}\,{}^{+5}_{-7})$ SNU. Subsequent runs have been in agreement with the GALLEX result and, including the 1990 result, they find for 1990–1993, $(74^{+13}_{-12}\,{}^{+5}_{-7})$ SNU.

It is concluded that while there may be many mysteries in the theory and experiments, overall, at this time, there is 'no compelling evidence' for any solar neutrino problem.

It is possible that there are solar neutrino anomalies which could be induced by neutrinos having mass, and if this mass were extremely small, then the only means of detecting any effect would be to use solar neutrinos where the L/E factor is very favourable, as the distance, L, is very long and energy is small. For this reason, a strong future programme is in preparation. One could consider the Chlorine experiment as being the first pioneering work and Ray Davis deserves great credit for his early work and perseverance. The second-generation experiments are then Kamiokande, GALLEX and SAGE. In about 1996, the third generation of experiments will start: Superkamiokande, SNO, Borexino, and Icarus. After these already large experiments, it is difficult at present to imagine still-larger experiments, but a fourth generation proposal is HELLAZ[50]. It would be significantly different from all previous experiments in its ability to measure the neutrino energy accurately (to about 5%). This would be an enormous advantage, as it would make it possible to obtain the detailed energy spectrum of the neutrinos and hence to measure the relative strengths of the different sources of neutrinos (^8B, ^7Be, ^{15}O, ^{13}N, pp). This would also allow those theories which claim that some type is anomalous to be tested directly and unambiguously.

10. Conclusions

Neutrino masses are a possibility and if observed, would open up entirely new fields of physics and astrophysics. They would take physics beyond the standard model. Neutrinos are a leading candidate for some or all of dark matter. Astrophysics gives a tough limit on the sum of the three masses of about 30 eV (assuming Omega is one). Direct measurements of the electron neutrino mass give an upper limit of about 5 eV, but the upper limits on muon and tau neutrinos masses are much higher than 30 eV.

Double-beta decay is widely studied in the search for evidence of Majorana neutrinos. Their existence would require that lepton number conservation be violated. So far the upper mass limit is about 1 eV.

If one neutrino has a mass, then it is expected that all three would be massive. It should then be possible for them to switch from one form to another — to oscillate as do K^0-mesons. Many experiments have been, and are being, performed using a variety of sources of neutrinos — accelerators, reactors, cosmic ray interactions in the atmosphere, and the Sun. Each source covers a wide range of mass differences and mixing angles. At present these ranges barely overlap, which makes confirmation of indications difficult. However, major new generations of experiments are beginning or are planned, which should extend the ranges and give overlap and some decisive checks on claims. A number of claims for evidence of neutrino masses have been made and most have been disproved or have faded, but at the time of writing, there remain several claims outstanding. In particular, major excitement surrounds atmospheric and solar neutrinos.

Another approach to deriving a conclusion about neutrino masses is to take all recent measurements and for each plot δm^2 with its error. The resulting plot would have approximately equal numbers of positive and negative values of δm^2. Also, many would be significantly different from zero. Some would say that negative values of δm^2 are non-physical and should be excluded, and hence the remaining results would all show a positive mass (though the ranges of these positive values is rather wide-spread). Others would say that this shows that neutrino experiments are difficult and the analyses tricky, as they often involve assumptions that are not well-established: they might conclude that there is 'no compelling evidence' for neutrinos having mass.

In this review, a checking of theory and experiments shows that the evidence provided in favour of neutrinos having mass is 'not compelling'.

The subject of neutrinos is so exciting that many pioneering experiments have been made. The pioneers who performed them are to be respected. However, there is the experimental result that a good fraction of the conclusions of these pioneering experiments have later had to be modified. This result is experimentally significant, though only a few examples have been given here. The history of the 17 keV neutrino is a good example of the scientific method in operation. A distinguished panel from the National Academy of Sciences, the National Academy of Engineering and the Institute of Medecine have issued a report[51] describing how Science should, and should not, be done. Questions of publishing policy, the publication all points of view and of non-abuse of the refereeing system, etc., are covered.

A wide range of neutrino masses and mixing angles remains to be explored and new generations of experiments are about to come on-line or are planned, for searching and for checking earlier claims. The consequences of finding definitive evidence for neutrino masses are so great that these major efforts with new generations of experiments are to be strongly encouraged.

11. References

1. G. Gyak and M.S. Turner, Nucl. Phys. B (Proc. Suppl) **38** (1995) 13–25.
2. S. Faber and J. Gallagher, Ann. Rev. Astron. Astrophys. **17** (1979) 135.
3. C. J. Copi, D.N. Schramm and M.S. Turner, Science **267** (1995) 192.
4. P. Coles and G. Ellis, Nature, **370** (1994) 609–615.
5. N. Bahcall, 'Large Scale Structure in the Universe', 11th Potsdam Cosmology Workshop, World Scientific, 1995.
6. D.R.O. Morrison, Nature, **366** (1993) 29–32.
7. H. Albrecht et al., Phys. Lett. **B292** (1992) 222.
8. R. Akers et al., sub. to Z. Phys. C and CERN–PPE/94–107.
9. ALEPH collaboration, sub. to Phys. Lett. B and CERN–PPE/95/03.
10. K. Assamagan et al., Phys. Lett. **B335** (1994) 21.
11. C. Weinheimer et al., Phys. Lett. **B300** (1993) 210.
12. A.I. Belesev et al., sub. to Nucl. Phys. and preprint INR–862/94.
13. E.W. Otten, Nucl. Phys. (Proc. Suppl.) **38** (1995) 26–35.
14. M. Moe and P. Vogel, Ann. Rev. Nucl. Sc., **44** (1994) 247, and Irvine preprint UCI-Neutrino 94–5.
15. T. Kirsten et al., Z. Phys. **C16** (1983) 189.
16. T. Bernatowski et al., Phys. Rev. **C47** (1993) 806.
17. A.L. Turkevitch et al., Phys. Rev. Lett. **67** (1991) 3211.
18. A. Pieke, Proc. XIII Moriond W'Shop, Villars, 1993.
19. B. Pontecorvo, Zh. Eksp. Teor. Fiz. **33** (1957) 549; ibid **34** (1958) 247.
20. Z. Maki, M. Nakagawa, and S. Sakata, Prog. Teor. Phys. **28** (1962) 870.
21. X. Shi, D.N. Schramm, and J.N. Bahcall, Phys. Rev. Lett. **69** (1992) 717.
22. A.I. Belesev et al., Inst. for Nuclear Research, Moscow preprint INR–862/94, and Phys. Lett. **B322** (1994) 176–181.
23. Achkar et al., Nucl. Phys. **B434** (1995) 503.
24. New York Times, 31 Jan. 1995, page 1.

25. Science, **267** (10 Feb. 1995) 789. I. Stancu, Moriond meeting, March 1995.
26. New Scientist, 11 Feb 1995, and Long Baseline newsletter, Jan. 1995. To be subm. to Phys. Lett. B. Moriond meeting, March 1995.
27. D.R.O. Morrison, 'Sciences: Raison et Deraisons', Payot, Lausanne (1994) p. 91–118.
28. M.G. Bowler and N.A. Jelley, Oxford report OUNP–94–04.
29. M.D. Jones, Proc. 15th Intl. Conf. on Neutrino Physics and Astrophysics, Granada, (1992).
30. M. Aglietta et al., Europhys. Lett. **8** (1989) 611.
31. H. Meyer, Proc. Baton Rouge W'shop on Atmos. Neutrinos, Ed. B. Svoboda, May 1993.
32. M. Goodman et al., Nucl. Phys. (Proc. Suppl.) **38** (1995) 337–342.
33. D. Casper et al., Phys. Rev. Lett. **66** (1991) 2561.
34. K.S. Hirata et al., Phys. Lett. **B280** (1992) 146.
35. T. Kajita, Inst. for Cosmic Ray Research preprint, ICRR–Report–332–94–27, (Dec. 1994) and Y. Fukuda et al., Phys. Lett. **B335** (1994) 237–245.
36. O. Ryazhskaya, Gran Sasso preprint, LNGS–94/110, (Nov. 1994).
37. D.H. Perkins, Oxford preprint OUNP–93–03 (1993).
38. D.R.O. Morrison, CERN preprint, CERN–PPE/94–125.
39. S. Turck-Chieze and I. Lopez, Ap. J., **408** (1993) 347.
40. A. Dar and G. Shaviv, Proc. VI Intl. W'shop on Neutrino Telescopes, ed. M. Baldo-Ceolin (1994) p. 303, and subm. to Ap. J.
41. A. Kovetz and G. Shaviv, Ap. J. **426** (1994) 787.
42. K. Rusager and A.S. Jensen, Aarhus preprint, IFA–93–1 (1993).
43. T. Motobayashi et al. subm to Phys. Rev. Lett.
44. P. Anselmann et al. Gran Sasso preprint 94/111 (Nov. 1994), and subm. to Phys. Lett. B.
45. J.N. Bahcall and M.H. Pinsonneault, Rev. Mod. Phys. **64** (1992) 885.
46. P. Anselmann et al., Phys. Lett. **B327** (1994) 377.
47. T. Kirsten, Report to XVII Texas W'shop on Relativistic Astrophysics, Munich, Dec. 1994.
48. V.N. Gavrin, et al. Phys. Rev. Lett. **67** (1990) 3332.
49. V.N. Gavrin, Nucl. Phys. (Proc. Suppl.) **38** (1995) 60–67.
50. T. Ypsilantis et al., College de France reports, LPC report 92–131 and D. Hatzifotiadou et al., LPC/94–29, and G. Bonvicini, Nucl. Phys. B (Proc. Suppl.) **35** (1994) 441.
51. Natl. Acad. of Sciences, Natl. Acad. of Engineering, Inst. of Medecine, 'Responsible Science; Ensuring the Integrity of the Research Process', 2 Vols., Natl. Acad. Press, 1992.

DOE/ER/40561-170-INT94-00-75

Core-Collapse Supernovae and Nucleosynthesis

W.C. Haxton
Institute for Nuclear Theory, NK-12 , and Dept. of Physics, NK-15,
University of Washington, Seattle, WA 98195 USA

Lectures presented at the Third School on Astrofundamental Physics
Erice, September, 1994

PREPARED FOR THE U.S. DEPARTMENT OF ENERGY
UNDER GRANT DE-FG06-90ER40561

CORE-COLLAPSE SUPERNOVAE AND NUCLEOSYNTHESIS

W.C. HAXTON
Institute for Nuclear Theory, NK-12, and Department of Physics, NK-15
University of Washington, Seattle, WA 98195

ABSTRACT

I discuss some of the physics that governs the collapse and explosion of a massive star, including issues such as lepton number losses in the infall stage and neutrino heating and convection following the core bounce. I review recent work on the neutrino process and the r-process, describing how the nucleosynthesis depends on the explosion mechanism. Some of the interesting possibilities for oscillations of closure mass ν_τs are discussed, along with their signatures in terrestrial detectors and in nucleosynthesis.

The challenge to understand the process by which a massive star, M$\sim$ 20 M$_\odot$, collapses and ejects its mantle involves a variety of issues in nuclear, particle, and astrophysics. Among these are the lepton number losses during the infall stage of the supernova; the equation of state of nuclear matter at several times nuclear density; the production and propagation of the shock wave; the types of convection that may be driven by the negative entropy and lepton number gradients imprinted on the core by the passage of the shock wave; the neutrino flux; and the heating of the core and mantle by neutrino reactions occurring outside the trapping radius. I will review some of this physics and its connections to galactic chemical evolution and to neutrino masses and mixing.

1. Core-collapse Supernovae

We begin with a massive star, M $\gtrsim$ 10 M$_\odot$, burning the hydrogen in its core under the conditions of hydrostatic equilibrium. When the hydrogen is exhausted, the core contracts until a density and temperature are reached where He burning, $3\alpha \rightarrow ^{12}$C, can commence. This pattern, fuel exhaustion, contraction, and ignition of the ashes of the previous burning cycle, repeats through the explosive burning of ^{28}Si to Fe. The result is the "onion skin" structure shown in Fig. 1, where the layers recount the evolutionary history of the presupernova star[1].

Since Fe is the most stable nucleus, no burning turns on to halt the collapse that begins when the core Si is exhausted. The collapse leads to rapid heating of the core, dissociation of Fe into nucleons and alphas, and enhanced electron capture on the freed protons, $e^- + p \rightarrow \nu_e + n$. Both the dissociation and electron capture pump energy out of the supporting electron gas, thereby assuring a very rapid collapse. Numerical simulations show that the iron core ($\sim$ 1.2 - 1.5 M$_\odot$) collapses at $\sim$0.6 of the free-fall velocity[2,3].

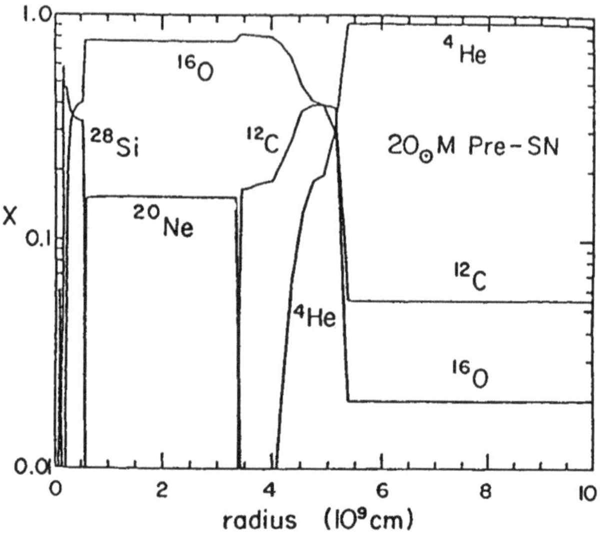

Figure 1. Mass fractions as a function of radius for a 20 $M_{\odot}$ pre-supernova star [1].

As the density increases, so does the electron chemical potential, so that electron capture on nucleons and nuclei becomes energetically favorable. The emission of the ν_es is the process by which the star's core begins to radiate its lepton number in the transformation to a neutron star. This loss of lepton number is halted, however, when densities reach $\rho \sim 10^{12} g/\text{cm}^3$: the coherent scattering of neutrinos off nuclei then becomes sufficiently strong that the neutrino diffusion time exceeds the time required to complete the collapse. That is, once the trapping density is reached, the trapped lepton fraction will remain constant until after core bounce.

The trapped lepton number Y_L is a crucial parameter of the explosion physics: a higher Y_L^{trapped} leads to a larger homologous core, a stronger shock wave, and easier passage of the shock wave through the outer iron core, as will be discussed below. The dashed curves in Fig. 2 show that most of the lepton number loss of an infalling mass element occurs as it passes through a narrow range of density near $\rho \sim 10^{12} g/\text{cm}^3$: at densities much below this value the matter temperature and electron capture rate, which varies as T^5, are low. As already noted, at higher densities the ν_es cannot escape.

The velocity of sound in nuclear matter rises with increasing density. The inner homologous core, with a mass $M_{HC} \sim 0.6$ - 0.9 $M_{\odot}$, is that part of the iron core where the sound velocity exceeds the infall velocity. This allows any pressure

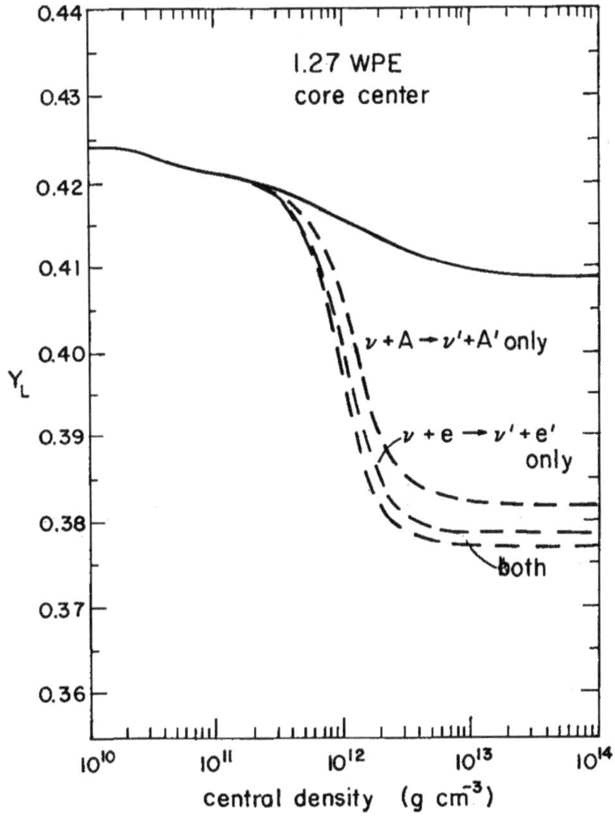

Figure 2. Trapped lepton number Y_L as a function of density for an infalling mass element prior to core bounce[5]. The solid curve shows the results for a collapse without neutrino downscattering (i.e., no inelastic neutral current scattering off nuclei and no neutrino-electron scattering), while the dashed lines show the effects of turning on one or both of the inelasticities.

variations that may develop in the homologous core during infall to even out before the collapse is complete. As a result, the homologous core collapses as a unit, retaining its density profile.

The collapse of the homologous core continues until nuclear densities are reached. The nuclear equation of state takes over, halting the collapse at about twice nuclear density, $\rho \sim (4 - 6) \cdot 10^{14} g/\mathrm{cm}^3$. The innermost shell of matter rebounds, producing a pressure wave that propagates out through the homologous core. Subsequent shells follow. The resulting pressure waves collect near the sonic point (the edge of the homologous core). As this point reaches nuclear density, the shock wave breaks out and begins its traversal of the outer core.

Initially the shock wave may carry an order of magnitude more energy than is needed to eject the mantle of the star ($\lesssim 10^{51}$ ergs). But as the shock wave travels through the outer iron core, it heats and melts the iron that crosses the shock front, at a loss of $\sim$ 8 MeV/nucleon. The enhanced electron capture that occurs off the free protons left in the wave of the shock, coupled with the sudden reduction of the neutrino opacity of this material (recall $\sigma_{\text{coherent}} \propto Z^2_{\text{weak}}$), greatly accelerates the neutrino emission. [Models predict a strong "breakout" burst of ν_es in the few milliseconds in which the shock wave travels from the edge of the homologous core to the neutrinosphere at $\rho \sim 10^{12} g/\text{cm}^3$ and $r \sim 50$ km. Note the enormous peak in luminosity in Fig. 3.] The summed losses from shock wave heating and neutrino emission are comparable to the initial energy carried by the shock wave. Thus it is far from clear that a successful hydrodynamic explosion occurs.

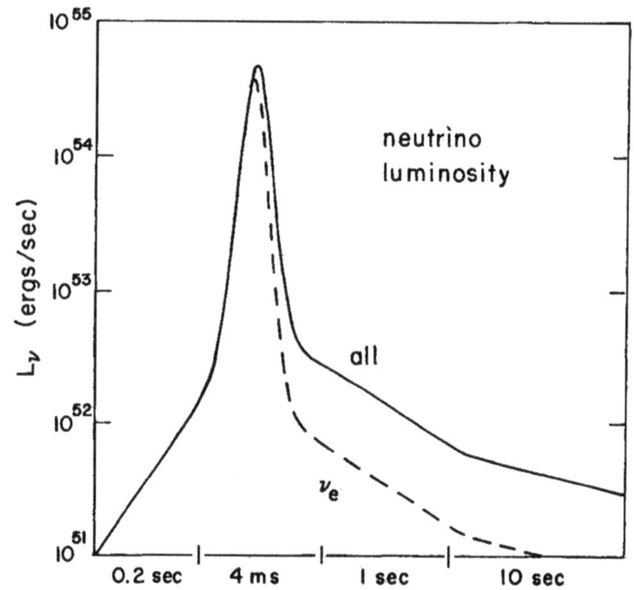

Figure 3. Schematic illustration of the time development of the neutrino luminosity from a core-collapse supernova, taken from Ref. 2. Note the broken time scale. The first 0.2 seconds corresponds to infall; the next 0.4 ms corresponds to the period between core bounce and shock breakout at the neutrinosphere; 1 second is a time characteristic of shock revival; while by 10 seconds most of the energy released in the collapse has been radiated in neutrinos.

Two explosion mechanisms were seriously considered in the past decade. In the prompt mechanism[2] the shock wave is sufficiently strong that it survives the passage through the outer iron core with enough energy to blow off the mantle. The most favorable results are achieved with smaller stars ($\sim 15\ M_\odot$) where there is less overlying iron, and with soft equations of state, which permit the core to rebound energetically from very high densities. While once this model was considered quite viable, it is now believed to fail for all but exceptionally soft equations of state.

The failure of the prompt explosion model is due in large part to the discovery[4] that neutrino-matter inelastic reactions would widen the density "window" where lepton number loss occurs during infall. The ν_es produced by electron capture tend to be hotter than the surrounding matter, and high energy neutrinos are more readily trapped, since $\sigma_{\text{coherent}} \propto E_\nu^2$. If these hot neutrinos only undergo elastic scattering, the "window" is essentially closed, yielding a $Y_L^{\text{trapped}} \sim 0.41$, as shown by the solid curve of Fig. 2. But the inclusion of $\nu_e + e^-$ inelastic scattering allows the ν_es to "downscatter" to unoccupied low-energy neutrino states, where they can more readily escape. The resulting $Y_L^{\text{trapped}} \sim 0.38$ (see Fig. 2) leads to a smaller homologous core, a weaker shock, and more overlying iron for the shock wave to penetrate. These are very difficult conditions for the prompt explosion model.

An alternative model, the delayed mechanism[3], begins with a hydrodynamic explosion that fails after $\sim$.01 sec, producing a stalled shock wave residing at $r \sim$ 200-300 km. However, after ~ 0.5 - 1.0 sec, the shock wave is revived due to neutrino heating (both charged current and quasielastic) of the nucleons left in the wave of the shock. One group[3] claims success with this model, and that success depends on the modeling of convection that helps to move hot matter away from the neutron star surface to larger radii.

However the explosion proceeds, modelers agree that the neutron star gradually cools in the ten seconds following core bounce, radiating $\gtrsim 99\%$ of the collapse energy ($3 \cdot 10^{53}$ ergs) in neutrinos of all flavors. The time scale over which the trapped neutrinos (and thus the core's lepton number) diffuse out of the cooling protoneutron star is a few seconds. (Fits to the $\bar{\nu}_e$ flux from SN1987A give, assuming exponential cooling $e^{-t/\tau}$, a time constant $\tau \sim 4.5$ sec.) Through most of their migration out of the star, the trapped neutrinos are in flavor equilibrium, e.g.,

$$\nu_e + \bar{\nu}_e \leftrightarrow \nu_\mu + \bar{\nu}_\mu \, . \tag{1}$$

As a result, there is an approximate equipartition of energy among the neutrino flavors. After weak neutral decoupling, the ν_es and $\bar{\nu}_e$s remain in equilibrium with the matter for a brief time due to the charged current reactions

$$\nu_e + n \leftrightarrow p + e^- \tag{2a}$$

$$\bar{\nu}_e + p \leftrightarrow n + e^+ \tag{2b}$$

and to the charged current enhancement of the cross section for $\nu_e + e \leftrightarrow \nu_e + e$. As a result, the peak of the heavy-flavor neutrino spectrum is characterized by a temperature $T \sim 8$ MeV, while $T_{\nu_e} \sim 4$ MeV and $T_{\bar{\nu}_e} \sim 5$ MeV. $T_{\nu_e} < T_{\bar{\nu}_e}$ because reaction (2a) is more effective than (2b) in the neutron-rich material near the neutrinosphere.

2. Convection

The Livermore group of Wilson and collaborators has produced successful simulations in the delayed explosion model. This model begins with a failed prompt explosion (but the closer to success the better!), producing a stalled shock wave at $\sim$ 200-300 km from the star's center. Approximately $\sim$ 0.5 sec later the shock wave is revived by the cumulative effects of neutrino heating of the nucleon soup that is left in the wake of the shock wave. The primary heating mechanism is the charged-current reactions of Eq. (2), though quasielastic neutrino scattering also deposits some energy.

The nucleon soup can also radiate ν_es and $\bar{\nu}_e$s by the reactions in Eq. (2). Thus the neutrino flux produces net heating only where the rate of neutrino energy deposition exceeds the cooling rate[3,6]

$$E = k(T_\nu)\left[\frac{L_\nu}{4\pi r_m^2} - \left(\frac{T_m}{T_\nu}\right)^2 acT_m^4\right] > 0 \tag{3}$$

Here r_m and T_m are the radius and temperature of a matter volume element lying outside the neutrinosphere, T_ν is the neutrinosphere temperature, and L_ν is the neutrino luminosity. The energy/unit volume of a backbody neutrino gas is $a = \frac{7}{16}(1.37 \cdot 10^{26}$ ergs/cm^3 MeV^4), and k(T_ν) is the neutrino absorption coefficient in cm^2/g. Assuming that the neutrinosphere luminosity is that of a blackbody, Eq. (3) then yields[3,7]

$$r_m T_m^3 \lesssim \frac{1}{2} r_\nu T_\nu^3 \tag{4}$$

where r_ν is the radius of the neutrinosphere. As the temperature in the hot nucleon soup drops sharply with increasing r_m, this defines a gain radius

$$r_m^{\text{gain}} \sim 200 \text{ km}$$

outside of which positive heating occurs. Thus neutrinos deposit net energy between this radius and that where the stalled shock wave resides. [More exactly, the heating extends initially up to the layer of undissociated He lying immediately behind the shock wave. As the heating proceeds, much of this He melts.]

Wilson and Mayle[7] have claimed that the success of this model depends on convection, particularly "neutron fingers", an analog of more familiar salt fingers. The neutron fingers carry neutrino-rich matter from below to above the neutrinosphere. This transport enhances the neutrino luminosity and thus the heating rate of Eq. (3), thereby decreasing the gain radius.

Several groups[7,8,9,10] have suggested that convection might be an integral part of the explosion model, enhancing neutrino heating and thereby explaining the difficulty in generating successful explosions in 1D calculations. The convection is presumably driven by the negative entropy and lepton number gradients that are imprinted on the star by the passage of the shock wave. There is substantial evidence from SN1987A, including the explosion asymmetry, the early emission of x-rays and gamma rays, and the outward mixing of ^{56}Ni, that convection is strong[11].

Neutron fingers[11,12] can arise in regions where negative lepton number gradients are stabilized against convection by positive entropy gradients, with a perturbed volume element coming to thermal equilibrium with the surrounding matter faster than it equilibrates chemically. One might assume that the thermal equilibration is rapid since there are six neutrino flavors to transport energy and only two to carry electron number. However a recent numerical study of the equilibration rates by Bruenn, Mezzacappa, and Dineva[11] concluded that the conditions for a neutron finger instability below the neutrinosphere are not realized in the post shock evolution of the stellar core. In particular, they argue that typical values of Y_ℓ found throughout most of the core at late times are about a factor of two below the critical values needed to produce a neutron finger instablity. Furthermore, they argue that at those times where Y_ℓ exceeds the critical value, neutron fingers still do not form because the lepton number equilibration rate of a displaced fluid element with the background is greater than the thermal equilibration rate, despite the hand-waving argument given above that suggested the opposite. Bruenn et al. observed that the heavy flavor neutrinos, which carry no lepton number, contribute little to the thermal transport because of their weaker cross sections, while the ν_e and $\bar{\nu}_e$ flows tend to be directed oppositely. Thus these flows rapidly transport lepton number, while only the difference contributes to thermal transport.

Another mode of convection, prompt convection arising near and below the neutrinosphere and occurring immediately after the passage of the shock wave, has been considered by several groups, with Burrows, Fryxell, and collaborators[8], in particular, advocating this as a "trigger" to initiate the explosion. This convection transports lepton number across the neutrinosphere, thereby increasing the ν_e and $\bar{\nu}_e$ luminosities governing the heating reactions of Eqs. (2a). As in the case of neutron fingers, an enhanced $\nu_e/\bar{\nu}_e$ luminosity moves the gain radius inward, allowing useful energy deposition over a larger portion of the star.

However, Bruenn, Mezzacappa, and Dineva[11,13] concluded that prompt convection has only a very modest effect on neutrino luminosities and root-mean-square energies, contradicting the results of Burrows and Fryxell[8]. They argue that such convection is short-lived ($\sim$ 10s of milliseconds), as the negative entropy gradients that drive the convection are not sustained by heating from below and thus are quickly destroyed by the convection; and that an increase in the neutrino luminosity as such early times will have minimal consequences for the explosion, as the narrow ν_e heating region above the gain radius and the large initial mass acretion rate across the shock front render this epoch unsuitable for shock revival.

Bethe and Herant, Benz, and Colgate[9,10] have discussed a third convection mechanism, entropy-driven convection between the gain radius and the shock front: hot matter rises and radiates, while cool matter flows downward and is efficiently heated near the gain radius. Thus this convection could be plausibly sustained in late-stage models. As little material is transported across the neutrinosphere, the neutrino luminosity is not modified significantly. However the heating efficiency is improved allowing rising high-entropy bubbles to maintain lower temperatures, thereby minimizing losses due to neutrino emission, while cold matter from large radii is swept closer to the neutron star, where it can be more efficiently heated[10].

3. Inelastic neutrino reactions

I noted earlier that the inclusion[4] of inelastic neutrino interactions (e.g., $\nu + e \rightarrow \nu' + e'$) led to the failure of prompt models: the resulting lepton number losses in the infall stage become too large. One might conclude that a search for further inelasticities, perhaps to enhance neutrino heating in the delayed model, would be counterproductive as it would further lower the trapped lepton number at core bounce. Interestingly, as illustrated in Fig. 2, this is not the case. The "downscattering" reactions $\nu + e \rightarrow \nu' + e'$ and $\nu + A \rightarrow \nu' + A$ are of comparable strength. The inclusion of either lowers the trapped Y_L by almost 10%. But the inclusion of both has very little additional effect. Evidently either of these reactions can fill low-energy holes in the neutrino sea as fast as they can be created by the escaping neutrinos.

Fig. 2 demonstrates that additional inelasticities might productively couple more energy into the star's mantle, without significantly altering prebounce evolution of the star. A few years ago Bruenn and I[5] investigated charged and neutral current inelastic scattering of neutrinos off nuclei. We found that neutrino heating of Fe nuclei lying in front of the stalled shock wave deposited $\sim 0.3 \cdot 10^{51}$ ergs. While this is a nonnegligible deposition, we concluded it was not sufficient to qualitatively change the explosion.

Some thought has also been given to the possibility of lower entropy collapses, with the goal of increasing the trapped lepton number fraction, so that the initial shock wave will be somewhat more energetic. (Clearly the easiest shock wave to revive is that corresponding to an almost prompt explosion.) One example of a mechanism that cools but radiates no net lepton number is

$$A^* \rightarrow A + \nu + \bar{\nu}. \qquad (5)$$

where A^* is a thermally excited nucleus. Unfortunately, the estimates14 that have been made seem to indicate that there is no significant lowering of the entropy.

There is a good deal of weak interaction physics that is not properly modeled in supernova collapse codes. Some of this physics is coupled to many-body effects such as the screening of coherent neutrino scattering, the excitation of plasma oscillations, etc. I believe there is still a possibility that some poorly treated aspect of current weak interactions modeling in supernova may be important to understanding the explosion mechanism.

4. Nucleosynthesis

Core-collapse supernovae are one of nature's major factories for producing nuclei with A > 6. Many of the most plentiful metals (He, C, O, Ne, Si, Fe, ...) are produced through hydrostatic burning and ejected into the interstellar medium by the explosion. Many other nuclei, including a large fraction of those of lower abundance, are synthesized in the explosive conditions associated with the shock wave.

I would like to focus here on two mechanisms that are especially dependent on the details of the explosion. The first is the neutrino process[15,16]. Supernova neutrinos can alter the chemistry of the mantle, thereby affecting the composition of the explosion ejecta. For example, the product of the neutral current inelastic cross section off ^{20}Ne and the neutrino fluence through the Ne shell, which resides ~30,000 km from the center of a 20 $M_\odot$ presupernova star, is ~1/300. That is, 0.3% of the matter is affected by such interactions. The cross sections are dominated by the higher temperature ν_μ and ν_τ neutrinos, which deposit ~20 MeV in the scattering. Thus the target is generally excited above the continuum,

$$^{20}\mathrm{Ne}(\nu,\nu')^{20}\mathrm{Ne}^* \rightarrow \begin{cases} ^{19}\mathrm{F}+\mathrm{p} \\ ^{19}\mathrm{Ne}+\mathrm{n} \\ \qquad \underset{\beta}{\hookrightarrow}\ ^{19}\mathrm{F} \end{cases} \tag{6}$$

leading to its breakup and the production of a new nucleus, ^{19}F. This is an important observation, since the origin of galactic ^{19}F has been unclear.

This process can be into incorporated more generally into explosive nucleosynthesis codes and its consequences explored. One has to account for the chemistry of the coproduced neutrons and protons in reactions like Eq. (6): for example, does the neutron capture back on ^{19}F, destroying the product of interest, or is it absorbed by some other neutron poison? One must consider which nuclei survive the heating associated with the passage of the shock wave, and which can be produced at late times and larger radii following the passage.

Such an exploration has been carried out[15,16]. It is found that elements like ^{7}Li, ^{11}B, and ^{19}F can be produced at about the correct galactic abundances for $T_{\nu_\mu/\nu_\tau} \sim 8$ MeV, and a value quite consistent with theory. Thus, if one accepts this as the correct production mechanism, nucleosynthesis constrains a neutrino temperature that we have not yet measured experimentally.

The idea of neutrino nucleosynthesis was first discussed by Domogatskii, Nadyozhin, and collaborators[17]. It was rediscovered by Woosley and Haxton[15], and this lead to the first realistic treatments of the reaction network and nuclear cross section aspects of this process. It is quite satisfying that the major productions may clarify some long-standing puzzles. I mentioned above the absence of a site for synthesizing ^{19}F. Traditionally ^{11}B and ^{7}Li have been considered, along with ^{10}B and ^{6}Li, cosmic ray (CR) spallation products. But conventional CR models that produce the requisite amount of ^{11}B overproduce ^{10}B (e.g. ratios of ^{10}B/^{11}B $\lesssim 2.5$ rather than the correct solar ratio of 4.6). The neutrino process work suggests that this mechanism is responsible for most of the ^{11}B (and possibly a significant fraction of the ^{7}Li) we observed, allowing the CR mechanism to be tuned to produce ^{10}B.

Fig. 4 shows a recent calculation of Timmes, Woosley, and Weaver[16] of ^{11}B galactic chemical evolution in the neutrino process. The required linearity of the boron abundance with Fe/H is clearly in accord with observation, while the neutrino production band nicely matches the data. The authors concluded that the neutrino process, CR spallation, and homogeneous big bang nucleosynthesis are complementary and together explain the evolution of the light elements ($A \leq 11$).

The second process I would like to mention is the work of Woosley and collaborators[18] on the r-process, the rapid capture of neutrons that we believe produces about half of the heavy elements and all of the transuranics. They have argued that a plausible site for the r-process is the expanding, high entropy bubble that forms between the neutrinosphere and the shock wave. The material in this region has experienced significant electron capture and is thus neutron rich. As this nucleon soup expands off the star and cools, the material undergoes an alpha-rich freezeout, with about 10% of the α-particles reassembling into heavier nuclei, up to masses A$\sim$100. There remains a signficant neutron mass fraction of about 100 neutrons per heavy nucleus. These neutrons then generate the r-process. Simultations of this process have yielded mass distributions that correspond closely to observation. Furthermore estimates of the amount of r-process material that would be ejected, 10^{-4} - 10^{-5}M$_\odot$, could reasonably account for the abundances we measure. This process appears to require very large entropies/baryon, and thus progress on the explosion mechanism is important to verify that such entropies are achieved. Present calculations do not fully include the effects of the neutrino process on the neutron-proton chemistry of the bubble or on the survival of the synthesized nuclei.

5. Neutrino oscillations in supernova

While there exists a long list of important particle physics constraints (on axions, Majorans, Dirac neutrino masses, neutrino magnetic moments, etc.) derived from the observation of the neutrinos that cooled SN1987A. However the limited time for these lectures prevents me from discussing this topic. Instead, I will briefly mention only one topic from particle astrophysics, the possibility that neutrino oscillations will occur due to an MSW level crossing just outside the neutrinosphere.

The favorite particle physics solution to the solar neutrino puzzle is MSW oscillations occurring for $\delta m_{12}^2 \sim 10^{-5}$ eV^2 and $\sin^2 2\theta \sim 10^{-2}$. If one assumes $m_2 \gg m_1$

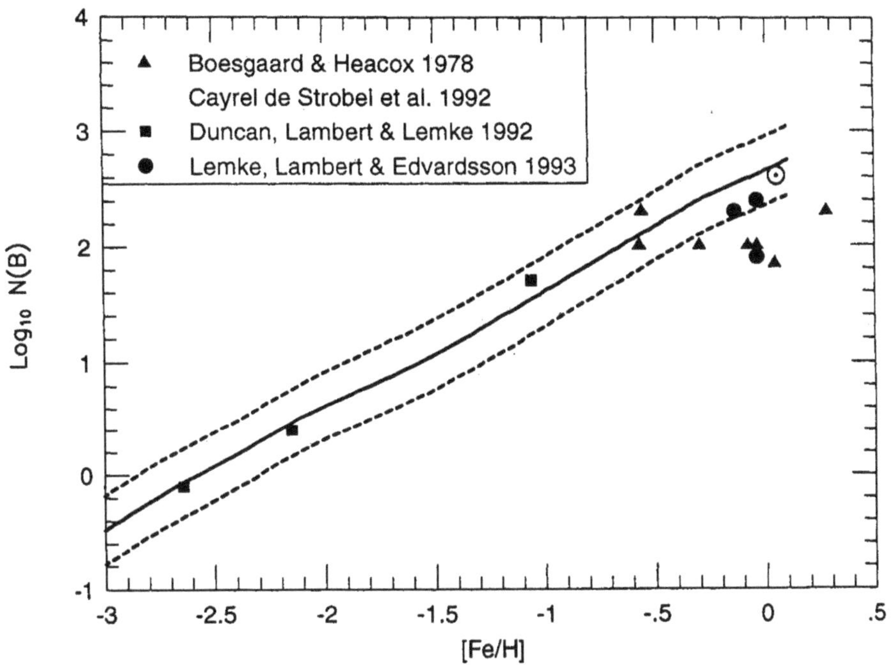

Figure 4. Evolution of boron relative to hydrogen as a function of the metallicity [Fe/H]. The calculated boron abundance (solid line) and a factor-of-two variation from this result (dashed lines) are taken from the galactic model calculations of Ref. 16 in which the neutrino process production of ^{11}B was estimated. References to the abundance determinations can also be found in Ref. 16.

as the seesaw mechanism suggests, this implies the existence of a massive neutrino $m_2 \sim$ few $\cdot 10^{-3}$ eV. If we furthermore assume that this crossing corresponds to $\nu_e \to \nu_\mu$ oscillations, the seesaw mechanism and the top quark mass predict $m_3 \sim$ 10 eV, a neutrino mass that would be of great significance because of its contributions to the closure mass and to the formation of large-scale structure.

The resulting three-flavor MSW level-crossing pattern is sketched in Fig. 5. The $\nu_e \leftrightarrow \nu_\tau$ crossing that would arise for a cosmologically interesting ν_τ typically occur at densities $\sim 10^8$ - 10^{10}g/cm^3, a point outside the neutrinosphere. This observation is very significant since, at this density, the supernova neutrinos are fully decoupled from the matter, so that their spectra are fixed. As noted in Section 1, the neutrino fluxes are characterized by temperatures $T_{\nu_e} \sim 4$ MeV, $T_{\bar{\nu}_e} \sim 5$ MeV, $T_{\nu_\mu/\bar{\nu}_\mu/\nu_\tau/\bar{\nu}_\tau} \sim$ 8 MeV.

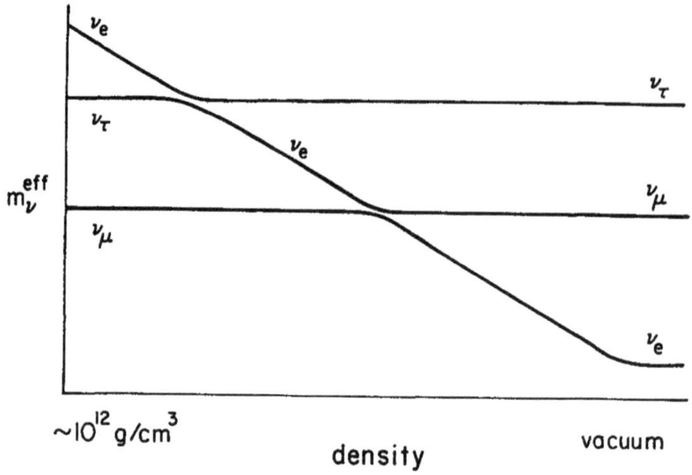

Figure 5. Schematic illustration of the $\nu_e \leftrightarrow \nu_\tau$ MSW level crossing that might occur at densities encountered in a supernova.

Adiabatic level crossings[19] occur for cosmologically significant ν_τs even for quite small mixing angles, $\sin^2 2\theta \gtrsim 10^{-5}$. For such crossings, we see immediately that the consequence is an anomalously high $T_{\nu_e}^{\rm MSW}$ temperature

$$T_{\nu_e}^{\rm MSW} \sim 8{\rm MeV} \gg T_{\bar{\nu}_e} \sim 5{\rm MeV}$$

due to the $\nu_e \leftrightarrow \nu_\tau$ crossing. The "'$\gg$" is particularly appropriate in the context of nuclear detection methods for supernova neutrinos, since nuclear transitions with high thresholds are typically characterized by cross sections that rise rapidly with T.

It follows that the detection of supernova ν_es and $\bar{\nu}_e$s could demonstrate that the ν_τ has a cosmologically interesting mass. This would be a result of great importance.

Such oscillations would also have the consequence of enhancing neutrino energy deposition outside the neutrinosphere. The net effect of the hot ν_es is to increase the rate of energy deposition due to the strong charged current reaction of Eq. (2a). Fuller and Qian[20] have pointed out that the attractive conditions for the hot-bubble r-process could be a casualty of this scenario, however. By enhancing the rate of Eq. (2a), one forces the matter in the hot bubble toward the proton-rich side. A demonstration that the hot-bubble r-process occurs in nature therefore would rule out a large range of δm^2 and $\sin^2 2\theta$ that might govern oscillations of cosmologically interesting ν_τs.

This work was supported in part by the U.S. Department of Energy and by NASA under grant #NAGW2523.

References

[1] See, for example, S.E. Woosley and T.A. Weaver, *Phys. Rep.* 163 (1989) 79.

[2] G.E. Brown, H.A. Bethe, and G. Baym, *Nucl. Phys.* A375 (1982) 481; S. Burrows and J.M. Lattimer, *Ap. J.* 270 (1983) 735; E. A. Baron and J. Cooperstein, *Ap. J.* 343 (1990) 597.

[3] J.R. Wilson, in *Numerical Astrophysics*, ed. J.M. Centrella, J.M. LeBlanc, and R.L. Bowers (Jones and Bartlett, Boston, 1985), p. 422; H.A. Bethe and J.R. Wilson, *Ap. J.* 295 (1985) 14; H.A. Bethe, *Ap. J.* 412 (1993) 192; S.A. Colgate and R.H. White, *Ap. J.* 143 (1966) 626.

[4] S.W. Bruenn, *Ap. J.* 340 (1989) 955 and 341 (1989) 385; E.S. Myra and S.A. Bludman, *Ap. J.* 340 (1989) 384.

[5] S.W. Bruenn and W.C. Haxton, *Ap. J.* 376 (1991) 678.

[6] R. Mayle, Ph.D. thesis, unpublished (UCRL preprint no. 53713).

[7] J.R. Wilson and R.W. Mayle, *Phys. Rep.* 227 (1993) 97 and *Phys. Rep.* 163 (1988) 63.

[8] A. Burrows and B.A. Fryxell, *Science* 258 (1992) 430 and *Ap. J. Lett.* 418 (1993) L33.

[9] H.A. Bethe, ref. [3]; M.Z. Herant, W. Benz and S.A. Colgate, *Ap. J.* 395 (1992) 642 and *Phys. Rep.* 227 (1993) 157; D.S. Miller, J.R. Wilson, and R.W. Mayle, *Ap. J.* 415 91993) 278; H.-Th. Janka, to appear in *Frontier Objects in Astrophysics and Particle Physics*, ed. F. Giovannelli and G. Mannocchi (Societá Italiana di Fisica, Bologna, 1993 MPA preprint).

[10] M.E. Herant, W. Benz, W.R. Hix, C.L. Fryer, and S.A. Colgate, Lick Observatory preprint No. 22 (March, 1994).

[11] S.W. Bruenn, A. Mezzacappa, and T. Dineva, University of North Carolina preprint TAR-046 (to appear in *Phys. Rep.*), and references therein.

[12] P.G. Baines and A.E. Gill, *J. Fluid Mech.* 37 (1968) 289; S.A. Grossman, R. Narayan, and W.D. Arnette, *Ap. J.* 407 (1993) 299.

[13] S.W. Bruenn and A. Mezzacappa, University of North Carolina preprint TAR-042 (submitted to *Ap. J.*) 1993.

[14] W.C. Haxton, talk presented at the APS Division of Nuclear Physics meeting, Asiloman, 1992; M. Aufderheide, talk presented at the Snowmass Workshop on Nuclear and Particle Astrophysics and Cosmology in the Next Millenium, 1994.

[15] S.E. Woosley and W.C. Haxton, *Nature* 334 (1988) 45; S.E. Woosley, D.H. Hartmann, R.D. Hoffman, and W.C. Haxton, *Ap. J.* 356 (1990) 272.

[16] F. T. Timmes, S.E. Woosley, and T.A. Weaver, Enrico Fermi Institute preprint (submitted to *Ap. J. Suppl.*)

[17] G.V. Domogatskii and D.K. Nadyozhin, M.N.R.A.S. 178 (1977) 33P; *Sov. Astr.* 22 (1978) 297; *Sov. Astr. Lett.* 6 (1980) 127; *Ap. Space Sci.* 70 (1980) 33.

[18] S.E. Woosley and R.D Hoffman, *Ap. J.* 395 (1992) 202; S.E. Woosley, J.R. Wilson, G.J. Mathews, R.D. Hoffman, and B.S. Meyer, to appear in *Ap. J.* (1994).

[19] G.M. Fuller, R.W. Mayle, B.S. Meyer, and J.R. Wilson, *Ap. J.* 389 (1992) 517; G.M. Fuller, *Phys. Rep.* 227 (1993) 149.

[20] Y.Z. Qian, G.M. Fuller, R. Mayle, G.J. Mathews, J.R. Wilson, and S.E. Woosley, *Phys. Rev. Lett.* 71 (1993) 1965; Y.Z. Qian and G.M. Fuller, submitted to *Phys. Rev. D.*

NEUTRINO ASTROPHYSICS AT HIGH ENERGIES—A BRIEF INTRODUCTION

MAURICE M. SHAPIRO
Department of Physics, University of Maryland
College Park, MD USA *

Abstract

Neutrinos have played a vital role in nuclear and elementary- particle physics. At high energies (HE), progress was facilitated by intense neutrino beams at large accelerators. At modest (MeV) energies, it has been possible to compare the predicted fluxes of solar neutrinos and of neutrinos from supernova explosions (SN 1987a) with those observed.

The deficit of solar neutrinos has been confirmed. By contrast, extraterrestrial neutrinos at high energies (HE ν), say $\gtrsim 10^{11}$ eV, have thus far eluded observation. The detection of cosmic HE ν requires a large array of sensors coupled to a massive target, since the interaction cross sections of neutrinos are exceedingly low, and backgrounds are troublesome. Nevertheless, major projects are under way to develop HEν astronomy with a view to opening a new window upon celestial regions where HE processes are prominent. Some promising arrays, and their prospect of detecting cosmic neutrinos from various sources are discussed. HE ν astronomy and HE gamma-ray astronomy are symbiotic, and both can shed light on puzzling phenomena observed in other channels in astronomy.

1 Introduction

Neutrinos are more abundant in nature than other elementary particles. Yet the only extraterrestrial neutrinos [†] detected hitherto—all in the energy range of $\approx$ 1-20 MeV—are solar neutrinos and those generated in supernova explosions. The solar neutrinos could yield information about suspected neutrino properties, e.g., oscillations, as well as about solar properties that are thus far only conjectural. As for supernovae, the few neutrinos so far observed from a single supernova burst have, in broad outline, confirmed our basic theoretical expectations of the critical role that neutrinos play in stellar collapse.

*address for correspondence: 205 Yoakum Pkwy., 1514, Alexandria, VA 22304, USA

[†] As employed here, the term neutrino will often refer to both neutrinos and antineutrinos, sometimes $\nu_e \bar{\nu}_e$, sometimes $\nu + \nu_\mu$ or $\bar{\nu}_\mu$, and occasionally ν_τ or $\bar{\nu}_\tau$.

There are compelling reasons to believe that the cosmos is full of objects that emit high-energy, e.g., multi-GeV neutrinos. Celestial regions, e.g., galaxies, stars, or the interstellar medium in which high-energy (HE) processes occur, generally involve cosmic-ray collisions. In these, neutrinos arise as secondary or tertiary progeny of interaction products such as pions and kaons. An adequate understanding of the "engines" that produce the cosmic rays will clarify the mechanisms by which they generate the various particles—pions, muons, electrons, neutrinos, gamma rays, X-rays, optical and radio photons, and other emissions. Because of their unique properties, HE ν's have a special role to play in astrophysics, and especially in HE astrophysics, which will be emphasized here. This lecture is intended as a brief elementary introduction to the subject.

In section 2 we present the motivation for developing the nascent science of neutrino astronomy and astrophysics. Section 3 outlines some important nuclear processes that produce neutrinos, and typical cross sections for their interaction with matter. Section 4 summarizes the features required in a telescope for HE neutrino astronomy. Section 5 describes the progenitor of deep undersea detectors, of HE ν, i.e. the DUMAND array, and the phenomenology of a ν_μ interaction in water. Also described are some basic features of the other underwater arrays, i.e., NESTOR in the Mediterranean, the Lake Baikal project, and AMANDA, embedded in Antarctic ice. Section 6 refers to underground neutrino telescopes. While some, e.g. MACRO, are impressively large, they are unlikely to detect discrete sources of HE ν's. Section 7 discusses diffuse sources of HE ν_μ—both astrophyical and atmospheric. The former are, thus far, candidate sources. Also discussed are some potential discrete sources. Section 8 explains the symbiosis between neutrino astronomy and gamma-ray astronomy. Finally, in Section 9, we discuss the present outlook for HE neutrino astronomy.

2 What Motivates The Effort to Develop Neutrino Astronomy?

Since neutrinos are notoriously difficult to detect, the question is not unreasonable. Some considerations suggest that the rewards will be commensurate with the great effort being expended: Neutrinos carry information of a very special kind. They reach us from sites otherwise inaccessible to observation. Like photons— and unlike the charged cosmic-ray particles—they travel in straight lines from their sites of origin. Mu neutrinos (ν_μ) are tracers of nuclear cosmic rays. They are inevitably produced in regions characterized by high-energy processes, as the secondary results of nuclear collisions. Gamma rays, on the other hand, while generated from neutral pions (π^0) born in nuclear collisions, are also copiously produced in electron interactions. Since, however, some 98 per cent of the primary cosmic rays are atomic nuclei, the ν_μ are unique tracers of this dominant component.

Although cosmic-neutrino astrophysics is a new discipline, the same can hardly be said of neutrino physics. Beams of HE ν have been available for many years, and their interactions have been the subject of intensive and fruitful study at the leading accelerator laboratories. This rich background of experimental information and expertise is in sharp contrast to the situation in another incipient—and very vital—field in astrophysics, i.e., that of gravitational wave astronomy, in which the observational difficulties may be even more formidable than in neutrino astronomy.

The apparatus and techniques developed for large neutrino telescopes will also be useful in exploring suspected physical properties of the particles themselves, e.g., the questions of oscillation and of possible neutrino masses. Finally, it should be recalled that as each new channel in astronomy has opened up, e.g., radio-, X-ray-, or gamma-ray astronomy, some of the richest dividends have been totally unexpected: e.g., pulsars, quasars, X- ray binaries, and gamma-ray bursts.

3 Production of Neutrinos, and Their Cross Sections

We have noted that ν_μ are unique tracers of the nuclear cosmic rays. Hence muons are the preferred secondaries in HE ν astronomy. Moreover, they are far more penetrating than electrons.

Hence their long ranges provide far better pointing back to a discrete source. A TeV muon, for example, has a range exceeding 2 km in water. The long ranges also enhance the effective target volume from which μ secondaries can be collected.

3.1 Interactions of HE Protons With Matter

Typically, these are of the following types:

$$\begin{array}{lclcl}
 & & \underline{\text{PionProduction}} & & \underline{\text{Decay}} \\
p+p & \rightarrow & p + n + \pi^+ & \rightarrow & \nu_{mu}, \nu_e \\
p+p & \rightarrow & p + p + \pi^+ + \pi^- & \rightarrow & \nu_{mu}, \nu_e \\
p+p & \rightarrow & p + p + \pi^o & \rightarrow & \nu_{mu}, \nu_e + \gamma - \text{rays} \\
p+p & \rightarrow & p + n + \pi^+ + \pi^o & \rightarrow & \nu_{mu}, \nu_e + \gamma - \text{rays}
\end{array}$$

Kaon production also occurs, but it need not be treated here.

The decays leading to ν_μ are:

$$\begin{array}{lcl}
\pi^\pm & \rightarrow & \mu^\pm + \nu_\mu(\bar{\nu}_\mu) \\
\mu^\pm & \rightarrow & e^\pm + \nu_\mu(\bar{\nu}_\mu) + \nu_e(\bar{\nu}_e)
\end{array}$$

3.2 Interaction of Protons With Photons

At sufficiently high energies, and in a regime of very high photon densities, the following interactions are also significant for ν_μ production:

$$p + \gamma \rightarrow \begin{cases} n + \pi^+ \\ p + \pi^+ + \pi^- \end{cases}$$

and, again, the pion decay yields neutrinos and leptons. Alternatively,

$$p + \gamma \rightarrow p + \pi^o, \ \pi^o \rightarrow 2\gamma$$

It is evident that the processes of ν production and of gamma-ray production are closely linked.

3.3 Neutrino Cross Sections

Sample cross sections for interaction of neutrinos with matter are:

At	1MeV,	10^{-43}
	1GeV,	$4 \times 10^{-39}\text{cm}^2$
	1TeV,	$4 \times 10^{-36}\text{cm}^2$

These may be compared with typical cross sections for high-energy nuclear reactions, $\approx 10^{-26} cm^2$. Thus, in order to have a reasonable chance of detecting a meaningful sample of neutrinos, we need very large targets.

4 Telescopes (Observatories) for HE Astronomy

What are the main requirements?

A large target mass (and volume)

An array of efficient sensors for directionality,

and for energy estimation

Adequate shielding against background radiation

Since the ultimate array must be large-scale, it is clear that the investment required will be comparable to that for a large optical telescope. However, design studies have shown that the cost will be much less than that of a TeV-class accelerator. Still, the feasibility would be judged by whether the price-tag would be justified by the possible benefits. If, for example, a promising detector array could be built for approximately the cost of a single large detector now required at Fermilab or CERN, then this would be a modest investment to make in return for the prospect of opening a new channel in astronomy and astrophysics.

5 Deep Underwater Observatories—And One Immersed In Ice

The central concept of an undersea neutrino detector originated in the 1960's. The idea was further elaborated under the leadership of F. Reines, J. Learned, S. Miyake, V. Peterson, A. Roberts, V. Stenger and others.[1] Over the past two decades the concept has evolved, with feasibility studies begun in two projects—DUMAND in the Pacific and at Lake Baikal in Siberia. The conceptual design requires a large volume of clear water deep undersea, instrumented with detectors of Cerenkov light produced by muon secondaries of neutrinos—all linked to a computer-data station on a nearby shore. More recently, project AMANDA, utilizing Antarctic ice, rather than sea water, has been under development. Among the advantages of an underwater (or ice) medium for housing the detector array are the following. Sea water is cheap, yet it serves three major functions: (a) as a shield of uniform density; (b) as a large target; (c) as a vital component of the detection process, i.e., as the transparent medium in which the Cerenkov light is generated by a passing muon.

5.1 HE ν_μ Interaction in Water

Figure 1 shows the main features of a ν_μ interactin with a nucleon in water. Emerging from the collision are a muon and a hadronic cascade which share the incident energy, say 10 TeV, about equally. The muon continues for a kilometer or further, suffering ionization loss of some 2 MeV/cm, and an occasional abrupt loss in a nuclear collision. At high energies the secondary muon proceeds in nearly the direction of the incident ν_μ (the angle θ in Figure 1 in grossly exaggerated.) The muon emits a cone of Cerenkov light which is detected by photomultipliers (PMT).

The hadronic cascade generates a burst of light along its more limited path of some meters. When the collision occurs inside the detector array (a "contained event") the hadronic shower is "seen" by the photosensors. The direction of the muon is well determined with a lattice of photomultipliers housed in pressurized modules. In ocean depths daylight is absorbed, and the Cerenkov light stands out. Good directional resolution is achieved thanks to the size of the array, the multiplicity of light sensors, and the long muon path. Thus, at energies of 1 TeV, the angular resolution is approximately $\pm$ 0.5 degree.

5.2 The DUMAND Array

I shall describe this observatory—Deep Underwater Muon and Neutrino Detector—in some detail, as it was the prototype of subsequent large-scale deep arrays.[2] It was designed by an international consortium of physicists, astrophysicists, engineers, and marine scientists. The original, ambitious plan envisaged a 1 km^3 array which, it was estimated, would have a good chance of detecting neutrinos from discrete celestial sources. The essential idea is illustrated in Figure 2 which depicts a set of vertical "strings" with flotation spheres at the top of each, and photomultipliers (PMT) spaced along the cables. To establish feasibility, more modest arrays were then designed, and the current telescope (DUMAND II)[3] is depicted in Figure 3. The array is modular, and can be expanded as resources permit. It is noteworthy that the effective volume of the telescope exceeds that of the contained array, since ν_μ-generated muons moving toward the telescope from outside the array, can also be "seen" by the sensors. If a significant number of the parent ν_μ come from a localized region in the celestial sphere, they may permit detection of a source.

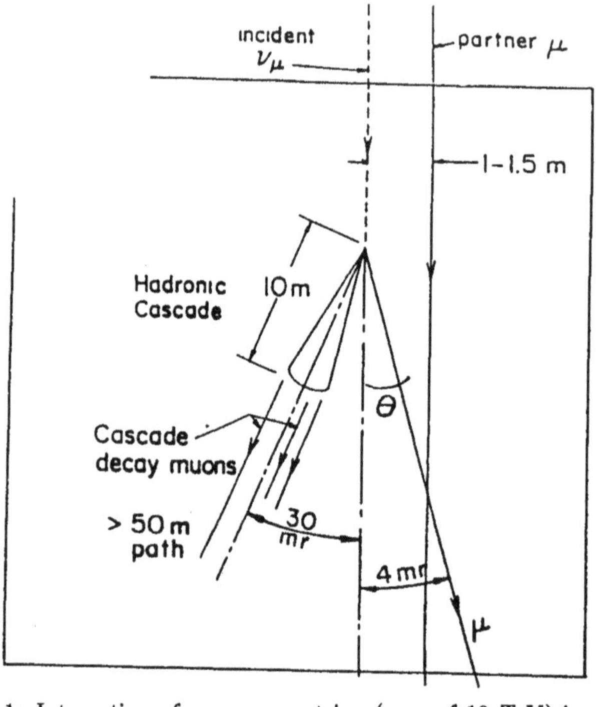

Figure 1: Interaction of a muon neutrino (e.g., of 10 TeV) in water, producing an hadronic cascade and a muon that preserves nearly the direction of the ν_μ ($\pm$ 0.5*deg.*). The muon can traverse the entire depth of the lattice, passing near several photomultipliers ; this permits good directonal resolution. (The angle θ, as shown, is much exaggerated.)

Figure 4 shows the location of the DUMAND II array in the Pacific waters off Keahole Point in Hawaii. The depth of the lattice is approximately 4.5 km. One advantage of the location is the roughly conical shape of the volcanic island. This puts the array reasonably close to the shore station, thereby keeping the length and

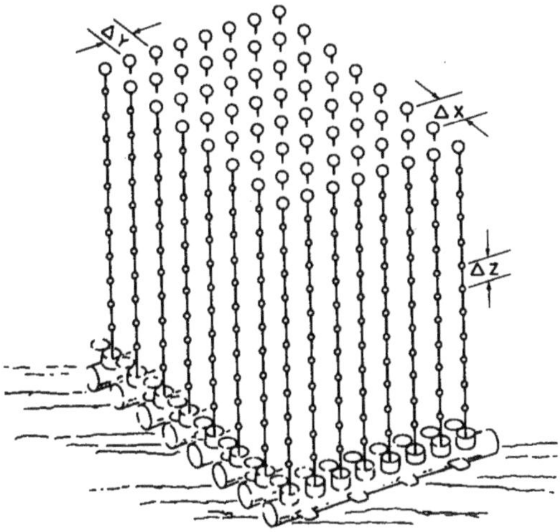

Figure 2: Configuration of the early DUMAND arrays (ca. 1978). The small circles denote phototubes for detection of Cerenkov light from the muon secondaries. The larger circles, at the top of the lattice, represent flotation spheres.

cost of the conecting cables within reason.

In principle, DUMAND is an omnidirectional (4π ster.) detector with a 100 per cent duty cycle. In practice, the cosmic-ray background makes it useful to employ the earth as a shield. Hence upward-moving ν_μ are preferred for source detection, and the PMT's are oriented downward to detect ν_μ coming up from well below the horizon.‡

Table 1 lists some important parameters of the DUMAND II array.

‡To reduce the effect of scattered light from the background of charged cosmic-ray particles coming from above the array, it is still desirable to maximize the depth of overlying water.

Table 1. Parameters of the DUMAND II array

Array dimensions	100 m diameter × 230 m high
String spacing	40 m side
Number of strings in array	8 in octagon, 1 in center
Sensor spacing along strings	10 m
Number of optical sensors/string	24
Total number of optical sensors	$9 \times 24 = 216$
Height of first sensor above bottom	100 m
Depth of bottom	4.8 km
Sensor pressure envelope	17" (43.2 cm) O.D. glass sphere
Optical sensor	16" photomultiplier
Volume of array, contained	$1.8 \times 10^6\,m^3$
Target area for through-going muons	23,000 m^2 horizontal, 7,850 m^2 vertical up-going, 2,500 m^2 down-going
Effective target volume for 2 TeV muons	$1.0 \times 10^8\,m^3$
Effective target volume for 1 TeV cascades	$7.0 \times 10^5\,m^3$
Muon energy threshold	20 – 50 GeV
Track reconstruction accuracy	0.5° – 1.0°
Cascade detection threshold	$\sim$ 1 TeV
Downgoing muon rate	3 / minute
Atmospheric neutrino rate for throughgoing muons	3500/yr
Atmospheric neutrino rate for contained events above 1 TeV	50/yr
Point source sensitivity	$4 - 7 \times 10^{-10}\,cm^{-2}sec^{-1}$ in a year above 1 TeV
Contained event sensitivity	$1 \times 10^{-8}\,cm^{-2}sec^{-1}$ in a year above 1 TeV

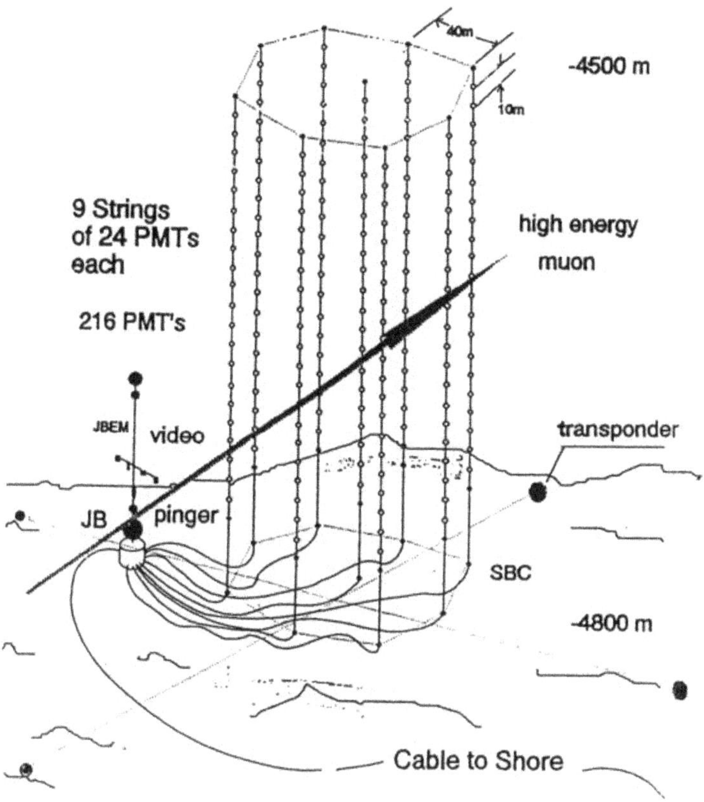

Figure 3: DUMAND II, the prototype HEν telescope. The small circles represent photomultiplier tubes (PMT). The octagonal array, 200 m high and $\sim$ 100 m in diameter, has a central "string" of PMT's surrounded by 8 peripheral strings.

5.3 The NESTOR Array

The Nestor§ telescope is situated in the Mediterranean near Pylos, southwest of Greece, at a depth of 3.5 km.[4] Under construction is a "tower" with 168 phototubes. The shore laboratory is located near the lighthouse of Sapienza. Based on the same principles and much the same methodology as DUMAND, the configuration of the sensors differs (see Figure 5). The tower consists of 12 floors, i.e., layers each

§Nestor was the king of Pylos who counselled the Greeks during the Trojan War. The acronym stands for Neutrinos from Supernovae and TeV Sources Ocean Range.

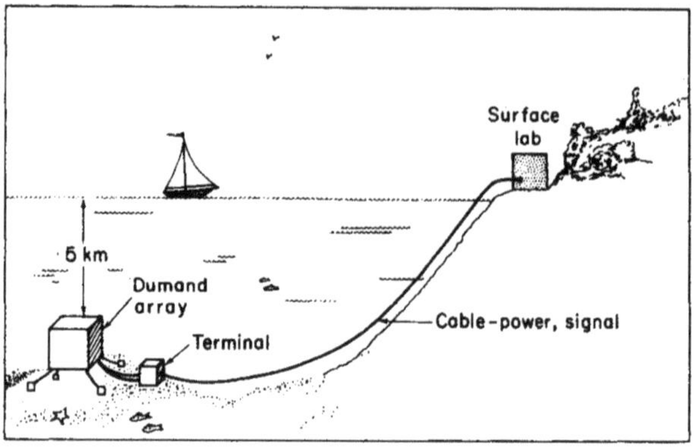

Figure 4: The DUMAND array in the waters off Keahole Point, Hawaii, near the Big Island, about 30 km from the shore station.

containing several pairs of phototubes; in a pair, one PMT faces up, the other one down. The contained volume has 3×10^5 tons.

To get a larger volume and aperture, the aim for the next stage of NESTOR is to build seven such towers, with a total sensitive area exceeding $10^5 m^2$. Collaborating in this enterprise are the University of Athens and the Institutes of Nuclear Research and Oceanology of the Russian Academy of Sciences. A Russian research vessel has been used for deployment of test strings of photomultipliers. At great depths the Mediterranean water is very clear, with an attenuation length for Cerenkov light of some 50 m. This makes it possible to space PMT's widely, thereby increasing the effective aperture at lower cost. Since the cable connecting the array to a shore station is relatively short, this provides another saving in cost.

5.4 The Lake Baikal Project

Under way for some years, the Baikal[5] array is situated near Irkutsk, Siberia. Baikal is one of the deepest lakes on earth, so that a neutrino telescope can be lowered to 1 km. As a sufficiently thick surface layer is frozen during the long winter months, this provides a convenient platform from which to deploy strings of sensors: holes are drilled through the ice, and the apparatus is lowered into the water below. The 1 km depth of the telescope, compared to the depths available to DUMAND and NESTOR, is a real limitation, as it provides relatively poorer shielding against charged cosmic rays. Like the DUMAND consortium, the Baikal team has pioneered in the underwater detection of HE ν.

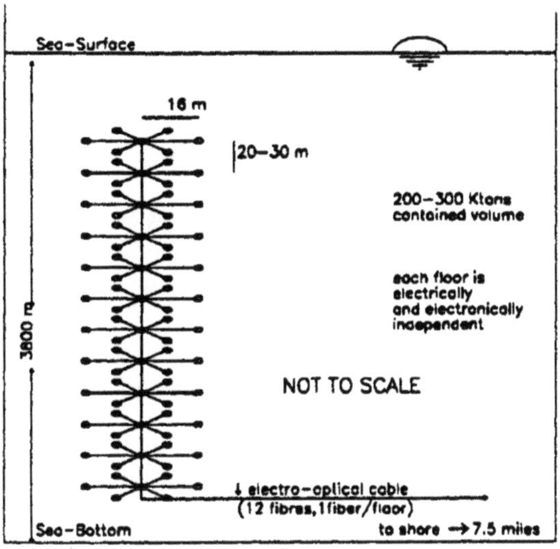

a. A NESTOR tower.

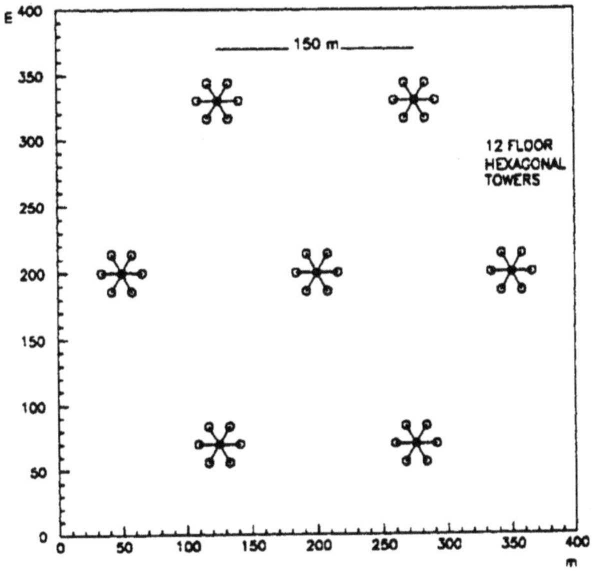

b. Full NESTOR array with 7 towers (top view); it will have a sensitive area $> 10^5 m^2$.

Figure 5: The NESTOR Array

5.5 AMANDA (Antarctic Muon and Neutrino Detector Array)[6]

This project is using Antarctic ice, rather than sea- water as a medium for HE ν_μ detection. The initial concept involves drilling holes in the ice down to a depth of 1.2 km. The drilling is accomplished with hot water.¶ The PMT's are lowered, and soon are frozen in. AMANDA benefits from an international infra-structure for science already in place near the South Pole, thanks partly to the U. S. National Science Foundation. Clearly, the ice is a more steady and convenient platform for deploying a telescope than is a ship at sea. Moreover, the South Polar ice has a reltively low background of radioactivity. Offsetting these advantages are some limitations. The modest depth results in a strong cosmic-ray background. The attenuation length λ of Cerenkov light is much shorter ($< 1/2$) that of sea water. Since the number of light sensors required in a given volume is proportional to λ^{-3}, this makes a telescope of a given size more expensive. In addition, the cost of drilling holes in the ice has hitherto restricted the hole diameter—hence the photocathode diameter—to 20 cm. So the phototubes are considerably less sensitive than the larger ones used in DUMAND and NESTOR. AMANDA will have 10 strings, each 200 m in length, and a total of 200 sensors.

5.6 Prospects of Collaboration

For some years hence, it is likely that all of the foregoing telescopes will suffer from meager statistics, especially in attempting to detect discrete sources. Of cosmic neutrinos streaking toward an observatory, only some 10^{-6} or less are likely to interact and give signals. Consequently there is much to be gained by joining forces. The desirability of pooling data has already been discussed by leading scientists in several of the projects.

6. Underground Arrays

Partly because of limited target size, these seem less likely than underwater telescopes to detect cosmic sources in the early foreseeable future. Yet they can contribute useful physics information. Many of them were, in fact, designed for detection of MeV-class neutrinos and various exotic particles. At least two of the projects were intended to search for proton decay. It was one of these, the Kamiokande detector in Japan, that has been recording muon and electron secondaries of atmospherically generated neutrinos on the far side of the earth (i.e., of upward moving neutrinos). It was also this array, as well as its U.S. counterpart, the IMB telescope, that signalled the birth

¶Fuel is scarce and costly in the South Polar regions; this has had an effect on PMT size; see below.

of extra- solar neutrino astronomy by detecting ν_e ($\approx$ 10-20 MeV) from SN 1987a in the Large Magellanic Cloud.

The reader who wishes to gain some idea of the widespread interest and activity in neutrino astronomy is referred to descriptive lists of most of the underground arrays capable of neutrino detection[7,8]. These observatories[7] are either extant or under construction in the last few years. Some are in the proposal stage.[8]

7. Cosmic Sources of High-Energy Muon Neutrinos

These may be classified as diffuse or discrete. Three principal types of diffuse emission of ν_μ have been identified; see Figure 6. Each creates a background, or noise, against which the ν_μ signal from discrete sources must be discerned. To be sure, one investigator's background may be another person's signal.

In the realm of neutrino physics, each type of diffuse emission is of intrinsic interest, as it may yield information about neutrino properties and about astrophysical processes. Nevertheless, as in the other astronomies, special interest attaches to individual, point-like sources. Neutrino astronomy will come into its own when these can be detected despite the diffuse backgrounds.

7.1 Diffuse Sources

The principal diffuse sources of HE ν_μ can be subdivided into terrestrial (i.e., atmospheric) and extraterrestrial ones. The latter comprise two classes, those originating in the interstellar medium (ISM) of our own Galaxy,and those produced by many of the active galactic nuclei (AGN) in the universe.

The atmospheric background arises from cosmic-ray collisions in the earth's atmosphere. It consists of neutrinos resulting from the decay of pions and kaons. The intensity and energy spectrum of these secondary HE ν_μ have been calculated using pion and kaon production models, and the measured muon spectrum.[9]

As was already learned by operating some of the early neutrino detectors in deep mines, atmospheric neutrinos from near-zenith directions are suppressed relative to those moving closer to the horizon. This is due mainly to competition between the rates of collison and decay of the parent pions. The nearly vertical ones plunge sooner into dense layers of the atmosphere; they suffer collision, and thus escape decay. In contrast to these, the near- horizontally moving pions produced at high altitudes can survive in the tenuous atmosphere long enough to decay into ν_μ and μ. Above 10^4 to 10^5 GeV, prompt neutrinos become a significant source of background even from the vertical direction.

It is well known that many of the Galactic cosmic rays are transformed by nuclear collisions in the ISM. A major difuse flux of HE ν_μ arises from interactions in the

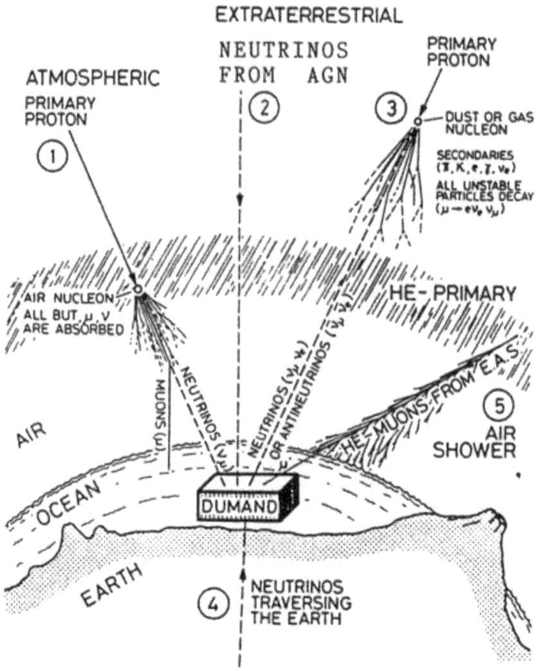

Figure 6: The main sources of diffuse HE ν_μ. (1) Neutrinos generated by cosmic rays striking the earth's atmosphere. (2) HE ν_μ from AGN distributed throughout the universe. (3) Neutrinos produced by cosmic rays colliding with the gas and dust of the ISM. Both the ocean above the array, and the earth below, shield the telescope, suppressing charged particles of non-ν origin. The ν_μ that collide produce secondary muons: $\nu_\mu + N \rightarrow \mu + N + \pi's$, where N denotes a nucleon.

ISM; these also yield HE gamma rays, mainly as secondaries of neutral pions. From observations of the latter, it has been deduced that most of the ν_μ generated in the ISM should reach us from the general direction of the Galactic Center,[10] within a solid angle of about 0.5 steradian. These constitute a considerable diffuse background of extraterrestrial ν_μ.

Another copious source of HE ν background is the ensemble of many active galactic nuclei (AGN) that produce intense fluxes of energetic protons (e.g., $\gtrsim 10^{15}$ eV). These generate HE ν_μ in regions with dense photon fields. The mechanism of photoproduction was discussed in Section 3.2. Stecker[11] and others[12] have shown that this flux of HE ν from AGN could be the dominant background. Of course, its observation will be of great intrinsic interest. However, to detect individual point sources (AGN or various objects less remote), we shall probably need larger targets than those available in the first-generation detectors now under construction.

7.2 Potential Discrete Sources of High-Energy Mu Neutrinos

Among candidate point-sources, we list some Galactic objects, then several extragalactic ones. Those proposed for the former class include young pulsars, certain stellar binaries—mainly compact ones—supernovae in dense gas clouds, and the region of the Galactic center (GC). Some extragalactic objects that have been considered as promising HE ν_μ sources are certain radiogalaxies, Seyferts, quasars, and blazars. The most plausible machine for powering an AGN is generally thought to be a supermassive black hole. Thus, most models of extragalactic HE ν_μ production have invoked such objects and their accretion disks in estimating their rates of HE ν_μ emission.

7.2.1 Galactic Sources

Neutron stars have been considered likely generators of HE cosmic rays since the discovery of pulsars.[13] Assume that the spectrum of emitted protons and other nuclei extends to energies sufficiently high so that they can produce extensive air showers (EAS) if they strike the earth's atmosphere. In that case, it was suggested by Berezinsky and Prilutsky,[14] such particles from a young pulsar could generate showers in the expanding shell of fresh debris expelled from a supernova explosion. Further development of this model[15,16] indicated that typical pulsars, with an incipient energy-loss rate of, say, $\approx 10^{38}$ ergs/sec would be observable by large telescopes if not too distant. More powerful—and more rare—pulsars with initial outputs of $\gtrsim 10^{40}$ ergs/sec might give detectable fluxes of HE ν_μ at distances of some 10 kpc in neutrino telescopes of sufficient aperture. The phenomenology of shower production and neutrino emission from the expanding shell is illustrated in Figure 7.

For a putative black hole of $\sim 10^3$ solar masses at the Galactic center (GC) and an output of 10^{37} erg/sec, the rate of ν_μ emission from the GC could be similar to that estimated for the much more distant Seyfert Galaxy 4151.[11]

X-ray binaries with an accreting neutron star or black hole are considered to be promising ν_μ sources. When energetic particles are accelerated at a neutron star, they spawn HE gamma rays and HE ν_μ in the accretion disk. These secondaries can appear to the observer as a beam of particles sweeping across the companion star. The gamma rays survive only at the stellar limbs, but are absorbed in between, producing an eclipse. The neutrinos, on the other hand, pass through the star, thereby reaching the observer much more copiously than the gamma rays; see Figure 8. Cyg X-3, which has been "seen" sporadically in gamma rays above 1 TeV, could be an example of this phenomenon. Cyg X-1, another accreting X-ray binary, is a favorite black hole candidate; it also seems to be a reasonable prospect for ν_μ emission.[17]

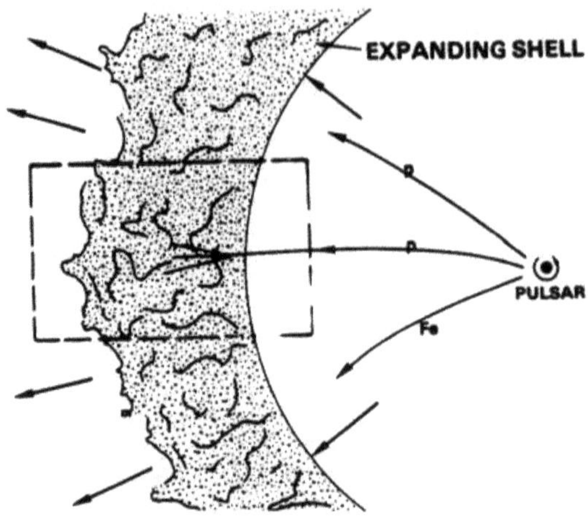

a. Very HE nuclei accelerated near a young pulsar, strike the shell of debris expanding from the site of the progenitor supernova.

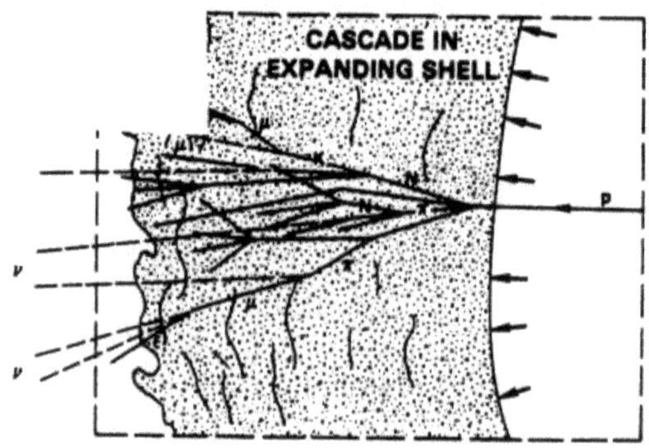

b. A shower of particles produced in a proton collision. The secondary neutrinos readily escape from the shell.

Figure 7: Shower Production and Neutrino Emission

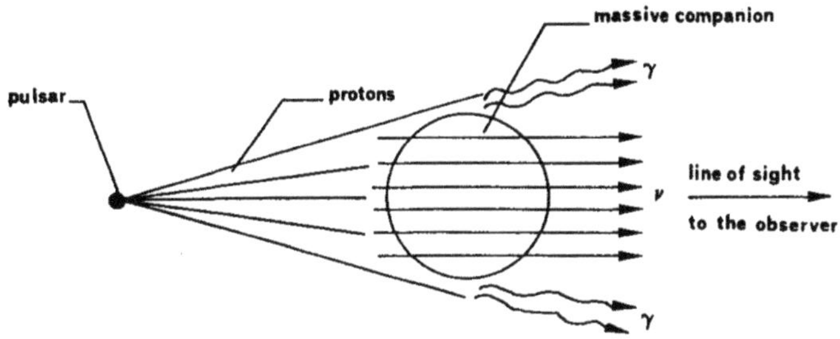

Figure 8: Particles accelerated at a neutron star impinge on a companion, generating gamma rays and neutrinos.

7.2.2 Extragalactic Sources—Mostly Active Galactic Nuclei (AGN)

We first cite a stellar-binary candidate—located not in the Milky Way, but in one of its satellite galaxies—the Large Magellanic Cloud (LMC). We refer to the X-ray binary LMC X-4. Since a rather intense HE gamma-ray flux reaches us from this object, we may be confident that the arriving ν_μ flux will be substantially higher.[18] This comes about because, as described in Section 7, the gamma-ray flux is much reduced by interactions with photons over the long path (160,000 light years) from the LMC.

The power output of an AGN is truly prodigious, like that of 10^{13} suns, and the entire emission comes from the compact galactic core. Such galactic nuclei occur in quasars, Seyfert galaxies, BL Lacs and radio galaxies. These objects are also prolific emitters of X-rays and gamma rays.[19] The nature of these energy sources has been the subject of intense inquiry. The most plausible explanation has been that a central supermassive black hole swallows up gas, dust, and even stars, thereby releasing energy from its accretion disk. Another model has involved a spinar, i.e., an ultramassive "star" of 10^6 solar masses or more. (It is thought that a spinar is fated ultimately to collapse to a supermassive black hole.) Yet another model proposes a dense conglomeration of stars which interact, speeding up the rate of a supernova formation to several per year. To discriminate among these models, new diagnostic observations may be crucial.

A Seyfert is a powerful spiral galaxy with a core so luminous that it looks star-like. The annual HE ν_μ event rate from the Seyfert galaxy NGC 1068 has been calculated[12(c)] under various assumptions. For a flat energy spectrum, say E^{-2} (differential), and a DUMAND-type telescope with area $\approx 10^5 m^2$ (for muons), an event rate of $10^2 - 10^3$ per year has been estimated. If one takes into account the (p, γ) process of neutrino production, this rate could be exceeded. Another Seyfert galaxy, NGC 4151, also in the Virgo supercluster, may be a detectable source of HE ν_μ as well.

Among the AGN that have been considered as candidate HE ν_μ sources are radiogalaxies like Cen A. Even more promising are some blazars and other quasars. Many of these galaxies have now been observed by EGRET, the Energetic Gamma Ray Experiment aboard the Compton GRO satellite at energies of $\approx$ 0.1 - 30 GeV. Important examples are 3C 273 and 3C 279.

Blazars are a set of AGN that include optically violently variable (OVV) quasars and BL Lacs. Their variability is sufficiently rapid so that these luminous sources seem very small (on a galactic scale). It has, however, been suggested that a relativistic beam oriented toward the earth can result in an apparent time compression.[20] The observation of "superluminal" motion of radio jets, and the lack of interstellar scintillations indicate that the lunimous cores may not be as small as was once supposed.

The HE gamma rays that are generated by blazars and other quasars are suppressed by pair production (i.e., $\gamma+\gamma \rightarrow e^+ + e^-$) against infra-red and other photons in the intergalactic medium. Accordingly, we would not expect such objects to be visible at much higher gamma-ray energies, say, $>$ 1 TeV. So it is not surprising that, in their diligent search for extragalactic sources of TeV gmma rays, T. Weekes and his collaborators at the Whipple Observatory[21] have detected only one such source, Mkn 421. This galaxy has a redshift of 0.03 which puts it closer to us than most of the other AGN observed by EGRET.

For distant sources, the loss of gamma rays by absorption en route will be very severe. On the other hand, the ν_μ that are also spawned by cosmic rays in the quasars and blazars will, in general, survive their transit toward a distant observer, and arrive with far greater intensity than the emitted gamma rays.

8 Neutrinos and Gamma Rays in Astrophysics—Contrast and Complementarity[22]

One of the major scientific triumphs of the space age has been the rapid development of gamma-ray astronomy. Initially, the new discipline got off to a slow start, owing mainly to atmospheric background and to the thin rain of cosmic gamma rays. Using balloons and relatively modest satellites (SAS II and COS B), the gamma-ray sky was explored, and many sources discovered. With the advent of the Compton Gamma-Ray Observatory, we have seen a veritable explosion of new results from the four telescopes aboard this large satellite. At the higher energies, EGRET and COMPTEL have yielded a rich harvest of information.

In celestial regions where high-energy processes are important, both gamma rays and neutrinos are copiously produced. Cosmic gamma rays (and also X-rays) are generated mainly by relativistic electrons interacting with magnetic fields and intense photon fields. Hence they are good tracers of electron acceleration processes in active galactic nuclei. Through synchrotron losses and inverse-Compton losses the electrons are efficiently degraded.

High energy gamma rays (cf. Section 3) are also generated (as secondaries of π^o) along with high-energy neutrinos, by high-energy proton and other nuclear reactions, with matter and dense photon fields. They are also readily degraded in both matter and photon fields. However, gamma rays are predominantly produced in electron processes. HE ν_μ, on the other hand, are uniquely generated in nuclear collisions. Therefore they can teach us about cosmic-ray sites and processes involving the principal component of the glactic cosmic rays reaching the heliosphere. In principle they carry information about the acceleration of energetic nuclei, their intensity, and the density of matter and photon fields in the vicinity of the sources.

Gamma-ray telescopes have been mainly used to probe the energy domains 10^6 – 10^{10} eV, while proposed HE ν_μ arrays will explore the region $\gtrsim$ 1 TeV.

Valiant searches are continuing for gamma-ray sources $\gtrsim$ 1 TeV, and it is important to pursue this research. To date only two TeV sources have been reliably observed, i.e., the Crab and Mkn 421. There is a strong obstacle to the detection of cosmic gamma rays at energies > 1 TeV; cf. Section 3. It is their severe attenuation, mostly by electron-pair production, either in the immediate vicinity of a powerful, compact source (e.e., a pulsar, a compact binary, or an AGN) or in the more tenuous but pervasive photon fields of interstellar and intergalactic space. Obviously, this absorption can work both near the source and en route to the observer.

Very energetic gamma rays ($E > 10^{14}$ eV) produced at distances exceeding 10^6 light years, cannot reach us at all; the same is true of photons with $E > 10^{12}$ eV generated beyond 10^9 light years. Thus, at sufficiently high photon energies, the universe is virtually opaque to gamma rays. Clearly, however, this does not diminish the great power of gamma-ray astronomy ove a wide range of lower energies.

As astrophysical probes, neutrinos have the advantage of prodigious penetrating power; they are very weakly interacting. An example of the symbiosis between gamma-ray and neutrino astronomies is the suggestion[12(d,e)] that it may be possible to discriminate betwen the black hole and spinar models of AGN. Suppose that one observes the energy in the AGN gamma-ray spectrum at which it steepens due to pair production, and also obsrves the luminosity of the AGN, then the neutrino flux should greatly exceed the gamma-ray flux if the source is a black hole. In the spinar model (say, of radius 10^{16} cm) the photon density is not sufficiently great to suppress gamma rays $\lesssim$ 1 GeV (though, at higher energies they are degraded). Near a black hole, the density of X-ray photons is high enough to absorb even 1 MeV gamma rays.

The acceleration of nuclei in the accretion disk of a central black hole in an AGN has been investigated by Lovelace[23] and by Kafatos et al.[24]

Altogether, it seems clear that neutrino astronomy and gamma-ray astronomy can provide complementary data of great—sometimes critical value. Together with observations from the other astronomies, they will allow various astrophysical models to be tested, and thereby enhance our understanding of many fascinating and exotic objects in the universe.

9 Prospects

In my view, the technical feasibility of HE ν_μ astronomy has been demonstrated. Strings of photomultipliers have been submerged in water (and ice), and these have recorded muons—at least those generated by cosmic rays in the earth's atmosphere. As of this writing, sources of cosmic HE muon neutrinos have not yet been demonstrably detected. However, the most plausible theoretical estimates of ν_μ fluxes really did not anticipate such detections with neutrino telescopes of aperture $\leq 2 \times 10^4 m^2$—the sensitive area (for muons) of the DUMAND II design. (A marginal exception might be the diffuse AGN background). This octagonal array has not yet been completed, nor have any other underwater (or under-ice) telescopes of comparable aperture.

Certainly, when any of the arrays now under construction becomes operational, invaluable experience will be gained toward the efficient operation of the larger arrays $\sim 10^5\ m^2$ that are envisaged. The latter class of telescopes should be able to detect the diffuse AGN HE ν_μ above other backgrounds. It is an open question whether any individual AGN or other discrete sources will then be discernible against the background of neutrinos emitted by the grand ensemble of AGN. The more distant goal of expanding the modular arrays by yet another order of magnitude, to $10^6 m^2$ may not be realized by any single one of the existing consortia for five to ten years. Support for a telescope of this size—one that is expected to "see" point sources—may require serious international collaboration among the leading groups.

Early in the twenty-first century, if not sooner,the dream of an operational neutrino astronomy at high energies will probably be realized.

10 Acknowledgements

The author thanks Professor Norma Sanchez for inviting him to lecture in the International School of Astrofundamental Physics. He is also indebted to Dr. Rein Silberberg for valuable comments, and to the members of the DUMAND collaboration for enthusiastic and stimulating discussions over the years.

11 References

1. F. Reines et al., discussions at the *International Cosmic Ray Conference* (*ICRC*), Denver, CO (1973, unpublished). Among earlier adumbrations of a deep underwater observatory for neutrinos, in the 1960's, were those of K. Greisen, M. A. Markov, and F. Reines.

2. For a detailed historical survey of the DUMAND project, see A. Roberts, *Rev. Mod. Phys.* **64** (1992) 259.

3. P Bosetti et al., *DUMAND II Proposal*, HDC- 288, Hawaii DUMAND Center (1988). For an update, see C M. Alexander et al. (for the DUMAND collaboration), 23rd ICRC **4** (1993) 515.

4. L. K. Resvanis, *High-Energy Neutrino Astrophysics*, Hawaii (1992, V. J. Stenger, J. G. Learned, S. Pakvasa and X. Tata, editors).

5. I. A. Belolaptikov et al., *The Lake Baikal Neutrino Project*, 23rd ICRC **4** (1993) 573.

6. T. Miller et al., for the AMANDA collaboration, op. cit., p. 557 ff.

7. M. M. Shapiro, in *The Astronomy and Astrophysics Encyclopedia*, S. Maran, editor (Van Nostrand Reinhold, New York, and Cambridge Univ. Press, Cambridge, 1992) 483.

8. J. G. Learned, *Proc. International Workshop on Neutrino Telescopes* (1990) HDC-8-90.

9. L. Volkova and G. T. Zatsepin, Proc. 9th ICRC **2** (1965) 1037. See also R. Silberberg, M. M. Shapiro, and F. W.Stecker, *Proc. DUMAND Summer Workshop* **2** (1978) 231.

10. R. Silberberg, M. M. Shapiro, and F. W. Stecker, op. cit.

11. F. W. Stecker et al. *Phys. Rev. Ltrs*, **66** (1991) 2697; Err. ibid. **69** (1992) 2738.

12. (a) A. P. Szabo and R. J. Protheroe, *Proc. High-Energy Astrophysics Workshop*, ed. V. Stenger et al., (World Scientific 1992) 24.
    (b) V. S. Berezinsky, *Proc. Neutrino 77* **1**, (Nauka Moscow, 1977) 177.
    (c) R. Silberberg and M. M. Shapiro, *Proc. 1th ICRC Kyoto* **10** (1979) 357.
    (d) D. Kazanas and D. C. Ellison, *Ap. J.* **304** (1986) 178.
    (e) V. S. Berezinsky and V. L. Ginzburg, *Monthly Notices R.A.S.* **194** (1981) 3.

13. T. Gold, *Nature* **221** (1969) 25.

14. V. S. Berezinsky and O. F. Prilutsky, *Astron. Astrophys.* **66** (1978) 325.

15. M. M. Shapiro and R. Silberberg, *Proc. DUMAND Workshop* **2** (1978) 183.

16. M M. Shapiro and R. Silberberg, *Summer Workshop at Khabarovsk and Lake Baikal*, J. G. Learned, ed. (Hawaii DUMAND Center, 1979) 269.

17. D. Eichler, *Ap. J.* **244** (1981) 1109.

18. G. Cocconi (unpublished).

19. M. M. Shapiro, *Neutrino 81 Conference*, Maui, HI (R. J. Cence, E. Ma, and A. Roberts, editors) **1** (1981) 44.

20. R. O. Blandford and M. J. Rees, *Testing the AGN Paradigm*, S. S. Holt, et. al., eds.. (American Inst. of Physics, New York, 1992) 3.

21. M. Punch, et al., *Nature* **358** (1992) 477.

22. M. M. Shapiro and R. Silberberg, *Space Sci. Rev.* **36** (1983) 51, and references therein.

23. R. V. E. Lovelace, *Nature* **262** (1976) 649.

24. M. Kafatos, M. M. Shapiro, and R. Silberberg, *Comm. Astrophys.* **9** (1981) 179.

III. X-RAY AND GAMMA-RAY ASTROPHYSICS

X-RAY BINARIES AND CONSTRAINTS ON GALACTIC EVOLUTION

Jonathan E. Grindlay
Harvard-Smithsonian Center for Astrophysics, 60 Garden Street, Cambridge, MA 02138

Abstract

X-ray binaries, containing a compact object (white dwarf, neutron star or black hole) accreting from a stellar companion, are luminous markers of stellar and dynamical evolution. Here we review the basic properties of x-ray binaries, with emphasis on the low mass x-ray binaries (LMXBs) containing accreting neutron stars and black holes and contrast these with accreting white dwarf systems (CVs). The new accretion disk theory of Narayan and Yi (1994, 1995), in which advection leads to a large inner disk radius (except at high accretion rates) and quasi-spherical accretion through a very hot corona, may constrain both source detectability and evolution. Globular clusters are prolific producers of LMXBs containing neutron stars but not black holes. Clusters may also be rich in CVs though these may be deficient in the dwarf novae types commonly found in the field. In contrast, the galactic disk is apparently over-abundant in black hole systems as indicated by the large x-ray transient source population. Recent studies of the cosmic x-ray background indicate that the galactic halo, or thick disk, may be dominated by accreting white dwarfs. Possible origins of these differing components, and some of the constraints they impose on the evolution of the Galaxy, are discussed in this tutorial review.

1 Introduction

X-ray binary systems, where luminous x-ray emission is produced by accretion onto a compact object from a binary companion star, have undergone a renaissance of sorts in the past decade and promise even greater returns in the next. The study of x-ray binaries with x-ray telescopes in space and optical and radio telescopes on the ground can provide powerful new clues for the origin and evolution of the Galaxy. Optical and uv studies with HST are now unravelling the nature of both the x-ray binaries in the dense cores of globular clusters and the mass functions (both initial and surviving) of stars in these oldest systems in the Galaxy. X-ray binaries (XRBs) are signposts for exotic endpoints of both stellar and dynamical evolution. They are few enough in total number and conspicuous enough in their x-ray vs. optical properties to stand out above stellar backgrounds (and foregrounds) and yet both numerous and long-lived enough to allow their distributions and thus origins to be traced.

With the advent of high spatial and spectral resolution studies possible with current instruments, it is now possible to study known sources in detail and to discover and locate many new sources. The ROSAT all-sky survey (see Trümper 1983 for general description), carried out in 1990-91 but not yet fully released, promises to provide thousands of new XRB sources (from the more than 70,000 sources detected, which are dominated by stellar coronal sources). Most of these will probably be accreting white dwarf systems, or cataclysmic variables (CVs). However at least several hundred or perhaps a few thousand accreting neutron star systems, with both low mass companions (LMXBs) and high mass companions (HMXBs), are also surely included in the survey. A comparable or perhaps even greater number of LMXB systems containing black holes as the primary are also likely to be present. Although distinguishing these from the vast number of stellar coronal sources would be greatly simplified eventually with much higher spatial and spectral resolution as well as higher energy coverage (e.g. with high resolution AXAF studies or the proposed (cf. Grindlay et al 1995) all-sky hard x-ray imaging survey mission EXIST), it will be a significant step forward in establishing the number density and distributions of XRBs in the Galaxy.

What are the relative numbers of these various types of XRBs and how are these related to stellar and binary evolution in the Galaxy ? This question requires a global perspective on the content of XRBs and their closely related systems, the cataclysmic variables (CVs), in the galactic disk and in globular clusters. In this tutorial review, we shall summarize properties of low mass x-ray binaries (LMXBs) although we shall make some comparisons with the XRBs containing high mass companions (HMXBs), particularly for those containing black holes. We shall attempt to connect the possible formation and evolution of several classes of LMXBs and CVs with the formation and evolution of millisecond pulsars (MSPs). This complex web of binary and stellar evolution has been summarized in parts by many previous authors as reviewed by Bhattacharya and Van den Heuvel (1991). Here we summarize some of the many interconnections and highlight what is likely to be a major new ingredient: the new accretion disk theory of Narayan and Yi (1994, 1995) which predicts that at low accretion rates ($\lesssim$ (0.001 – 0.01)$\dot{M}_{Edd}$) the standard thin accretion disk of Shakura and Sunyaev (1973) is likely to apply only at $\gtrsim$ 3000 Schwarzschild radii from the central object (with mass perhaps 1-10$M_{\odot}$). These truncated accretion disks, or doughnuts, will transition to quasi-spherical inner flows of very high temperature and low efficiency. We note that this more radial flow must also decrease the angular momentum coupling to the central object. Thus the formation of MSPs from LMXBs by the so-called standard model may be limited still further (to only the relatively few objects at high accretion rate) and the role of accretion induced collapse (AIC) as a direct channel for MSP production (Michel 1987, Grindlay and Bailyn 1988) is perhaps more pressing.

How might the distribution of LMXBs containing neutron stars vs. black holes offer clues for the initial mass functions of stars in and out of globular clusters ? Do the LMXBs containing neutron stars and main sequence companions, which are the typical burst sources, arise in globular clusters which have largely been disrupted ? Are the possible CVs in the thick disk or halo of the Galaxy, which we have suggested (Maoz and Grindlay 1995 and Grindlay and Maoz 1995) could be a significant contribution to the cosmic x-ray background, from this same primordial cluster distribution or must a still larger number of primordial clusters be invoked ? Could these CVs and the large scale-height population of MSPs be related and also give rise to an extended halo population of NSs as required for galactic models (e.g. Lamb 1995) for gamma-ray bursts ?

These are some of the questions to be explored in this review. We first provide introductory summaries of basic properties of CVs and then LMXBs, distinguishing between those containing neutron stars (NSs) vs. black holes (BHs). We also briefly consider the total populations and scenarios for formation of LMXBs (both NSs and BHs) in the disk. We then summarize our recent reviews of the globular cluster x-ray sources and their distinct populations of high luminosity (LMXB) and dim (CV ?) source types. We shall compare the LMXB vs. CV populations in globulars with those in the thick disk and halo to conclude that any primordial cluster distribution must have been relatively deficient in stars more massive than about 6$M_{\odot}$.

2 Basic Properties of CVs

Cataclysmic variables (CVs) are the oldest known type of compact binary. Kraft (1962) first pointed out the likely binary interpretation of the strange blue flickering variable stars which sometimes undergo cataclysmic outbursts. We now recognize a variety of types of CV although the relative numbers and luminosity functions of each are still poorly known.

2.1 CV Types

Without reviewing the long and rich history of the study of these systems (see Patterson 1984 for an early review and Ritter and Kolb 1995 for more recent results), we simply summarize the basic types of CV and their primary distinguishing features:

Dwarf Novae
Perhaps the most common type of CV are the dwarf novae (DN), or those systems which occasionally (typically every few months) undergo sudden outbursts lasting typically days to weeks and reaching relatively fixed maximum luminosities with absolute magnitudes $M_V \sim 4.5$ (Warner 1987). These systems are exemplified by SS-Cyg or U-Gem and have typical x-ray luminosities in the ROSAT band L_x(0.2-2.4 keV) $\sim 10^{31-32}$ erg/s (Ritter and Kolb 1995). These systems typically have binary period in the 4-8 hour range and thus have K or M-dwarf low mass secondary companions. As such, they are the white dwarf analogues of both the x-ray bursters and the soft x-ray transients for LMXBs containing NSs and BHs, respectively.

DQ Her or Intermediate Polars
Those CVs with moderately magnetic white dwarf (WD) primaries, such as DQ Her, are often called intermediate polars. They are reviewed by Chanmugam (1992, 1995) and Cropper (1990). As opposed to the DN in which the accretion disk was thought to extend down to the WD to a boundary layer (but see below), the DQ Her systems must have an inner disk radius at the magnetospheric radius (where magnetic pressure from the central dipole magnetic field of the WD matches the ram pressure of the accretion flow). Thus the accretion onto the WD is via the magnetic field and the systems are often detected as pulsed at the WD rotation rate (typically 30-100 sec) and with x-ray luminosities usually $\sim 10^{32-33}$ erg/s. Thus these systems (e.g. GK Per) can be more luminous than the DN systems and also have larger values of their x-ray to optical flux, f_x/f_{opt}. These systems are the analogues of the disk-fed x-ray pulsars like Her X-1.

AM Her or Polars
The AM Her type CVs are those with the largest magnetic fields, typically $\lesssim 3 \times 10^7$ gauss, as measured directly by circular polarization, and have their WD rotation periods (nearly) tidally locked at the binary period. Again, reviews are given by Chanmugam (1992, 1995) and Cropper (1990). The x-ray luminosities are comparable to the DQ Her systems, but the spectra are different, with significant emission in a very soft x-ray component which extends down into the XUV with cyclotron emission components as recently studied with EUVE. These systems are mostly found with ultra-short periods just below the so-called period gap (~2-3 hours) so that their binary companions are likely to be evolved. It is significant that the ROSAT sky survey has recently found many new AM Her systems including the first one with significantly longer binary period indicating a main sequence or possibly sub-giant companion as well as several within the

period gap (Chanmugam 1995) so that their space density may be much higher than previously suspected.

Nova-like and Classical Novae

CVs are of course the progenitors of classical novae and are sometimes still detected as XRBs after nova outbursts. The so-called nova-like CVs are those with relatively weak emission lines (resembling DNs in outburst) and are undoubtedly accreting at relatively high accretion rates, with $\dot{M} \sim 10^{-8}\ M_{\odot}$/year. Thus they may be expected to have sub-giant companions (in order to drive the high accretion rates by nuclear evolution and expansion of the secondary) and correspondingly longer binary periods (~1 day) than the other CV types listed above. They should also be the most luminous soft x-ray sources and indeed GK Per (which is also an intermediate polar) is particularly luminous with $L_x \sim 10^{33}$ erg/s (cf. Patterson and Raymond 1985).

In contrast, the classical novae are usually relatively faint well after their nova outburst implying relatively lower accretion rates for triggering nova outbursts. GK Per, which was an historical nova (1903) is an exception, perhaps because it is an intermediate polar with accretion from the disk via the magnetic field. However, classical novae may be luminous supersoft x-ray sources (with temperatures kT ~ 30 eV), as discovered by ROSAT for the nova GQ Muscae (cf. Ögelman et al 1993). Indeed this case provided strong confirmation of the model (Van den Heuvel et al 1992) for supersoft sources (SS) being accreting white dwarfs on which steady (or quasi-steady) nuclear burning is possible for a narrow range of $\dot{M}$.

2.2 CV Statistics

Distances to CVs are usually only poorly known and thus their absolute space density is not well established. Their space density is further complicated by their uncertain duty cycles for accretion and total system lifetimes. As will be discussed below, these are similar problems for accreting NS systems, the LMXBs. In addition, truly complete surveys for CVs are still lacking. Although optical surveys for uv-excess objects (e.g. the Palomar-Green (PG) survey down to m_v =16) have provided ostensibly complete samples), they are limited in source numbers and sample very different volumes for different CV types: AM Her objects, for example, are so intrinsically faint (with typical $M_v \sim 10$) as to be detectable only out to ~100 pc in this survey. The ROSAT all-sky survey will surely greatly enhance the CV statistics (and has in fact discovered a number of new AM Her systems, for example) but is again still volume limited: a typical CV with $L_x \sim 10^{31}$ erg/s would exceed the survey flux limit of $10^{-12.5}$ erg/cm^2 only for distances $\lesssim$ 600 pc.

Two main possibilities have been advanced for the space density and thus total population of disk CVs: either the ~100 known CVs (with their uncertain distances typically ~100-1000 pc) are most of the local population or there is a much larger number in quiescence (the hibernation model of Shara 1989). The conventional model (e.g. Patterson 1984) would have the space density of CVs about $10^{-6} - -10^{-5}$ pc^{-3} whereas the Shara model, with most CVs quiescent, would predict densities about $\gtrsim$ 10 times higher. The discovery of faint CVs in the Einstein galactic plane survey and its optical followup (Hertz et al 1990) supports the higher space density value.

3 Basic Properties of LMXBs Containing NS Primaries

LMXBs have (by definition) low mass secondary stars but primaries that are either neutron stars (NS) or black holes (BH). Here we briefly review the NS-primary systems, saving the BH systems for the next section.

3.1 NS System Types

The LMXBs containing NS primaries appear to be divided into at least two groups: the so-called Z sources vs. Atoll sources (cf. Hasinger and Van der Klis 1989). The division of LMXBs into at least two primary classes has been long been suggested (e.g. Parsignault and Grindlay 1978) by the differences in source spectra and variability characteristics. There is also evidence that the small class of LMXB pulsars, such as the 7 sec pulsar 4U1626-67, may form yet another class of LMXBs (Mereghetti and Stella (1995).

Z Sources
The so-called Z sources are the very luminous sources ($10^{37.5-38}$ erg/s) in the galactic bulge and disk that have well-defined correlations in their spectral variability. In the plot of x-ray color (i.e. ratio of fluxes in two distinct bands) vs. flux (in a given band), or also in the color-color plane; these sources exhibit a Z-like trajectory of possible states. Only 6 such sources have been identified out of a total of some 100 LMXBs, indicating they are relatively rare or short-lived. The three different branches of the Z have well-defined source variability characteristics (e.g. differing quasi-periodic oscillation behavior) and suggest that $\dot{M}$ increases from the top to the bottom of the Z, with the Eddington value reached at the lower vertex. Thus these objects are accreting at the highest rates and must have sub-giant or giant companions with binary periods of ~1 day to supply the mass transfer. This is directly confirmed for the (only) two Z sources with optical c: Sco X-1 and Cyg X-2. As reviewed by Van der Klis (1989), the characteristic QPO characteristic of the Z sources is still most consistent with the beat frequency model (Alpar and Shaham 1985) for at least these intensity-dependent QPOs. This, in turn, requires the NSs in these systems to have a moderate magnetic field ($B \gtrsim 10^{9-10}$ gauss) in order for the magnetospheric radius to extend out to several stellar radii where the Keplerian frequencies would be ~30 Hz as observed in the QPO for such high $\dot{M}$ systems.

Thus the Z sources appear to be magnetic NSs in relatively young (companion masses are $\gtrsim 0.8\ M_\odot$) systems. Their NSs must be spinning up, since the accretion disk extends relatively close to the NS. However the maximum observed QPO frequencies of ~40 Hz may imply a rotational frequency cutoff for the NSs (and thus millisecond pulsars) at a comparable value if they are near their equilibrium spin periods. No Z sources are known in globular clusters, although the LMXB AC211, with 17 hour binary period, in M15 shows some interesting parallels (but also important differences).

Atoll Sources and Bursters
Hasinger and Van der Klis (1989) made a clear case for a separate class of sources based on their distinctly different color-flux or color-color diagrams. This group, which typically showed clumped islands (atolls) of characteristic behavior in these diagrams include the bursters and lower flux (and luminosity - typically $\sim 10^{36-37}$ erg/s) sources previously recognized as different (e.g. Parsignault and Grindlay 1978, Grindlay 1984) from the luminous galactic bulge sources

(and Z sources). Although there are only 11 atoll sources identified as such by detailed studies of their spectral variability, it is likely that most or all of the 42 presently known bursters (cf. Van Paradijs 1995) are in this class.

The bursters, then, must have lower mass companions usually still on the main sequence and supplying their characteristic $\dot{M} \sim 10^{-10}$ $M_{\odot}$/year accretion rate values by magnetic braking as the source of loss of angular momentum to maintain binary (and Roche lobe overflow) contact. However some, like the 11-min binary 4U1820-30 in the globular cluster NGC 6624, clearly have degenerate (WD) companions and so must represent end-phases of the common-envelope evolution of a companion star near the turnoff mass (e.g. Bailyn and Grindlay 1987). In any case, the NSs in atoll sources must have very low magnetic fields ($\lesssim 3 \times 10^8$ gauss) since they do not have apparently magnetospheric-related QPOs and since no pulsations have been detected. Recent understanding of accretion flows onto BHs (or NSs with low magnetic fields) at comparably low $\dot{M}$ values by Narayan and Yi (1994, 1995) shows that the inner disk is truncated at a large value (perhaps 3000 Schwarzshild radii) and accretion down onto the star is quasi-spherical through a thin corona. We shall return to this important new development in our discussion of BH LMXBs below and the formation/evolution of LMXBs vs. MSPs.

LMXB Pulsars

Finally, the LMXBs include a (very) few pulsars, such as the 7 sec pulsar 4U1626-67. Clearly there is a route for LMXB formation with magnetic NSs, with magnetic fields $\gtrsim 10^{11}$ gauss or much stronger than in the Z sources. As pointed out recently by Mereghetti and Stella (1995) the 3-5 members of this class are remarkably similar in their pulse periods (5.4-8.7 sec) and extremely low mass companions (therefore degenerate) as required by typical limits on their binary periods of $\lesssim 1$ hour. Thus they probably have a common formation mechanism, as discussed below.

3.2 Statistics of NS/LMXBs

Apart from the 12 known LMXBs in globular clusters (see discussion below), the distances and therefore luminosities and space densities of the approximately 125 field LMXBs listed in the catalogue of Van Paradijs (1995) are poorly known. The best distance estimates for field LMXBs come from peak fluxes in x-ray bursts, which are likely to be Eddington limited luminosities and have been used to devive a reasonable galactic center distance, $R_o \sim 7$ kpc, from the centroid of the galactic bulge distribution of bursters (Ebisuzaki et al 1984). Similarly, the emission lifetimes are controversial and yet are (of course) needed to establish the total density and birthrate of LMXBs. It is likely that the atoll (assumed here to include all the bursters) sources, with low $\dot{M}$ values (typically $10^{-10}/M_{\odot}$/year) have lifetimes $\tau_{atoll} \sim \delta m/\dot{M} \sim 10^9$ years, where δm is the mass transferred onto the NS and is likely to be $\gtrsim 0.1$ $M_{\odot}$, particularly if the NSs are to be spun up as millisecond pulsars (MSPs) as described below. However, the lifetimes could be either shorter due to x-ray heating effects (e.g. Tavani 1991) or longer if the sources are highly variable and spend a significant fraction of their time in quiescence. Thus a compromise value (highly uncertain) is still provided by the value for the birthrate $B_{atoll} \sim N_{atoll}/\tau_{atoll} \sim 10^2/10^9 = 10^{-7}$/year. For comparison, the luminous Z sources with $\dot{M}$ values typically within a factor of 3 of the Eddington value probably only sustain this for the typical AGB lifetimes of their giant companions, or $\tau_Z \sim 3 \times 10^7$ years, so that $B_Z \sim N_Z/\tau_Z = 6/(3 \times 10^7) = 2 \times 10^{-7}$/year, or within a factor of two of the burster/atoll rate. Thus although there is evidence (cf. Van der Klis 1994) that the atoll and Z sources may differ in their NS magnetic field (since Z sources display QPOs generally well described by the beat frequency model), the similar birth rates may simply suggest that the Z sources are the descendents of atoll sources when their companion stars have evolved off the main sequence.

3.3 Relation to Millisecond Pulsars ?

The millisecond pulsars (MSPs) are a new and rich source of constraints on the nature, origin and evolution of LMXBs in the Galaxy. A recent Aspen Workshop (January 1994) and the published proceedings (Fruchter, Backer and Tavani 1995 and papers therein) provide a wealth of new observational and theoretical results. Here we extract only a few of these whch are most relevant to the central question of whether MSPs are the descendents of LMXBs via the so-called standard model or whether the evolution can go the other way (cf. Bhattacharya and Van den Heuvel 1991 for references). Several new results were reported in Aspen, including:

1. With much larger samples of galactic MSPs in the disk (i.e. not in globular clusters, which until 1992 dominated the samples), the scale height of MSPs is probably significantly larger (e.g. $\gtrsim$ 500 pc) than that for ordinary pulsars ($\sim$150 pc), which would also suggest a high velocity ($\gtrsim$ 100 km/sec) for the underlying MSP population (Bailes and Lorimer 1995).

2. Some of newly discovered disk pulsars are extremely close and low luminosity (e.g. PSR J0108-143 - cf. Manchester 1995), suggesting that the luminosity function extends to very low luminosity and that the total MSP population is therefore large. The total galactic population estimated by Bailes and Lorimer (1995) is $\sim 10^5$, although these estimates are still very uncertain due to poorly known beaming and scale height values.

3. The number of NS-NS MSPs, or relativistic MSPs such as PSR1534+12 which allow tests of general relativity, has *not* grown despite factor of $\gtrsim$ 10 increases in survey depths. Thus the total number of these systems, which surely do not descend from LMXBs (but rather HMXBs), and thus the number of NS-NS coalescence events/galaxy which could give rise to gamma-ray bursts and to gravitational wave events possibly detectable by future upgrades of LIGO, is correspondingly reduced by an order of magnitude (Bailes and Lorimer 1995).

The birthrate for MSPs, which have characteristic spin-down lifetimes of $\sim 10^{10}$ years, must be $B_{MSP} \sim N_{MSP}/\tau_{MSP} \sim 10^{-5}$/year, or significantly larger than the rate derived above for either Z or atoll sources. With earlier estimates for a smaller total number N_{MSP}, the rates were within factors of 3-5 of each other. However, the more recent estimates (which result in part from the high velocity distribution of MSPs) suggest again there is the LMXB deficit originally pointed out by Kulkarni and Narayan (1988) for the field MSPs and by Grindlay and Bailyn (1988) for the globular cluster MSPs. The deficit may be further enhanced since the atoll source contribution may be smaller still: with the advection accretion solutions of Narayan and Yi (1994) for low $\dot{M}$ (as appropriate for atoll sources), the accretion is quasi-spherical through an inner corona. Thus the angular momentum transport to the NS must be less and the spinup of LMXBs into MSPs (the "standard model) must be less effective.

Van den Heuvel (1995) has pointed out that a large fraction (perhaps $\gtrsim$ 80%) of the MSPs may arise from the possibly large population of very short-lived LMXBs like Her X-1, which is probably just coming into Roche-lobe contact and will soon undergo unstable mass transfer (since its secondary is $\sim 2M_\odot$, and thus more massive than the NS). However, the total population of these systems is obviously very uncertain (given small number statistics) and thus this evolutionary channel must also be regarded as tentative. The "standard model would predict that the LMXBs being spun up have accreted significant mass ($\gtrsim 0.5\ M_\odot$if the Ns are initially born with high fields that are only decreased to the low values typical of MSPs by an extended accretion phase — cf. Van den Heuvel 1995). However the NS masses derived for MSPs (cf. contributions in Fruchter et al 1995) are still consistent with $1.4M_\odot$, as are the LMXB masses in globulars (cf. Grindlay 1995 and references therin).

Since there are now several examples of NS-NS pulsar binaries, it is reasonable that pulsar with a black hole (BH) companion should be found since it would only require a pulsar companion star perhaps twice as massive (in order to collapse to a BH) as in the NS-NS progenitor systems. Such a system would enable novel tests of general relativity (cf. Narayan, Piran and Shemi 1991) and may be found in pulsar surveys if they target specifically regions of massive star formation. Although pulsar-BH systems are not yet found, BHs are now almost certainly seen in x-ray binaries with both low mass and high mass companions.

4 Black Hole XRB Systems

Within the past decade, BHs have been convincingly seen in at least 6 x-ray binary systems. Although indicated early on for HMXBs (e.g. Cyg X-1), the great surprise has been that BHs with low mass companions may in fact be the most common x-ray binary systems in the Galaxy.

4.1 BH System Types

Both HMXBs and LMXBs contain, or are strongly suspected to contain, BHs. The evidence for BHs comes most directly from measurements of the mass functions, $f_c(M/M_\odot)$, of the binary systems by measurement of the radial velocity amplitudes of the optical companion stars (cf. Van Paradijs and McClintock 1995 for a critical review). Any system for which the compact object mass can be established to be in excess of $3M_\odot$, or approximately the causality limit for the maximum mass of a NS, is almost certainly a BH. Six systems have been so measured: three HMXBs and three LMXBs. Several more LMXB systems are currently being measured and many more are likely be found at larger distances and thus fainter magnitudes (making them, of course, even more difficult to measure mass functions). A discussion with references is given in Van Paradijs and McClintock (1995) and a summary of the measured system parameters is given in Tables 1 and 2 below.

X-ray signatures for BHs vs. NSs are of major interest and are summarized by Van der Klis (1994). No signatures (e.g. flickering or hard x-ray spectra) appear to be unique, although the combined presence of an ultra-soft x-ray component (probably blackbody from the accretion disk) and a hard x-ray power law component extending out to beyond 100-200 keV (from an inner accretion disk corona ?) comes closest. The x-ray spectra of atoll-LMXBs at relatively low accretion rates $\dot{M}$ (i.e. bursters in low states) can appear very similar to BHCs at hard x-ray energies although their spectra usually break at lower energies ($\lesssim 100$ keV) and have steeper spectral indices (Barret and Grindlay 1995). These possible spectral differences are consistent with the advection-flow accretion disk models of Narayan and Yi (1994, 1995) which would suggest low $\dot{M}$ flows onto NSs (non-magnetized) would still be partially dominated by blackbody emission from the central NS. Both more sensitive hard x-ray spectra and more dynamical mass measurements of BHCs are needed to further test the current paradigm that harder spectral components (possibly also with 511 keV annihilation line radiation) are BH signatures.

An enticing, but still tentative, BH signature may be the formation and existence of relativistic jets from BHs (both high and low mass) accreting at very high rates. Two recently discovered super-luminal radio sources (the x-ray transient, X1655-40, and the peculiar variable GRS1915+10) both show hard x-ray spectra characteristic of BHCs. No confirmed NS systems show such jets. Of course both Cyg X-3 and SS433, both HMXBs, show relativistic jets but at least SS433 is better "fit" by BH models (cf. Van Paradijs and McClintock 1995 for references).

High Mass Systems

The HMXBs containing BHs are epitomized by Cyg X-1, the original BH candidate (BHC) identified by Bolton (1972). It is somewhat surprising that only 2 more such systems are likely BHCs since massive companions would seem favored for BH production (i.e. binary mass ratios are strongly peaked at unity). Although several other HMXBs are suspected to be BHCs (e.g. SS433 and possibly even Cyg X-3), only the two sources LMC X-1 and X-3 have relatively clean radial velocity measures for their (primary) companions and thus mass functions. Results for the HMXB BHCs are given in Table 1.

Table 1

Galactic Black Hole Candidates With High Mass Companions [1,2,3]

Object	Opt. ID	RA (2000)	Dec (2000)	m_v	Type	P (days)	$f_c(M/M_\odot)$	$M_x/M_\odot$
LMC X-1	star R148	05 39 38.7	-69 44 36	14	OB	4.2	0.14 ±0.05	
LMC X-3	star WP	05 38 56.4	-64 05 01	17	B3V	1.7	2.3 ±0.3	>7
Cyg X-1	HDE 226868	19 58 21.7	35 12 06	9	O9.7Iab	5.6	0.24 ±0.01	>7

(References given at end of Table 2 below)

It is likely that many more HMXB systems harbor BH secondaries and are recurrent transients with Be primaries. In fact it is at first surprising that none have been detected thus far, since Van Paradijs and McClintock (1995) estimate from the population of such systems containing NS secondaries that there must be at least 2000-6000 Be/HMXBs in the Galaxy. The NS/Be systems are easily recognized as pulsars (e.g. the transient A0535+26) which have characteristic spectra with breaks at energies of ~50 keV (itself dependent on the magnetic field strength and thus cyclotron opacity of the NS). Thus detection of a recurrent transient with the characteristic ~1 year recurrence times typical of Be systems (e.g. their typical orbital periods), but either no detection of x-ray pulsations or detection of an x-ray spectrum with hard spectral index out to 100-200 keV would be a plausible BH/Be system. Obviously, discovery of a Be optical counterpart would make the case clear.

It is possible one such system has in fact been found: the recurrent transient GROJ1849-03 discovered by us with BATSE (cf. Zhang et al 1994). This system appears to be a 241 day period system (which would imply an eccentric orbit since Be systems usually do not have strictly periodic outbursts at the orbital period) and a hard x-ray spectrum that is flatter than NS/Be systems and without an obvious break. A limited search for an optical counterpart after one outburst was not centered on what later emerged as the best (but still uncertain) source position; a wider search is planned for the next outburst.

Because NS/Be transients typically reach outburst luminosities of only $\sim 10^{36}$ erg/s, their peak $\dot{M}$ values are probably still $\lesssim 0.01$ of the Eddington value. Thus the accretion onto a BH/Be for the same outburst may be governed by the advection flows found by Narayan and Yi (1994,1995) for $\dot{M}$ values below about 0.001-0.01 of the Eddington value (and depending on the α value for the disk). If so, the resulting x-ray luminosity will be much lower for the BH/Be system (e.g. perhaps only $\sim 10^{33}$ erg/s) than for the NS case, and these systems will be indeed difficult to find.

Low Mass Systems

Perhaps the most dramatic development in the study of x-ray binaries in the past decade has been the discovery of the almost certain BH nature of several LMXB transient sources: the so-called ultra-soft transients. The archetype is the bright "x-ray nova" A0620-00, which in 1975 became the brightest source ever detected (still) in x-rays and in 1986 was found by McClintock and Remmilard (cf. review by Van Paradijs and McClintock 1995) to be a BHC by virtue of their measurements of its mass function. Two other such x-ray novae, Nova Muscae 1991 and V404 Cyg, were then found to be excellent candidates. (Note these transients are not novae in the classical sense – nuclear burning on the surface of a white dwarf – but are due to accretion instabilities, probably in the accretion disk. Unfortunately the term "x-ray nova" seems to have been adopted and we use it here). All three x-ray novae have mass function values that are near or in excess of the $3M_{\odot}$limit (and dramatically so for V404 Cyg !), thus providing the most conservative requirements that even for zero companion mass the compact object must be a BH. Results are summarized in Table 2.

Table 2

Galactic Black Hole Candidates With Low Mass Companions [1,2,3]

Object	Opt. ID	RA (2000)	Dec (2000)	m_v	Type	P (days)	$f_c(M/M_{\odot})$	$M_x/M_{\odot}$
A0620-00	Nova Mon 75	06 22 44.5	-00 20 45	18	K5V	0.32	2.90 ±0.08	>3.4
GS1124-68	Nova Mus 91	11 26 26.7	-68 40 33	20	K5V	0.43	3.07 ±0.40	>2.9
GS2023+33	V404 Cyg	20 24 03.8	33 52 04	19	G9V	6.47	6.26 ±0.31	>5.6

[1] McClintock, J.E. 1992 in ***Relativistic Astrophysics, Cosmology, and Fundamental Physics***, eds. J.D. Barrow, L. Mestel, and P.A. Thomas, The New York Academy of Science (New York), Vol. 647, p. 495.
[2] Tanaka, Y. 1992, in *Ginga Memorial Symp.*, (F. Makino and F. Nagase, eds.), ISAS Publications (Tokyo), p. 19.
[3] Van Paradijs, J. and McClintock, J.E. 1995, in *X-ray Binaries*, (W.H.G. Lewin, J. van Paradijs and E.P.J. van den Heuvel, eds.), Cambridge Univ. Press (Cambridge).

Many more of these x-ray novae have been discovered (see below) and several are being studied optically in an effort to measure their mass functions. Preliminary results on these indicate they too are very strong BHCs. Thus the LMXB/BH systems have emerged as the most convincing and probably also most numerous (see below) BHCs in the Galaxy.

4.2 Statistics of BH/LMXBs

A significant number of BH/LMXB transients are now being discovered with the BATSE all-sky monitor instrument on the Compton Gamma Ray Observatory (CGRO). All are so bright as to suggest that if they reach similar peak x-ray luminosities (which are, somewhat surprisingly, only about 10^{37} erg/sec or ~0.01 of their Eddington limit values) they are typically within ~3 kpc, and indeed such are the approximate distances indicated by their optical counterparts. The rate of detection of these x-ray novae is about 1-2/year for peak fluxes $\gtrsim$ 1 Crab (it is convenient to measure fluxes in units of that of the Crab nebula, a standard which can cross calibrate different detectors). Extrapolating to the entire galactic disk population, assumed here to be a simple uniform disk of radius 10 kpc, a total of ~10/year is expected in the Galaxy. Because each transient is typically detectable for only about 3 months (the outbursts decay with similar time constant values of ~30 days), and recur with typical intervals of (only) 30 years, their effective duty cycle for detection is only ~0.01. Thus the total number active at any one time in the Galaxy is probably of order 1000. A similar total is derived by Tanaka (1992) on the basis of the similar rates of detection of x-ray novae with the Ginga satellite at lower x-ray energies. It is interesting that the BATSE detections at higher energies, where BH/LMXBs are expected to be more readily distinguished from NS/LMXBs, yields about the same rates and total statistics.

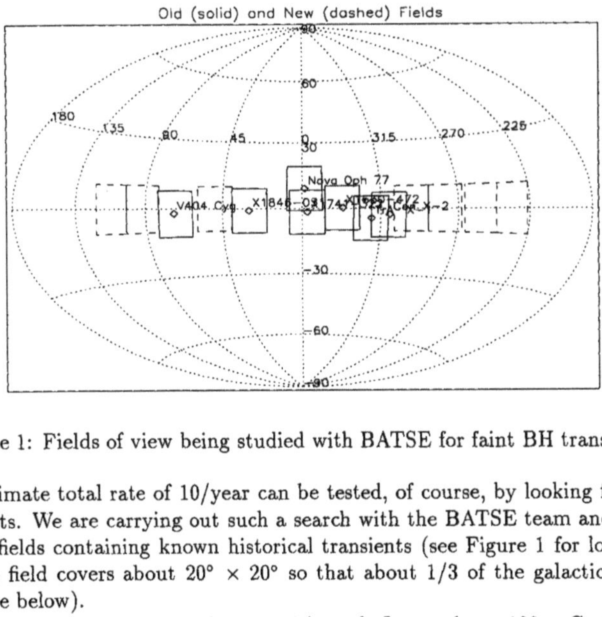

Figure 1: Fields of view being studied with BATSE for faint BH transients

The approximate total rate of 10/year can be tested, of course, by looking for more distant fainter transients. We are carrying out such a search with the BATSE team and have begun to search some 7 fields containing known historical transients (see Figure 1 for locations of fields studied). Each field covers about 20° × 20° so that about 1/3 of the galactic plane could be covered (but see below).

The sensitivity reached is for outbursts with peak fluxes about 100 mCrab (or 0.1 of the typical bright novae) so that transients out to ~10 kpc should be detectable. The search will yield improved estimates of the recurrence time for known systems as well as search for new sources. Preliminary results (Grindlay et al 1995) indicate probable detection of faint outbursts from 1-2 of the sources and 1-2 possible new sources for some 3 years of data. Because of our offset position from the galactic center (at 7 kpc), a survey out to about 10 kpc could sample about half of the assumed galactic disk distribution with radius 10 kpc. However, the actual distribution of galactic longitudes and latitudes of the 7 fields used (cf. Figure 1) reduces this to about 1/5 of the assumed disk. Thus in 3 years we would expect about 6 new transients in these fields but have instead detected only 1-2. This might suggest that the total BH/LMXB population is a factor of ~3 less than the 1000 estimated above. However, much more analysis remains and the 7 fields chosen initially are limited from optimum sensitivity by the presence of many nearby bright sources. An extension of the survey to an additional 7 fields (cf. Figure 1), which are relatively blank, will therefore be studied with BATSE and should allow the total population of BHCs to be better measured.

5 Populations and Formation of Disk LMXBs

5.1 Populations

Within the assumptions, and still uncertain parameters, the preliminary results of our BATSE survey are consistent with a galactic population of BH/LMXBs of $\gtrsim$ 300. Although obviously in need of confirmation with more data, this suggests that the number of BH/LMXBs in the Galaxy is larger than the number of NS/LMXBs that are detected as persistent sources since the total number of NS/LMXBs (see above discussion) is ~100 for the galactic disk. Clearly a correction for the number of NS/LMXB transients is required to compare the total number of LMXBs containing NSs vs. BHs. Here it is interesting that the number of historical NS/LMXB transients is significantly less (only ~2-3) than the number of BH/LMXB transients (now $\gtrsim$ 10). Note we define transients as systems which are usually in a quiescent state and not usually bright (although variable by factors of 3 perhaps) but which occasionally and only briefly fade to extremely low luminosity. The burster 4U1608-52, which was recently found by ASCA (Inoue 1995) to be in a very low luminosity state (with $L_x \sim 10^{33}$ erg/s), is such an example. A similar excess of BH/LMXBs would also be derived from the simple fact that while 1-2 new BH/LMXB transients have been discovered each year for the past 5 years when all-sky monitors with hard x-ray sensitivity have been available (first on GRANAT, with the WATCH experiment, and then BATSE on CGRO), no new systems like Cen X4 (a classical NS/LMXB transient – cf. Van Paradijs and McClintock 1995) have been found. Thus surely the BH systems outnumber the NS systems among the LMXB transients. Conversely, the number of persistent BH/LMXBs (defined as detectable at least half of the time) is much smaller: only 5 (cf. list given by White 1994). Clearly the combination of the more sensitive all-sky monitor (2-10 keV) on XTE (soon to be launched) with BATSE, as well as followup hard x-ray pointed observations, will allow better measurements of the apparent excess of BHs over NSs among the LMXB transients.

The galactic distributions of the BH/LMXBs appear to be much more confined to the disk (and thus essentially Pop I; cf. White 1994) than are the NS/LMXBs, which have long been thought to be a bulge or thick disk distribution. Indeed, Naylor and Podsiadlowski (1993) show that the field NS/LMXB are consistent with a thick disk with scale height of ~500 pc. Similar conclusions were reached by Eikenberry and Grindlay (1995), who attempted to differentiate between the burster (atoll source) and Z source distributions. It is now also clear that the NS/LMXBs must be further differntiated to distinguish the emeging class of LMXB pulsars, which appears to contain 5 systems with remarkably similar periods (5-7 sec), as described recently by Mereghetti and Stella (1995). These NS/LMXBs are clearly even more confined to the plane than the the total NS/LMXB distribution (to which they also contribute). Since 2 of these 5 also appear to be associated with supernova remnants, it seems clear they are young and fundamentally a Pop I distribution.

Thus several conclusions can be drawn for total populations of BH vs. NS LMXBs in the disk:

1. The BH/LMXBs are primarily transient systems, which suggests their companions are generally under-filling their Roche lobes. It is interesting that no ultra-compact binaries with degenerate secondaries (like the NS/LMXB and 11 min binary 4U1820-30) have been found among the BH/LMXBs; all known BH/LMXB periods are $\gtrsim 4.8$ hours.

2. The total number of BH/LMXBs is probably at least as large, and possibly significantly larger, than the NS/LMXB population (~100).

3. The BH/LMXB galactic distribution appears to be more of a Pop I distribution than does the NS/LMXB distribution (atoll and Z sources), which are a thick disk distribution. However, the NS/LMXB pulsars are also Pop I.

5.2 Formation Scenarios

The formation of LMXBs in the disk (i.e. not in globular clusters) remains a particularly challenging problem. With a birth rate of $\sim 10^{-7} - 10^{-6}$/year in the entire Galaxy (see discussion above), they are certainly rare (indeed, LMXBs have been described as the "freaks" of binary evolution !) but hardly can be dismissed as irrelevant. Webbink and Kalogera (1994) provide a recent review of constraints on formation scenarios. No satisfactory model yet exists for the formation of at least the NS/LMXBs in the disk. This is in sharp contrast to the formation of these systems in globular clusters, which is certainly driven by either tidal capture of isolated NSs or exchange collisions of NSs with primordial binaries (cf. Hut et al 1992 and references therein). Thus we have suggested (Grindlay 1984, 1988) the field LMXBs might have been formed in globular clusters which have since disrupted by tidal shocking on Giant Molecular Clouds (GMCs) in the galactic disk. A thick disk distribution would be expected since those globulars on initial orbits closest to the galactic plane would spend the most time encountering GMCs and be most readily disrupted. Their debris, including field LMXBs, would thus constitute the presently observed thick disk population in the Galaxy. This theory has yet to be worked out in detail, although cluster disruption on GMCs has been investigated in part by Chernoff and Weinberg (1990). Note also that this model would not be expected to account for the now apparently distinct class of NS/LMXBs that are accreting pulsars which, as noted above, appear to be of Pop I origin.

The formation of NS/LMXBs has been made all the more urgent by the discovery of the very much larger population of millisecond pulsars (MSPs) in both the galactic disk (and halo) and globular clusters. The "standard model" for MSP formation invokes a progenitor LMXB which has been spun up to millisecond periods by the accretion of $\sim 0.1 M_{\odot}$(cf. Bhattacharya and Van den Heuvel 1991 for references). This requires $\sim 10^7$ years if the LMXB is accreting at maximum rate (i.e. is a Z source; see discussion above) or $\sim 10^9$ years if it is a typical atoll source or burster. For either, the birth rate derived above is only $\sim 10^{-7} - -10^{-6}$/year, or insufficient to produce the number of MSPs observed. Kulkarni and Narayan (1988) originally recognized this probable birth rate discrepancy between MSPs and NS/LMXBs in the disk, with MSPs a factor of $\gtrsim$ 10 higher. Grindlay and Bailyn (1988) and Bailyn and Grindlay (1990) pointed out the same problem for the MSPs vs. LMXBs in globular clusters and suggested it could possibly be solved if MSPs were formed directly from the accretion induced collapse (AIC) of a massive white dwarf in a CV type system rather than an LMXB progenitor. Although early MSP surveys seemed to suggest a limited MSP population that, with shortened LMXB lifetimes, could be produced by LMXB progenitor, the most recent MSP surveys (cf. Bailes and Lorimer 1995) suggest a still larger MSP population comparable with the total pulsar population (i.e. $\sim 10^5$). In this case, the MSP vs. LMXB birth rate discrepancy is still a factor of 10 (though see Iben and Tutakov 1995 and Kulkarni 1995 for claims of possible consistency). Thus MSPs apparently must form more efficiently than LMXBs (perhaps by AIC), and LMXBs in the field (and perhaps even in globulars) may form from MSPs.

The need for an alternative to the standard model for production of MSPs from LMXBs is also suggested by the recent discoveries of low quiescent luminosities from recurrent transients such as Aql X-1 and Cen X-4 with ROSAT (Verbunt et al 1994) or from the usually bright (though variable, with high and low states) persistent source 4U1608-52 with ASCA (Inoue 1995). There is a potentially fatal problem for producing MSPs from the bulk of the LMXB population such as might be represented by these classes (NS/LMXB transients and bursters in very low states): for the systems to accrete at the very low rates inferred in quiescence ($\sim 10^{-13} M_\odot$/year $\sim 10^{13}$ g/sec), they must be either very low magnetic field NSs or slow rotators since otherwise the accretion cannot penetrate the centrifugal barrier at the Alfven radius. Inoue (1995) writes the usual equilibrium spin period condition with appropriate scaling as

$$P \gtrsim 63 \text{ msec } (B_8 \cdot \dot{M}_{13}^{-3/7} \cdot (M/M_\odot)^{-5/7} \cdot R_6^{18/7}$$

This suggests that quiescent transients could not be spun up to the ~5 msec periods found in MSPs and thus cannot be the hidden reservoir sometimes invoked to solve the LMXB vs. MSP birthrate discrepancy. Thus if AIC is needed to make the MSPs in low mass systems, then it may also be responsible for making the LMXBs themselves since accretion can resume onto the MSP as the orbit of the secondary decays (by gravitational radiation) so that the secondary can once again fill its Roche lobe.

Surprisingly, the BH/LMXBs may be easier to understand. Romani (1992) has outlined an evolutionary scheme where they could evolve via a common envelope phase from massive HMXBs (probably containing a Wolf Rayet star). The fact that the distribution of BH/LMXBs appears to be more Pop I and less confined to the galactic bulge, like the NS/LMXBs, supports this picture.

The LMXB pulsars (NS-PSR/LMXBs), which also appear to be Pop I, may therefore be related to the BH/LMXBs. [Certainly the fact that these systems now have such low mass companions suggests this companion could not have driven the high mass transfer rates needed to produce them by AIC, although AIC for these was also suggested by Mereghetti and Stella (1995). Furthermore, their apparent association with SNR as pointed out by these authors would itself point to their NSs being formed in the SN event, and not by AIC]. The difference would be the pre-supernova (or pre-collapse) stellar mass: the LMXBs would arise from less massive progenitors (perhaps 15 $M_\odot$) which collapse to NSs through a supernova phase, whereas the BH/LMXBs arise from wider systems with larger (perhaps 30 $M_\odot$) primary mass. Collapse to the BH/LMXB would thus occur sooner, leaving the low mass (~1 $M_\odot$) secondary at larger radius and longer orbital period than for the NS-PSR/LMXBs. The pulsars, in contrast, would undergo a longer pre-SN common-envelope phase and thereby have secondaries considerably more stripped in mass when the NS was formed.

6 Globular Cluster X-ray Sources

We have provided numerous reviews of the key problems of the nature and origin of the XRBs in globular clusters. The most recent of these summarize the bright vs. dim sources in globulars (Grindlay 1994a), the dim sources in detail and their probable identification with CVs (Grindlay 1994b) and the relationships between the bright and dim sources and MSPs in globulars (Grindlay 1995). In the following two sections we briefly summarize some of the main points, adding some recent observational and theoretical developments.

6.1 LMXBs and MSPs

The study of low mass x-ray binaries and millisecond pulsars in globular clusters continues to be an endlessly fascinating problem. Since our last review (Grindlay 1995), at least two notable developments have occurred: the ROSAT survey results for clusters sources have been reported by Verbunt et al (1995) and the MSP population in 47 Tuc has been described in detail by Robinson et al (1995). Both results offer tantalizing astrometry for source positions within globulars, with the promise that more precise positions can provide for source and cluster evolution models.

The ROSAT all sky survey measured fluxes and thus luminosities for 10 of the 12 cluster LMXBs (cf. Grindlay 1994a for complete list); only the presumed transient (see below) in NGC 6440 and the Rapid Burster in Liller 1 were not detected. A significant conclusion drawn by the authors is that some 4 of the 12 cluster LMXBs may be regarded as transients in the sense that their luminosities have been observed at some time to be at least a factor of 10 below normal quiescent values. This changes somewhat the statistics of persistent vs. transient sources in globulars, which had been set by the previous recognition that only 1-2 of the 12 were transients (the sources in NGC 6440 and Liller 1). In addition, source positions accurate to typical PSPC values (~10 arcsec) were presented, so that one is now tempted to derive radial offsets for the 3 new LMXBs as discussed by Grindlay (1995). However this is still premature (and thus not presented here) since the positions are not accurate enough to better constrain the radial offset distribution and thus either source mass or (more likely) cluster internal dynamics. This will be the subject of forthcoming HRI proposals for ROSAT.

At long last, the timing positions for the MSPs in the globular cluster 47 Tuc have started to appear. Although Manchester et al (1991) originally shocked the MSP (and LMXB !) community with the discovery of the then (and still) record number of 11 MSPs in a single cluster, 47 Tuc, timing positions for the individual sources were not possible to measure initially. They have now done this for the first two (brightest) of the 47 Tuc cluster zoo: 47 Tuc C and D. The surprising result is that these first two are not located in the core but rather at radial offsets of 1.25 and 0.62 arcmin, respectively. These offsets are to be compared to the nominal core radius of 47 Tuc of 23 arcsec or the (more likely) HST core radius of only 8 arcsec recently reported by Calzetti et al 1993) and indicative of 47 Tuc having undergone (or now undergoing) core collapse. Thus these first two MSPs are at radial offsets of $\gtrsim$ 3-9 and 1.5-4.5 core radii (depending on which core radius value is used) and are therefore similar to the large radial offsets found for some MSPs in M15 (e.g. M15 D). Note, as with all such projected radial offset values, that the true radial offset is of course larger than this value projected onto the plane of the sky. In the case of these first two 47 Tuc MSPs, both of which have negative P-dot values, they are therefore on the back side of the cluster and are being accelerated back into the core. The obvious interpretation is as with the M15 D system: these 47 Tuc MSPs have also been ejected from the core. The fact that both of these are not (obviously) still in binary systems (unlike M15 D) makes them easier to eject. However it also implies a very high central density and/or central peak in the cluster binary distribution (since binaries are even more effective scattering targets). Robinson et al (1995) use the negative P-dot values to derive a minimum central average density of 5 $\times 10^4 M_\odot/pc^3$; the true value is likely to be much higher still to account for the ejected fraction.

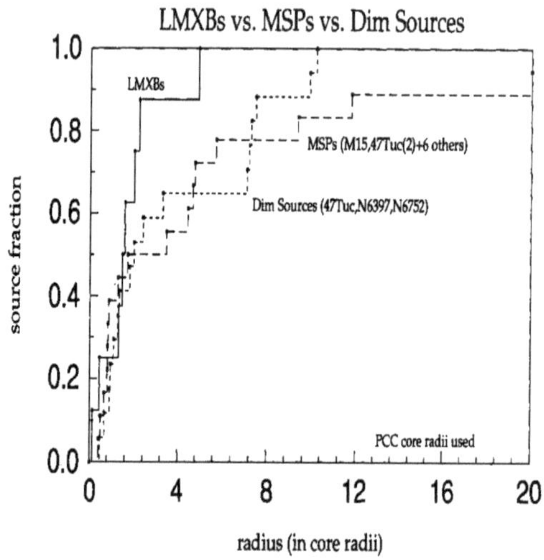

Figure 2: Radial distributions of XRBs vs. Dim sources vs. MSPs in globulars

These new results, limited as they are, would seem to further support our suggestion (Grindlay 1995) that the MSPs and CVs (or, more exactly, the dim x-ray sources) in globulars are more extended in their radial distribution than are the LMXBs. In Figure 2 we show the cumulative distributions for the radial offsets of the three types of compact objects in globular clusters as in Grindlay (1995) but now updated with the two new MSP positions within 47 Tuc.

We shall return to this question of ejection of MSPs, and possibly also LMXBs and CVs, in the Summary discussion below.

6.2 CVs and Transients

The nature of the dim sources in globulars, or those with soft x-ray luminosities in the range $\sim 10^{31-33}$ erg/s which were originally discovered in the Einstein globular cluster survey (Hertz and Grindlay 1983; hereafter HG), remains controversial. HG suggested that the apparent gap in the luminosity distribution of the cluster sources, with the dim sources $\sim 10^3$ times fainter than the bright LMXBs, was most naturally explained if most of the dim sources were CVs. Recent models for CV production in globulars (DeStefano and Rappaport 1993) have also borne out the early prediction of HG for a significant population of CVs in globulars. The limited sensitivity of the short Einstein observations limited the source detections to luminosities $\gtrsim 10^{32.5}$ erg/s, which motivated Verbunt et al (1984) to conclude the sources were too luminous to be CVs but rather were LMXBs in quiescence (as also pointed out by HG for at least the dim source in NGC 6440). The problem of CVs vs. quiescent LMXBs is important for at least two other reasons: i) where *are* the cluster CVs (if not in the dim sources), since they ought to form (for sufficiently massive WDs) as easily as the cluster LMXBs yet be enhanced by the far greatter numbers of WDs (even allowing for the competing effects of cluster mass segregation—cf. also DeStefano and Rappaport (1993)) and ii) the relative populations of each are important for understanding the formation and evolution of the cluster MSPs, as discussed above .

With our more sensitive (much longer exposure) and higher resolution (HRI vs. the IPC on Einstein or the PSPC on ROSAT, as used by Verbunt et al (1995)) studies of several globulars, we have argued that the dim sources are dominated by CVs (Grindlay 1994a,b). The supporting evidence for our HST discovery of CV candidates in the error circles of three (of the 5) dim sources we discovered with ROSAT (Cool et al 1993) has now been published (Cool et al 1995). The followup spectra with the FOS on HST are scheduled to be obtained in the current (Cycle 4) phase. Thus the debate over whether these objects (at least) are CVs will soon be possible to test with optical spectra. These spectra should also test our suggestion (Grindlay 1994b, 1995) that the cluster CVs and dim sources may be dominated by magnetic CVs since that would be more consistent with the probable L_x/L_{opt} values for the dim sources/CV candidates in NGC 6397 (Cool et al 1995) . A DQ-Her type CV as the counterpart for the brightest of the dim sources in 47 Tuc has also been possibly found with HST by Paresce et al (1992) (though FOS spectra are sorely needed). And finally, the lack of dwarf novae (DN) found by Shara et al (1995) in the globular NGC 6752 (and also the discovery of DeMarchi and Paresce (1994) of only one DN in 47 Tuc) points to the CVs, if present, being dominated by magnetic systems since these do not show DN outbursts. Thus the search for CVs in globulars may well constrain the origin of magnetic WDs which could be enhanced in Pop II systems.

However, as pointed out in Grindlay (1995), the spectroscopic difference between a CV and quiescent LMXB may not be that dramatic in the visible region (the FOS spectra will cover the wavelength range from approximately Hβ to Hα) at least. The quiescent LMXB, with apparent luminosity (e.g. Verbunt et al 1984) $\sim 10^{32}$ erg/s in the ROSAT band, will then have $\dot{M} \sim 10^{12}$ g/sec unless there are large bolometric corrections for luminosity not measured in the ROSAT band. Thus it is deep within the low $\dot{M}$ regime of advection-dominated accretion (cf. Narayan and Yi 1994, 1995), and its accretion disk will be truncated at $\sim$3000 Schwarzschild radii or $\sim 10^9$ cm. Interior to this, the accretion is expected to be quasi-spherical and through a hot corona (which may give rise to a significant hard continuum, extending out to perhaps 100 keV, as seen for bursters like 4U1608-52 in their "low state" of quiescent (not transient) emission - cf.

Barret and Grindlay 1995), as discussed below). The optical spectrum (though not yet calculated in detail) might then look like that from a truncated disk and thus be deficient in uv continuum for the quiescent LMXB case (as in fact anticipated by Grindlay 1995). The continuum and emission lines in the visible, which arise from farther out in the (still thin) disk, may be rather similar, however, to that from a CV. Clearly better spectra for the known NS/LMXB transients (Aql X-1 and Cen X-4), as well as additional examples, are very much needed.

The x-ray spectrum expected for a quiescent transient (NS) is, to first order, the very soft (kT $\sim$ 0.3 keV) blackbody emission expected for the thermalization of the accretion flow on the NS surface. Indeed this seems consistent with the ROSAT (Verbunt et al 1994) and ASCA (Inoue 1995) spectra of several quiescent field LXMBs as mentioned above. However, current (in progress) calculations (Narayan, private communication) suggest that some fraction of the luminosity may indeed emerge in the hard continuum (contributing to, or providing energy input for, the possible hard coronal emission mentioned above) since the very low $\dot{M}$ flows through the corona may in fact be settling solution flows. In this case, our claim (Grindlay 1995) that CVs could be distinguished from quiescent LMXBs by their (typically) harder x-ray spectra in the ROSAT band may be weakened.

7 Halo CVs and the Cosmic X-ray Background ?

The origin of the cosmic x-ray background (CXB) is still not completely understood more than 30 years after its discovery. Although it is now clear that it must be largely due to the integrated emission from active galactic nuclei (see, for example, Fabian and Barcons 1992), the lack of clustering strongly suggests that in fact no more than about 70% of the CXB can arise from known classes of AGNs. A new population of weak sources may be required. Thus it was extremely interesting when in an epochal study of the CXB and source counts with the Deep Survey observations undertaken with ROSAT, Hasinger et al (1993; hereafter H93) concluded that in fact about 60% of the CXB at fluxes $\geq 2.5\times10^{-15}$ cgs (hereafter cgs = erg cm^{-2} s^{-1}) in the 0.5-2 keV band could indeed be resolved into discrete sources but that the source counts are inconsistent with models for the x-ray luminosity functions (XLF) of AGNs (Boyle et al 1993).

7.1 Constraints on Halo/Thick Disk Sources

We have suggested (Maoz and Grindlay 1995, hereafter MG and Grindlay and Maoz 1995) that a *galactic* population of faint x-ray sources in the halo or thick disk could be the new source population required and could satisfy all the new constraints from ROSAT as well as the earlier CXB flux and fluctuation constraints. We have considered whether either a spherical halo ($\propto 1/R^2$) or a thick disk (exponential in both R and Z) distribution of discrete sources of fixed (standard candle) luminosity L and spatial density n_0 (in the solar neighborhood) can satisfy all the constraints imposed by the source counts in the 27 deep fields surveyed by H93 as well as the fluctuation and flux constraints imposed by the CXB. Our models are normalized to a possible galactic contribution to the source counts of 120 deg^{-2}, at the flux limits of H93, which allows for the measured contributions of stars and all known classes of extragalactic sources (galaxies, clusters, BL Lacs and QSOs with the XLF evolution of Boyle et al 1993). Details of the calculation are given in MG and our results are shown below in Figure 3, which presents all the constraints on the allowed values of L and n_0 for two spherical halo models with small and large core radii, and for two thick disk models with modest and large scale heights. For each model, the solid line represents the combinations of L and n_0 that would exactly resolve the discrepancy found in the faint source counts; the shaded area corresponds to combinations which would produce a number density of unidentified bright sources which is consistent with observations; the two inclined dashed lines confine the parameter space which would not violate constraints on the contribution to the CXB intensity; the horizontal dash-dot lines confine the region which would be consistent with the results of fluctuation analyses of the unresolved background, and the region below the dotted line corresponds to models which contribute $\leq 10^{40}$ erg/s to the Galaxy's total x-ray luminosity.

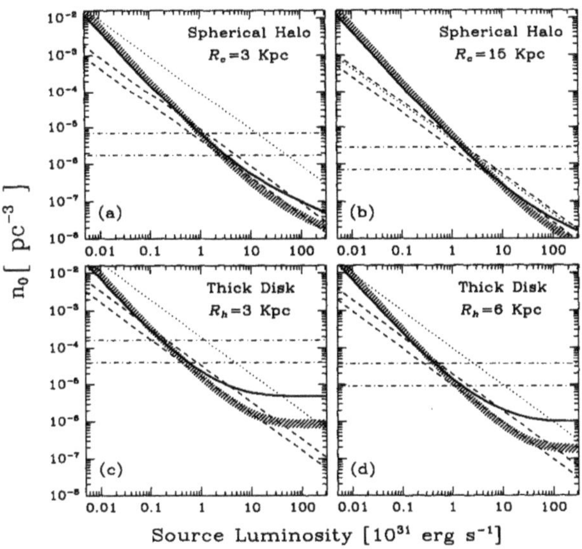

Figure 3: Combined observational constraints (see text) on the typical x-ray luminosity of the sources, L, and their number density in the Solar vicinity, n_0, for two different halo models (a,b) and thick disk models (c,d). (From Maoz and Grindlay 1995).

The striking result is that for either halo or thick disk models a relatively well defined region of parameter space is defined by all the constraints. In particular, the typical luminosity of the sources is $10^{31\pm0.5}$ erg/s (0.5-2 keV), and is not very sensitive to changes in the scale of the spatial distribution. The total number of sources required is about $10^{8.5}$. Thus although Kashyap *et al.* (1994) have recently suggested that MACHOs may have x-ray coronae with typical luminosities of order $\sim 10^{27}$ erg/s, and may produce a small fraction ($\sim$ 10%) of the soft CXB *flux*, their luminosity is far too low to satisfy the constraints imposed by number counts and the excess of faint sources detected by H93.

All our models have an approximately Euclidean logN-logS relation at fluxes above $\sim 10^{-14}$ cgs, and which flattens considerably below a flux of a few times 10^{-16} cgs. In the flux range of 10^{-15}-10^{-14} cgs, the logN-logS relation is steeper than that expected for an evolving population of QSOs (Boyle *et al.* 1993), thus providing an increasing relative contribution of the Galactic sources with decreasing flux.

We have also calculated the constraints imposed by the spectrum and anisotropy (limits) of the CXB on our proposed source population. H93 found that the average spectrum of the resolved sources becomes harder with decreasing flux. Since the extragalactic sources have a steeper average spectrum in the 0.5 - 2 keV band, we conclude that the average spectral index of the proposed population of sources must be < 1 in this band. Allowing for the Galactic emission components of the CXB which already indicated thin *and* thick disk components at higher flux levels, we conclude that the source population proposed here most probably has a flattened distribution and thus should have a significant contribution to the detected anisotropy. The fractional contribution of the proposed population to the CXB in the 2-60 keV band, averaged over the entire sky, is $\sim 8\%$ but can be 20-30% near the galactic center depending on the model. Anisotropy constraints would allow a slightly higher source temperature in more spherical halo models than in flattened thick disk models.

7.2 Possible Identification of the Sources

Our primary requirements are sources with typical $L_x \sim 10^{31}$ erg/s and a relatively hard spectrum. These are not consistent with dM stars (which are only this bright during brief flares, and are typically soft), with LMXBs in quiescence (the ROSAT studies of Verbunt et al 1994 have shown these sources to be typically very soft, with kT $\sim$ 0.3 keV) or with isolated neutron stars (e.g. old pulsars, for which thermal emission is also expected to be very soft - cf. Ögelman et al 1993). However pulsars also appear to have harder emission components (Ögelman 1994), but with much lower L_x than we require. Magnetized neutron stars (NSs) accreting from the ISM could produce hard cyclotron emission (cf. Nelson et al 1995) with $L_x \sim 0.05 L_{acc} \sim 3 \times 10^{25} n_{-3}/v_{100}^3$ erg/s for a typical pulsar at 100 km/s moving through a halo interstellar medium of density $n_H \sim 10^{-3}$ cm^{-3}, which is again far too low. Thus although (old) pulsars would nicely satisfy the requirement for some $10^{8.5}$ total sources in a thick disk or halo, it seems they could not satisfy the luminosity constraint.

Similarly, we now consider $\sim 10^8$ galactic black holes (BHs) with mass $\lesssim 10^4 M_\odot$ (so as not to exceed the halo mass) and moving on halo orbits with v $\sim$ 200 km/s. Note these are probably not the $\sim 10^6 M_\odot$ halo BHs invoked by Lacey and Ostriker (1985) to stabilize the disk although it is conceivable that our lower mass could be accomodated with changes in the primordial fluctuation spectrum (Ostriker, private communication). To obtain the required L_x by accretion from the halo ISM would imply a radiation efficiency $\epsilon \sim 10^{-3}$ (v_{200}^3/n_{-3}). However at the low Bondi accretion rates implied ($\sim 10^{-9} n_{-3}/v_{200}^3$ of Eddington), the recent results of Narayan and Yi (1994) for the effective accretion efficiencies onto BHs when advection is included would suggest much lower efficiencies (typically $\lesssim 10^{-5}$, but dependent on the assumed viscosity parameter α). Thus BHs are unlikely but perhaps cannot be totally excluded.

Cataclysmic variables (hereafter CVs) can most naturally fit our requirements for luminosity and local space density. Hertz *et al.* (1990), from optical identifications of the x-ray selected Einstein galactic survey (Hertz and Grindlay 1984), find a local space density for low $\dot{M}$ CVs which is consistent with our prediction. Howell and Szkody (1990) discuss the halo CV population and report that their significantly lower mean orbital period implies that these systems are relatively old. As discussed above, magnetic CVs (e.g. AM-Her type) are dominated by very short period systems (i.e. they are predominantly below (and within) the period gap at 2-3 hours). This is probably due to effects of the WD magnetic field both locking the secondary into synchronous rotation and enhancing $\dot{M}$ for those systems in the period gap where normally magnetic braking is thought to have ceased (for such low mass secondaries) to be able to drive the mass transfer (Chanmugam 1995). Since the AM Her types are also optically fainter than nonmagnetic CVs (larger L_x/L_v ratio) because they do not have well developed accretion disks, it is possible that our proposed population of x-ray sources are some kind of magnetic CVs. Their low $\dot{M}$ could push their characteristically soft x-ray spectral components into the XUV band leaving just their hard components.

As discussed above, globular clusters contain dim sources with typical soft x-ray luminosity of $\sim 10^{31.5-32.5}$ erg/s in the ROSAT band, and the bulk of these sources are fully consistent with being CVs (Grindlay 1994a, and references therein). X-ray and optical observations indicate a presence of $\gtrsim 10$ CVs per cluster (at least for (near-) core collapsed globulars) with $L_x \sim 10^{30.5-31.5}$ erg/s. These may be optically fainter than disk CVs for a given x-ray luminosity (Grindlay 1994a), deficient in dwarf novae, and contain a significant fraction of magnetic CVs (Grindlay 1994a,b). Although magnetic Cvs have been predicted to be in globulars (cf. Chanmugam, Ray and Singh 1991), they have not yet been directly observed optically. Our forthcoming HST/FOS spectra of the Hα objects in the cluster NGC 6397 may provide the first tests.

Thus the thick disk or halo CV population proposed by MG resembles the globular cluster CV population, suggesting they may have been ejected from globular clusters. Indeed, Grindlay (1995) shows that the dim sources in globulars appear to have an extended radial distribution which extends to radial offsets well beyond what purely relaxed objects (CVs) should obtain and may thus imply ejection. Although this process may contribute to the proposed Galactic population, it obviously cannot provide the required number of $\sim 10^{8.5}$ CVs in the halo from the present ~150 globular clusters. Thus the required halo CVs, if from globulars, must imply a population of some 10^{6-7} primordial globulars which have been disrupted and which had a flatter radial distribution than the $\sim 1/R^3$ distribution of the survivors. In particular, a thick disk distribution of CVs (and Pop II stars) may form from globular clusters on low-inclination orbits which have been disrupted by encounters with giant molecular clouds, as suggested by Grindlay (1984, 1988) for the origin of the field LMXBs.

7.3 Tests for CV/CXB Model and Halo CVs

The CXB may partially originate in a large population of intrinsically faint (and presumably old) CVs which are distributed within a thick Galactic disk (or an oblate halo) with a scale height of a few kpc. The inferred $\sim 3 \times 10^8$ CVs would have luminosity of $\approx 10^{30-31}$ erg/s and a number density of $\sim 10^{-5}$-10^{-4} pc^{-3} sources in the Solar vicinity. Our model can be tested by searching for a large-scale variation in the surface *number density* of resolved x-ray sources. We predict that the density of sources with flux below 10^{-14} cgs will decrease either with an increasing galactic latitude ($|b|$) or with an increasing angular distance to the Galactic center (θ). The data presented by H93 are insufficient for performing this test since they do not represent a broad range in either b or θ and the number of sources is limited. Deeper HRI surveys (for more precise source positions) at several θ and $|b|$ values are needed.

The most direct test of our extended halo or thick disk CV population and contribution to the CXB is to optically identify the predicted ~20% of the faintest ROSAT survey sources. Especially if these are magnetic CVs (e.g. AM-Her or DQ-Her CVs), they should appear almost as "pure" emission line sources and could thus be found by very deep images in broad band (R) vs. Hα. We emphasize that *all* faint x-ray sources in the ROSAT fields should be examined, including ones already "identified" as AGNs since some fraction of these may have been misidentified given an average AGN angular separation of only $\sim 4'$ at the typical fluxes of interest.

8 Summary: XRBs and Constraints on Evolution of the Galaxy

We conclude with a brief summary of the populations of x-ray binaries and how they can be used to constrain the evolution of the Galaxy. The CVs in the disk are likely to arise from the evolution of primordial binaries in which the more massive member has evolved to a giant and the binary has undergone common envelope evolution to leave the WD and lower mass secondary. (cf. Ritter and Kolb 1995 for references). In contrast, the nearly comparable density of CVs in the halo or thick disk that may account for part of the CXB would be unlikely to have the same origin since the density of halo binaries is less by at least the factor of halo/disk stellar density ratio. If these CVs are instead formed in globular clusters (by tidal capture, as well as from primordial cluster binaries), then a significant population ($\sim 10^{6-7}$) of clusters must have been disrupted or have evaporated. Since the surviving globular cluster distribution is steeply concentrated ttoward the galactic center (with radial distribution $\propto 1/R^3$), the primordial clusters are most likely to have been a disk distribution. Because the number of LMXBs in the thick disk is lower than the inferred CV population by a factor of at least 10^4, this primordial globular distribution must have been composed of stars with high mass cuttoff $\lesssim 6\ M_{\odot}$so that NSs were not formed.

In contrast, the disk LMXBs could have formed from a much smaller number ($\sim 10^4$) globulars containing higher mass stars. The simplest interpretation is that the primordial globulars contained stars with a single IMF with mass cutoff at about 4-6 $M_{\odot}$. Only those (very few) stars with masses above $\sim 8 M_{\odot}$, or a minimum mass for direct production of NSs by Type II SNe, would produce NSs that could be captured to form cluster LMXBs that in turn could produce the survivving disk LMXBs when these clusters were disrupted by encounters with the galactic disk. If the primordial cluster IMF mass cutoff as well as slope were themselves a function of galactic radius, with higher cutoffs at smaller radii, then a qualitative understanding of the thick disk may be as follows: the $\sim 10^{4-5}$ most massive globulars with highest mass stars were disrupted by encounters with the central bar in the galactic bulge. This produced (with perhaps 1-10% efficiency) the distribution of disk LMXBs presently observed. Their derived average birthrate B $\sim 10^{-6}$/year (see above) would then governed by the rates of binary and stellar evolution which govern their possible mass transfer rate, $\dot{M}$. Since the derived birthrates of atoll and Z sources are very comparable, yet the Z sources are limited to shorter lifetimes, the rate-limiting step must be the decay of binary orbits into Roche-lobe contact. This would imply that "successful" tidal captures in the primordial globulars (i.e. those not consumed to a common envelope merger by tidal heating and expansion of the captured secondary) are at the largest radii (e.g. the outermost annulus for possible capture) such that the secondary does not fill its Roche lobe and initiate mass transfer immediately after the capture. Such a "delayed mass transfer" was in fact invoked by Bailyn and Grindlay (1987) for the formation of the system leading to the 11-min binary 4U1820-30 in the globular NGC 6624.

Alternatively, and perhaps more consistent with the formation of MSPs, the cluster LMXBs and thus the disk LMXBs (from the disrupted cluster population) could arise from the high-$\dot{M}$ accretion onto a massive WD ($\gtrsim$ 1.1 $M_{\odot}$, as formed from primordial 3-4 $M_{\odot}$stars in the cluster) which can undergo AIC into a NS if the rate is $\dot{M} \gtrsim 3 \times 10^{-8}$ $M_{\odot}$/year (cf. Nomoto and Kondo 1991). In this case, the requirement for an extension of the primordial cluster IMF above the 6-8$M_{\odot}$limit likely to be needed for direct production (via SN-II) of NSs is eased; all primordial globulars could be limited to just low mass stars below this limit. This is attractive since then the cluster (and eventually the disk) LMXBs form from cluster MSPs (which also are scattered into the disk) that have been re-cycled via a second tidal capture. This delays the LMXB phase as well as reduces (perhaps greatly) the efficiency of making LMXBs (two tidal captures are now required), so that perhaps just one population of primordial globulars can now account for both the halo or thick disk LMXBs as well as CVs. Obviously this model needs to be worked out in detail.

We note that the HMXBs and their progeny (e.g. the BH/LMXBs ?) raise interesting questions for galactic evolution. The fact that no accretion powered pulsars are found in the globular clusters today is consistent with our suggestion above that these objects among the LMXBs in the disk (e.g. 4U1626-67, as discussed above) formed from high mass stars and that such high mass stars were lacking in globulars. Similarly, the lack of BHCs in globulars (cf. Grindlay 1993, Kulkarni, Hut and McMillan 1993) shows that the cluster IMF probably was indeed limited to cuttoffs not much above that needed for NS (but not BH) collapse and that stellar mass ($\lesssim$ 10 $M_{\odot}$) BHs are thus not expected in the halo or thick disk. Rather, the BH/XRBs (including the BH/LMBs) are probably Pop I (as essentially suggested by Romani 1992) or thin disk systems.

Finally, we comment briefly on the likely escape of CVs, dim sources and probably also LMXBs from globulars into the halo (or thick disk), as suggested by Figure 2 above. Although the case is (like so much of that discussed here) not proven, it appears likely from the large tail of the radial offset distribution that all three classes of compact binary (or binary products) are escaping from globulars. Clearly the escape alone could not populate the thick disk or halo with enough CVs or MSPs to account for the entire galactic population (inferred to be $\gtrsim 10^{8.5}$ and 10^5, respectively, as discussed above) from the present population of globulars. There are simply not enough globulars, and thus the disrupted primordial distribution is invoked. However, the binary heating of globulars, which causes core bounce from core collapse (cf. Hut et al 1992), is probably central to cluster disruption when combined with the galactic (bar or bulge) or discrete GMC tidal field. Thus if the large primordial cluster distribution invoked here evaporated many of its NSs and WDs (and associated compact binaries) on the way to its dissolution and total disruption, an extended halo would result. That is, the evaporating stars (and binaries), with by definition random velocities of $\gtrsim$ 30 km/sec (or typical cluster escape velocity) superimposed on the cluster velocity, will have diffused out to galactocentric radii of ~300 kpc in the $\sim 10^{10}$ years since the formation of the Galaxy and the primordial cluster distribution. Such a large radius halo of NSs (and WDs) is just that invoked for current extended halo models for gamma-ray bursts (GRBs), as reviewed by Lamb (1995). Thus the otherwise (and still !) challenging problem of how to populate the extended halo required for galactic models of GRBs with NSs may also be solved by the general considerations of the nature and origin of XRBs as summarized in this Review.

This work was supported in part by NASA, grants NAGW-3280, NAG5-1624 and NAG5-2763. I thank D. Barret and R. Narayan for discussions and the organizers of this Summer School for their support.

9 References

Alpar, A. and Shaham, J. 1985, *Nature*, **316**, 239.

Bailes, M. And Lorimer, D. 1995, in *Millisecond Pulsars: A Decade of Surprise*, (A. Fruchter, D. Backer and M. Tavani, eds.), ASP Conf. Series, vol. 72, 17.

Bailyn, C. and Grindlay , J. 1987, *Ap.J. Letters*, **316**, L25.

Bailyn, C. and Grindlay, J. 1990, *Ap. J.*, **353**, 159.

Barret, D. and Vedrenne, G. 1994, *Ap.J. Suppl.*, **92**, 505.

Barret, D. and Grindlay, J. 1995, in preparation.

Bhattacharya, D. and Van den Heuvel, E. 1991, *Physics Reports*, **203**, Nos. 1 and 2, 1

Boyle, B.J., Griffiths, R.E., Shanks, T., Stewart, G.C., & Georgantopoulos, I. 1993, *M.N.R.A.S.*, **260**, 49

Calzetti, D., De Marchi, G., Paresce, F., and Shara, M. 1993, *Ap.J. Letters*, **402**, L1.

Chanmugam, G., Ray, A. and Singh, K. 1991, *Ap. J.*, **375**, 600.

Chanmugam, G. 1992, *Ann. Rev. Astron. Ap.*, **30**, 143

Chanmugam, G. 1995, in *Proc. CV Workshop, Padua*, in press.

Chernoff, D. And Weinberg, D. 1990, *Ap. J.*, **351**, 121.

Cool, A., Grindlay, J., Krockenberger, M. and Bailyn, C. 1993, *Ap.J. Letters*, **410**, L103.

Cool, A., Grindlay, J., Cohn, H., Lugger, P., and Slavin, S. 1995, *Ap. J.*, **439**, 695.

Cropper, M. 1990, *Space Sci. Rev.*, **54**, 195

Di Stefano, R. and Rappaport, S. 1993 *Ap. J.*, **423**, 274

Ebisuzaki, T., Hanawa, T. and Sugimoto, D. 1984, *PASJ*, **36**, 551.

Eikenberry, S. and Grindlay , J. 1995, in preparation.

Fabian, A. and Barcons, X. 1992, *Ann. Rev. Astron. Ap.*, **30**, 429.

Fruchter, A., Backer, D. And Tavani, M. 1995, *Millisecond Pulsars: A Decade of Surprise*, (A. Fruchter, D. Backer and M. Tavani, eds.), ASP Conf. Series, vol. 72.

Giacconi, R. et al 1979, *Ap. J.*, **234**, L1

Grindlay, J.E., 1984, it Adv. Space.Res., **3**, 19

Grindlay, J. 1988, in *Globular Cluster Systems in Galaxies, Proc. IAU Symp. 126*, (J. Grindlay and A. G. D. Philip, eds.), Dordrecht:Reidel, p. 347

Grindlay, J. 1993, in *The Globular Cluster–Galaxy Connection*, (G. Smith and J. Brodie, eds.), ASP Conf. Series, Vol. 48, p. 156.

Grindlay, J.E. 1994a, in *Evolutionary Links in the Zoo of Interactive Binaries*, (eds. F. D'Antona, V. Caloi, C. Maceroni and F. Giovanelli), *Mem. della Soc. Astron. Ital.*, vol. 65, no. 1, 259.

Grindlay, J.E. 1994b, in *Evolution of X-Ray Binaries*, (S. Holt & C. Day, eds.), AIP Conf.Proc., 308, 339

Grindlay, J.E. 1995, in *Millisecond Pulsars: A Decade of Surprise*, (A. Fruchter, D. Backer and M. Tavani, eds.), ASP Conf. Series, vol. 72, 57.
Grindlay, J. and Bailyn, C. 1988, *Nature*, **336**, 48
Grindlay, J. and Maoz, E. 1995, in *Proc. IAU Symp. 168*, in press.
Grindlay, J.E. et al 1995, in *Proc. SPIE*, vol. 2518, in press.
Hasinger, G. and Van der Klis, M. 1989 *Astr. Ap.*, **225**, 79.
Hasinger, G., Burg, R., Giacconi, R., Hartner, G., Schmidt, M., Trümper , J., & Zamorani G. 1993, *Astr. Ap.*, **275**, 1 (H93)
Hertz, P., and Grindlay, J.E. 1984, *Ap. J.*, **275**, 105
Hertz, P., and Grindlay, J.E. 1983, *Ap. J.*, **278**, 137
Hertz, P., Bailyn, C.D., Grindlay, J.E., Garcia, M.R., Cohn, H., & Lugger, P.M. 1990, *Ap. J.*, **364**, 251
Howell, S.B., & Szkody, P. 1990, *Ap. J.*, **356**, 623
Hut, P. et al 1992, *PASP*, **104**, 981.
Iben, I. and Tutakov, A. 1995, in *Millisecond Pulsars: A Decade of Surprise*, (A. Fruchter, D. Backer and M. Tavani, eds.), ASP Conf. Series, vol. 72, 119.
Inoue, H. 1995, preprint.
Kashyap, V., Rosner, R., Schramm, D. & Truran, J. 1994, submitted to *Ap. J.*,
Kulkarni, S. 1995, in *Millisecond Pulsars: A Decade of Surprise*, (A. Fruchter, D. Backer and M. Tavani, eds.), ASP Conf. Series, vol. 72, 79.
Kulkarni, S. and Narayan, R. 1988, *Ap. J.*, **335**, 755.
Kulkarni, S., Hut, P., and McMillan, S. 1993, *Nature*, **364**, 421.
Kraft, R. 1962, *Ap. J.*, **135**, 408
Lacey, C.G. and Ostriker, J.P. 1985, *Ap. J.*, **299**, 633
Lamb, D.Q. 1995, *PASP*, in press.
Manchester, R. 1995, in *Millisecond Pulsars: A Decade of Surprise*, (A. Fruchter, D. Backer and M. Tavani, eds.), ASP Conf. Series, vol. 72, 3.
Manchester, R., Lyne, A., Robinson, C., D Amico, N., Bailes, M., and Lim, J. 1991, *Nature*, **352**, 219.
Maoz, E. and Grindlay, J.E. 1995, *Ap. J.*, in press (May 1, 1995)
Mereghetti, S. and Stella, L. 1995, preprint.
Michel, F.C. 1987, *Nature*, **329**, 310.
Narayan, R., Piran, T. and Shemi, A. 1991, *Ap.J. Letters*, 379, L17.
Narayan, R. and Yi, I. 1994, *Ap.J. Letters*, **428**, L13.
Narayan, R. and Yi, I. 1995, *Ap. J.*, **444**, 231.
Naylor, T. and Podsiadlowski, Ph. 1993, *M.N.R.A.S.*, **262**, 929.
Nelson, R. et al 1995, *Ap. J.*, in press.
Nomoto, K. and Kondo, Y. 1991, *Ap.J. Letters*, **367**, L19.
Ögelman, H. et al 1993, *Nature*, **361**, 331.
Paresce, F. and De Marchi, G. 1994, *Ap.J. Letters*, **427** L33
Paresce, F. et al 1992, *Nature*, **360**, 46.
Parsignault, D. and Grindlay, J. 1978, *Ap. J.*, **225**, 970.
Patterson, J. 1984, *Ap. J. Suppl.***54**, 443
Patterson, J. and Raymond, J. 1985, *Ap. J.*, **292**, 535

Ritter, H. and Kolb, U. 1995, in *X-ray Binaries*, (W. Lewin, J. Van Paradijs, and E. Van den Heuvel, eds.), Cambridge Univ. Press, in press.

Robinson, C., Lyne, A., Manchester, R., Bailes, M., D'Amico, N., and Johnston, S. 1995, *M.N.R.A.S.*, in press.

Romani, R. 1992, *Ap. J.*, **399**, 621.

Shakura, N. and Sunyaev, R. 1973, *Astr. Ap.*, **24**, 337.

Shara, M. 1989, *PASP*, **101**, 5.

Tanaka, Y. 1992, in *X-ray Binaries and Recycled Pulsars*, (E. Van den Heuvel and S. Rappaport, eds.), Kluwer, NATO ASI Series, **377**, 37

Tavani, M. 1991, *Nature*, **351**, 39.

Trümper , J. 1983, *Adv. Space Res.*, **2**, 241.

Van Paradijs, J. 1995, in *X-ray Binaries*, (W. Lewin, J. Van Paradijs, and E. Van den Heuvel, eds.), Cambridge Univ. Press, in press.

Van Paradijs, J. and McClintock, J. 1995, in *X-ray Binaries*, (W. Lewin, J. Van Paradijs, and E. Van den Heuvel, eds.), Cambridge Univ. Press, in press.

Van den Heuvel, E.P.J. et al 1992, *Astr. Ap.*, **262**, 97

Van der Klis, M. 1989, *Ann. Rev. Astron. Ap.*, **27**, 515.

Van der Klis, M. 1994, in *Evolution of X-Ray Binaries*, (S. Holt & C. Day, eds.), AIP Conf.Proc., 308, 505.

Verbunt, F., van Paradijs, J. and Elson, R. 1984, *M.N.R.A.S.*, **210**, 899

Verbunt, F. *et al.* 1994, *Astr. Ap.*, **285**, 903.

Verbunt, F. *et al.* 1995, *Astr. Ap.*, in press

Warner, B. 1987, *M.N.R.A.S.*, **227**, 23

Webbink. R. and Kalogera, V. 1994, in in *Evolution of X-Ray Binaries*, (S. Holt & C. Day, eds.), AIP Conf.Proc., 308, 321.

White, N. 1994, in in *Evolution of X-Ray Binaries*, (S. Holt & C. Day, eds.), AIP Conf.Proc., 308, 53.

Zhang, S. et al 1994, *IAUC No. 6096*

Selected Highlights from COMPTEL

Giselher Lichti
Max-Planck-Institut für extraterrestrische Physik, Giessenbachstrasse,
Postfach 1603, 85740 Garching, Germany

ABSTRACT

Since its launch in April 1991 COMPTEL has performed a wealth of observations all over the sky. It did not only perform for the first time a complete sky survey in the so far unexplored 1-30 MeV energy range, but also carried out a variety of observations devoted to special objects. In general the main topics of interest for the COMPTEL observations were the performance of γ-ray line spectroscopy, the measurement of γ-ray emission from point-like sources, the investigation of the diffuse galactic γ-ray emission and the location of γ-ray bursts. Since not all the results obtained so far can be presented this lecture concentrates on the measurement of the 1.809 MeV γ-ray line from galactic ^{26}Al, the detection of the 1.157 MeV line from ^{44}Ti from Cas-A and the detection of the 4.4 and 6.1 MeV lines from Orion. Then results on quasar observations will be presented with emphasis given to the spectral break at low-MeV energies which is observed in all energy spectra from quasars seen by COMPTEL. The results of a search for black-hole candidates and pulsars are adressed and finally a distance limit for a specific burst will be given.

1. Introduction

1.1 General

With the launch of the Compton Gamma-Ray Observatory (CGRO) on the 5th of April 1991 a new era in γ-ray astronomy was opened. Especially the low-energy γ-ray range (1 - 30 MeV) was nearly unexplored because only few percent of the sky had been observed by balloon telescopes. Now the first complete sky survey at these energies has been performed by COMPTEL which has lead to new insights to the appearance of the γ-ray sky. The most spectacular results are the γ-ray line measurements, the detection of γ-ray emission from point-like sources and the location of γ-ray bursts. Here the most important results from these measurements are reported.

1.2 Description of the COMPTEL Instrument

COMPTEL is a Compton telescope consisting of two detectors which are 1.5 m apart, an upper one filled with a liquid scintillator (NE 213A) and a lower one made of NaI(Tl) crystals. A γ-ray normally interacts via a Compton scattering in the upper detector and the scattered γ-ray will, if it travels to the lower detector, deposit there in most cases its final energy via photoelectric absorption. The time sequence of the process (first interaction in the upper detector and the second one in the lower detector) is guaranteed via a delayed coincidence measurement. The energy of the infalling γ-ray is given by the sum of the two measured energy losses from which also the scatter angle φ can be calculated by using the kinematics of the Compton process. Furthermore the interaction locations in the two detectors are measured by applying the Anger-camera principle so that the direction of the scattered γ-ray is known yielding the arrival direction of the infalling γ-ray. Compton telescopes, however, suffer from one inherent disadvantage: no azimuthal information on the direction of the incoming γ-ray can be determined and therefore the arrival direction is only known to lie on a circle in the sky. This makes the data analysis a rather complicated process.

The main properties of COMPTEL can be characterized as follows: it measures within a large field of view (FoV) of ≈ 1 sr γ-rays in the energy range 0.75 - 30 MeV. The angular resolution at 1 MeV is $\approx 2.4^0$ and at 5 MeV it is $\approx 1.3^0$. The energy resolution lies between 5% and 10%, the lower value being valid for higher energies. The effective area depends on the energy and the selection criteria and ranges from 20 - 50 cm^2. Strong point sources can be located with an accuracy of 0.5^0. A detailed description of the COMPTEL instrument is given in Schönfelder et al.[1].

2. γ-Ray Line Spectroscopy

2.1 Introduction

γ-ray lines are generated in the galaxy by two different processes: first by the decay of radioactive elements which were synthesized by nucleosynthesis processes in the interior of stars and second by excitation of atomic nuclei via high-energetic interactions between atomic nuclei and elementary particles. The energies of all these lines range from ~ 0.5 MeV to ~ 7 MeV. This coincides nicely with the energy range of COMPTEL (0.75 - 30 MeV) which is therefore well suited to measure the radiation from these processes. The measurements of these lines yield therefore information about the creation processes and about the locations where these processes take place. The radioactive elements are synthesized in novae and supernovae explosions and in nucleosynthesis processes in the interior of massive stars (e. g. Wolf-Rayet (WR) stars and Asymptotic Giant Branch (AGB) stars). The excitation of atomic nuclei takes place in the interstellar medium and in solar flares. The most important radionuclides of γ-ray astronomy are listed in Table 1.

Table 1: Important radionuclides of γ-ray astronomy with their characteristic features.

Decay Chain	mean lifetime	Emission [MeV]	Study Object	observed from
$^{56}Ni \rightarrow ^{56}Co \rightarrow ^{56}Fe$	0.31 y	e^+, 0.847, 1.2138	single SN	SN 1987A
$^{57}Ni \rightarrow ^{57}Co \rightarrow ^{57}Fe$	1.1 y	0.014, 0.122	single SN	SN 1987A
$^{22}Na \rightarrow ^{22}Ne$	3.8 y	e^+, 1.275	single nova	-
$^{44}Ti \rightarrow ^{44}Sc \rightarrow ^{44}Ca$	78 y	e^+, 0.068, 0.078 1.156	single SN	Cas-A
$^{60}Fe \rightarrow ^{60}Co \rightarrow ^{60}Ni$	$4.3 \cdot 10^5$ y	0.059, 1.173, 1.332	SN distribution	-
$^{26}Al \rightarrow ^{26}Mg$	$1.1 \cdot 10^6$ y	e^+, 1.809	novae, SN and/or massive star distribution	galactic plane Vela SNR

2.2 The 1.809 MeV Line Emission from Galactic ^{26}Al

The 1.809 MeV γ-ray line emitted by the decay of radioactive ^{26}Al was first detected by the HEAO-C instrument[2]. Since then it has been measured by several other instruments (SMM and several balloon experiments) and its existence is a clear sign that recent nucleosynthesis has taken place in the Galaxy. Although the existence of the ^{26}Al line had been clearly established before COMPTEL, the origin of this radioactive element was not known by then. From the nucleosynthesis theories it was known that ^{26}Al is produced in the interior of stars and during explosive nucleosynthesis processes[3,4,5]. But where and how much ^{26}Al is produced during these events was highly uncertain. The attempts to get an answer to this question via imaging of the line emission did not lead to conclusive results because of the bad angular resolutions of the former instruments. Due to its high energy resolution of 8.5% at 1.8 MeV and its good imaging capability COMPTEL, however, is well suited to tackle the problem of the origin of galactic ^{26}Al.

^{26}Al decays with a half-life of about 10^6 years to ^{26}Mg. It is due to this long life time that ^{26}Al is widely distributed and accumulated in the galaxy thus leading to a large-scale distribution with a continuum emission. Another result of the long life time is that the ^{26}Al atoms come nearly to rest before they decay. Therefore the 1.8 MeV line is only broadened by the galactic rotation which results in a sharp line profile (intrinsic line width $\approx$ 3 keV). The observation of the distribution of the 1.8 MeV line emission from our galaxy should therefore yield information about the sources of ^{26}Al, about the distribution of these sources in our galaxy and about a better understanding of the underlying nucleosynthesis processes.

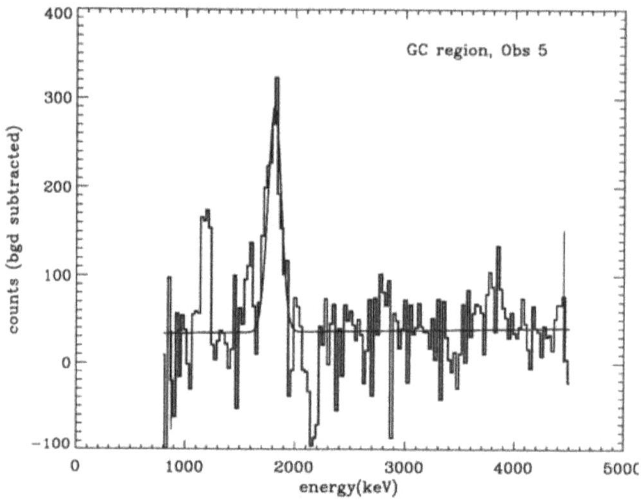

Figure 1: The background-subtracted energy spectrum of γ-rays from the GC region.

The good energy resolution of COMPTEL allows already the detection of the 1.8 MeV line in the raw data. In Figure 1 a background-subtracted energy spectrum of γ-rays from the galactic-center region is shown. The line at 1.8 MeV is clearly seen with a statistical significance of $\sim 10\ \sigma$.

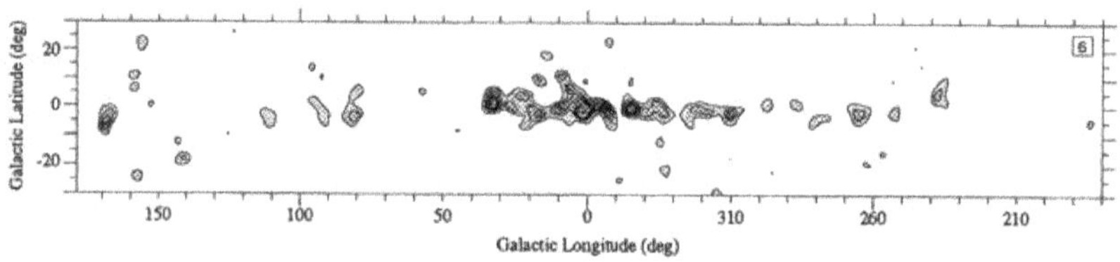

Figure 2: A skymap of the Galaxy in the light of the 1.8 MeV γ-ray line.

If one maps the emission from the galaxy in the light of the 1.8 MeV line one obtains the skymap shown in Figure 2. An inspection of this map reveals a clear concentration of the emission towards the galactic plane and towards the galactic center. Furthermore one notices that the emission is asymmetric with respect to the galactic center (i. e. much less γ-rays are received from the first and second quadrant than from the third and fourth one). A closer look reveals that the emission maximum is not identical with the galactic center (GC). The maximum is located at l≈ 2^0, b≈ -1.5^0. Another striking fact is that the distribution is not smooth but clumpy. Hot spots far from the GC exist in the Cygnus region (at l≈ 80^0), in the Carina region (at l≈ 310^0) and in the Vela region (at l≈ 265^0).

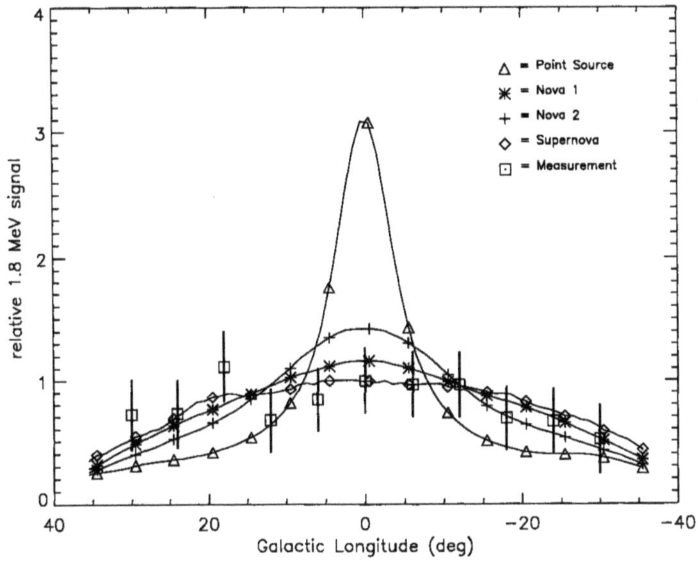

Figure 3: Comparison of the measured 1.8 MeV excess-counts profile with various model profiles.

The measured emission was compared with the distribution from different source populations. The result is shown in Figure 3. From the comparison one learns that a single point source at the GC is not compatible with the data and can clearly be excluded as origin of the line emission[6]. The extended smooth distributions (novae and supernovae distributions) are in better agreement with the data, but do not fit them satisfactorily. Especially the nova distribution can hardly be reconciled with the data. From this one can conclude that a single-site origin of ^{26}Al can be excluded[7]. The COMPTEL data even favour a composite origin of ^{26}Al with massive-star formation sources (SN, WR stars) as main contributors.

A blow up of the hot spot in the Vela region is shown in Figure 4. It is obvious that the maximum γ-ray emission coincides nicely with the Vela supernova remnant (SNR) indicated by the circle but not with the position of the WR star closest to the sun (WR 11). The statistical significance of the γ-ray emission (at a level of $\sim 4 \cdot 10^{-5}\ \gamma/(\mathrm{cm}^2\ \mathrm{s})$)

is 5σ. This observation is interpreted as γ-ray emission from ^{26}Al which was synthesized in the progenitor star of the Vela SNR and expelled into the interstellar space by the corresponding SN explosion. The missing enhanced γ-ray emission from WR 11 is not in conflict with the aforemade statement. But if one determines the amount of ^{26}Al produced in a WR star according to the standard nucleosynthesis models one finds that not enough ^{26}Al is produced to make this star visible to COMPTEL[8].

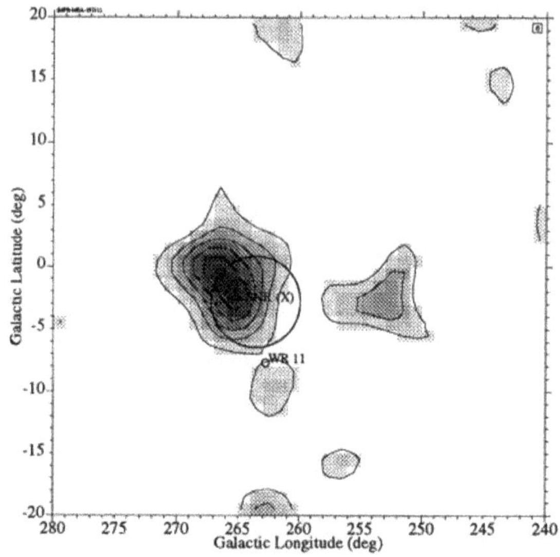

Figure 4: A skymap of the Vela region in the light of the 1.8 MeV γ-ray line.

From the fact that the model fits of novae, SN, WR and AGB stars can explain only a part of the observed intensity and from the fact that nearby sources contribute significantly to the total intensity one can conclude that the total mass of ^{26}Al in the galaxy must be less than the canonical value of $3M_\odot$. This reduced mass helps in explaining the origin of galactic ^{26}Al because the yields in the standard nucleosynthesis models were not sufficient to account for the $3M_\odot$ as required before.

It should be noted at the end of this section that the true nature of the ^{26}Al sources is still not undoubtedly known. But it is the hope that the analysis of the excesses observed from the near Carina and Cygnus regions can help in clarifying these open questions.

2.3 Observation of ^{44}Ti from Cas-A

^{44}Ti is besides ^{26}Al another potential probe of nucleosynthesis sites in our galaxy.

This isotope decays according to the scheme $^{44}Ti_{22} \longrightarrow {}^{44}Sc_{21} \longrightarrow {}^{44}Ca_{20}$ to ^{44}Ca. The half life of the first decay is ~ 54 years and the one of the second 2.7 hours. In the course of the latter decay a γ-ray line at 1.157 MeV is emitted. Because of the short (compared to ^{26}Al) mean life of about 78 years ^{44}Ti is a good tracer for recent supernovae. One of those, the SNR Cas-A was in the FoV of COMPTEL from July 16 till August 6, 1992 and the analysis of the data obtained during this observation period yielded for the firts time the proof for the existence of the 1.157 MeV γ-ray line from the ^{44}Ti decay.

Cas-A is the remnant of a supernova whose explosion was possibly observed by the English astronomer John Flamsteed in the year 1667. Its distance is about 2.8 kpc and the diameter of the remnant is about 3.7 pc. The supernova explosion was very probably of the type Ib. Supernovae of this type have progenitor stars which have lost most of their hydrogen envelope via a strong stellar wind or via Roche-lobe overflow. It is believed therefore that the progenitor star of Cas-A was a WR star with 15-25 $M_\odot$.

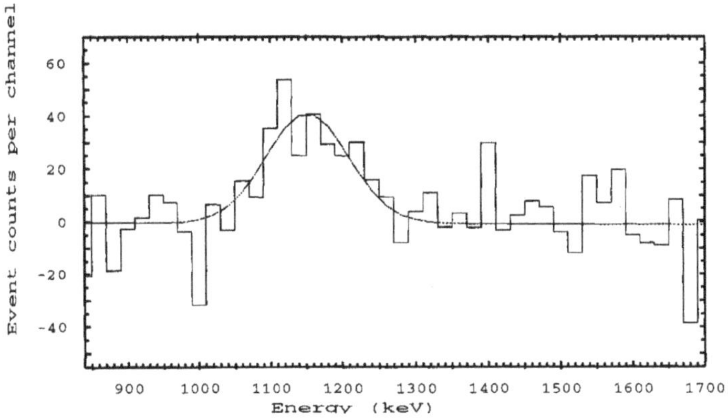

Figure 5: The background-subtracted energy spectrum of Cas-A.

In Figure 5 a background-subtracted energy spectrum of γ-rays coming from Cas-A is shown. The line at an energy of 1.157 MeV is clearly visible. The statistical significance of the line detection comes to about 5σ. A more refined analysis indicates that the line is broader than expected from the instrumental energy resolution at that energy by about 30%. This additional line broadeneing is consistent with the expansion of the supernova shell[9].

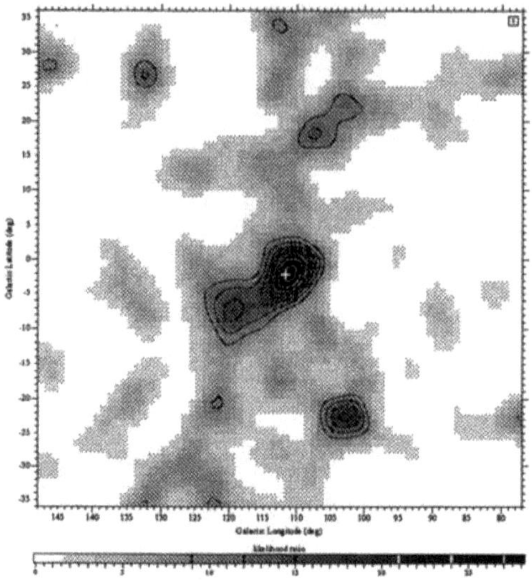

Figure 6: Skymap from the region around Cas-A in the light of the 1.157 MeV line.

That the line quanta are indeed coming from Cas-A is illustrated by the skymap of Figure 6. The maximum of the measured γ-ray intensity coincides nearly exactly with the position of Cas-A. If one calculates from the measured line flux of $(7 \pm 1.7) \cdot 10^{-5}$ $\gamma/$ (cm^2 s) the mass of ^{44}Ti one obtains $(3.2 \pm 0.8) \cdot 10^{-4}$ $M_\odot$. This result is in good agreement with the theoretical expectations of $\sim 2 \cdot 10^{-4}$ $M_\odot$ per SN explosion of this SN type[1]0. The large error in the mass estimate comes from the uncertainties in the mean life time of ^{44}Ti (70 ± 20 years), in the distance (2.8 ± 0.5 kpc) and in the time of the explosion (A. D. 1680 ± 15 years).

2.4 Observation of Line Emission from Excited Atomic Nuclei

The existence of nuclear interaction lines was first predicted by Ramaty and Lingenfelter[11]. These lines are produced by the interaction of cosmic-ray particles with the atoms of the interstellar medium. The most prominent lines are the de-excitation lines of ^{12}C and ^{16}O at 4.44 MeV and 6.13 MeV, respectively, which lie amidst the energy range of COMPTEL of 0.75-30 MeV. A search for these lines in the COMPTEL data from regions where these processes may take place (regions with dense interstellar clouds) was therefore obvious and has been performed.

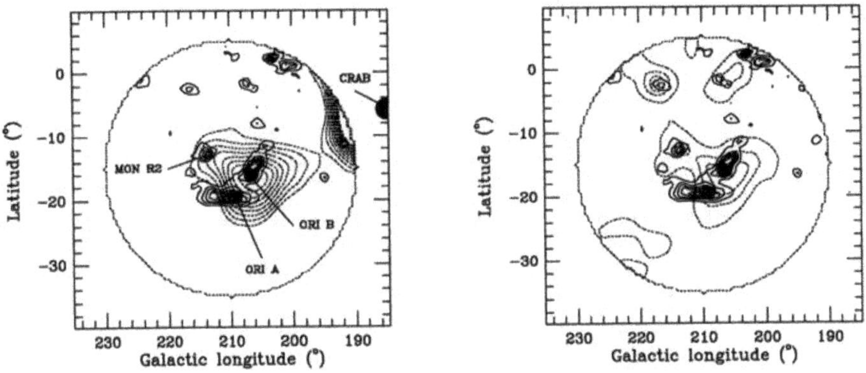

Figure 7: Maximum-likelihood map (left) and maximum entropy map of the Orion region in the energy range 3-7 MeV. The Columbia CO observations are superimposed for comparison (solid lines).

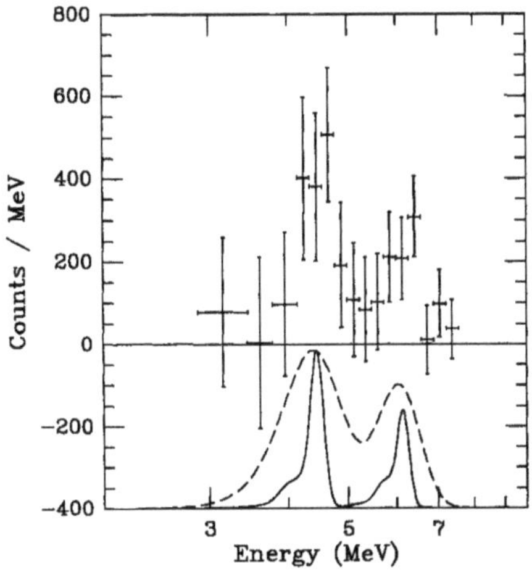

Figure 8: Background-subtracted count-rate spectrum of Orion for a selected region of 3^0 radius centered on $l = 208^0$ and $b = -18^0$.

One of these regions is the Orion-cloud complex which was in the field-of-view of COMPTEL during five observation periods. Unfortunately the angular distance of Orion to the pointing axis of COMPTEL was always $> 17^0$ and therefore this object was only observed with moderate sensitivity. Nevertheless emission from this region in the energy range 3-7 MeV has been detected with a statistical significance of $\sim 7.5\sigma$ when combining all these observations. In Figure 7 maximum-likelihood and maximum-entropy skymaps of the Orion region for the energy range 3-7 MeV are shown. Also shown by the solid contour lines are the Columbia CO measurements as an indicator for regions with dense interstellar matter. Clearly the maximum of the γ-ray emissivity coincides with one (Orion-B) of the two CO maxima. If one performs a spectral analysis and generates a background-subtracted energy spectrum of γ-arys coming from Orion-B one obtains the spectrum of Figure 8. Two lines around 4.4 MeV and 6.1 MeV are present. In the lower panel of Figure 8 it is shown how narrow lines and intrinsically broadened lines (FWHM = 1 MeV) would appear in the data. We interpret this result as a proof for the existence of the predicted de-excitation lines of excited ^{12}C and ^{16}O at 4.44 MeV and 6.13 MeV, respectively. The flux measured in the energy range 3-7 MeV amounts to $(1.0 \pm 0.15) \cdot 10^{-4}$ $\gamma/(\mathrm{cm}^2$ s). Such a high flux requires a $\sim$ 10-times higher emissivity per H-atom than is observed from the galactic plane[12].

What do we now learn from this observation? The stimulation of the nuclei is possible via two processes:

1. by bombardment of C, O atoms of the interstellar matter by low-energy protons and α particles
2. by bombardment of H and He of the interstellar matter by medium-heavy nuclei (C, O, N etc.) of the cosmic radiation.

The measured line profiles should allow a distinction between these two processes because in the first case narrow, in the second case boadened lines should be observed. The measurement, however, does not yet allow this decision.

Yet astrophysical considerations lead to the conclusion that the second process is more likely, because the first process would need too a high energy density ($\sim$ 500 eV/cm^3) of the cosmic radiation. This, however, is on the one side not in accordance with the ionization rate of the interstellar medium [$\leq 10^{-16}$ /(H-atom s)] (500 eV/cm^3 would result in a 1000-times larger ionization rate) and on the other side in conflict with the missing enhanced bremsstrahlungs- and π^0 γ-radiation from Orion as derived from COS-B[13] and COMPTEL[14] observations. In turn the second process requires for an explanation of the measurement an abundance enhancement of the medium-heavy elements in the cosmic radiation by a factor of 30! Bykov and Bloemen[15] have shown that in the Orion complex with its massive star-formation rate the second process seems to be possible.

3. Observation of γ-Ray Emission from Point Sources

3.1 Introduction

A search for point sources in the COMPTEL data is complicated by three facts:

- COMPTEL's point-spread function has a conical shape in a 3-dimensional data space[16]
- few counts per bin result in statistical problems

- a high intrinsic background (signal-to-noise ratio < 5%) makes the recognition of a source signal difficult.

Therefore sophisticated deconvolution methods like the maximum-entropy[17] or maximum-likelihood[18] methods have to be used to find γ-ray sources and derive their properties. In addition consistency checks are always necessary to validitate a result. Nevertheless the following point sources have been detected by COMPTEL: the active galactic nuclei (AGN) 3C 273, 3C 279, PKS 0528+134, CTA 102, 3C 454.3 and an yet unidentified source near the LMC, the radiogalaxy Cen-A, the Crab pulsar, the Vela pulsar and PSR 1509-08 and the two X-ray binaries Cyg X-1 and Nova Persei. It is interesting to note that no Seyfert galaxy has been detected by COMPTEL so far[19].

3.2 Observation of γ-Ray Emission from AGNs

In the pre-launch era of the Compton Gamma-Ray Observatory only one γ-ray quasar was known: 3C 273[20]. EGRET has now detected about 40 radio-loud quasars of the blazar type and BL Lac objects[21]. Only few of them have been seen by COMPTEL. From those I shall pick up 3C 273 as an exemplary source.

3C 273 is with a redshift of z = 0.158 one of the nearest and most luminous quasars. It has two radio lobes and an optical jet with a high polarization. The jets show superluminal motion and one has observed at all wavelengths time variability with different time scales. Thus this source possesses all the characteristics of a typical γ-ray blazar which are:

- strong radio emission from a compact region with a flat radio spectrum
- time variability on short($\approx$ hours-days) and long ($\approx$ years) time scales
- high degree of polarization at optical and IR wavelengths
- a significant fraction of these objects shows superluminal motion
- the γ-ray luminosity often exceeds the luminosity observed in other spectral ranges
- nearly all have breaks in the energy spectra in the low-MeV energy range.

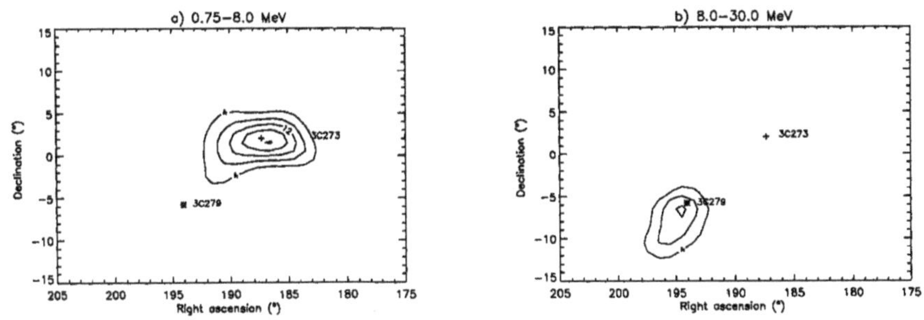

Figure 9: Skymaps of the Virgo region with the detections of 3C 273 and 3C 279.

In Figure 9 two skymaps of the Virgo region for two different energy ranges are shown. In the energy interval 0.75 - 8 MeV one sees a γ-ray source which coincides with the position of 3C 273. At higher energies (8 - 30 MeV) this source is not seen anymore, but a new source shines up which coincides with the position of 3C 279. The detection significances are $\sim 4\sigma$ for 3C 273 and $\sim 3.2\sigma$ for 3C 279[22]. From this observation one learns that γ-ray blazars can have quite different energy spectra.

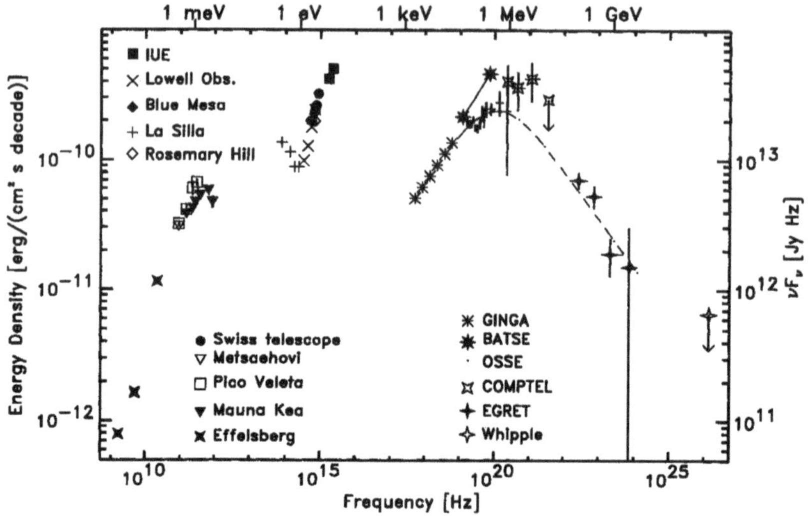

Figure 10: Energy-density spectrum of 3C 273 of quasi-simultaneous observations.

Since multiwavelength aspects are essential for the understanding of the processes which power the AGNs an international observation campaign had been initiated to observe 3C 273 at all other wavelengths at the same time when CGRO looked at it (from June 15 to June 28, 1991). The data of all these observations were collected and an overall energy-density spectrum was constructed. This spectrum is shown in Figure 10. The most striking features of this spectrum are three maxima at $\sim 10^{13}$ Hz, $2.3 \cdot 10^{15}$ Hz and $1.2 \cdot 10^{21}$ Hz. If one calculates the luminosity for isotropic emission and a Hubble constant of 60 (km/s)/Mpc one obtains $1.2 \cdot 10^{46}$ erg/s, $3.1 \cdot 10^{46}$ erg/s and $3.4 \cdot 10^{46}$ erg/s, respectively[23]. These three maxima can basically be associated with three different emission mechanisms: the radio and IR radiation is mainly produced by relativistic electrons via the synchrotron process, the optical and UV photons are thermally emitted from the hot accretion disk and its associated corona and the X- and γ-rays are generated via the inverse Compton scattering of relativistic electrons on the synchrotron and/or UV radiation from the accretion disk.

The most striking feature in the X-γ-ray regime is a spectral break at low-energy γ-rays of $\sim$ 0.9. Such a break in the energy spectrum at MeV energies seems to be a common feature of all γ-ray blazars. The physical processes which lead to these breaks are not completely understood. In the inverse-Compton model of Dermer and Schlickeiser[24] this break could either be due to an energy-dependent escape time of the electrons from the emission region or due to adiabatic energy losses of the electrons due to radial variation of the magnetic fields (in their model a constant magnetic field was assumed).

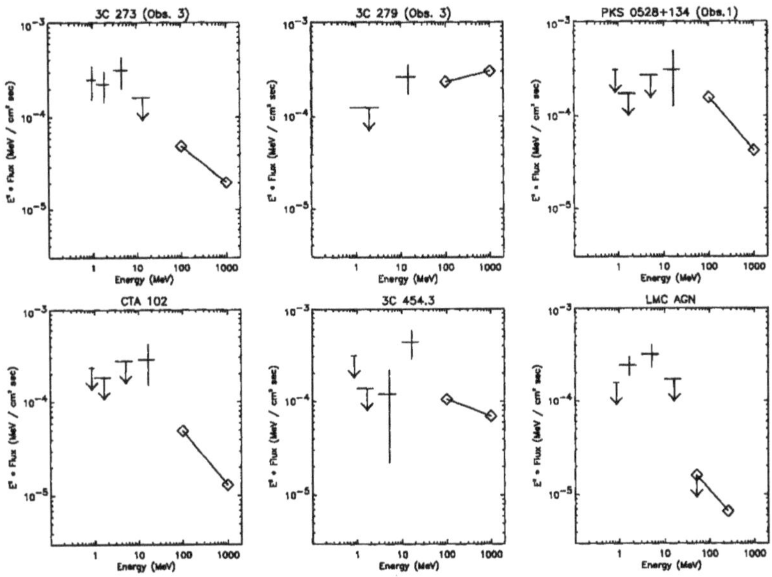

Figure 11: Energy-density spectra of six γ-ray blazars.

That such spectral breaks are indeed a common feature of γ-ray blazars becomes obvious from Figure 11. There the luminosity spectra of the six COMPTEL blazars are compared with the EGRET spectra. With the exception of the AGN near the LMC the COMPTEL data require always a spectral break[25].

3.3 COMPTEL Observations of Black-Hole Candidates

Black-hole (BH) candidates are X-ray binaries whose mass function suggest that the compact companion is a black hole. In these systems matter is accreted by the black hole either via Roche-lobe overflow or via a stellar wind from the companion. Therefore X- and γ-ray spectra can be expected which are similar to those observed from AGNs[26]. These sxpectra can best be explained by Compton upscattering of UV photons and low-energy X-rays by electrons of a hot electron gas[27]. But sometimes multi-temperature gas models are necessary to give adequate fits to the measured spectra.

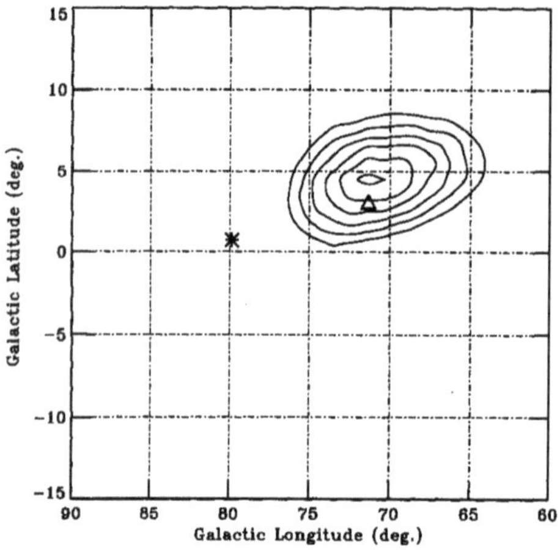

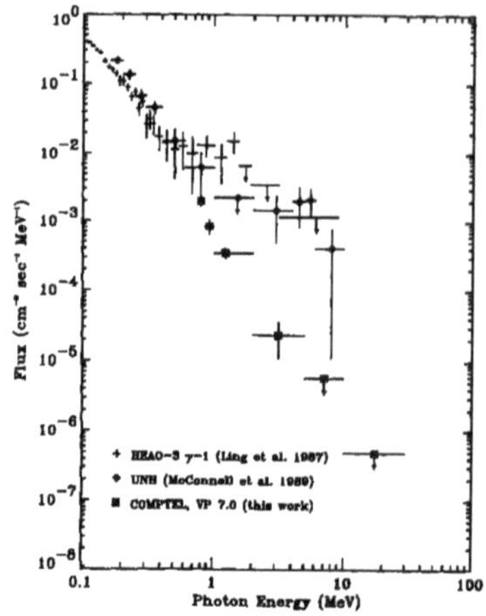

Figure 12 (left): A maximum-likelihood skymap of the Cygnus region in the energy range 0.75-1 MeV (triangle: Cyg X-1, asterisk: Cyg X-3).

Figure 13 (right): The γ-ray spectrum of Cygnus X-1.

A search for BH candidates has been performed with the COMPTEL data. So far two of those candidates have been detected: Cyg X-1 and Nova Persei. Other BH candidates are Nova Musca, 1E 1740.7-2942 and GX 339-4. Figure 12 shows a skymap of the Cygnus region in the energy range 0.75 - 1 MeV. The positions of Cyg X-1 and Cyg X-3 are indicated. A clear signal from Cyg X-1 is seen. The significance of the detection is 7.7σ[28]. In contrary no signal from Cyg X-3 is detected.

A spectral analysis indicates that the emission extends at least to 2 MeV. If one compares the COMPTEL data points with the measurements of BATSE and OSSE (see Figure 13) then one sees that they are consistent with an extrapolation of the spectra measured by these two instruments. The COMPTEL upper limits are not low enough to give an answer to the question whether the spectrum breaks off above 2 MeV as is expected from thermal emission processes involving electrons or whether there exists a power-law extension towards higher energies. In the latter case an additional non-thermal emission process would be required.

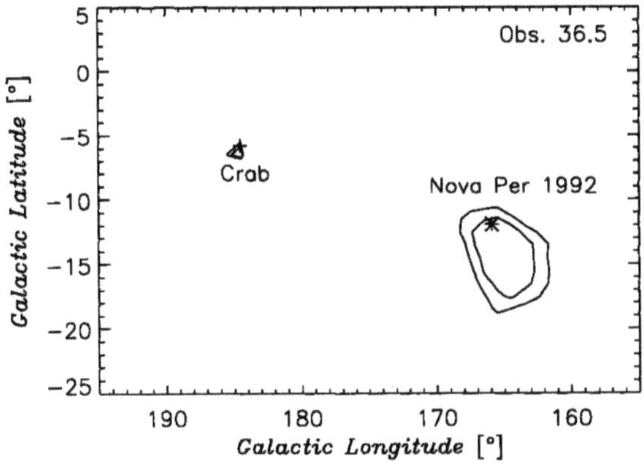

Figure 14: A skymap of the anticenter region in the energy range 1-2 MeV with the Crab and Nova Persei.

If one compares the COMPTEL measurement with the HEAO-3 data[29] collected several years earlier one sees that there is a discrepancy by a factor of ~ 10. So COMPTEL does not see the "MeV-bump". Whether this is a sign of time variability of a non-thermal process or an incorrect measurement is unclear because the statistical significance of the data above 1 MeV is rather low. Future COMPTEL observations of Cygnus X-1 will certainly help in solving this puzzle!

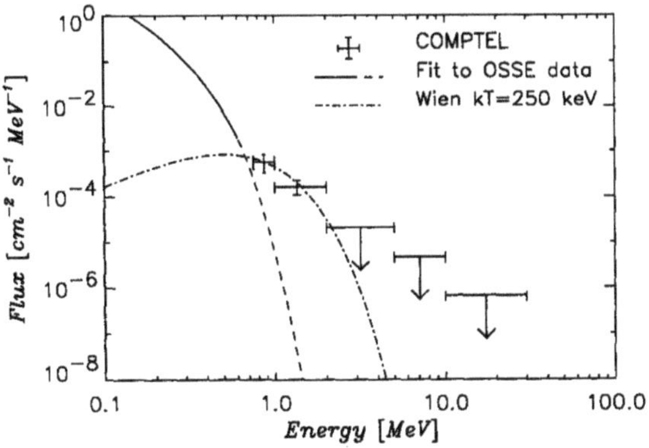

Figure 15: The γ-ray emission spectrum of Nova Persei.

On Aug. 5, 1992 a nova was discovered by the BATSE instrument[30,31]. This nova was declared a target of opportunity for CGRO and therefore this source was in the FoV of COMPTEL from Aug. 11-20, 1992 and again from Sept. 1-17, 1992. An analysis of the COMPTEL data showed significant emission up to 2 MeV. A skymap of the August observation for event energies between 1 and 2 MeV is shown in Figure 14. A significant signal ($\sim 3.5\sigma$) from the position of Nova Persei is present[32].

If one performs a spectral analysis one finds in contrary to Cyg X-1 a second emission component above 1 MeV (see Figure 15). The extrapolation of the X-ray spectrum measured by OSSE according to the Sunyaev-Titarchuk model falls short with respect to the COMPTEL data. A separate component is needed to fit the data. The COMPTEL data can best be described by a Wien-type spectrum with a temperature of kT = 250^{+150}_{-50} keV. But the upper limits do not rule out a power law. A hot electron-positron plasma cloud in the inner accretion disk for example can produce such a component[33].

In summary further observations by COMPTEL are needed to settle the following two questions:

- Do all BH candidates show a spectral hardening above 1 MeV?

- Is such a hard component time variable?

3.4 γ-Ray Pulsars

Before CGRO only two γ-ray pulsars were known to emit radiation above 1 MeV: the Crab and Vela pulsars[34]. Both of these have, of course, been seen by COMPTEL. Since then four more γ-ray pulsars have been found by BATSE and EGRET: Geminga, PSR 1055-52, PSR B1509-58 and PSR 1706-44[21,35]. Although an extensive systematic search for γ-ray pulsars was performed with the COMPTEL data only one of these pulsars (PSR B1509-58) was found at low energies only (0.7 - 1 MeV)[36].

In Figure 16 the lightcurves of the three COMPTEL pulsars are shown. It should be noted, however, that the energy intervals are different: 1-10 MeV for the Crab, 10-30 MeV for Vela and 0.7-1 MeV for PSR B1509-58. It is interesting to note further that all three lightcurves are different, too. Whereas Crab and Vela have a double-peak lightcurve with a peak separation of $\sim$ 0.4 of the phase PSR B1509-58 has only one broad peak. Crab shows in addition a strong interpulse emission which is not (or only marginally) present for Vela. The peak widths are also different: the Crab pulses are much broader (6% and 10%) than the ones of Vela (1% and 3.3%). From the energy intervals in which the pulsars have been seen one can deduce that the spectral behaviour of these pulsars is also quite different. The pulsed Crab spectrum can be well described above 130 keV over five decades of γ-ray energies by a power law with a spectral index of -2.2[37]. In contracst the Vela spectrum has a break at low-energy γ-rays[34]. For PSR B1509-58 no spectral analysis could be performed because of the marginal detection.

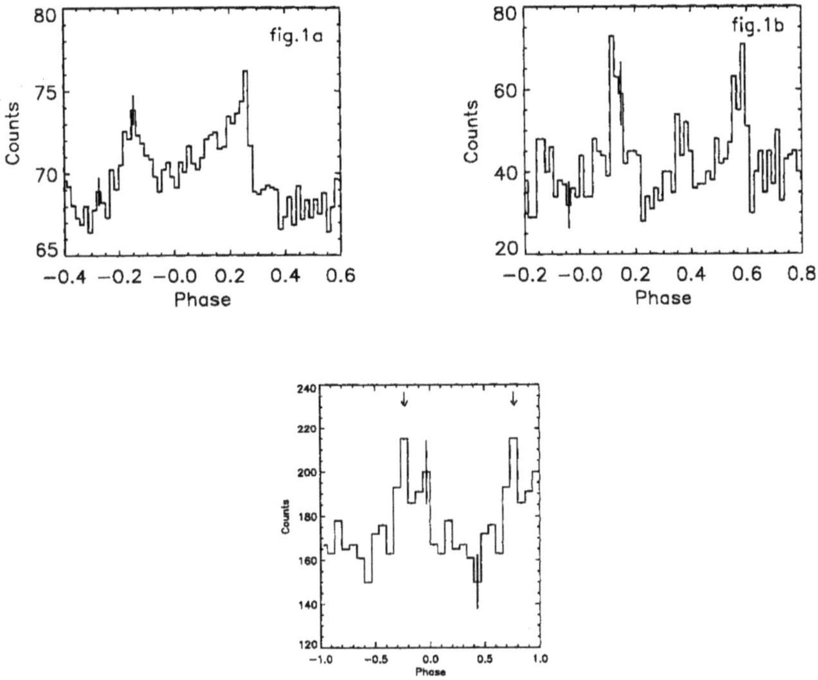

Figure 16: The γ-ray lightcurves of the three COMPTEL pulsars.

4. Observation of γ-Ray Bursts

COMPTEL measures γ-ray bursts in two modes, the burst-module mode and the telescope mode. The first mode is quasi-omnidirectional sensitive to γ-ray bursts and yields no directional information, whereas the second mode is sensitive only to γ-ray bursts which occur in the FoV of ≈ 1 sr. In the burst-module mode COMPTEL investigates the spectral and timing behaviour of γ-ray bursts and in the telescope mode COMPTEL has in addition the capability to locate γ-ray bursts (strong ones with an accuracy better than 1^0). Since the sources of γ-ray bursts are still unknown a search for counterparts is very important. With COMPTEL such searches are promising because

- the triangulation rings of CGRO and ULYSSES on the one side and the error region of COMPTEL on the other side yield narrow elongated error boxes ($\approx$ few arcmin to $\approx 1^0$)
- the rapid burst-response facility of COMPTEL allows the location of a burst within about one hour after the event[38].

So far COMPTEL has localized 17 cosmic γ-ray bursts in the energy range 0.75 - 30 MeV with an average location accuracy of $\sim 1.3^0$. From those the results of the strong γ-ray burst GRB 930131 will be presented as an example. In Figure 17 the intensity profile of this burst using COMPTEL telescope data is shown. One sees that the burst consisted of two peaks separated by ~ 0.8 s. The fact that the first peak appears weaker than the second one is due to a large dead-time effect[39]. Actually the two peaks have nearly the same strength.

Employing the maximum-likelihood method[17] the position of the burst was determined using telescope data. Figure 18 shows the result. The COMPTEL position is compared with the position determined by EGRET and with the triangulation arc from the arrival-time measurements of CGRO and the Ulysses spacecraft. The right panel of Figure 18 shows the burst position as derived from the combined COMPTEL/EGRET data. This best position is $\alpha_{2000} = 12^h\ 12^m$ and $\delta_{2000} = -10^0\ 21'$ (1σ error $\pm 32'$ in both directions).

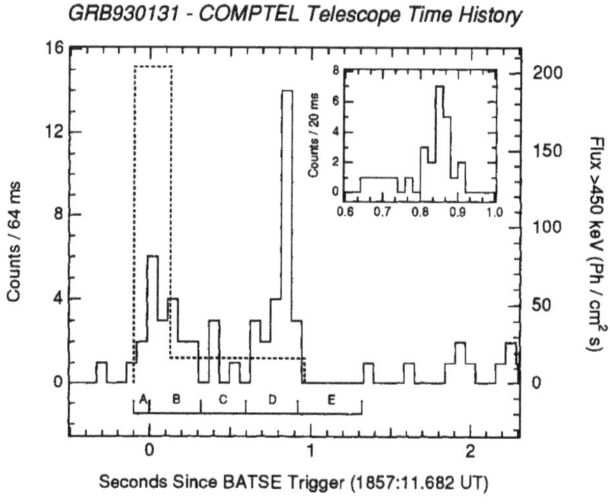

Figure 17: The intensity-time profile of GRB 930131 from COMPTEL telescope data.

The facts that the emission extends beyond 1 MeV and rapid intensity variations of these high-energy photons have been observed allow the determination of an upper limit for the burst distance. The reasoning is as follows: from the ~ 50 ms timing structure seen in the inlet of Figure 17 a maximum diameter ($2c\Delta t$) of 30000 km on the burst photosphere can be placed. In order to leave this photosphere the optical thickness τ for

γ-γ absorption of these high-energy photons must be < 1. This requires a compactness parameter l$\leq$ 10 ($l = \frac{\sigma \cdot L}{mc^3 \cdot R}$; σ = Thomson cross section, L = luminosity, R = extension of emission region). The luminosity can be calculated from the measured energy flux and the distance: $L = 4\pi \cdot D^2 \cdot F$ [with F $\approx 10^{-4}$ erg/(cm^2 s)]. Putting all this together one obtains for isotropic emission an upper limit on the distance:

$$D\ [cm] \leq 1.716 \cdot 10^{14} \cdot \sqrt{\frac{R}{F}}$$

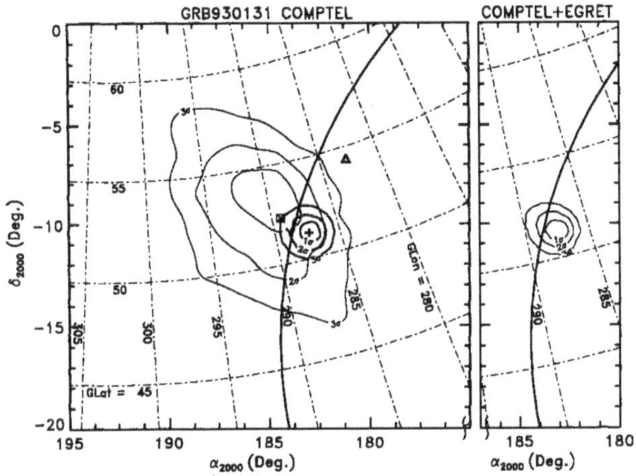

Figure 18: Maximum-likelihood skymaps of GRB 930131 (for explanation see text).

Inserting numbers (in cgs units) one obtains a value for the distance of 250 pc. The distance limit will be larger if beaming plays a role. In order for a cosmological object at 10^3 Mpc to emit a $\sim$ 50 ms spike with the observed intensity at 1 MeV, it must have a Doppler-Lorentz beaming factor greater than the unprobably high value of 2800 (the emission cone in this case would be $< 4 \cdot 10^{-7}$ sr!). For a distance of $\sim$ 30 kpc (galactic halo distance) this factor reduces to a reasonable value of $\sim$ 10. It should be noted that only the rapid intensity variations of high-energetic γ-rays (E $>$ 1 MeV) lead for the first time to the derivation of an upper limit on a burst-source distance[39].

The models in which the γ-ray bursts originate at cosmological distances are the neutron star (NS)-NS and BH-NS merging models. Although in this merger scenario beaming effects play a role[40,41] the high beaming factor derived for GRB 930131 questions the validity of these models. The merger rate estimated from the observations of NS-NS binaries of $\sim 10^{-4} - 10^{-6}$ events/(year & galaxy)[42,43,44,45] is just in agreement with the observed burst rate. Now due to the beaming effect the true rate of merger events must be much higher (by a factor of $\sim 3 \cdot 10^7$) than the observed burst rate which have problems to explain the COMPTEL observation.

5. Acknowledgements

I would like to thank all my colleagues of the COMPTEL team who supported me in completing this work.

6. References

1. V. Schönfelder et al., *Ap. J. Suppl. Ser.* **86** (1993) 657
2. W. A. Mahoney et al., *Ap. J.* **286** (1984) 578
3. D. D. Clayton, *Ap. J.* **280** (1984) 144
4. D. D. Clayton and M. D. Leising, *Phys. Rep.* **144** (1987) 1
5. N. Prantzos, *Ap. J.* **404** (1993) L55
6. R. Diehl et al., *Ap. J. Suppl. Ser.* **92** (1994) 429
7. R. Diehl et al., *AA* in press
8. U. Oberlack et al., *Ap. J. Suppl. Ser.* **92** (1994) 433
9. A. Iyudin et al., *AA* **284** (1994) L1
10. T. Shigeyama et al., *Ap. J.* **286** (1990) 644
11. R. Ramaty et al., *Ap. J. Suppl. Ser.* **40** (1979) 487
12. H. Bloemen et al., *AA* **281** (1994) L5
13. H. Bloemen et al., *AA* **139** (1984) 37
14. A. Strong et al., *AA* in press
15. A. Bykov and H. Bloemen, *AA* in press
16. R. Diehl et al., in *Data Analysis in Astronomy IV*, eds. Di Gesù (Plenum, New York) Vol. **59** (1992) 201
17. A. Strong et al., in *Data Analysis in Astronomy IV*, eds. Di Gesù (Plenum, New York) Vol. **59** (1992) 251
18. H. de Boer et al., in *Data Analysis in Astronomy IV*, eds. Di Gesù (Plenum, New York) Vol. **59** (1992) 241
19. M. Maisack et al., *AA* in press
20. B. N. Swanenburg et al., *Nature* **275** (1978) 298
21. C. E. Fichtel et al., *Ap. J. Suppl. Ser.* in press
22. O. R. Williams et al., *AA* in press
23. G. G. Lichti et al., *AA* in press
24. C. D. Dermer and R. Schlickeiser, *Ap. J.* **416** (1993) 458
25. V. Schönfelder et al., *Advances in Sp. Res.* in press (1995)
26. A. Owens and M. McConnell, *Comm. Astrophys.* **16** (1992) 205
27. R. A. Sunyaev and L. G. Titarchuk, *AA* **86** (1980) 121
28. M. McConnell et al., *Ap. J.* **424** (1994) 933
29. J. C. Ling et al., *Ap. J.* **321** (1987) L117

30. W. S. Paciesas et al., *IAU circular* **5580** (1992)
31. B. A. Harmon et al., *IAU circular* **5584** (1992)
32. R. van Dijk et al., *AA* in press
33. E. P. Liang and C. D. Dermer, *Ap. J.* **325** (1988) L39
34. W. Hermsen et al., *Ap. J. Suppl. Ser.* **92** (1994) 559
35. R. Wilson et al., *Bull. Am. Astr. Soc.* **23** (1991) 1440
36. K. Bennett et al., *Ap. J. Suppl. Ser.* **90** (1994) 823
37. M. Ulmer, *Ap. J. Suppl. Ser.* **90** (1994) 789
38. R. M. Kippen et al., *AIP Conf. Proc.* **280** (1993) 823
39. J. Ryan et al., *Ap. J.* **422** (1994) L67
40. B. Paczyński, *Ap. J.* **308** (1986) L43
41. J. Goodman, *Ap. J.* **308** (1986) L47
42. J. P. Clark et al., *AA* **72** (1979) 120
43. D. Eichler et al., *Nature* **340** (1989) 126
44. R. Narayan et al., *Ap. J.* **379** (1991) L17
45. E. S. Phinney, *Ap. J.* **380** (1991) L17

BATSE OBSERVATIONS OF GAMMA-RAY BURSTS

CHRYSSA KOUVELIOTOU
USRA, NASA/MSFC ES-84, Huntsville, AL 35812, USA

ABSTRACT

Over 1200 Gamma-Ray Bursts (GRBs) have been observed with the Burst and Transient Source Experiment (BATSE) onboard the Compton Gamma Ray Observatory (CGRO) during the first 3.5 years of its operation. I present here a review of these observations and summarize the results on temporal (morphology, time variability), spatial (sky distribution, log N - log P diagram) and spectral (spectral evolution, spectral shapes) characteristics of the events.

1. Introduction

Gamma-Ray Bursts (GRBs) were discovered accidentally in 1967 by a group of Los Alamos National Laboratory scientists, who had launched the VELA satellite series as a monitor of terrestrial gamma-ray emission within the frame of the 1963 Nuclear Test Ban Treaty. The VELAs recorded bursts of gamma-rays from outer space, a quite unexpected phenomenon. It took a while for this new phenomenon to be unambiguously recognised and accepted in the scientific community: the paper announcing the discovery was only published in 1973[1].

For several years GRB studies depended on "piggy–back" experiments: small detectors added in already designed payloads that were approved for flight. As a result, until 1977, GRBs were being discovered with an average rate of only $\sim$10 events per year. The first instrument dedicated to GRB studies went into orbit in 1977 (Prognoz 6); it was launched by an international group (French-Soviet collaboration). Many more followed: Pioneer Venus Orbiter, International Sun Earth Explorer - 3, the Venera series of 4 satellites, Prognoz - 7, Solar Maximum Mission, Kvant on MIR, GINGA, GRANAT, Ulysses, to name a few. Between 1980 and 1991, GRBs were discovered on an average rate of about 100 per year. I should parenthetically mention here one of the major uncertainities involved in GRB studies: since gamma-ray emission within the GRB energy range can be associated with a variety of transient phenomena, such as magnetospheric particles, solar flares, *etc.* , a GRB was only considered "confirmed", when it had been observed with more than one instrument. This requirement, together with the then existing instrument sensitivity, severely limited the number of observed events.

On April 5, 1991, one of the NASA Great Observatories, the Compton Gamma-Ray Observatory (CGRO) was launched. One of the four experiments onboard CGRO is the Burst And Transient Source Experiment (BATSE), especially designed to detect GRBs with unprecedented sensitivity[2]. Like the *KONUS* experiment on the Veneras, BATSE can provide directional information for each transient event it detects, thus determining whether it is terrestrial, solar or cosmic. By the end of 1994 BATSE had detected over 1200 GRBs; it has accumulated sufficient statistics to define the observational constraints of the GRB problem.

In the following sections, I will describe the GRB observational characteristics using mainly BATSE results. I will divide these characteristics in three main types: temporal, spectral and spatial. BATSE has provided a wealth of information in each one of them. In analyzing temporal characteristics of GRBs we have attempted classifications based on their morphologies and studied the behaviour of their time variability. Gamma-ray burst sky (maps) and intensity (log N - log P diagram) distributions provide constraints on their spatial properties. GRB spectral behaviour is best studied by tracking their spectral evolution and fitting their spectral shapes. These studies have significantly enhanced our understanding of the GRB phenomenon. Nevertheless, today we still find ourselves debating a fundamental issue, namely, the nature of Gamma Ray Bursts: are they cosmological or are they galactic ?

2. Temporal Properties

2.1. Morphology

In the past 27 years there have been several attempts to classify GRB lightcurves into different types according to their common characteristics[3–5]. All have been unsuccessful, insofar that there is always the odd event: the event that will create its own unique class. Statistically speaking, when such "loner" cases dominate the sample, classification is useless. In Fig. 1 I try to provide an overview of the morphologies that are often encountered in GRBs. I have separated and examined the events according to duration into two subsets divided at 2 s; the choice of this separator is discussed in Section 2.2. The diagram reveals that the same profile varieties are present at all time scales. GRBs can consist of single pulses or have many episodes, each of them either single or multi-pulsed. Each pulse can have a smooth overall envelope or it may contain rich substructure. The substructure can be entirely random (described as chaotic in the diagram) or it may exhibit quasi-periodicity. A particular pulse-form, exhibiting Fast Rise and Exponential-like Decay, has been termed FRED on the diagram; a word coined by the BATSE group to afford quick GRB description.

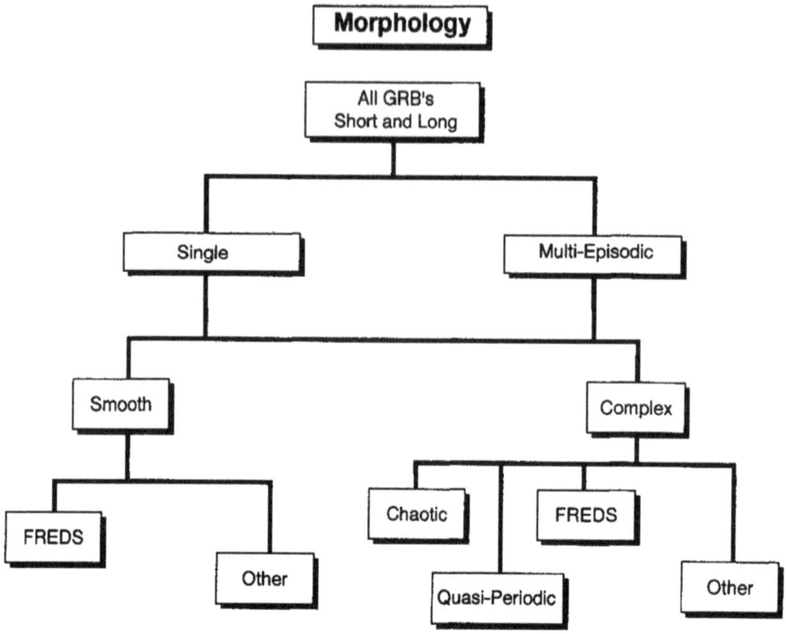

Figure 1. Scheme of GRB morphological classification.

The majority of bursts can be classified in the non-committal way described above; the remainder fall in the "other" box. Figures 2 and 3 display typical examples.

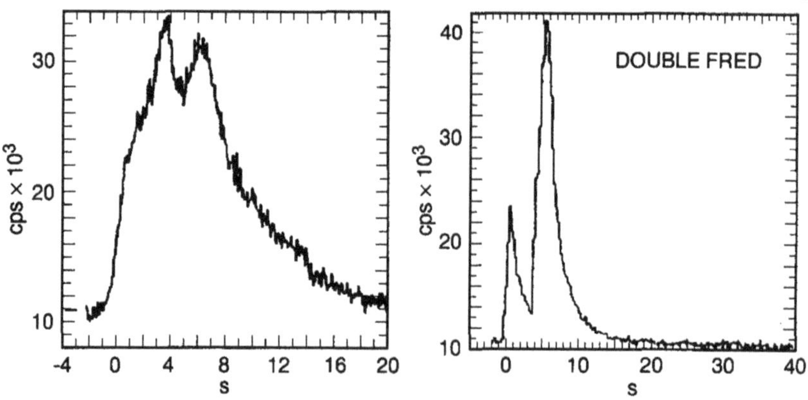

Figure 2. Typical smooth GRB light-curves.

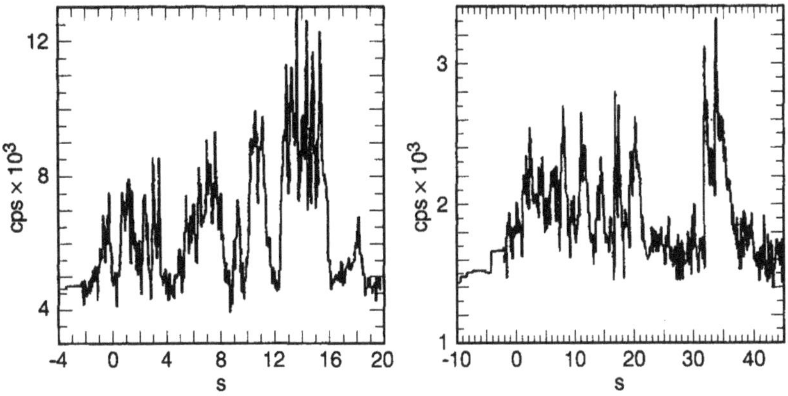

Figure 3. Typical complex GRB light-curves.

Yet, there are light-curves that are unique; as examples, Figure 4a shows the only symmetric GRB and Figure 4b the only anti-FRED detected so far.

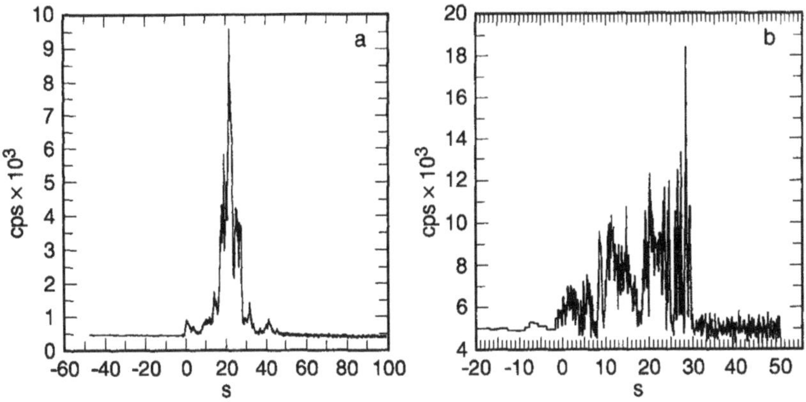

Figure 4a) The symmetric GRB of 910601. 4b) The anti-FRED GRB of 940803.

Although it is difficult to classify highly resolved GRB light-curves, when we look at an average GRB profile we can identify significant overall trends. Mitrofanov *et al.* [6] have created such a profile from a set of GRBs by first coarsely binning each event, then aligning all GRBs at their highest peak and adding the profiles after normalizing to the sample size; they found that the leading part of this average GRB (rise-to-peak) is shorter and has a harder spectrum than the trailing part (peak-to-decay). The asymmetry of the average GRB profile has also been quantitatively established by Nemiroff *et al.* [7], who have devised a simple test they call the Υ

statistic. This parameter is the ratio of the number of times that the counts in the previous time bin are higher to the number where the counts are lower. Figure 5 shows Υ for three individual GRBs and for the complete sample studied by Nemiroff *et al.* [7] for the energy band 50 - 300 keV. Their results indicate that GRBs show significant time-asymmetry, on all time scales from 64 ms to 4096 ms.

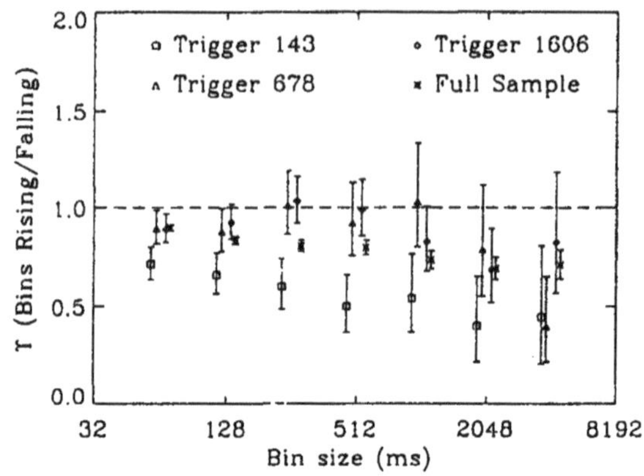

Figure 5. Distribution of the Υ statistic for three individual GRBs and for the entire GRB sample from Nemiroff *et al.* [7].

Summarizing, GRB light-curves vary from very simple, single pulse structures to extremely complex, multi-peaked events and this behaviour manifests itself at **all** time scales from ms to hundreds of seconds. After looking at GRB light-curves for the last 16 years, I feel that the most appropriate summary for this section should be: "when you've seen one GRB, you've seen *one* GRB !"

2.2. Time scales

Each GRB light-curve is a set of pulses that result from a random convolution of three distributions: individual pulse widths, separations between pulses and number of pulses per GRB. Given each of these distributions, one can reproduce an entire light-curve by assuming a few, simple functions for the pulse shapes, such as triangular, or FREDs. Determining the *total* duration of a transient event has always been a controversial issue. The criteria for "duration" have widely varied over the past and no consensus was reached for an algorithm. We introduced in the first BATSE catalog[8] an unbiased and reproducible way of estimating durations. We have used the sum of the counts above 25 keV in all triggered (GRB) detectors to reduce the effect of Earth scattering and the dependence of detector response on the burst arrival direction. We define T_{90} as the time during which the cumulative counts increase from 5% to 95% above background, thus encompassing 90% of the total GRB counts. The times thus defined are an intensity-independent measure of duration, unlike previous definitions[9]. Figure 6a shows the distribution of T_{90} for the 3B catalog [10–11]. As reported previously, we find that there is a strong bimodality in the distribution; the two modes are separated at ~ 2 s. The resulting two GRB subsets, with (logarithmic) mean T_{90} values of 0.57 ± 0.08 and 30.6 ± 1.9 s respectively, have been checked extensively for additional differences. We find that they are indistinguishable, except for their spectral characteristics.

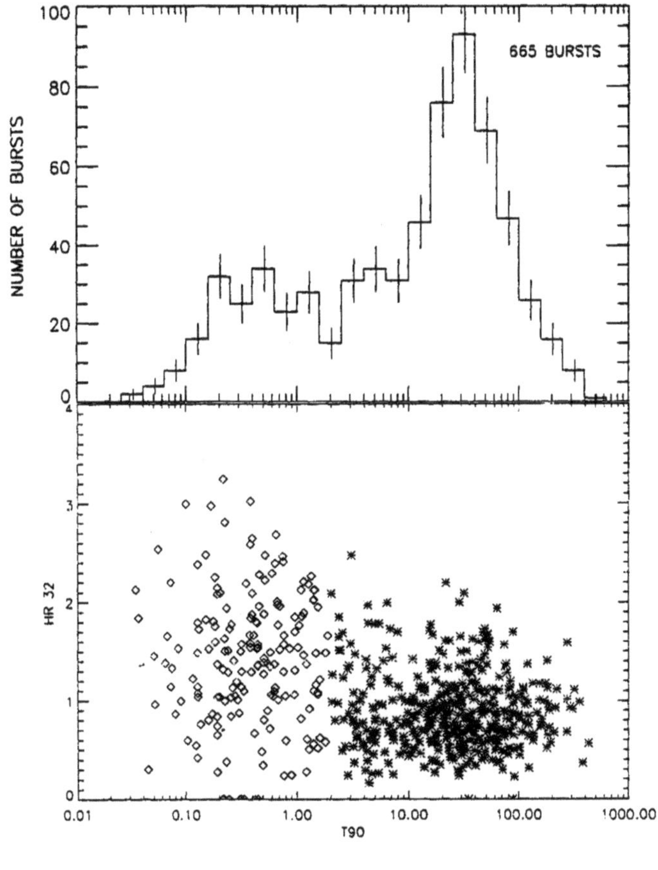

Figure 6. *upper pannel:* Distribution of T_{90} for 665 GRBs from the 3rd BATSE catalog. *lower pannel:* Distribution of HRs *versus* T_{90} for the same set of 665 GRBs.

We define as an integral Hardness Ratio (HR) the ratio of total burst counts above background in the 100 – 300 keV and 25 – 100 keV bands. The distribution of HRs is illustrated in Figure 6b. We see that there is a strong trend for the short events to have relatively hard spectra while the spectra of the longer events are predominantly softer. Their corresponding average HRs are 1.44 ± 0.08 and $0.90 \pm$

0.03, respectively. This trend has been reported in the past[12], albeit not linked with a duration bimodality.

A very important test for the origin of GRBs would be the discovery of cosmological signatures (such as time dilation effects) in their light-curves. If GRBs are cosmological, and if they can be considered standard candles, the weaker events would originate in more distant sources and their durations should be time dilated, (*i.e.*, longer on the average), relative to these of the brighter events. Norris *et al.* [13] have searched for such an effect and reported that they found significant evidence of light-curve stretching by a factor of ~ 2.3 between the extreme 10% dim and 10% bright GRBs (Figure 7). This amount of dilation would correspond to the dim events being at cosmological distances corresponding to $z \sim 1$. Their results have been disputed by Mitrofanov *et al.* [14] who claim that the measured stretching is a result of selection criteria for the two extreme (dim and bright) populations of GRBs. This is currently one of the hotly debated issues in the field.

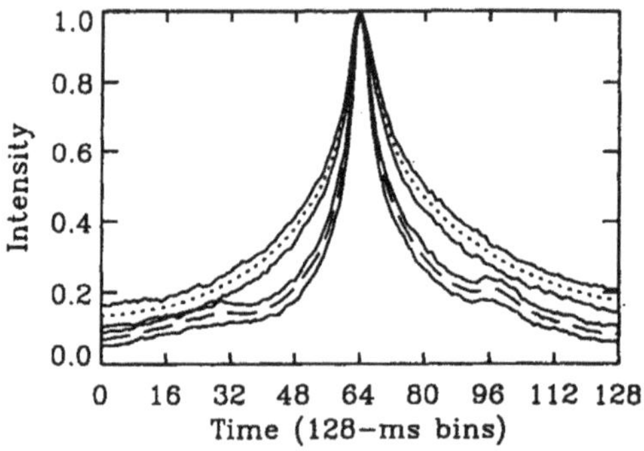

Figure 7. Average profiles of BATSE bursts in three brightness groups, aligned at the peaks. Dimmest (solid, outer profile), dim (dotted), and bright groups (solid, inner profile). (From Norris *et al.* [13].)

2.3. Substructure

The subset of GRBs that are rich in substructure has been extensively searched for coherent pulsations. To prove the existence of coherent oscillations in short, transient phenomena has always been a challenging task. Before BATSE very few events have been identified to exhibit pulsations that are periodic and last for several cycles[15–17]. In a preliminary search using 22 out of the first 137 long (> 2 s) GRBs detected with BATSE, Kouveliotou *et al.* [18] reported that they found no significant coherent oscillations in any of the events. They also find that the power in the red noise component of the Fourier spectrum of GRBs does not extend above 1.5 Hz in 90% of the cases; this would mean that the characteristic GRB time scales are longer than ~ 0.7 s.

Bhat *et al.* [19] have studied the short (< 2 s) GRBs for evidence of variability. Their results from an analysis of 37 events show that more complex bursts have higher average HRs and sharper temporal features at higher energies. A search for photon bunching within the bursts had negative results.

Although no coherent oscillations have been unambiguously found in GRBs there is evidence in several cases for peaks in the GRB Fourier power spectrum around 2 - 7 s, reminiscent of the quasi-periodic oscillations (QPO) found in the intensity variations of X-ray binaries[20]. The interpretation of these peaks is currently unclear.

3. Spatial Properties

GRB distances cannot be measured directly. With self-sufficient experiments like KONUS or BATSE, which have directional capability, one can only determine the direction of arrival of an event and project it on the sky as a circle (with radius reflecting the location uncertainty). All other experiments depend on simultaneous detection of the same event by detectors on widely separated spacecraft and subsequent triangulation (see Section 3.2.) of the burst arrival front, producing arrival rings on the sky. The resulting sky-map, combined with the intensity distribution of bursts, provides constraints on the radial density distribution of GRB sources. Distances can then only be inferred, or modelled, if two basic questions could be answered: are GRBs isotropic and/or homogeneously distributed in space ? Two measured burst properties, locations and intensities, are traditionally employed in a statistical way to find answers to these questions. There are, however, several caveats in these studies. Single-instrument locations depend on the assumed burst input spectrum and the modelled detector response matrices; intensities depend on detector sensitivity and the - unknown - GRB intrinsic luminosity function. I will describe in the following the most commonly used tests that probe the GRB population for evidence of homogeneity and isotropy.

3.1. GRB Homogeneity

3.1.1. The V/V_{max} test

This is a direct statistical test[21-22] for homogeneity (or the lack thereof) of the spatial distribution of a large sample of objects. It employs the ratio of the volume in which the burst was found (V) to that in which it could have been observed (V_{max}) with the sensitivity of a given instrument. In Euclidean space, the source distance, (r), its intrinsic luminosity (L) and the flux measured with a detector (P) are related by $L = 4\pi r^2 P$. The volume (V) to the source, is proportional to r^3. It follows that V is then proportional to $(L/P)^{3/2}$; similarly, V_{max} is proportional to $(L/P_{det})^{3/2}$ (where P_{det} is the the detector threshold flux). Clearly, the ratio V/V_{max} is independent of L. One can also envision the concept as follows: the farther out the sources are, the more the observed space, V, expands to fill the observable space, V_{max}, and the higher their ratio becomes. In a Euclidean space, and if the sources are distributed homogeneously, V/V_{max} is uniformly distributed between 0 (near-by GRBs) and 1 (distant GRBs) with an average value of 0.5.

In practice, the test is associating the count rate threshold for each burst and the peak rate recorded during the event with V_{max} and V, respectively. In this way one avoids a varying threshold bias and bypasses one of the caveats, *i.e.*, the assumed GRB input spectrum for the flux conversion. BATSE observations of 577 GRBs have yielded so far an average value of $V/V_{max} = 0.33 \pm 0.01$, a clear deviation of homogeneity. Values of $V/V_{max} < 0.5$ have been reported in the past[23–25], albeit with much smaller sample sizes.

3.1.2. The log N – log P distribution

Using the arguments of Section 3.1.1., one finds a very simple functional form that describes the cumulative distribution of observed GRB peak intensities (P) for a constant source density in Euclidean space. Brightness (or peak intensity P) decreases proportionally to the inverse square of the distance (r) to the source. Volume to that

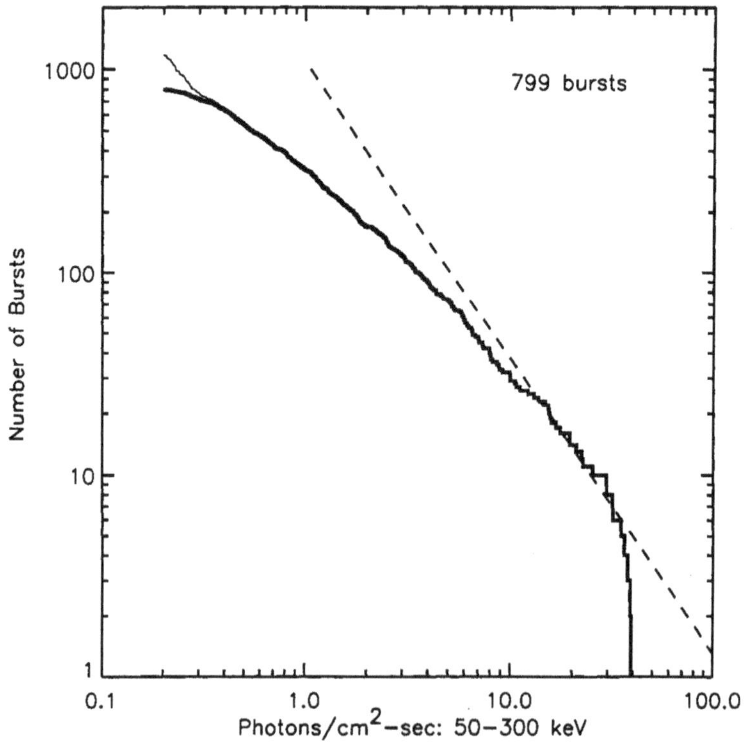

Figure 8. Log N – log P distribution for 799 GRBs. The peaks are integrated with 1024 ms over 50 – 300 keV. The dashed line represents a $-3/2$ power law expected for a homogeneous distribution. (From Pendleton *et al.* [27].)

distance increases as the cube of r. Under the above assumptions, the integral number of bursts observed above a given peak intensity P is: $N(>P) \propto P^{-3/2}$. The $-3/2$ power law expected for such a distribution is independent of the intrinsic burst luminosity function, *if* the burst density is constant; the detector dependence is alleviated by correcting for varying thresholds and burst geometries (trigger efficiency).

Before BATSE, the log N – log P distribution was found to be inconsistent with a $-3/2$ power law. It was widely believed[26], however, that the relative lack of weak events was caused by the sample incompleteness: existing detector sensitivity was good enough to sample the high intensity part of the distribution, but most of the expected weaker events should have been missed. One crucial and unexpected result obtained with BATSE was that the low (faint) end of the distribution significantly deviates from the $-3/2$ power law; in fact, it follows a -0.8 power law. Figure 8 shows the BATSE log N-log P distribution for 799 GRBs[27]. The BATSE range covers three orders of magnitude, substantially larger than for all previous instruments. I emphasize, that the relative dearth of weak events is not caused by a selection effect: if they would have been there they would have been detected with BATSE.

One has to conclude that GRB sources are **not** distributed homogeneously in Euclidean space. The farther out we probe, the less events we find, or in other words : we appear to have reached the end of the GRB distribution. We have no direct information, however, for the distance scale of GRBs.

3.2. GRB Locations

Single–instrument locations of GRBs have, so far, only been possible with multi–detector experiments (KONUS, BATSE). They are derived by comparing and combining responses of the detectors that have detected an event. The BATSE localization algorithm[28] takes into account corrections for scattering of burst photons by the Earth's atmosphere, the angular dependence of the detector sensitivity and the burst attenuation through the spacecraft body. Currently, the systematic error of the BATSE locations is 4°; the statistical error depends on the event intensity, and varies between 1° (for strong events) and 10° (for weak events).

Another method of GRB location is the triangulation of the arrival times of the burst wavefront at multiple detectors located in space. Once an event has ben recorded simultaneously by instruments separated by a large distance R (long baseline), one can use the difference dt between the signal arrival times, to locate the source, following: $\cos\theta = c\,dt/R$ and $\theta = \arccos\,(c\,dt/R)$. Her θ is the angle between the vector that separates the two satellites,and the direction from which the GRB came. In this way we can define a cone of opening angle θ and thickness $d\theta$ (depending on the accuracy of the measurements). The source should lie at the projection of the cone on the sky, which is a ring of thickness $d\theta$. Two satellite detections result in a ring in the sky, three, in the crossing areas of two rings, a forth detector eliminates one of the areas and results in a single location.

So far there have been 3 InterPlanetary Networks (IPN)[29–35] of satellites with GRB detectors, with mutual distances of up to several Astronomical Units (AU). The accuracy of locations derived via IPN triangulation depends on the baselines between satellites and the number of instruments that detected an event. IPN locations can be as small as several tens of square arcmin in size if the satellites are separated by long baselines; so far the number of such accurate locations is limited and thus cannot provide insight in the overall sky distribution of GRBs. The only way to study burst locations today is to evaluate their distributions statistically using isotropy tests.

3.2.1. Isotropy tests

BATSE had detected 1005 GRBs by early June 1994. Figure 9 shows the distribution of their locations in the sky in galactic coordinates. Several tests for isotropy have been applied to this distribution; they differ in the type of the anisotropy each can detect.

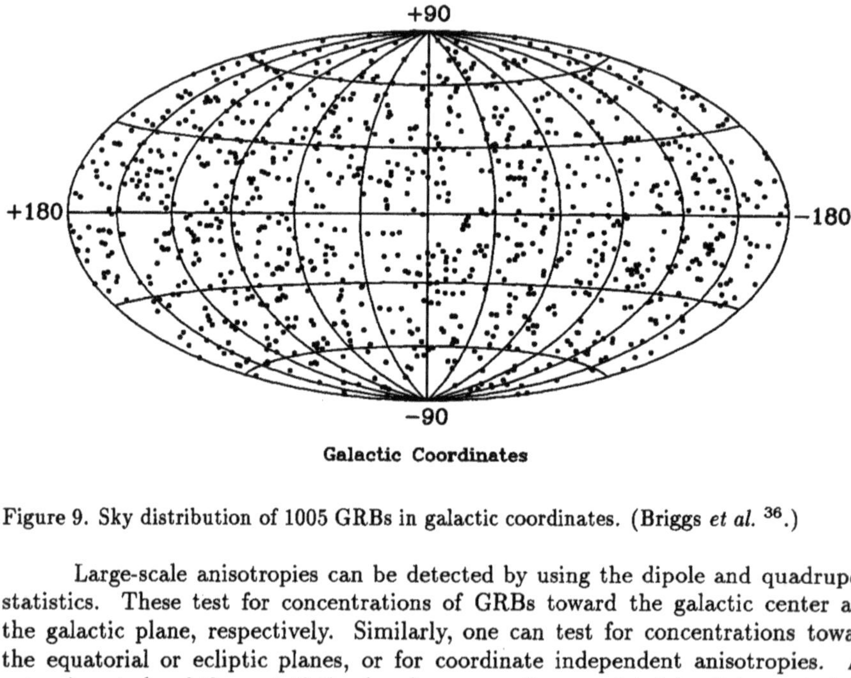

Figure 9. Sky distribution of 1005 GRBs in galactic coordinates. (Briggs *et al.* [36].)

Large-scale anisotropies can be detected by using the dipole and quadrupole statistics. These test for concentrations of GRBs toward the galactic center and the galactic plane, respectively. Similarly, one can test for concentrations toward the equatorial or ecliptic planes, or for coordinate independent anisotropies. An extensive study of these statistics has been recently completed by Briggs *et al.* [36]. Table 1 contains the values of the galactic and equatorial tests. The deviations from large-scale isotropy are at most 0.9σ for both coordinate systems: the distribution of GRBs is extremely isotropic.

TABLE 1

Statistics for the first 1005 Bursts detected with BATSE[36]

Statistic[a]	Moment	Coordinate System	Value expected for isotropy with uniform sky exposure	Value expected for isotropy with BATSE sky exposure	Observed value
$\langle \cos\theta \rangle$	Dipole	Galactic	0±0.018	−0.013±0.018	−0.004
$\langle \sin^2 b - \frac{1}{3} \rangle$	Quadrupole	Galactic	0±0.0093	−0.005±0.0093	−0.008
$\langle \sin\delta \rangle$	Dipole	Equatorial	0±0.019	0.026±0.019	0.016
$\langle \sin^2 \delta - \frac{1}{3} \rangle$	Quadrupole	Equatorial	0±0.0097	0.026±0.0097	0.019

[a] θ is the angle between a burst and the Galactic Center, b is galactic latitude and δ is declination.

3.2.2. Small-scale anisotropies, and GRB Repetition tests

Two main tests have been used to probe possible small-scale anisotropies of the GRB sky distribution, and/or repetitions of GRBs. These are practically indistinguishable, in that a concentration of GRB sources around a point in the sky cannot be differentiated from one source repeating several times. The first test is the two point angular correlation function: it studies the distribution of the number of pairs with angular separation θ, within a solid angle dΩ for a given point. The second is the nearest-neighbour analysis: it tests whether the separations between bursts are consistent with the separations expected for an isotropic distribution.

Before BATSE there had been only upper limits of ~0.1 burst/year for burst repetition rates based on few locations derived by IPN[37–38]. After the first BATSE catalog was published[8] several analyses reported detection of burst repetition or clustering at small angles (~5°)[39] at small and large angles (4°and 180°)[40] and in space *and* time (4°and 4 day-intervals)[41]. Meegan *et al.* [42] an Brainerd *et al.* [43] have recently performed an extensive study, utilizing the above tests for the 1005 GRBs of the third BATSE catalog. Their results are consistent with the hypothesis that GRB sources do not repeat. They also set an upper limit to the repeater fraction of up to about 20% of the observed bursts, assuming a simple model of two observed bursts per repeating source. One has to stress here though, that such limits are highly model-dependent on assumptions for the number of repeaters and the number of repetitions per source.

Conclusion: BATSE data have not solved the problem of the GRB distances but have enabled us to envision the GRB universe around us, as spherically symmetric and bound.

4. Spectral Properties

GRBs have very hard, non–thermal spectra, with the bulk of their power being emitted above 100 keV, up to several tens of MeV. Very few cases have been reported[44–46], where low–level X-ray (1–10 keV) emission, comprising $\lesssim 1-2$ % of the total power, has been seen before and during a GRB. An even smaller number of events (~ 5) has been recently[47] associated with very high–energy photons in the GeV range, the outstanding example being GRB940217, in which photons up to 18 GeV were recorded $\sim$ 100 min after its onset.

Spectral studies comprise fits of the GRB continua, statistical studies of spectral parameters and their evolution, and searches for lines. In the following I will describe the current results obtained with BATSE data and compare them to previously reported studies.

4.1. Spectral continua

Several functions (see Hurley[48] and references therein) have at times been fit to burst spectral continua: optically thin thermal bremsstrahlung (OTTB), power–laws (single and/or broken), log–normal[49] and thermal synchrotron. Average values for the characteristic spectral parameters derived were: kT's between 200 – 400 keV, and power–law indices, $\alpha = 1.5 - 2.5$, clearly indicating hard spectra.

Recently, Band *et al.* [50] have established the following four–parameter functional form, to characterize GRB spectra:

$$N_E(E) = A\ E^{\alpha}\ e^{-(E/E_0)} \text{ for } E < (\alpha-\beta)\ E_0, \text{ and } N_E(E) = B\ E^{\beta} \text{ for } E > (\alpha-\beta)\ E_0$$

where B is chosen so that N_E and its derivative are continuous. The parameters α, β and E_0 vary, with characteristic values[50] of $\alpha \sim -0.5, \beta \sim -2.0$ and $E_0 \sim 150$ keV.

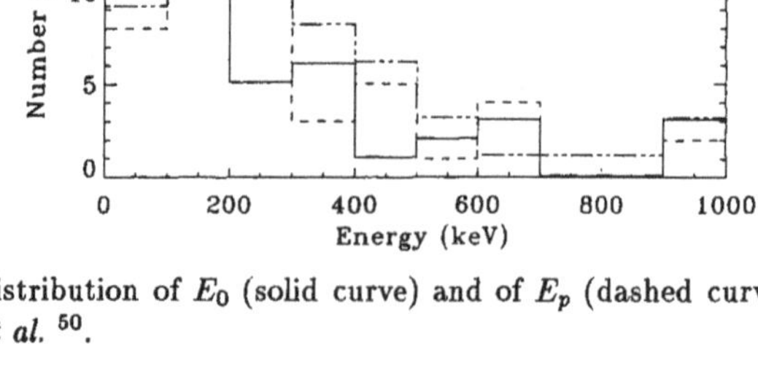

Figure 10. Distribution of E_0 (solid curve) and of E_p (dashed curve) for 54 spectral fits. From Band *et al.* [50].

This spectral form has been found to fit the data of the BATSE spectral detectors very well over their entire energy range of $\sim$ 20 keV to several MeV. The break energy E_0, and both α and β vary significantly from burst to burst and within bursts, thus not affording a single parameter set characterization of a GRB spectrum.

A popular trend[51] today, is to plot GRB spectra in the form of $E^2 N_E$ vs. E, *i.e.*, giving power per logarithmic energy band. This plot is also known as "the theoretician's plot" because it depicts clearly where most of the energy of a GRB is radiated. Figure 10 shows the distribution of the energy peak E_p of the $E^2\ N_E$ plot and the distribution of the turnover energy mentioned above, E_0, for fits of 54 GRB spectra[50]: both vary widely between 0 – 600 keV. E_p seems to peak around 200 keV, another confirmation that GRBs are predominantly hard transients.

4.2. Spectral Evolution

As we mentioned above, the spectrum of a GRB may change significantly during the event. Several trends have been reported in the literature in the form of a correlation between spectral hardness and event intensity: hard–to–soft evolution[52–53], both within a pulse and from pulse to pulse in a burst (anti–correlation of hardness and intensity), or no evolution at all[54] (no correlation).

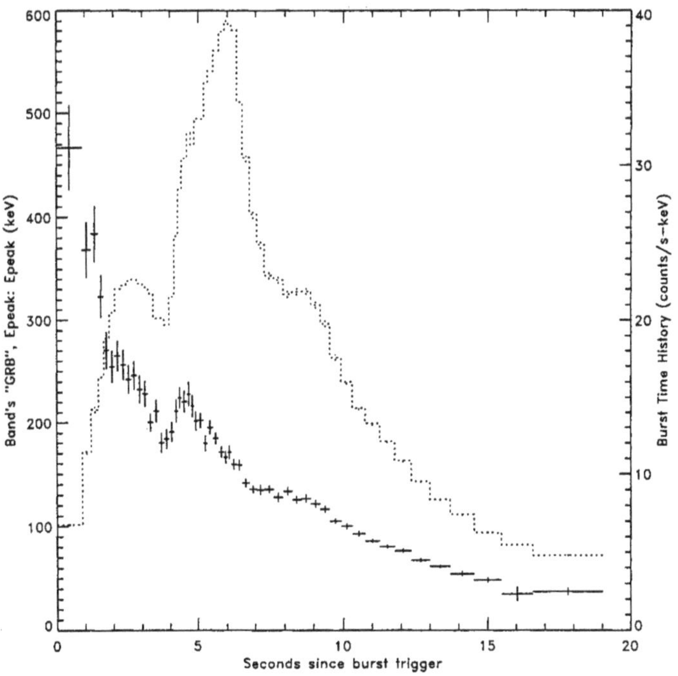

Figure 11. Histogram of the GRB911118 light curve superposed on the E_p values for the same event.

Ford *et al.* [53] have studied the evolution of E_p during 37 bright BATSE bursts. In most cases they find that E_p either rises with or slightly precedes major intensity increases and softens for the remainder of the pulse. Inter–pulse emissions are generally harder early in the burst, while in multi–pulsed GRBs they find that late spikes tend to be softer than earlier ones. Figure 11[55], shows a characteristic hard–to–soft evolution of a GRB: the crosses are the values of E_p during the burst, superposed on the burst light curve (dotted line).

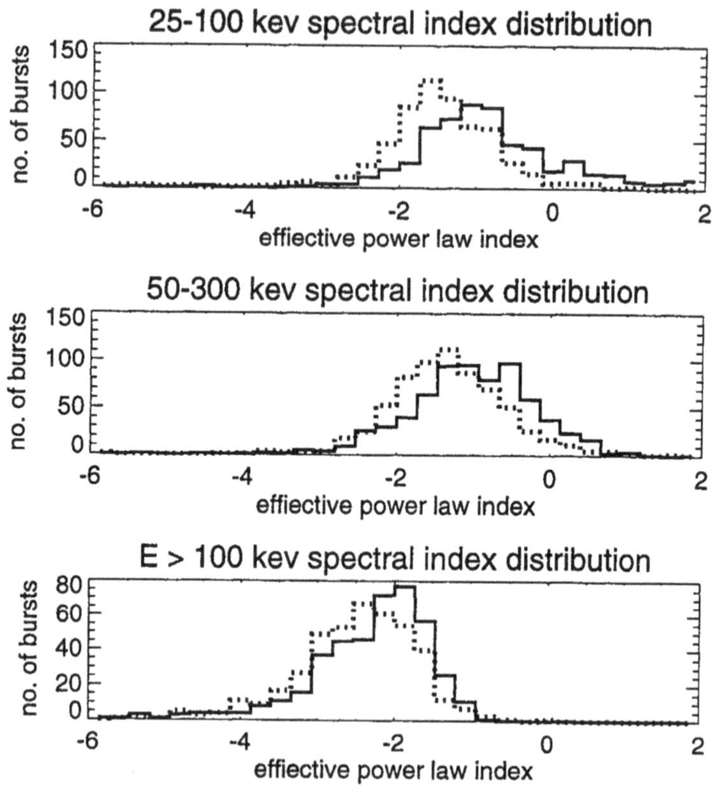

Figure 12. Histograms of the spectral indices over the three energy ranges indicated on the plot. Solid lines are distributions for peak rate spectra and dashed lines represent total fluence spectra. From Pendleton *et al.* [49].

Pendleton *et al.* [49] have published distributions of the continuum spectral characteristics of 596 GRBs derived from flux ratios calculated from the BATSE Large Area Detector (LAD) four-channel discriminator data. These data have been converted from counts to photons using a direct spectral inversion technique to remove effects of atmospheric scattering and energy dependence of the detector angular response. Figure 12 shows the histograms of the spectral indices of the fits over three energy ranges: 25 – 100, 50 – 300 and > 100 keV. Solid lines represent distributions for peak rate spectra and dotted lines show the distribution for the total fluence spectra. We see clearly that the fluence spectra tend to be softer than those for the peak flux, particularly in the 20 – 100 keV range. Pendleton *et al.* [49] interpret their results as indications that the more intense portions of bursts usually have harder spectra than the less intense portions, and that the spectra change significantly within individual bursts.

Finally, another technique (adapted[56] from X-ray astronomy) employes color–color diagrams to study GRB spectral evolution. Figure 13 shows the evolution of the hard versus the soft color of GRB911118; colors are the ratios of counts in different energy bands. Time evolves from the upper left corner of the plot to the lower left one, and clearly shows the hard–to–soft trend depicted in Figure 11.

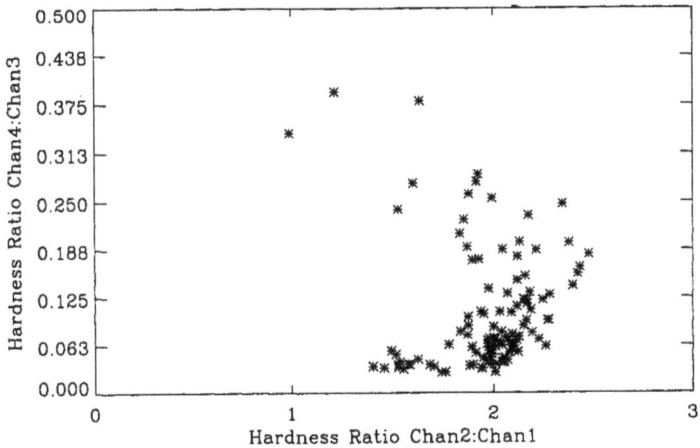

Figure 13. Color–color diagram of GRB911118. From Kouveliotou *et al.* [56].

4.3. Spectral Lines

An important argument that supported the pre-BATSE idea that GRBs originate from galactic neutron stars was the detection of discrete features in some GRB spectra, with energies of several tens of keV (interpreted as cyclotron transition lines) and near 400 keV (interpreted as gravitationally red–shifted electron–positron annihilation lines). For a review of these line reports see Hurley[48] and references therein. In spite of intensive studies, so far no lines have been detected with BATSE[57–60]. A consistency study has been made for two data sets: GINGA and BATSE[61]. This leads to the conclusion[61] that the probability that the BATSE non–detection is consistent with the GINGA detection, is of the order of 10%. New software has been developed to facilitate an automated line search, which examines each GRB in numerous time intervals and combinations of such over large energy intervals[62]. This analysis is currently in progress.

5. Counterparts at other Wavelengths

The search for counterparts to GRBs in other wavelengths started very soon after the discovery of the phenomenon. These efforts have been concentrated mainly in the optical region, although there are ongoing projects that monitor the GRB error boxes in the radio[63], and TeV ranges[64] of the electromagnetic spectrum, and with neutrino[65] detectors. The searches have been both of archival nature, utilizing large archives of optical photographic plates[66], and simultaneous with gamma-ray observations[67], initiated by an alert activated by a GRB instrument that triggered on a burst. So far no GRB counterpart has been identified and unambiguously accepted; there have been several controversial claims that have been discounted by others as instrumental effects or as of low significance[68–70].

In addition there have been searches for associations of the GRB sky map with known celestial object distributions[71–72]. GRB locations have been compared with those of flare stars, AGNs, the nearest 2809 galaxies, Abell clusters of galaxies, Seyfert galaxies, globular clusters, BL Lac's, Quasars, *etc.* No correlations were found so far; in other words, there is no galactic or extragalactic population of objects that can be associated with GRBs.

The discovery of gravitationally lensed GRBs by intervening large masses between their sources and the Earth, would provide undisputable evidence for cosmological distances to GRB sources. Nemiroff *et al.* [73–74] have searched the BATSE data base for lensed events, and have found none. Here we have another null result that cannot lead to conclusions for the nature of GRBs.

6. Summary

The BATSE results have unambiguously shown that GRBs are distributed isotropically in the sky. The most intense bursts are consistent with homogeneity[75], but the weak bursts are clearly inhomogeneous. In Euclidean space these results indicate that we have probed to the end of the GRB universe. The above combined with a BATSE non–detection of repeaters and lines in GRB spectra have serious consequences for the GBR models; some of the previously favoured galactic models are now falling out of grace. Instead, a consensus is forming slowly within the community towards an extragalactic origin for GRB sources.

It somewhat discomforting that so fundamental a question cannot yet be answered: well into the third decade after their discovery GRBs remain to date one of the unsolved mysteries of high–energy astrophysics. The majority opinion of the GRB community has oscillated over the years between a galactic and an extragalactic nature of the events.

BATSE obervations have led to a thickening of the plot of the GRB mystery, and its results have created more controversies than solutions. To unravel this mystery, future experiments should be focused on providing accurate (of order of arcsecond) GRB locations that would help to identify e.g., repeaters and burst counterparts in other wavelength bands. In particular, high–resolution spectra should be obtained, most desirably covering the low energy (0.5 – 10 keV) part of the spectrum, which would provide stringent constraints for the GRB emission mechanism.

7. Acknowledgements

I thank my colleagues of the BATSE Team for many stimulating discussions and for supplying a major part of the results presented here. I also thank Jan van Paradijs for critically commenting on this manuscript.

8. References

1. R. W. Klebesadel, I. B. Strong, R. A. Olson, *Ap. J. Letters* **182** (1973) L85.
2. G. J. Fishman *et al.* , *in Proc. of the GRO Science Workshop*, ed. W. N. Johnson (NASA/GSFC, 1989), p. 39.
3. U. Desai, *Astrophys. Space Sci.* **75** (1981) 15.
4. E. Mazets, S. Golenetskii, *Astrophys. Space Sci.* **75** (1981), 47.
5. C. Barat *et al.* , *Astrophys. Space Sci.* **75** (1981), 83.
6. I. Mitrofanov *et al.* , *in Gamma-Ray Bursts, Huntsville AL 1993*, eds. G. Fishman, J. Brainerd and K. Hurley (New York:AIP 307), p. 187.
7. R. Nemiroff *et al.* , *in Gamma-Ray Bursts, Huntsville AL 1993*, eds. G. Fishman, J. Brainerd and K. Hurley (New York:AIP 307), p. 237.
8. G. J. Fishman *et al.* , Ap. J. Suppl. Series **92** (1994), 229.
9. C. Kouveliotou *et al.* , *Ap. J. Letters* **413** (1993), L101.
10. C. A. Meegan *et al.* , *in preparation.*
11. C. Kouveliotou *et al.* , *in preparation.*
12. J-P. Dezalay *et al.* , *in Gamma-ray Bursts, Huntsville AL 1991*, eds. W. Paciesas and G. J. Fishman, (New York:AIP 265), p. 304.
13. J. P. Norris *et al.* , *Ap. J.* **424** (1994), 540.
14. I. Mitrofanov *et al.* , *in preparation.*
15. E. Mazets *et al.* , *Nature* **282** (1979), 587.
16. C. Barat *et al.* , *Ap. J. Letters* **286** (1984), L5.
17. C. Kouveliotou *et al.* , *Ap. J. Letters* **330** (1988), L101.
18. C. Kouveliotou *et al.* , *in Gamma-ray Bursts, Huntsville AL 1991*, eds. W. Paciesas and G. J. Fishman, (New York:AIP 265), p. 299.
19. N. Bhat *et al.* , *in Gamma-Ray Bursts, Huntsville AL 1993*, eds. G. Fishman, J. Brainerd and K. Hurley (New York:AIP 307), p. 197.
20. G. Hasinger and M. van der Klis, *A & A* **225** (1989), 79.
21. M.Schmidt, *Ap. J.* **151** (1968), 393.
22. M. Schmidt, J. Higdon, G. Hueter, *Ap. J. Letters* **329** (1988), L85.
23. J. Higdon *et al.* , *in Gamma-ray Bursts, Huntsville AL 1991*, eds. W. Paciesas and G. J. Fishman, (New York:AIP 265), p. 89.

24. Y. Ogasaka, T. Murakami, J. Nishimura, A. Yoshida, E. Fenimore, *Ap. J. Letters* **383** (1991), L61.
25. O. Terekhov, D. Denishenko, V. Lobachev, R. Sunyaev, A. Kovtun, *Astron. Letters* **20** (1994), 265.
26. K. Hurley, *in Gamma-Ray Bursts 1986*, eds. E. P. Liang & V. Petrosian, (New York: AIP 141), p. 1.
27. G. Pendleton *et al.* , *Ap. J.*, submitted.
28. M. N. Brock *et al.* , *in Gamma-ray Bursts, Huntsville AL 1991*, eds. W. Paciesas and G. J. Fishman, (New York:AIP 265), p. 383.
29. T.L. Cline *et al.* , *in Gamma-ray Bursts, Huntsville AL 1991*, eds. W. Paciesas and G. J. Fishman, (New York:AIP 265), p. 72.
30. K. Hurley *et al.* , *A & A Suppl.* **97** (1993), 39.
31. K. Hurley *et al.* , *in Gamma-Ray Bursts, Huntsville AL 1993*, eds. G. Fishman, J. Brainerd and K. Hurley (New York:AIP 307), p. 369.
32. K. Hurley *et al.* , *in Gamma-Ray Bursts, Huntsville AL 1993*, eds. G. Fishman, J. Brainerd and K. Hurley (New York:AIP 307), p. 27.
33. K. Hurley *et al.* , *in Gamma-Ray Bursts, Huntsville AL 1993*, eds. G. Fishman, J. Brainerd and K. Hurley (New York:AIP 307), p. 364.
34. K. Hurley *et al.* , *Ap. J. Suppl. Series* **92** (1994), 655.
35. K. Hurley *et al.* , *Ap. J. Letters* **431** (1994), L31.
36. M. Briggs *et al.* , *Ap. J.*, submitted.
37. B. Schaefer T. L. Cline, *Ap. J.* **289** (1985), 490.
38. J-L. Atteia *et al.* , *Ap. J. Suppl. Series* **64** (1987), 305.
39. J.M. Quashnock, D. Q. Lamb, *MNRAS* **265** (1993), L45.
40. R. Narayan, T. Piran, *MNRAS* **265** (1993), L65.
41. V. Wang, R. E. Lingenfelter, *Ap. J. Letters* **416** (1993), L 13.
42. C. A. Meegan *et al.* , Ap. J. Letters, submitted.
43. J.J. Brainerd *et al.* , Ap. J. Letters, accepted.
44. A. Yoshida *et al.* , *Publ. Astron. Soc. of Japan* **41** (1989), 509.
45. T. Murakami *et al.* , *Nature* **350** (1991), 592.
46. T. Murakami *et al.* , *in Proc. Los Alamos Workshop, Taos NM 1990*, eds. C. Ho, R. Epstein, E. Fenimore, (Cambridge University Press), p. 239.
47. K. Hurley *et al.* , *Nature* **372** (1994), 652.
48. K. Hurley *in Gamma-ray Bursts, Huntsville AL 1991*, eds. W. Paciesas and G. J. Fishman, (New York:AIP 265), p. 3.
49. G. Pendleton *et al.* , *Ap. J.* **431** (1994), 416.
50. D. Band *et al.* , *Ap. J.* **413** (1993), 281.
51. E. Liang, *in Proc. GRO Science Workshop, GSFC 1989*, ed. W.N. Johnson, p. 4.
52. J.P. Norris *et al.* , *Ap. J.* **301** (1986), 213.
53. L. Ford *et al.* , Ap. J., accepted.
54. V. Kargatis *et al.* , *Ap. J. Letters* **421** (1994), L83.
55. R. Preece, *private communication.*
56. C. Kouveliotou *et al.* , *A & A Suppl. Series* **97** (1993), 55.

57. D. Palmer *et al.* , *in Gamma-Ray Bursts, Huntsville AL 1993*, eds. G. Fishman, J. Brainerd and K. Hurley (New York:AIP 307), p. 247.
58. L. Ford *et al.* , *in Gamma-Ray Bursts, Huntsville AL 1993*, eds. G. Fishman, J. Brainerd and K. Hurley (New York:AIP 307), p. 298.
59. R. Preece *et al.* , *in Gamma-Ray Bursts, Huntsville AL 1993*, eds. G. Fishman, J. Brainerd and K. Hurley (New York:AIP 307), p. 266.
60. B. Schaefer *et al.* , *in Gamma-Ray Bursts, Huntsville AL 1993*, eds. G. Fishman, J. Brainerd and K. Hurley (New York:AIP 307), p. 271.
61. D. Band *et al.* , *Ap. J.* **434** (1994), 560.
62. M. Briggs, *private communication.*
63. D. Frail *et al.* , *Ap. J.Letters*
64. V. Connaughton *et al.* , *in Gamma-Ray Bursts, Huntsville AL 1993*, eds. G. Fishman, J. Brainerd and K. Hurley (New York:AIP 307), p. 470.
65. J.M. LoSecco, *Ap. J.* **425** (1994), 217.
66. B. Schaefer, *Ap. J.* **364** (1990), 590.
67. B. Schaefer *et al.* , *Ap. J. Letters* **422** (1994), L71.
68. A. Zytkow, *Ap. J.* **359** (1990), 138.
69. J. Greiner, *A & A* **264** (1992), 121.
70. J. Greiner, E. Moskalenko, *A & A* **283** (1994), 693.
71. S. Howard *et al.* , *in Proc. CGRO Symposium St. Louis 1992*, eds. M. Friedlander, N. Gehrels, D. Macomb, (New York: AIP 280), p. 793.
72. F. Hack *et al.* , *in Gamma-Ray Bursts, Huntsville AL 1993*, eds. G. Fishman, J. Brainerd and K. Hurley (New York:AIP 307), p. 359.
73. R. Nemiroff *et al.* , *Ap. J.* **414** (1993), 36.
74. R. Nemiroff *et al.* , *Ap. J.* **423** (1994), 432.
75. E. Fenimore *et al.* , *Nature* **366** (1993), 40.

IV. GRAVITY, COSMOLOGY AND STRINGS

STRING THEORY IN COSMOLOGICAL SPACETIMES

H.J. de Vega[(a)] **and N. Sánchez**[(b)]

(a) Laboratoire de Physique Théorique et Hautes Energies, Université Pierre et Marie Curie (Paris VI) et Université Denis Diderot (Paris VII), Tour 16, 1er. étage, 4, Place Jussieu 75252 Paris, cedex 05, France.

Laboratoire Associé au CNRS URA280.

(b) Observatoire de Paris, DEMIRM, 61, Avenue de l'Observatoire, 75014 Paris, France. Laboratoire Associé au CNRS URA336, Observatoire de Paris et École Normale Supérieure.

(April 1995)

Abstract

Progress on string theory in curved spacetimes since 1992 is reviewed. After a short introduction on strings in Minkowski and curved spacetimes, we focuse on strings in cosmological spacetimes.

The classical behaviour of strings in FRW and inflationary spacetimes is now understood in a large extent from various types of explicit string solutions. Three different types of behaviour appear in cosmological spacetimes: **unstable, dual** to unstable and **stable**. For the unstable strings, the energy and size grow for large scale factors $R \to \infty$, proportional to R. For the dual to unstable strings, the energy and size blow up for $R \to 0$ as $1/R$. For stable strings, the energy and proper size are bounded. (In Minkowski spacetime, all string solutions are of the stable type).

Recent progress on self-consistent solutions to the Einstein equations for string dominated universes is reviewed. The energy-momentum tensor for a gas of strings is then considered as source of the spacetime geometry and from the above string behaviours the string equation of state is determined. The self-consistent string solution exhibits the realistic matter dominated behaviour $R \sim (X^0)^{2/3}$ for large times and the radiation dominated behaviour $R \sim (X^0)^{1/2}$ for early times.

Finally, we report on the exact integrability of the string equations plus the constraints in de Sitter spacetime that allows to systematically find **exact** string solutions by soliton methods and the multistring solutions. **Multistring solutions** are a new feature in curved spacetimes. That is, a single world-sheet simultaneously describes many different and independent strings. This phenomenon has no analogue in flat spacetime and appears as a consequence of the coupling of the strings with the spacetime geometry.

I. INTRODUCTION

Since the previous Erice School on 'String Quantum Gravity' a host of impressive developments has taken place on strings in curved spacetime. For the state of the art in 1992 we refer to the 1992 Proceedings [1] where the programme on string quantization in curved spacetimes initiated by us in 1987 [2] is reviewed.

A consistent quantum theory of gravity is the strongest motivation for string theory and hence to study strings in curved spacetime [1]. As stressed in the sec. 1 of the 1992 lectures, a quantum theory of gravity must be a theory able to describe all physics below the Planck scale $M_{Planck} = \hbar c/G = 1.22\,10^{16}$Tev. That means that a sensible theory of quantum gravity is necessarily part of a unified theory of all interactions. **Pure gravity** (a model containing only gravitons) cannot be a physical and realistic quantum theory. To give an example, a theoretical prediction for graviton-graviton scattering at energies of the order of M_{Planck} must include all particles produced in a real experiment. That is, in practice, all existing particles in nature, since gravity couples to all matter.

String theory is therefore a serious candidate for a quantum description of gravity since it provides a unified model of all interactions overcoming at the same time the nonrenormalizable character of quantum fields theories of gravity.

The present lectures deal mainly with strings in cosmological spacetimes; substantial results were achieved in this field since 1992. The classical behaviour of strings in FRW and inflationary spacetimes is now understood in a large extent [3]. This understanding followed the finding of various types of exact and numerical string solutions in FRW and inflationary spacetimes [5] - [9]. For inflationary spacetimes, the exact integrability of the string propagation equations plus the string constraints in de Sitter spacetime [4] is indeed an important help. This allowed to systematically find **exact** string solutions by soliton methods using the linear system associated to the problem (the so-called dressing method in soliton theory) and the multistring solutions [5] - [8].

In summary, three different types of behaviour are exhibited by the string solutions in cosmological spacetimes: **unstable, dual** to unstable and **stable.** For the unstable strings, the energy and size grow for large scale factors $R \to \infty$, proportional to R. For the dual to unstable strings, the energy and size blow up for $R \to 0$ as $1/R$. For stable strings, the energy and proper size are bounded. (In Minkowski spacetime, all string solutions are of the stable type). The equation of state for these string behaviours take the form

- (i) **unstable** for $R \to \infty$ $p_u = -E_u/(d-1) < 0$
- (ii) **dual to unstable** for $R \to 0$, $p_d = E_d/(d-1) > 0$.
- (iii) **stable** for $R \to \infty$, $p_s = 0$.

Here E_u and E_d stand for the corresponding string energies and $d-1$ for the number of spatial dimensions where the string solutions lives. For example, $d-1 = 1$ for a straight string, $d-1 = 2$ for a ring string, etc. This number d is obviously less or equal than the number of spacetime dimensions D.

As we see above, the dual to unstable string behavior leads to the same equation of state than radiation (massless particles or hot matter). The stable string behavior leads to the equation of state of massive particles (cold matter). The unstable string behavior is a purely 'stringy' phenomenon. The fact that in entails a negative pressure is however physically acceptable. For a gas of strings, the unstable string behaviour dominates in inflationary universes when $R \to \infty$ and the dual to unstable string behavior dominates for $R \to 0$.

The unstable strings correspond to the critical case of the so called *coasting universe* [15,29]. That is, classical strings provide a *concrete* realization of such cosmological models. The 'unstable' behaviour is called 'string stretching' in the cosmic string literature [16,17].

It must be stressed that while time evolves, a **given** string solution may exhibit two and even three of the above regimes one after the other (see sec. III). Intermediate behaviours are also observed in ring solutions [7,9]. That is,

$$P = (\gamma - 1)\, E \quad \text{with} \quad -\frac{1}{d-1} < \gamma - 1 < +\frac{1}{d-1}$$

Another new feature appeared in curved spacetimes: **multistring solutions**. That is a single world-sheet simultaneously describes many different and independent strings. This phenomenon has no analogue in flat spacetime. This is a new feature appearing as a consequence of the interaction of the strings with the spacetime geometry.

The world-sheet time τ turns out to be an multi-valued function of the target string time X^0 (which can be the cosmic time T, the conformal time η or for de Sitter universes it can be the hyperboloid time q^0). Each branch of τ as a function of X^0 corresponds to a different string. In flat spacetime, multiple string solutions are necessarily described by multiple world-sheets. Here, a single world-sheet describes one string, several strings or even an infinite number of different and independent strings as a consequence of the coupling with the spacetime geometry. These strings do not interact among themselves; all the interaction is with the curved spacetime. One can decide to study separately each of them (they are all different) or consider all the infinite strings together.

Of course, from our multistring solution, one *could* just choose only one interval in τ (or a subset of intervals in τ) and describe just one string (or several). This will be just a **truncation** of the solution.

The really remarkably fact is that all these infinitely many strings come **naturally together** when solving the string equations in de Sitter spacetime as we did in refs. [5] - [7].

Here, interaction among the strings (like splitting and merging) is neglected, the only interaction is with the curved background.

The study of string propagation in curved spacetimes provide essential clues about the physics in this context but is clearly not the end of the story. The next step beyond the investigation of **test** strings, consist in finding **self-consistently** the geometry from the strings as matter sources for the Einstein equations or better the string effective equations (beta functions). This goal is achieved in ref. [3] for cosmological spacetimes at the classical level. Namely, we used the energy-momentum tensor for a gas of strings as source for the Einstein equations and we solved them self-consistently.

To write the string equation of state we used the behaviour of the string solutions in cosmological spacetimes. Strings continuously evolve from one type of behaviour to another, as is explicitly shown by our solutions [4] - [7]. For intermediate values of R, the equation of state for gas of free strings is clearly complicated but a formula of the type:

$$\rho = \left(u_R\, R + \frac{d}{R} + s \right) \frac{1}{R^{D-1}} \tag{1.1}$$

where

$$\lim_{R\to\infty} u_R = \begin{cases} 0 & \text{FRW} \\ u_\infty \neq 0 & \text{Inflationary} \end{cases} \tag{1.2}$$

This equation of state is qualitatively correct for all R and becomes exact for $R \to 0$ and $R \to \infty$. The parameters u_R, d and s are positive constants and the u_R varies smoothly with R.

The pressure associated to the energy density (1.1) takes then the form

$$p = \frac{1}{D-1} \left(\frac{d}{R} - u_R\, R \right) \frac{1}{R^{D-1}} \tag{1.3}$$

Inserting this source into the Einstein-Friedman equations leads to a self-consistent solution for string dominated universes (see sec. IV) [3]. This solution exhibits the realistic matter dominated behaviour $R \sim (X^0)^{2/(D-1)}$ for large times and the radiation dominated behaviour $R \sim (X^0)^{2/D}$ for early times.

For the sake of completeness we analyze in sec. IV the effective string equations [3]. These equations have been extensively treated in the literature [25] and they are not our central aim.

It must be noticed that there is no satisfactory derivation of inflation in the context of the effective string equations. This does not mean that string theory is not compatible with inflation, but that the effective string action approach *is not enough* to describe inflation. The effective string equations are a low energy field theory approximation to string theory containing only the *massless* string modes. The vacuum energy scales to start inflation are typically of the order of the Planck mass where the effective string action approximation breaks down. One must also consider the *massive* string modes (which are absent from the effective string action) in order to properly get the cosmological condensate yielding inflation. De Sitter inflation does not emerge as a solution of the the effective string equations.

In conclusion, the effective string action (whatever be the dilaton, its potential and the central charge term) is not the appropriate framework in which to address the question of string driven inflation.

Early cosmology (at the Planck time) is probably the best place to test string theory. In one hand the quantum treatment of gravity is unavoidable at such scales and in the other hand, observable cosmological consequences are derivable from the inflationary stage. The natural gravitational background is an inflationary universe as de Sitter spacetime. Such geometries are not string vacua. This means that conformal and Weyl symmetries are broken at the quantum level. In order to quantize consistently strings in such case, one must enlarge the physical phase space including, in particular, the factor $\exp \phi(\sigma, \tau)$ in the world-sheet metric [see eq.(2.6)]. This is a very interesting and open problem. Physically, the origin of such difficulties in quantum string cosmology comes from the fact that one is not dealing with an **empty** universe since a cosmological spacetime necessarily contains matter. In the other hand, conformal field theory techniques are only adapted to backgrounds for which the beta functions are identically zero, i. e. sourceless geometries.

The outline of these lectures is as follows. Section II presents an introduction to strings in curved spacetimes including basic notions on classical and quantum strings in Minkowski spacetime and introducing the main physical string magnitudes: energy-momentum and invariant string size. Section III deals with the string propagation and the string energy-momentum tensor in cosmological spacetimes. (In sections III.A, III.B and III.C we treat the straight strings, ring strings and generic strings respectively and derive the corresponding string equations of state). In section IV we treat self-consistent string cosmology including the string equations of state. (Section IV.A deals with general relativity, IV.B with the string thermodynamics). Section V discuss the effective (beta functions) string equations in the cosmological perspective and the search of inflationary solutions. Finally, in sec. VI, we briefly review the systematic construction of string solutions in de Sitter universe *via* soliton methods and the new feature of multistring solutions.

II. INTRODUCTION. STRINGS IN CURVED AND MINKOWSKI SPACETIMES.

Let us consider bosonic strings (open or closed) propagating in a curved D-dimensional spacetime defined by a metric $G_{AB}(X), 0 \leq A, B \leq D-1$. The action can be written as

$$S = \frac{1}{2\pi\alpha'} \int d\sigma d\tau \sqrt{g}\, g_{\alpha\beta}(\sigma,\tau)\, G_{AB}(X)\, \partial^\alpha X^A(\sigma,\tau)\, \partial^\beta X^B(\sigma,\tau) \tag{2.1}$$

Here $g_{\alpha\beta}(\sigma,\tau)$ ($0 \leq \alpha, \beta \leq 1$) is the metric in the worldsheet, α' stands for the string tension. As in flat spacetime, $\alpha' \sim (M_{Planck})^{-2} \sim (l_{Planck})^2$ fixes the scale in the theory. There are no other free parameters like coupling constants in string theory.

We will start considering given gravitational backgrounds $G_{AB}(X)$. That is, we start to investigate *test* strings propagating on a given spacetime. In section IV, the back reaction problem will be studied. That is, how the strings may act as source of the geometry.

String propagation in massless backgrounds other than gravitational (dilaton, antisymmetric tensor) can be investigated analogously.

The string action (2.1) classically enjoys Weyl invariance on the world sheet

$$g_{\alpha\beta}(\sigma,\tau) \to \lambda(\sigma,\tau)\, g_{\alpha\beta}(\sigma,\tau) \tag{2.2}$$

plus the reparametrization invariance

$$\sigma \to \sigma' = f(\sigma,\tau) \quad , \quad \tau \to \tau' = g(\sigma,\tau) \tag{2.3}$$

Here $\lambda(\sigma,\tau), f(\sigma,\tau)$ and $g(\sigma,\tau)$ are arbitrary functions.

The dynamical variables being here the string coordinates $X_A(\sigma,\tau)$, $(0 \leq A \leq D-1)$ and the world-sheet metric $g_{\alpha\beta}(\sigma,\tau)$.

Extremizing S with respect to them yields the classical equations of motion:

$$\partial^\alpha[\sqrt{g}\, G_{AB}(X)\, \partial_\alpha X^B(\sigma,\tau)] = \frac{1}{2}\sqrt{g}\, \partial_A G_{CD}(X)\, \partial_\alpha X^C(\sigma,\tau)\, \partial^\alpha X^D(\sigma,\tau) \tag{2.4}$$
$$0 \leq A \leq D-1$$

$$T_{\alpha\beta} \equiv G_{AB}(X)[\,\partial_\alpha X^A(\sigma,\tau)\, \partial_\beta X^B(\sigma,\tau)$$

$$- \frac{1}{2} g_{\alpha\beta}(\sigma,\tau)\, \partial_\gamma X^A(\sigma,\tau)\, \partial^\gamma X^B(\sigma,\tau)\,] = 0 \quad , \quad 0 \le \alpha, \beta \le 1. \tag{2.5}$$

Eqs. (2.5) contain only first derivatives and are therefore a set of constraints. Classically, we can always use the reparametrization freedom (2.3) to recast the world-sheet metric on diagonal form

$$g_{\alpha\beta}(\sigma,\tau) = \exp[\phi(\sigma,\tau)]\ \mathrm{diag}(-1,+1) \tag{2.6}$$

In this conformal gauge, eqs. (2.4) - (2.5) take the simpler form:

$$\partial_{-+}X^A(\sigma,\tau) + \Gamma^A_{BC}(X)\, \partial_+ X^B(\sigma,\tau)\, \partial_- X^C(\sigma,\tau) = 0\,, \quad 0 \le A \le D-1, \tag{2.7}$$

$$T_{\pm\pm} \equiv G_{AB}(X)\, \partial_\pm X^A(\sigma,\tau)\, \partial_\pm X^B(\sigma,\tau) \equiv 0\,, \quad T_{+-} \equiv T_{-+} \equiv 0 \tag{2.8}$$

where we introduce light-cone variables $x_\pm \equiv \sigma \pm \tau$ on the world-sheet and where $\Gamma^A_{BC}(X)$ stand for the connections (Christoffel symbols) associated to the metric $G_{AB}(X)$.

Notice that these equations in the conformal gauge are still invariant under the conformal reparametrizations:

$$\sigma + \tau \to \sigma' + \tau' = f(\sigma+\tau) \quad , \quad \sigma - \tau \to \sigma' - \tau' = g(\sigma-\tau) \tag{2.9}$$

Here $f(x)$ and $g(x)$ are arbitrary functions.

The string boundary conditions in curved spacetimes are identical to those in Minkowski spacetime. That is,

$$\begin{aligned} X^A(\sigma+2\pi,\tau) &= X^A(\sigma,\tau) \quad \text{closed strings} \\ \partial_\sigma X^A(0,\tau) &= \partial_\sigma X^A(\pi,\tau) = 0 \quad \text{open strings.} \end{aligned} \tag{2.10}$$

A. A brief review on strings in Minkowski spacetime

In flat spacetime eqs.(2.7) become linear

$$\partial_{-+}X^A(\sigma,\tau) = 0\,, \quad 0 \le A \le D-1, \tag{2.11}$$

and one can solve them explicitly as well as the quadratic constraint (2.8) [see below]:

$$\left[\partial_\pm X^0(\sigma,\tau)\right]^2 - \sum_{j=1}^{D-1} \left[\partial_\pm X^j(\sigma,\tau)\right]^2 = 0 \tag{2.12}$$

The solution of eqs.(2.11) is usually written for closed strings as

$$X^A(\sigma,\tau) = q^A + 2p^A\alpha'\tau + i\sqrt{\alpha'} \sum_{n\neq 0} \frac{1}{n} \{\alpha^A_n \exp[in(\sigma-\tau)] + \tilde{\alpha}^A_n \exp[-in(\sigma+\tau)]\} \tag{2.13}$$

where q^A and p^A stand for the string center of mass position and momentum and α_n^A and $\tilde{\alpha}_n^A$ describe the right and left oscillator modes of the string, respectively. Since the string coordinates are real,

$$\bar{\alpha}_n^A = \alpha_{-n}^A \quad , \quad \bar{\tilde{\alpha}}_n^A = \tilde{\alpha}_{-n}^A$$

This resolution is no more possible in general for curved spacetime where the equations of motion (2.7) are non-linear. In that case, right and left movers interact with themselves and with each other.

In Minkowski spacetime we can also write the solution of the string equations of motion (2.11) in the form

$$X^A(\sigma, \tau) = l^A(\sigma + \tau) + r^A(\sigma - \tau) \tag{2.14}$$

where $l^A(x)$ and $r^A(x)$ are arbitrary functions. Now, making an appropriate conformal transformation (2.9) we can turn any of the string coordinates $X^A(\sigma, \tau)$ (but only one of them) into a constant times τ. The most convenient choice is the light-cone gauge where

$$U \equiv X^0 - X^1 = 2p^U\alpha'\tau. \tag{2.15}$$

That is, there are no string oscillations along the U direction in the light-cone gauge. We have still to impose the constraints (2.12). In this gauge they take the form

$$\pm\, 2\alpha' p^U\, \partial_{\pm} V(\sigma, \tau) = \sum_{j=2}^{D-1} \left[\partial_{\pm} X^j(\sigma, \tau)\right]^2 \tag{2.16}$$

where $V \equiv X^0 + X^1$. This shows that V is not an independent dynamical variable since it expresses in terms of the transverse coordinates $X^2, \ldots, X^{D-1}$. Only q^V is an independent quantity.

The physical picture of a string propagating in Minkowski spacetime clearly emerges in the light-cone gauge. The gauge condition (2.15) tells us that the string 'time' τ is just proportional to the physical null time U. Eqs.(2.13) shows that the string moves as a whole with constant speed while it oscillates around its center of mass. The oscillation frequencies are all integers multiples of the basic one. The string thus possess an infinite number of normal modes; $\alpha_n^A, \tilde{\alpha}_n^A$ classically describe their oscillation amplitudes. Only the modes in the direction of the transverse coordinates $X^2, \ldots, X^{D-1}$ are physical. This is intuitively right, since a longitudinal or a temporal oscillation of a string is meaningless. In summary, the string in Minkowski spacetime behaves as an extended and composite relativistic object formed by a $2(D-2)$-infinite set of harmonic oscillators.

Integrating eq.(2.16) on σ from 0 to 2π and inserting eq. (2.13) yields the classical string mass formula:

$$m^2 \equiv p^U p^V - \sum_{j=2}^{D-1} (p^j)^2 = \frac{1}{\alpha'} \sum_{j=2}^{D-1} \sum_{n=1}^{\infty} \left[\alpha_n^j\, \alpha_{-n}^j + \tilde{\alpha}_n^j\, \tilde{\alpha}_{-n}^j\right] \tag{2.17}$$

We explicitly see how the mass of a string depends on its excitation state. The classical string spectrum is continuous as we read from eq.(2.17). It starts at $m^2 = 0$ for an unexcited string: $\alpha_n^j = \tilde{\alpha}_n^j = 0$ for all n and j.

The independent string variables are:
the transverse amplitudes $\{\alpha_n^j, \tilde{\alpha}_n^j, n\varepsilon \mathcal{Z}, n \neq 0,\ 2 \leq j \leq D-1\}$,
the transverse center of mass variables $\{q^j, p^j,\ 2 \leq j \leq D-1\}$,
q^V and p^U.
Up to now we have considered a classical string.
At the quantum level one imposes the canonical commutation relations (CCR)

$$[\alpha_n^i, \alpha_m^j] = n\,\delta_{n,-m}\,\delta^{i,j}\ ,$$

$$[\tilde{\alpha}_n^i, \alpha_m^j] = n\,\delta_{n,-m}\,\delta^{i,j}\ ,$$

$$[\tilde{\alpha}_n^i, \alpha_m^j] = 0\ ,$$

$$[q^i, p^j] = i\,\delta^{i,j} \quad , \quad [q^V, p^U] = i \tag{2.18}$$

All other commutators being zero. An order prescription is needed to unambiguously express the different physical operators in terms of those obeying the CCR. The symmetric ordering is the simplest and more convenient.

The space of string physical states is the the tensor product of the Hilbert space of the $D-1$ center of mass variables $q_V, p_U, \{q^j, p^j, 2 \leq j \leq D-1\}$, times the Fock space of the harmonic transverse modes. The string wave function is then the product of a center of mass part times a harmonic oscillator part. The center of mass can be taken, for example, as a plane wave. The harmonic oscillator part can be written as the creation operators $\alpha_n^{j\,\dagger}, \tilde{\alpha}_n^{j\,\dagger}$, $n \geq 1,\ 2 \leq j \leq D-1$ acting on the oscillator ground state $|0>$. This state is defined as usual by

$$\alpha_n^j\,|0> = \tilde{\alpha}_n^j\,|0> =, \quad \text{for all } n \geq 1,\ 2 \leq j \leq D-1$$

Notice that a string describes **one particle**. The kind of particle described depends on the oscillator wave function. The mass and spin can take an infinite number of different values. That is, there is an infinite number of different possibilities for the particle described by a string.

Let us consider the quantum mass spectrum. Upon symmetric ordering the mass operator becomes,

$$m^2 = \frac{1}{2\alpha'} \sum_{j=2}^{D-1} \sum_{n=1}^{\infty} \left[\alpha_n^j\,\alpha_{-n}^j + \alpha_{-n}^j\,\alpha_n^j + \tilde{\alpha}_n^j\,\tilde{\alpha}_{-n}^j + \tilde{\alpha}_{-n}^j\,\tilde{\alpha}_n^j\right]\ . \tag{2.19}$$

Using the commutation rules (2.18) yields

$$m^2 = \frac{D-2}{\alpha'} \sum_{n=1}^{\infty} n + \frac{1}{2\alpha'} \sum_{j=2}^{D-1} \sum_{n=1}^{\infty} \left[\alpha_n^{j\,\dagger}\,\alpha_n^j + \tilde{\alpha}_n^{j\,\dagger}\,\tilde{\alpha}_n^j\right] \tag{2.20}$$

The divergent sum in the first term can be defined through analytic continuation of the zeta function

$$\zeta(z) \equiv \sum_{n=1}^{\infty} \frac{1}{n^z} \tag{2.21}$$

One finds $\zeta(-1) = -1/12$ [27]. Thus,

$$m^2 = -\frac{D-2}{12\alpha'} + \frac{1}{2\alpha'} \sum_{j=2}^{D-1} \sum_{n=1}^{\infty} \left[\alpha_n^{j\,\dagger} \alpha_n^j + \tilde{\alpha_n^j}^{\dagger} \tilde{\alpha}_n^j \right] \tag{2.22}$$

Hence, the string ground state $|0>$ has a negative mass squared

$$m_0^2 = -\frac{D-2}{12\alpha'} \tag{2.23}$$

Such particles are called tachyons and exhibit unphysical behaviours. When fermionic degrees of freedom are associated to the string the ground state becomes massless (superstrings) [26].

Notice that the appearance of a negative mass square yields a dispersion relation $E^2 = p^2 - |m_0^2|$ similar to waves with Jeans unstabilities [28].

Let us consider now excited states.

The constraints (2.12) integrated on σ from 0 to 2π impose

$$\sum_{j=2}^{D-1} \sum_{n=1}^{\infty} (\alpha_n^j)^\dagger \alpha_n^j = \sum_{j=2}^{D-1} \sum_{n=1}^{\infty} (\tilde{\alpha_n^j})^\dagger \tilde{\alpha}_n^j \tag{2.24}$$

on the physical states. This means that the number of left and right modes coincide in all physical states.

The first excited state is then described by

$$|i,j> = (\tilde{\alpha_1^i})^\dagger (\alpha_1^j)^\dagger \, |0> \tag{2.25}$$

times the center of mass wave function. We see that this wavefunction is a symmetric tensor in the space indices i, j. It describes therefore a spin two particle plus a spin zero particle (the trace part).

From eqs.(2.22-2.25) follows that

$$m^2 \, |i,j> = -\frac{D-26}{12\alpha'} |i,j> \tag{2.26}$$

This state is then a massless particle only for $D = 26$. In such critical dimension we have then a graviton (massless spin 2 particle) and a dilaton (massless spin 0 particle) as string modes of excitation. For superstrings the critical dimension turns to be $D = 10$.

We shall consider, as usual, that only four space-time dimensions are uncompactified. That is, we shall consider the strings as living on the tensor product of a curved four dimensional space-time with lorentzian signature and a compact space which is there to cancel the anomalies. Therefore, from now on strings will propagate in the curved (physical) four dimensional space-time. However, we will find instructive to study the case where this curved space-time has dimensionality D, where D may be 2, 3 or arbitrary.

B. The string energy-momentum tensor and the string invariant size

The spacetime string energy-momentum tensor follows (as usual) by taking the functional derivative of the action (2.1) with respect to the metric G_{AB} at the spacetime point X. This yields,

$$\sqrt{-G}\, T^{AB}(X) = \frac{1}{2\pi\alpha'}\int d\sigma d\tau \left(\dot{X}^A\dot{X}^B - X'^A X'^B\right)\delta^{(D)}(X - X(\sigma,\tau)) \tag{2.27}$$

where dot and prime stands for $\partial/\partial\tau$ and $\partial/\partial\sigma$, respectively.

Notice that X in eq.(2.27) is just a spacetime point whereas $X(\sigma,\tau)$ stands for the string dynamical variables. One sees from the Dirac delta in eq.(2.27) that $T^{AB}(X)$ vanishes unless X is exactly on the string world-sheet. We shall not be interested in the detailed structure of the classical strings. It is the more useful to integrate the energy-momentum tensor (2.27) on a volume that completely encloses the string. It takes then the form [14]

$$\begin{aligned}\Theta^{AB}(X) &= \frac{1}{2\pi\alpha'}\int d\sigma d\tau \left(\dot{X}^A\dot{X}^B - X'^A X'^B\right)\delta(X^0 - X^0(\tau)) \\ &= \frac{1}{2\pi\alpha'|\dot{X}^0(\tau)|}\int_0^{2\pi} d\sigma \left[\dot{X}^A\dot{X}^B - X'^A X'^B\right]_{\tau=\tau(X^0)}\end{aligned}$$

Another relevant physical magnitude for strings is the invariant size. We define the invariant string size ds^2 using the induced metric on the string world-sheet:

$$ds^2 = G_{AB}(X)\, dX^A\, dX^B \tag{2.28}$$

Inserting $dX^A = \partial_+ X^A\, dx^+ + \partial_- X^A\, dx^-$, into eq.(2.28) and taking into account the constraints (2.8) yields

$$ds^2 = G_{AB}(X)\,\partial_+ X^A\,\partial_- X^B\left(d\tau^2 - d\sigma^2\right) \tag{2.29}$$

Thus, we define the string size as the integral of $\sqrt{G^{AB}(X)\partial_+ X^A\partial_- X^B}$ over σ at fixed τ.

Let us now consider a circular string as a simple example of a string solution in Minkowski spacetime.

$$\begin{aligned}&X^0(\sigma,\tau) = \alpha' E\tau \quad , \quad X^3(\sigma,\tau) = \alpha' p\tau \\ &X^1(\sigma,\tau) = \alpha' m\cos\tau\cos\sigma = \frac{\alpha' m}{2}\left[\cos(\tau+\sigma) + \cos(\tau-\sigma)\right] \\ &X^2(\sigma,\tau) = \alpha' m\cos\tau\cos\sigma = \frac{\alpha' m}{2}\left[\cos(\tau+\sigma) - \cos(\tau-\sigma)\right]\end{aligned} \tag{2.30}$$

This is obviously a solution of eqs.(2.11) where only the modes $n = \pm 1, j = 1,2$ are excited. The constraints (2.12) yields

$$E^2 = p^2 + m^2$$

Eqs.(2.30) describe a circular string in the X^1, X^2 plane, centered in the origin and with an oscillating radius $\rho(\tau) = \alpha' m\cos\tau$. In addition the string moves uniformly in the z-direction with speed p/E. (That is, p is its momentum in the z-direction). The oscillation amplitude m can be identified with the string mass and E with the string energy. Notice that the string time τ is here proportional to the physical time X^0 (this solution is not in the light-cone gauge).

It is instructive to compute the integrated energy-momentum tensor (2.28) for this string solution. We find in the rest frame ($p = 0$) that it takes the fluid form

$$\Theta_A^B = \begin{pmatrix} \rho & 0 & 0 & 0 \\ 0 & -p & 0 & 0 \\ 0 & 0 & -p & 0 \\ 0 & 0 & 0 & 0 \end{pmatrix} \tag{2.31}$$

where

$$\rho = E = m \quad , \quad p = -\frac{m}{2}\cos(2\tau) \tag{2.32}$$

We see that the total energy coincides with m as one could expect and that the (space averaged) pressure oscillates around zero. That is the string pressure goes through positive and negative values. The time average of p vanishes. The string behaves then as cold matter (massive particles).

The upper value of p equals $E/2$. This is precisely the relation between E and p for radiation (massless particles). (Notice that this circular strings lives on a two-dimensional plane). The lower value of p correspond to the limiting value allowed by the strong energy condition in General Relativity [11]. We shall see below that these two extreme values of p appear in the general context of cosmological spacetimes.

The invariant size of the string solution (2.30) follows by inserting eq.(2.30) into eq.(2.29). We find

$$ds^2 = (\alpha' m)^2 \left(d\tau^2 - d\sigma^2\right) \tag{2.33}$$

Therefore, this string solution has a constant size $2\pi\alpha' m$.

One obtains the invariant size for a generic solution in Minkowski spacetime inserting the general solution (2.13) into $\partial_+ X^A \partial_- X_A$. This gives constant plus oscillatory terms. In any case the invariant string size is always **bounded** in Minkowski spacetimes. We shall see how differently behave strings in curved spacetimes.

III. STRING PROPAGATION IN COSMOLOGICAL SPACETIMES

We obtain in this section physical string properties from the exact string solutions in cosmological spacetimes.

We consider strings in spatially homogeneous and isotropic universes with metric

$$ds^2 = (dT)^2 - R(T)^2 \sum_{i=1}^{D-1} (dX^i)^2 \tag{3.1}$$

where T is the cosmic time and the function $R(T)$ is called the scale factor. In terms of the conformal time

$$\eta = \int^T \frac{dT}{R(T)}\,, \tag{3.2}$$

the metric (3.1) takes the form

$$ds^2 = R(\eta)^2 \left[(d\eta)^2 - \sum_{i=1}^{D-1} (dX^i)^2 \right] \tag{3.3}$$

The classical string equations of motion can be written here as

$$\begin{aligned} &\partial^2 T \;\; - R(T)\, \tfrac{dR}{dT}\, \textstyle\sum_{i=1}^{D-1} (\partial_\mu X^i)^2 = 0\,, \\ &\partial_\mu \left[R^2 \partial^\mu X^i \right] = 0\,, \qquad 1 \le i \le D-1\,, \end{aligned} \tag{3.4}$$

and the constraints are

$$T_{\pm\pm} = (\partial_\pm T)^2 - R(T)^2\, (\partial_\pm X^i)^2 = 0\,. \tag{3.5}$$

The most relevants universes corresponds to power type scale factors. That is,

$$R(T) = a\, T^\alpha = A\, \eta^{k/2} \tag{3.6}$$

where $\alpha = \frac{k}{k+2}$. For different values of the exponents we have either FRW or inflationary universes.

$$\mathrm{FRW}: 0 < k \le \infty,\; 0 < \alpha \le 1 = \begin{cases} \alpha = 1/2,\; k = 2, & \text{radiation dominated} \\ \alpha = 2/3,\; k = 4, & \text{matter dominated} \\ \alpha = 1,\; k = \infty, & \text{`stringy'} \end{cases}$$

$$\text{Inflationary}: -\infty < k \le 0,\; \alpha < 0 \text{ and } \alpha > 1 = \begin{cases} \alpha = \infty,\; k = -2, & \mathrm{R(T)} = \mathrm{e}^{HT},\ \text{de Sitter}, \\ \alpha > 1,\; k < -2, & \text{power inflation} \\ \alpha < 0,\; -2 < k < 0, & \text{superinflationary} \end{cases}$$

As we will see below, $T^{AB}(X)$ takes the fluid form for string solutions in cosmological spacetimes, allowing us to define the string pressure p and energy density ρ :

$$T_A^B = \begin{pmatrix} \rho & 0 & \cdots & 0 \\ 0 & -p & \cdots & 0 \\ \cdots & \cdots & \cdots & 0 \\ 0 & 0 & \cdots & -p \end{pmatrix} \tag{3.7}$$

Notice that the continuity equation

$$D^A\, T_A^B = 0$$

takes here the form

$$\dot\rho + (D-1)\, H\, (p + \rho) = 0 \tag{3.8}$$

where $H \equiv \frac{1}{R}\frac{dR}{dT}$.
For an equation of state of the type of a perfect fluid, that is

$$p = (\gamma - 1)\, \rho \qquad , \qquad \gamma = \text{ constant}, \tag{3.9}$$

eqs.(3.8) and (3.9) can be easily integrated with the result

$$\rho = \rho_0\, R^{\gamma(1-D)}\quad . \tag{3.10}$$

For $\gamma = 1$ this corresponds to cold matter ($p = 0$) and for $\gamma = \frac{D}{D-1}$ this describes radiation with $p = \frac{\rho}{D-1}$.

A. 1+1 Dimensional Universes. Straight Strings in D-Dimensional Universes

Let us start by considering strings in this simpler case. For a $D = 1+1$ dimensional universe or for a straight string parallel to the X-axis in a D-dimensional universe, the metric (3.1) takes the form

$$ds^2 = (dT)^2 - R(T)^2\,(dX)^2$$

It is convenient to start by solving the constraints (3.5)

$$(\partial_\pm T)^2 = R(T)^2\,(\partial_\pm X)^2\,. \tag{3.11}$$

They reduce to

$$\partial_\pm T = \epsilon_\pm \;\; R(T)\,\partial_\pm X\,. \tag{3.12}$$

where $\epsilon_\pm^2 = 1$. Using the conformal time (3.2) , eq.(3.12) yields

$$\partial_\pm(\eta - \epsilon_\pm X) = 0\,.$$

We find a first family of solutions choosing $\epsilon_\pm = \pm 1$. Then

$$\eta + X = \phi(\sigma+\tau) \quad , \quad \eta - X = \chi(\sigma - \tau) \tag{3.13}$$

Where ϕ and χ are arbitrary functions of one variable. It is easy now to check that eq.(3.13) fulfills the string equations of motion (3.4).

The solution (3.13) is analyzed in detail for de Sitter spacetime ($R(T) = e^{HT}$) in ref. [4] where the global topology of the space is taken into account.

Since one can always perform conformal transformations

$$\sigma + \tau \to f(\sigma+\tau) \quad , \quad \sigma - \tau \to g(\sigma - \tau) \quad ,$$

with arbitrary functions f and g , the solution (3.13) has no degrees of freedom other than topological ones.

Let us compute the energy momentum tensor for the string solution (3.13). We find from eqs.(2.27) and (3.13),

$$\begin{aligned}\sqrt{-G}\;T^{00}(X) &= \frac{1}{2\pi\alpha'}\int d\sigma d\tau\,\left(\dot\eta^2 - \eta'^2\right)\delta(\eta - \eta(\sigma,\tau))\delta(X - X(\sigma,\tau)) \\ &= \frac{1}{2\pi\alpha'}\frac{\dot\eta^2 - \eta'^2}{J}\end{aligned} \tag{3.14}$$

where $J = \frac{\partial(X,\eta)}{\partial(\sigma,\tau)}$ is the jacobian. From eq.(3.13) we find $J = -\chi'\,\phi'$ and $\dot\eta^2 - \eta'^2 = -\chi'\,\phi'$. Then,

$$\sqrt{-G}\;T^{00}(X) = \frac{1}{2\pi\alpha'}\,.$$

We analogously find $\dot X^2 - X'^2 = -\chi'\,\phi'$. Then

$$\sqrt{-G}\, T^{11}(X) = -\frac{1}{2\pi\alpha'} \quad , \quad T^{01}(X) = 0 \; .$$

That implies,

$$\rho = \frac{1}{2\pi\alpha'} \quad , \quad p = -\frac{1}{2\pi\alpha'} \quad , \quad p + \rho = 0. \tag{3.15}$$

We find a constant energy density and a constant **negative** pressure. They exactly fulfill the continuity equation (3.8). These results hold for **arbitrary** cosmological spacetimes in $1+1$ dimensions. That is, for arbitrary factors $R(T)$. In particular they are valid for strings winded n-times around the de Sitter universe [4].

A second family of string solutions follows from eq.(3.11) by choosing

$$\eta = \pm X + C_{\pm} \tag{3.16}$$

where $C_{\pm}$ is a constant. Then, the string equations of motion (3.4) become

$$\partial_\mu \left[R(\eta)^2\, \partial^\mu \eta \right] = 0 \tag{3.17}$$

Using eq.(3.16), we find that the energy-momentum tensor (2.27) is traceless for this string solution:

$$T^{00} = T^{11} \quad , \quad \text{that is } \; p = \rho \tag{3.18}$$

Let us call

$$V(\eta) = \int^{\eta} R^2(x) dx$$

Then, eq.(3.17) implies that

$$\left(\frac{\partial^2}{\partial \tau^2} - \frac{\partial^2}{\partial \sigma^2} \right) V(\eta) = 0.$$

The general solution $\eta = \eta(\sigma, \tau)$ is implicitly defined by:

$$\int^{\eta} R^2(x) dx = A(\sigma - \tau) + B(\sigma + \tau) \; .$$

where $A(x)$ and $B(x)$ are arbitrary functions. Upon a conformal transformation, without loss of generality we can set

$$A(\sigma - \tau) + B(\sigma + \tau) \;\; \Rightarrow \;\; \tau \; .$$

Hence, η is in general a function solely of τ with

$$\tau = \int^{\eta} R^2(x) dx \;\; \text{or} \;\; \frac{d\eta}{d\tau} = \frac{1}{R^2(\eta)} \; .$$

The energy-momentum tensor (2.27) takes for this solution the form:

$$T_A^B = \frac{1}{\alpha' R^2}\,\delta(\eta \mp X - C_\pm)\begin{pmatrix} 1 & 0 \\ 0 & -1 \end{pmatrix}$$

The $\delta(\eta \mp X - C_\pm)$ characterizes a localized object propagating on the characteristics at the speed of light. This solution describes a massless point particle since it has been possible to gauge out the σ dependence .

In summary, the two-dimensional string solutions in cosmological spacetimes generically obey perfect fluid equations of state with either

$$p = -\rho \quad (\gamma = 0) \qquad \text{or} \qquad p = +\rho \quad (\gamma = 2) \tag{3.19}$$

The respectives energy densities being

$$\rho = \rho_u \quad (\rho_u = \text{constant}) \text{ for } \gamma = 0 \quad \text{or} \quad \rho = \frac{u}{R^2} \quad (u = \text{constant}) \text{ for } \gamma = 2. \tag{3.20}$$

These behaviors fulfill the continuity equation (3.8) for $D = 2$.

B. 2+1 Dimensional Universes. Ring Strings in D-Dimensional Universes

A large class of exact solutions describing one string and multistrings has been found in $2+1$ dimensional de Sitter universe [5] - [8]. For power-like expansion factors $R(\eta)^2 = A\eta^k$, $(k \neq -2)$ only ring solutions are known [9]. ($k = -2$ corresponds to de Sitter spacetime). These string solutions are on a two-dimensional plane that can be considered embedded on a D-Dimensional universe.

Ring solutions correspond to the Ansatz [9]:

$$\begin{aligned} T &= T(\tau) \\ X^1 &= f(\tau)\cos\sigma \\ X^2 &= f(\tau)\sin\sigma \end{aligned} \tag{3.21}$$

The total energy of one string is then given by (recall $G^{00} = 1$)

$$E(T) = \int d^{D-1}X\sqrt{-G}\;T^{00}(X) = \frac{1}{\alpha'}\frac{dT}{d\tau}$$

More generally, the energy-momentum tensor integrated on a volume that completely encloses the string, takes the form [14]

$$\begin{aligned} \Theta^{AB}(X) &= \frac{1}{2\pi\alpha'}\int d\sigma d\tau\left(\dot{X}^A\dot{X}^B - X'^A X'^B\right)\delta(T - T(\tau)) \\ &= \frac{1}{2\pi\alpha'|\dot{X}^0(\tau)|}\int_0^{2\pi} d\sigma\left[\dot{X}^A\dot{X}^B - X'^A X'^B\right]_{\tau=\tau(T)} \end{aligned}$$

For multistring solutions, one must sum over the different roots τ_i of the equation $T = T(\tau)$, for a given T.

We find for the ring ansatz eq.(3.22):

$$\Theta^{00}(X) = E(T)$$
$$\Theta^{11}(X) = \Theta^{22}(X) = \frac{1}{2\alpha'|\dot{X}^0(\tau)|}[\dot{f}^2 - f^2]$$
$$\Theta^{01}(X) = \Theta^{02}(X) = \Theta^{12}(X) = 0\ .$$

That is,

$$\Theta_A^B = \begin{pmatrix} E & 0 & 0 \\ 0 & -P & 0 \\ 0 & 0 & -P \end{pmatrix}$$

where,

$$E = \frac{1}{\alpha'}\,\dot{T}(\tau) \quad , \quad P = \frac{R(\tau)^2}{2\alpha'|\dot{X}^0(\tau)|}\ [\dot{f}^2 - f^2] \tag{3.22}$$

Let us first consider ring strings in de Sitter universe [5] - [7]. We find there three different asymptotic behaviors: stable, unstable and its dual. The unstable regime appears for $\eta \simeq \frac{\tau}{H} \to 0$. That is, when the scale factor tends to infinity. From eqs.(3.22) and ref. [5], we find

$$E(\tau) \overset{\tau\to 0^-}{=} \frac{1}{\alpha' H}\left(\frac{1}{|\tau|} + 1\right) \simeq \frac{1}{\alpha' H}\,[R(\tau) + 1] \to +\infty$$
$$P(\tau) \overset{\tau\to 0^-}{=} -\frac{1}{2\alpha' H|\tau|} \simeq -\frac{R(\tau)}{2\alpha' H} \to -\infty$$

(Here H stands for the Hubble constant). The invariant string size grows as R/H in this unstable regime. Notice that the pressure is **negative** for the unstable strings and proportional to the expansion factor R. In this regime we also see that

$$P \overset{R\to\infty}{=} -\frac{E}{2} + \frac{1}{2\alpha' H} + O(1/R) \to -\infty \tag{3.23}$$

with $E = \frac{1}{\alpha' H}\,[R(\tau) + 1]$. The regime dual to the unstable regime appears when the conformal time η tends to infinity. For the solution $q_-(\sigma, \tau)$ of ref. [5] in de Sitter universe, η diverges for finite $\tau \to \tau_0$ as

$$\eta \overset{\tau\to\tau_0}{=} \frac{6\ e^{-\tau_0}}{\tau - \tau_0} + O(1) \to +\infty$$

Here

$$\tanh\frac{\tau_0}{\sqrt{2}} = \frac{1}{\sqrt{2}} \quad , \quad \tau_0 = \sqrt{2}\,\log(1 + \sqrt{2}) = 1.246450....$$

Then,

$$
\begin{aligned}
R(\tau) &\overset{\tau\to\tau_0}{=} \frac{e^{\tau_0}}{6}\,(\tau-\tau_0) \to 0^+ \\
E(\tau) &\overset{\tau\to\tau_0}{=} \frac{1}{\alpha' H(\tau-\tau_0)} = \frac{0.5796...}{\alpha' H R} \to +\infty \\
P(\tau) &\overset{\tau\to\tau_0}{=} \frac{1}{2\alpha' H(\tau-\tau_0)} = E/2 \to +\infty
\end{aligned}
$$

We call dual to this regime since it appears related to the unstable regime (3.23) through the exchange $R \leftrightarrow 1/R$. The invariant string size tends to $\frac{1}{H}$ in this regime.

In the stable regime, $\tau \to \infty$, (and the cosmic time $T \simeq \frac{\tau}{H} \to \infty$), from eq.(3.22) and ref. [5], we find

$$
E(\tau) \overset{\tau\to\infty}{=} \frac{1}{\alpha' H} \quad , \quad P(\tau) \overset{\tau\to\infty}{=} \frac{1+\sqrt{2}}{2H} e^{-\tau\sqrt{2}} \to 0. \tag{3.24}
$$

For stable strings in de Sitter universe, the pressure is positive and vanishes asymptotically, and the invariant string size tends to $\frac{1}{\sqrt{2}H}$.

The solution $q_-(\sigma,\tau)$ in ref. [5] describes two ring strings:

a stable string for $q_0 \to +\infty$ ($\tau \to \infty$) (q_0 being the hyperboloid time-like coordinate) and

a unstable one for $q_0 \to -\infty$ ($\tau \to \tau_0$). The pressure P depends on τ ; it is negative for $q_0 < l$ and positive for $q_0 > l$, where $l = -1.385145....$

It can be noticed that the behaviour eq.(3.23) for the energy can be interpreted as an unstable piece $R/(\alpha' H)$ *plus* a stable one $1/(\alpha' H)$. The constant term $1/(\alpha' H)$ is precisely the energy for the stable solution eq.(3.24).

Let us now study the energy and pressure for the ring string solutions in power-type inflationary universes considered in [9]. In terms of the conformal time η , we have as expansion factor $R^2(\eta) = A\,\eta^k$ with $k < 0$.

Near $\eta = 0$, two types of behavior were found [9]. The first one is a linear behavior

$$
\begin{aligned}
\eta &\overset{\tau\to\tau_0}{=} \tau-\tau_0+O(\tau-\tau_0)^2 \\
f(\tau) &\overset{\tau\to\tau_0}{=} 1-\frac{(\tau-\tau_0)^2}{2(k+1)} \quad \text{for } k<-1 \\
\text{and } f(\tau) &\overset{\tau\to\tau_0}{=} 1+c\,(\tau-\tau_0)^{1-k} \quad \text{for } -1<k<0.
\end{aligned} \tag{3.25}
$$

Eqs.(3.25) describe a expanding string for $k < 0$ with invariant size $\simeq (\tau-\tau_0)^{k/2}$. That is, in inflationary universes ($k < 0$) , the string size grows indefinitely for $\eta \simeq \tau-\tau_0 \to 0$ as the universe radius $R \simeq (\tau-\tau_0)^{k/2} \to +\infty$. The growing of the proper string size for $\eta \simeq \tau-\tau_0 \to 0$ is a typical feature of string unstability.

Using eqs.(3.22) and (3.25), the energy and pressure take the form,

$$
\begin{aligned}
E(\tau) &\overset{\tau\to\tau_0}{=} \frac{\sqrt{A}}{\alpha'}\,(\tau-\tau_0)^{k/2} = R/\alpha' \quad , \\
P(\tau) &\overset{\tau\to\tau_0}{=} -\frac{\sqrt{A}}{2\alpha'}\,(\tau-\tau_0)^{k/2} = -R/(2\alpha') \quad .
\end{aligned} \tag{3.26}
$$

That yields

$$P \overset{\tau\to\tau_0}{=} -E/2 \tag{3.27}$$

Eq.(3.27) is also valid in de Sitter universe for unstable strings [eq.(3.23)]. In all these cases strings exhibit **negative** pressure with an equation of state $P = -E/2$. This equation of state exactly saturates the strong energy condition in general relativity. This **unstable** string behavior dominates in all inflationary universes for $R \to \infty$.

The second behavior present near $\eta = 0$ is [9]

$$\eta \overset{\tau\to\tau_0}{=} (\tau-\tau_0)^{1/(k+1)} \quad , \quad f(\tau) \overset{\tau\to\tau_0}{=} f_0 \pm (\tau-\tau_0)^{1/(k+1)} \tag{3.28}$$

where f_0 must be set equal to zero for $-1 < k < 0$. The invariant string size behaves for $\tau \to \tau_0$ as,

$$S(\tau) \overset{\tau\to\tau_0}{=} \sqrt{A}\,(\tau-\tau_0)^{\frac{k+2}{2(k+1)}} \overset{\tau\to\tau_0}{=} A\,T \quad \text{for } k < 0$$

This solution describes a string that collapses for $k < -2$ (power inflation) and blows up for $-2 < k < 0$ (super inflation). In both cases the string size is proportional to the horizon size which is of order T (cosmic time).

Here, the expansion factor tends to zero as

$$R(\tau) \overset{\tau\to\tau_0}{=} \sqrt{A}\,(\tau-\tau_0)^{\frac{k}{2(k+1)}} \to 0.$$

when $k < -1$ and blows up for $0 < k < -1$.

From eq.(3.22) we find that

$$P = \frac{A}{2(k+1)\alpha' R} \to +\infty \quad , \quad E = \frac{A}{(k+1)\alpha' R} \to +\infty$$

That is,

$$P \overset{\tau\to\tau_0}{=} E/2 \tag{3.29}$$

Notice that the pressure is here **positive**. This second behaviour is related by duality ($R \leftrightarrow 1/R$) to the first behaviour described by eqs.(3.26)-(3.27). This is the **dual to unstable** regime. In such regime the strings behave as massless particles (radiation).

Let us now consider ring string solutions in FRW universes considered in [9]. These solutions start ex-nihilo at the big bang ($R = 0$). In terms of the conformal time η , we have as expansion factor $R^2(\eta) = A\,\eta^k$ with $k > 0$.

Near $\eta = 0$, two types of behavior were found [9]. The first one is a linear behavior

$$\begin{aligned} \eta &\overset{\tau\to\tau_0}{=} \tau - \tau_0 + O(\tau-\tau_0)^2 \\ f(\tau) &\overset{\tau\to\tau_0}{=} 1 - \frac{(\tau-\tau_0)^2}{2(k+1)} \quad . \end{aligned} \tag{3.30}$$

Eqs.(3.30) describe a collapsing string for $k > 0$ with invariant size $\simeq (\tau-\tau_0)^{k/2}$. That is, in FRW universes ($k > 0$), the string size goes to zero for $\eta \simeq \tau - \tau_0 \to 0$ as the universe radius $R \simeq (\tau-\tau_0)^{k/2} \to 0$.

Using eqs.(3.22) and (3.30), the energy and pressure take the form,

$$E(\tau) \overset{\tau\to\tau_0}{=} \frac{\sqrt{A}}{\alpha'}\,(\tau-\tau_0)^{k/2} = R/\alpha' \quad ,$$
$$P(\tau) \overset{\tau\to\tau_0}{=} -\frac{\sqrt{A}}{2\alpha'}\,(\tau-\tau_0)^{k/2} = -R/(2\alpha') \quad . \tag{3.31}$$

That yields

$$P \overset{\tau\to\tau_0}{=} -E/2 \tag{3.32}$$

As we shall see below, this behaviour is subdominant in FRW universes. [Notice that eq.(3.32) is identical eq.(3.27) that holds for $R \to \infty$ in power-like inflationary universes $(k < 0)$].

The second behavior present near $\eta = 0$ is [9]

$$\eta \overset{\tau\to\tau_0}{=} (\tau-\tau_0)^{1/(k+1)} \quad , \quad f(\tau) \overset{\tau\to\tau_0}{=} f_0 \pm (\tau-\tau_0)^{1/(k+1)} \tag{3.33}$$

The invariant string size behaves for $\tau \to \tau_0$ as,

$$S(\tau) \overset{\tau\to\tau_0}{=} \sqrt{A}\,(\tau-\tau_0)^{\frac{k}{2(k+1)}} \to 0 \quad \text{for } k > 0$$

This solution describes a string that collapses for $k > 0$.

Here, the expansion factor tends to zero as

$$R(\tau) \overset{\tau\to\tau_0}{=} \sqrt{A}\,(\tau-\tau_0)^{\frac{k}{2(k+1)}} \to 0.$$

since $k > 0$.

From eq.(3.22) we find that

$$P = \frac{A}{2(k+1)\alpha' R} \to +\infty \quad , \quad E = \frac{A}{(k+1)\alpha' R} \to +\infty$$

That is,

$$P \overset{\tau\to\tau_0}{=} E/2 \tag{3.34}$$

Notice that the pressure is here **positive**. This second behaviour is related by duality $(R \leftrightarrow 1/R)$ to the first behaviour described by eqs.(3.31)-(3.32).

Hence, we see from eqs.(3.31-3.34) that near the big bang $(R \to 0)$ the **dual** behavior dominates over the unstable behavior.

It should also be noticed that for this dual behavior, the energy redshifts exactly as $1/R$ for $R \to 0$.

For large τ, the ring strings exhibit a stable behaviour [9],

$$\eta(\tau) \overset{\tau\to\pm\infty}{=} \tau^{2/(k+2)} \, ,$$
$$T(\tau) \overset{\tau\to\pm\infty}{=} \frac{2\sqrt{A}}{k+2}\,\tau \, ,$$
$$f(\tau) \overset{\tau\to\pm\infty}{=} \frac{2}{k+2}\,\tau^{-k/(k+2)} \cos(\tau+\varphi)\, , \tag{3.35}$$

where φ is a constant phase and the oscillation amplitude has been normalized. For large τ, the energy and pressure of the solution (3.35) behave as

$$E(\tau) \overset{\tau\to\pm\infty}{=} \frac{2\sqrt{A}}{\alpha'(k+2)} = \text{constant} \quad ,$$
$$P(\tau) \overset{\tau\to\pm\infty}{=} -\frac{\sqrt{A}}{\alpha'(k+2)} \cos(2\tau + 2\varphi) \to 0 \quad . \tag{3.36}$$

This is the analog of the stable behaviour (3.24) in de Sitter universe for $\tau \to \infty$. Notice that eqs.(3.36) hold both for FRW ($k > 0$) and inflationary spacetimes ($k < 0$).

The behaviour described by eqs.(3.36) is similar to Minkowski spacetime. Notice that the factor $\tau^{-k/(k+2)}$ in the comoving radius [eq.(3.36)] is just the scale factor. That is, the physical string radius oscillates with constant amplitude as in Minkowski spacetime. The fact that the spacetime curvature tends to zero (except for de Sitter universe) when $T(\tau) \to \infty$ explains here the presence of a Minkowski-type behaviour.

For power-like inflationary universes with $k < -1$ a special exact ring solution exist [9] with

$$\eta = C \exp \frac{\tau}{\sqrt{-k-1}} \quad , \quad f(\tau) = \frac{C}{\sqrt{-k}} \exp \frac{\tau}{\sqrt{-k-1}} \tag{3.37}$$

where C is an arbitrary constant. For $k = -2$ (de Sitter universe) this is the solution $q^{(o)}(\sigma, \tau)$ in ref. [5] which has constant string size.

For this solution we find

$$E = \frac{1}{\alpha'}\sqrt{\frac{A}{-k-1}} \exp\left[\frac{\tau(k+2)}{2\sqrt{-k-1}}\right] = K\, R^{1+2/k}$$
$$P = -(\frac{1}{2} + \frac{1}{k})\; E$$

where K is a constant.

This is a fluid-like equation of state with $\gamma = \frac{1}{2} - \frac{1}{k}$. Notice that $\frac{1}{2} < \gamma < \frac{3}{2}$. For this solution, the energy grows with R as R to a power $1 + 2/k$ where, since $k < -1$,

$$-1 < 1 + \frac{2}{k} < 1.$$

(E and P of this solution are constants in de Sitter spacetime [$k = -2$]). This means that these special strings are subdominant both for $R \to \infty$ and for $R \to 0$ where the unstable strings ($E \simeq R$) and their dual ($E \simeq R^{-1}$) dominate respectively.

In ref. [7] the exact general evolution of circular strings in $2 + 1$ dimensional de Sitter spacetime is described closely and completely in terms of elliptic functions. Such solutions follow the form of eq.(3.21). The evolution depends on a constant parameter b, related to the string energy, and falls into three classes depending on whether $b < 1/4$ (oscillatory motion), $b = 1/4$ (degenerated, hyperbolic motion) or $b > 1/4$ (unbounded motion). The novel feature here is that one single world-sheet generically describes *infinitely many* (different and independent) strings. The world-sheet time τ is an infinite-valued function of the string physical time, each branch yields a different string. This has no analogue in flat spacetime.

We computed in ref. [7] the string energy E as a function of the string proper size S, and analyze it for the expanding and oscillating strings. For expanding strings ($\dot{S} > 0$): $E \neq 0$ even at $S = 0$, E decreases for small S and increases $\propto S$ for large S. For an oscillating string ($0 \leq S \leq S_{max}$), the average energy $< E >$ over one oscillation period is expressed as a function of S_{max} as a complete elliptic integral of the third kind [7].

Let us briefly describe here the periodic solutions found in ref. [7]. The scale factor takes the form

$$R(T) = e^{HT_+(\tau)} = e^{\tau[\frac{k}{1+k^2}+\Omega\frac{\vartheta_4'}{\vartheta_4}(\Omega y)]}\ \frac{\Omega\vartheta_1'(0)}{\pi\vartheta_1(\Omega y)}\ \mid \frac{\vartheta_4(\Omega(\tau-y))}{\vartheta_4(\Omega\tau)} \mid \tag{3.38}$$

where $0 \leq k \leq 1$ stands for the elliptic modulus,

$$\Omega \equiv \frac{\pi}{2\sqrt{k^2+1}\,K(k)} \quad , \quad y \equiv \sqrt{k^2+1}\,F(\frac{1}{\sqrt{k^2+1}}, k)$$

and $F(\phi, k)$ stands for a 1st. class elliptic integral. Notice that $1 \geq \Omega \geq 0$ for $0 \leq k \leq 1$. b relates with k through $\sqrt{b} = \frac{k}{1+k^2}$. We see that $R(T)$ is an oscillatory function of τ.

The cosmic time can be explicitly written as a Fourier series:

$$\begin{aligned} HT(\tau) &= \tau\,[\frac{k}{1+k^2} + \Omega\frac{\vartheta_4'}{\vartheta_4}(\Omega y)] + \log \mid \frac{\Omega\vartheta_1'(0)}{\pi\vartheta_1(\Omega y)} \mid \\ &-4\sum_{n=1}^{\infty}\frac{q^n}{n(1-q^{2n})}\sin(ny\Omega)\sin[n\pi\Omega(2\tau-y)]\,. \end{aligned}$$

The comoving string radius $f(\tau)$ [see eq.(3.21)] takes here the form

$$\begin{aligned} f(\tau) &= e^{-HT(\tau)}\ \frac{k}{\sqrt{1+k^2}}\ \mathrm{sn}\left(\frac{\tau}{\sqrt{1+k^2}}, k\right) \\ &= \frac{2\pi}{K(k)\sqrt{1+k^2}}\sum_{n=1}^{\infty}\frac{q^{n-1/2}}{1-q^{2n-1}}\sin[(2n-1)\Omega\tau]\,. \end{aligned} \tag{3.39}$$

The cosmic time and the comoving string radius are therefore completely regular functions of τ, and it follows that this string solution which is *oscillating regularly* as a function of world-sheet time τ, is also oscillating regularly when expressed in terms of hyperboloid time or cosmic time. This solution represents one *stable* string.

The invariant string size results:

$$S(\tau) = \frac{k}{H\sqrt{1+k^2}}\ \mathrm{sn}\left(\frac{\tau}{\sqrt{1+k^2}}, k\right) \tag{3.40}$$

We see that the string oscillates **inside** the horizon:

$$0 < S(\tau) < \frac{1}{H}$$

We see from eqs. (3.39-(3.39) that the string oscillations do not follow a pure harmonic motion as in flat Minkowski spacetime, but they are precise superpositions of all frequencies

$(2n-1)\Omega$ $(n = 1, 2, ..., \infty)$ with uniquely defined coefficients. The non-linearity of the string equations in de Sitter spacetime fixes the relation between the mode coefficients. In the present case the basic frequency Ω depends on the string energy, while in Minkowski spacetime the frequencies are merely n.

The energy (3.22) of this solution is given by [7]:

$$E = \frac{1}{H\alpha'}\left[\frac{k}{1+k^2} + \frac{1}{\sqrt{1+k^2}}\,\mathrm{zn}\left(\frac{y}{\sqrt{1+k^2}}, k\right)\right] + \textit{oscillating terms}, \tag{3.41}$$

where $\mathrm{zn}(x,k)$ stands for the Jacobi zeta-function [27]. Averaging over a period $2K\sqrt{1+k^2}$, the average energy $< E >$ is just the square bracket term.

The pressure (3.22) averaged over a period can be expressed here as a combination of complete elliptic integrals of the third kind. Numerical evaluation gives zero with high accuracy. We have checked in addition that the first orders in the expansion in the k identically vanish. We conclude that the average pressure is zero as in Minkowski spacetime. Thus, in average the *oscillating strings* describe *cold matter.*

In the $k \to 0$ limit the amplitude of the string size oscillation and the energy goes to zero [see eq.(3.40-3.41)]. Actually the solution disappears in such limit.

The other interesting limit is $k \to 1$ where the elliptic solutions turn into hyperbolic solutions. Then $< E > \to 1/(H\alpha')$.

More generally a numerical analysis shows that $< E >$ is a monotonically increasing function of b for $b \in [0, 1/4]$.

In summary, three asymptotic behaviors are exhibited by ring solutions.

i) Unstable for $R \to \infty$ and $R \to 0$:
$E =$ (constant) $R \to +\infty$, $P = -E/2 \to -\infty$, $S \simeq R \to \infty$.

ii) Dual to unstable for $R \to 0$ and $R \to \infty$:
$E =$ (constant) $R^{-1} \to +\infty$, $P = +E/2 \to +\infty$,
$S \simeq R \to 0$ (except for de Sitter where $S \to \frac{1}{H}$).

iii) Stable for $R \to \infty$:
$E =$ constant , $P = 0$, $S =$ constant .

Recall that the unstable behaviour dominates for $R \to \infty$ in inflationary universes whereas the dual behaviour dominates in FRW universes for $R \to 0$.

In addition we have the special behavior (3.38) for $k < -1$. Notice that the three behaviors i)-iii) appear for all expansion factors $R(\eta)$. The behaviours i) and ii) are related by the duality transformation $R \leftrightarrow 1/R$, the case iii) being invariant under duality . In the three cases we find perfect fluid equations [see eq.(3.9)] with different γ :

$$\gamma_u = 1/2 \ , \quad \gamma_d = 3/2 \ , \quad \gamma_s = 1 \ ,$$

where the indices u, d and s stand for 'unstable', 'dual' and 'stable', respectively. Assuming a perfect gas of strings on a volume R^2 , the energy density ρ will be proportional to E/R^2. This yields the following scaling with the expansion factor using the energies from i)-iii):

$$\rho_u = \text{constant } R^{-1} \ , \quad \rho_d = \text{constant } R^{-3} \ , \quad \rho_s = \text{constant } R^{-2}. \tag{3.42}$$

All three densities and pressures obey the continuity equation (3.8), as it must be.

The factor 1/2 in the relation between P and E for the cases i) and ii) is purely geometric. Notice that this factor was one in $1+1$ dimensions [eq.(3.19)].

C. D-Dimensional Universes

The solutions investigated in secs. II.A and II.B can exist for any dimensionality of the spacetime. Embedded in D-dimensional universes, the $(1+1)$ solutions of section II.A describe straight strings, the $(2+1)$ solutions of section II.B are circular rings. In D-dimensional spacetime, one expects string solutions spread in $D-1$ spatial dimensions. Their treatment has been done asymptotically in ref. [13]. One finds for $\eta \simeq \tau - \tau_0 \to 0$, $R \to \infty$,

$$P_u = -\frac{1}{D-1}\,\rho_u \quad , \quad \gamma_u = \frac{D-2}{D-1} \quad \text{for unstable strings.} \tag{3.43}$$

This relation coincides with eq.(3.15) for $D=2$ and with eqs.(3.23) and (3.27) for $D=3$.

The energy density scales with R as

$$\rho_u = u\,R^{2-D} \quad , \quad R \to \infty \tag{3.44}$$

(where u is a constant) in accordance with eq.(3.20) for $D=2$ and with eq.(3.42) for $D=3$.

For the dual regime, $R \to 0$, we have:

$$P_d = +\frac{1}{D-1}\,\rho_d \quad , \quad \gamma_d = \frac{D}{D-1} \quad \text{with } \rho_d = d\,R^{-D} \quad , \quad R \to 0 \ , \tag{3.45}$$

(where d is a constant). Eq.(3.45) reduces to eq.(3.18) for $D=2$ and to eqs.(3.24) and (3.29) for $D=3$. In this dual regime strings have the same equation of state as massless radiation.

Finally, for the stable regime we have

$$P_s = 0 \quad , \quad \gamma_s = 1 \quad \text{with } \rho_s = s\,R^{1-D} \tag{3.46}$$

(where s is a constant). This regime is absent in $D=2$ and appears for $D=3$ solutions in eqs.(3.24) and (3.36). The lack of string transverse modes in $D=2$ explains the absence of the stable regime there. The equation of state for stable strings coincides with the one for cold matter.

In conclusion, an ideal gas of classical strings in cosmological universes exhibit three different thermodynamical behaviours, all of perfect fluid type:

1) Unstable strings : negative pressure gas with $\gamma_u = \frac{D-2}{D-1}$
2) Dual behaviour : positive pressure gas similar to radiation , $\gamma_d = \frac{D}{D-1}$
3) Stable strings : positive pressure gas similar to cold matter, $\gamma_s = 1$.

Tables I and II summarize the main string properties for any scale factor $R(X^0)$.

TABLE 1. String energy and pressure as obtained from exact string solutions for various expansion factors $R(X^0)$.

STRING PROPERTIES FOR ARBITRARY $R(X^0)$

	Energy	Pressure	Equation of State: $p = (\gamma - 1)\rho$
D = 1 + 1: two families of solutions			
(i) $\eta \pm X = f_\pm(\sigma \pm \tau)$ (ii) $\eta \pm X =$ constant	$E = u\,R$ $E = d/R$	$P = -E$ $P = +E$	$\gamma = 0$ $\gamma = 2$
$D = 2 + 1$: Ring Solutions, three asymptotic behaviours (u, d, s)	$E = \frac{1}{\alpha'}\,\dot{X}^0(\tau)$	$P = \frac{R(\tau)^2}{2\alpha' \lvert \dot{X}^0(\tau) \rvert}\,[\dot{f}^2 - f^2]$	
(i) unstable for $R \to \infty$ (ii) dual to (i) for $R \to 0$ (iii) stable for $R \to \infty$	$E_u \overset{R\to\infty}{=} u\,R \to \infty$ $E_d \overset{R\to 0}{=} d/R \to \infty$ $E_s =$ constant	$P_u = -E_u/2 \to -\infty$ $P_d = +E_d/2 \to \infty$ $P_s = 0$	$\gamma_u = 1/2$ $\gamma_d = 3/2$ $\gamma_s = 1$
D-Dimensional spacetimes: general asymptotic behaviour			
(i) unstable for $R \to \infty$ (ii) dual to (i) for $R \to 0$ (iii) stable for $R \to \infty$	$E_u \overset{R\to\infty}{=} u\,R \to \infty$ $E_d \overset{R\to 0}{=} d/R \to \infty$ $E_s =$ constant	$P_u = -\frac{E_u}{D-1} \to -\infty$ $P_d = +\frac{E_d}{D-1} \to \infty$ $P_s = 0$	$\gamma_u = \frac{D-2}{D-1}$ $\gamma_d = \frac{D}{D-1}$ $\gamma_s = 1$

TABLE 2. The string energy density and pressure for a gas of strings can be summarized by the formulas below which become exact for $R \to 0$ and for $R \to \infty$.

STRING ENERGY DENSITY AND PRESSURE FOR ARBITRARY $R(X^0)$

	Energy density: $\rho \equiv E/R^{D-1}$	Pressure
Qualitatively correct formulas for all R and D	$\rho = \left(u\,R + \frac{d}{R} + s\right) \frac{1}{R^{D-1}}$	$p = \frac{1}{D-1} \left(\frac{d}{R} - u\,R\right) \frac{1}{R^{D-1}}$

TABLE 3. The **self-consistent** cosmological solution of the Einstein equations in General Relativity with the string gas as source.

STRING COSMOLOGY IN GENERAL RELATIVITY

Einstein equations (no dilaton field)	Expansion factor $R(X^0)$	Temperature $T(R)$
$X^0 \to 0$	$\frac{D}{2} \left[\frac{2d}{(D-1)(D-2)}\right]^{\frac{1}{D}} (X^0)^{\frac{2}{D}}$	$\frac{dD}{S(D-1)}\, 1/R$
$X^0 \to \infty$ (without string decay)	$\left[\frac{(D-2)u}{2(D-1)}\right]^{\frac{1}{D-2}} (X^0)^{\frac{2}{D-2}}$	$\frac{(D-2)u}{(D-1)S}\, R$
$X^0 \to \infty$ (with string splitting)	$\left[\frac{(D-1)s}{2(D-2)}\right]^{\frac{1}{D-1}} (X^0)^{\frac{2}{D-1}}$	usual matter dominated behaviour

The unstable string behaviour corresponds to the critical case of the so-called coasting universe [15,29]. In other words, classical strings provide a *concrete* matter realization of such cosmological model. Till now, no form of matter was known to describe coasting universes [15].

Finally, notice that strings continuously evolve from one type of behaviour to the other two. This is explicitly seen from the string solutions in refs. [5] - [7] . For example the string described by $q_-(\sigma, \tau)$ for $\tau > 0$ shows unstable behaviour for $\tau \to 0$, dual behaviour for $\tau \to \tau_0 = 1.246450...$ and stable behaviour for $\tau \to \infty$.

The equation of state for strings in four dimensional flat Minkowski spacetime is discussed in ref. [21]. One finds the values 4/3, 2/3 and 1 for γ by choosing appropriate values of the average string velocity in chap. 7 of ref. [21].

IV. SELF-CONSISTENT STRING COSMOLOGY

In the previous section we investigated the propagation of test strings in cosmological spacetimes. Let us now investigate how the Einstein equations in General Relativity and the effective equations of string theory (beta functions) can be verified **self-consistently** with our string solutions as sources.

We shall assume a gas of classical strings neglecting interactions as string splitting and coalescing. We will look for cosmological solutions described by metrics of the type (3.1). It is natural to assume that the background will have the same symmetry as the sources. That is, we assume that the string gas is homogeneous, described by a density energy $\rho = \rho(T)$ and a pressure $p = p(T)$. In the effective equations of string theory we consider a space independent dilaton field. Antisymmetric tensor fields wil be ignored.

A. String Dominated Universes in General Relativity (no dilaton field)

The Einstein equations for the geometry (3.1) take the form

$$\frac{1}{2}\,(D-1)(D-2)\;H^2 = \rho \quad ,$$
$$(D-2)\dot{H} + p + \rho = 0 \quad . \tag{4.1}$$

where $H \equiv \frac{dR}{dT}/R$. We know p and ρ as functions of R in asymptotic cases. For large R, the unstable strings dominate [eq.(3.44)] and we have

$$\rho = u\,R^{2-D} \quad , \quad p = -\frac{\rho}{D-1} \quad \text{for } R \to \infty \tag{4.2}$$

For small R, the dual regime dominates with

$$\rho = d\,R^{-D} \quad , \quad p = +\frac{\rho}{D-1} \quad \text{for } R \to 0 \tag{4.3}$$

We also know that stable solutions may be present with a contribution $\sim R^{1-D}$ to ρ and with zero pressure. For intermediate values of R the form of ρ is clearly more complicated but a formula of the type

$$\rho = \left(u_R\, R + \frac{d}{R} + s \right) \frac{1}{R^{D-1}} \tag{4.4}$$

where

$$\lim_{R\to\infty} u_R = \begin{cases} 0 & \text{FRW} \\ u_\infty \neq 0 & \text{Inflationary} \end{cases} \tag{4.5}$$

This equation of state is qualitatively correct for all R and becomes exact for $R \to 0$ and $R \to \infty$. The parameters u_R, d and s are positive constants and the u_R varies smoothly with R.

The pressure associated to the energy density (4.4) takes then the form

$$p = \frac{1}{D-1} \left(\frac{d}{R} - u_R\, R \right) \frac{1}{R^{D-1}} \tag{4.6}$$

Inserting eq.(4.4) into the Einstein-Friedmann equations [eq.(4.1)] we find

$$\frac{1}{2}\,(D-1)(D-2) \left(\frac{dR}{dT} \right)^2 = \left(u_R\, R + \frac{d}{R} + s \right) \frac{1}{R^{D-3}} \tag{4.7}$$

We see that R is a monotonic function of the cosmic time T. Eq.(4.7) yields

$$T = \sqrt{\frac{(D-1)(D-2)}{2}} \int_0^R dR\, \frac{R^{D/2-1}}{\sqrt{u_R\, R^2 + d + s\, R}} \tag{4.8}$$

where we set $R(0) = 0$.

It is easy to derive the behavior of R for $T \to 0$ and for $T \to \infty$.

For $T \to 0$, $R \to 0$, the term d/R dominates in eq.(4.7) and

$$R(T) \overset{T\to 0}{\simeq} \frac{D}{2} \left[\frac{2d}{(D-1)(D-2)} \right]^{\frac{1}{D}} (T)^{\frac{2}{D}} \tag{4.9}$$

For $T \to \infty$, $R \to \infty$ and the term $u_R\, R$ dominates in eq.(4.7). Hence,

$$R(T) \overset{T\to\infty}{\simeq} \left[\frac{(D-2)u_\infty}{2(D-1)} \right]^{\frac{1}{D-2}} (T)^{\frac{2}{D-2}} \tag{4.10}$$

[u_R tends to a constant u_∞ for $R \to \infty$]. This expansion is faster than (cold) matter dominated universes where $R \simeq [T]^{\frac{2}{D-1}}$. For example, for $D = 4$, R grows linearly with T whereas for matter dominated universes $R \simeq [T]^{2/3}$. However, eq.(4.10) **is not** a self-consistent solution. Assuming that the term $u_R\, R$ dominates for large R we find a scale factor $R(T) \sim (T)^{\frac{2}{D-2}} \sim \eta^{\frac{2}{D-4}}$ for $D \neq 4$ and $R(T) \sim T \sim e^\eta$ at $D = 4$. This **is not an inflationary universe** but a FRW universe. The term $u_R R$ is absent for large R in FRW universes as explained before. Therefore, we must instead use for large R

$$\rho = \left(\frac{d}{R} + s \right) \frac{1}{R^{D-1}} \tag{4.11}$$

Now, for $T \to \infty, R \to \infty$ and we find a matter dominated regime:

$$R(T) \stackrel{T\to\infty}{\simeq} \left[\frac{(D-1)s}{2(D-2)}\right]^{\frac{1}{D-1}} (T)^{\frac{2}{D-1}} \tag{4.12}$$

For intermediate values of T, $R(T)$ is a continuous and monotonically increasing function of T.

In summary, the universe starts at $T = 0$ with a singularity of the type dominated by radiation. (The string behaviour for $R \to 0$ is like usual radiation). Then, the universe expands monotonically, growing for large T as $R \simeq [T]^{\frac{2}{D-1}}$. In particular, this gives $R \simeq [T]^{2/3}$ for $D = 4$.

It must be noticed that the qualitative form of the solution $R(T)$ does not depend on the particular positive values of u_R, d and s.

We want to stress that we achieve a **self-consistent** solution of the Einstein equations with string sources since the behaviour of the string pressure and density given by eqs.(4.4)-(4.6) precisely holds in universes with power like $R(T)$.

In ref. [3] similar results were derived using arguments based on the splitting of long strings.

B. Thermodynamics of strings in cosmological spacetimes

Let us consider a comoving volume R^{D-1} filled by a gas of strings. The entropy change for this system is given by:

$$TdS = d(\rho\, R^{D-1}) \; + \; p\, d(R^{D-1}) \tag{4.13}$$

The continuity equation (3.8) and (4.13) implies that dS/dt vanishes. That is, the entropy per comoving volume stays constant in time. Using now the thermodynamic relation [20]

$$\frac{dp}{dT} = \frac{p+\rho}{T} \tag{4.14}$$

it follows [21] that

$$S = \frac{R^{D-1}}{T}(p+\rho) + \text{constant} \tag{4.15}$$

Let us first ignore the possibility of string decay. Then, eq.(4.15) together with eqs.(4.4) and (4.6) yields the temperature as a function of the expansion factor R. That is,

$$T = \frac{1}{S}\left\{s + \frac{1}{D-1}\left[\frac{D\,d}{R} + (D-2)\,u\,R\right]\right\} \tag{4.16}$$

where S stands for the (constant) value of the entropy.

Eq.(4.16) shows that for small R, T scales as $1/R$ whereas for large R it scales as R. The small R behaviour of T is the usual exhibited by radiation.

For large R, in FRW universes $u_R \to 0$ and the constant term in s dominates. We just find a cold matter behaviour for large R.

For large R in inflationary universes, $u_R \to u_\infty$ and eq.(4.16) would indicate a temperature that **grows** proportionally to R. However, as stressed in ref. [3], the decay of long strings (through splitting) makes u_R exponentially decreasing with R.

V. EFFECTIVE STRING EQUATIONS WITH THE STRING SOURCES INCLUDED

Let us consider now the cosmological equations obtained from the low energy string effective action including the string matter as a classical source. In D spacetime dimensions, this action can be written as

$$
\begin{aligned}
S &= S_1 + S_2 \\
S_1 &= \frac{1}{2}\int d^D x \,\sqrt{-G}\, e^{-\Phi}\, \left[R + G_{AB}\, \partial^A\Phi\, \partial^B\Phi \;+ 2\, U(G,\Phi) - c \right] \\
S_2 &= -\frac{1}{4\pi\alpha'} \sum_{strings} \int d\sigma d\tau\; G_{AB}(X)\; \partial_\mu X^A\; \partial^\mu X^B \qquad , \qquad (5.1)
\end{aligned}
$$

Here $A, B = 0, \ldots, D-1$. This action is written in the so called 'Brans-Dicke frame' (BD) or 'string frame', in which matter couples to the metric tensor in the standard way. The BD frame metric coincides with the sigma model metric to which test strings are coupled.

Eq.(5.1) includes the dilaton field (Φ) with a potential $U(G,\Phi)$ depending on the dilaton and graviton backgrounds; c stands for the central charge deficit or cosmological constant term. The antisymmetric tensor field was not included, in fact it is irrelevant for the results obtained here. Extremizing the action (5.1) with respect to G_{AB} and Φ yields the equations of motion

$$
\begin{aligned}
R_{AB} + \nabla_{AB}\Phi + 2\,\frac{\partial U}{\partial G_{AB}} - \frac{G_{AB}}{2}\left[R + 2\,\nabla^2\Phi - (\nabla\Phi)^2 - c + 2\,U \right] &= e^{\phi}\, T_{AB} \\
R + 2\,\nabla^2\Phi - (\nabla\Phi)^2 - c + 2U - \frac{\partial U}{\partial\Phi} &= 0 \;, \qquad (5.2)
\end{aligned}
$$

which can be more simply combined as

$$
\begin{aligned}
R_{AB} + \nabla_{AB}\Phi + 2\,\frac{\partial U}{\partial G_{AB}} - G_{AB}\,\frac{\partial U}{\partial\Phi} &= e^{\Phi}\, T_{AB} \\
R + 2\,\nabla^2\Phi - (\nabla\Phi)^2 - c + 2\,U - \frac{\partial U}{\partial\Phi} &= 0 \qquad (5.3)
\end{aligned}
$$

Here T_{AB} stands for the energy momentum tensor of the strings as defined by eq.(2.27). It is also convenient to write these equations as

$$
R_{AB} - \frac{G_{AB}}{2}\, R = T_{AB} + \tau_{AB} \qquad (5.4)
$$

where τ_{AB} is the dilaton energy momentum tensor :

$$
\tau_{AB} = -\nabla_{AB}\Phi + \frac{G_{AB}}{2}\left[2\,\frac{\partial U}{\partial\Phi} - R \right]
$$

The Bianchi identity

$$
\nabla^A\left(R_{AB} - \frac{G_{AB}}{2}\, R \right) = 0
$$

yields, as it must be, the conservation equation,

$$\nabla^A (T_{AB} + \tau_{AB}) = 0 \tag{5.5}$$

It must be noticed that eqs.(5.3) do not reduce to the Einstein equations of General Relativity even when $\Phi = U = 0$. Eqs. (5.3) yields in that case the Einstein equations *plus* the condition $R = 0$.

A. Effective String Equations in Cosmological Universes

For the homogeneous isotropic spacetime geometries described by eq.(3.1) we have

$$\begin{aligned} R_0^0 &= -(D-1)(\dot{H} + H^2) \\ R_i^k &= -\delta_i^k \, [\dot{H} + (D-1)\, H^2] \\ R &= -(D-1)(2\,\dot{H} + D\,H^2). \end{aligned} \tag{5.6}$$

where $H \equiv \frac{1}{R}\frac{dR}{dT}$.

The equations of motion (5.3) read

$$\begin{aligned} \ddot{\Phi} - (D-1)(\dot{H} + H^2) - \frac{\partial U}{\partial \Phi} &= e^{\Phi}\, \rho \\ \dot{H} + (D-1)\, H^2 - H\,\dot{\Phi} + \frac{\partial U}{\partial \Phi} + \frac{R}{D-1}\frac{\partial U}{\partial R} &= e^{\Phi}\, p \\ 2\,\ddot{\Phi} + 2(D-1)\, H\,\dot{\Phi} - \dot{\Phi}^2 - (D-1)(2\,\dot{H} + D\,H^2) - 2\,\frac{\partial U}{\partial \Phi} - c + 2\,U &= 0 \end{aligned} \tag{5.7}$$

where dot $\cdot$ stands for $\frac{d}{dT}$, and

$$\rho = T_0^0 \quad , \quad -\delta_i^k \, p = T_i^k \, . \tag{5.8}$$

The conservation equation takes the form of eq.(3.8)

$$\dot{\rho} + (D-1)\, H\,(p + \rho) = 0 \quad . \tag{5.9}$$

By defining,

$$\begin{aligned} &\Psi \equiv \Phi - \log\sqrt{-G} = \Phi - (D-1)\,\log R \\ &\bar{\rho} = e^{\Phi}\, \rho \quad , \quad \bar{p} = e^{\Phi}\, p \, , \end{aligned} \tag{5.10}$$

equations (5.7) can be expressed in a more compact form as

$$\begin{aligned} \ddot{\Psi} - (D-1)\, H^2 - \left.\frac{\partial U}{\partial \Psi}\right|_R &= \bar{\rho} \\ \dot{H} - H\,\dot{\Psi} + \frac{R}{D-1}\left.\frac{\partial U}{\partial R}\right|_\Psi &= \bar{p} \\ \dot{\Psi}^2 - (D-1)\, H^2 - 2\,\bar{\rho} - 2\,U + c &= 0 \, , \end{aligned} \tag{5.11}$$

The conservation equation reads

$$\dot{\bar{\rho}} - \dot{\Psi}\,\bar{\rho} + (D-1)\,H\,\bar{p} = 0 \tag{5.12}$$

As is known, under the duality transformation $R \longrightarrow R^{-1}$, the dilaton transforms as $\Phi \longrightarrow \Phi + (D-1)\log R$. The shifted dilaton Ψ defined by eq.(5.10) is invariant under duality.

The transformation

$$R' \equiv R^{-1} \quad , \tag{5.13}$$

implies

$$\Psi' = \Psi \quad , \; H' = -H \quad , \; \bar{p}' = -p \quad , \; \bar{\rho}' = \bar{\rho} \tag{5.14}$$

provided $u = d$, that is, a duality invariant string source. This is the duality invariance transformation of eqs.(5.11).

Solutions to the effective string equations have been extensively treated in the literature [25] and they are not our main purpose. For the sake of completeness, we briefly analyze the limiting behaviour of these equations for $R \to \infty$ and $R \to 0$.

It is difficult to make a complete analysis of the effective string equations (5.11) since the knowledge about the potential U is rather incomplete. For weak coupling (e^{Φ} small) the supersymmetry breaking produces an effective potential that decreases very fast (as the exponential of an exponential of Φ) for $\Phi \to -\infty$.

Let us analyze the asymptotic behavior of eqs.(5.11) for $R \to \infty$ and $R \to 0$ assuming that the potential U can be ignored. It is easy to see that a power behaviour Ansatz both for R and for e^{Ψ} as functions of T is consistent with these equations. It turns out that the string sources do not contribute to the leading behaviour here, and we find for $R \to 0$

$$\begin{aligned} R_{\mp} &= C_1\,(T)^{\pm 1/\sqrt{D-1}} \to 0 \quad , \\ e^{\Psi_{\mp}} &= \quad C_2\,(T)^{-1} \to \begin{cases} \infty \\ 0 \end{cases} \end{aligned} \tag{5.15}$$

Where C_1 and C_2 are constants. Here the branches $(-)$ and $(+)$ correspond to $T \to 0$ and to $T \to \infty$ respectively. In both regimes $R_{\mp} \to 0$ and $e^{\Phi_{\mp}} \to 0$.

The potential $U(\Phi)$ is hence negligible in these regimes. In terms of the conformal time η , the behaviours (5.15) result

$$\begin{aligned} R_{\mp} &= \quad C_1'\,(\eta)^{\pm\frac{1}{\sqrt{D-1}\mp 1}} \to 0 \\ e^{\Psi_{\mp}} &= C_2'\,(\eta)^{-\frac{\sqrt{D-1}}{\sqrt{D-1}\mp 1}} \to \begin{cases} \infty \\ 0 \end{cases} \end{aligned} \tag{5.16}$$

Where C_1' and C_2' are constants. The branch $(-)$ would describe an expanding non-inflationary behaviour near the initial singularity $T = 0$, while the branch $(+)$ describes a 'big crunch' situation and is rather unphysical.

Similarly, for $R \to \infty$ and $e^{\Phi} \to \infty$, we find

$$R_{\mp} = D_1\,(T)^{\mp 1/\sqrt{D-1}} \to \infty \quad ,$$

$$e^{\Psi_\mp} = \quad D_2\,(T)^{-1} \to \begin{cases} \infty \\ 0 \end{cases} \tag{5.17}$$

Where D_1 and D_2 are constants. Here again, the branches $(-)$ and $(+)$ correspond to $T \to 0$ and to $T \to \infty$ respectively, but now in both regimes $R_\mp \to \infty$ and $e^{\Phi_\mp} \to \infty$. (In this limit, one is not guaranteed that U can be consistently neglected). In terms of the conformal time, eqs.(5.17) read

$$\begin{aligned} R_\mp = &\quad D_1'\,(\eta)^{\mp\frac{1}{\sqrt{D-1}\pm 1}} \to \infty \\ e^{\Psi_\mp} = &\; D_2'\,(\eta)^{-\frac{\sqrt{D-1}}{\sqrt{D-1}\pm 1}} \to \begin{cases} \infty \\ 0 \end{cases} \end{aligned} \tag{5.18}$$

The branch $(+)$ describes a noninflationary expanding behaviour for $T \to \infty$ faster than the standard matter dominated expansion, while the branch $(-)$ describes a super-inflationary behaviour $\eta^{-\alpha}$, since $0 < \alpha < 1$, for all D.

The behaviours (5.15) for $R_\mp \to 0$ and (5.17) for $R_\mp \to \infty$ are related by duality $R \leftrightarrow 1/R$.

B. String driven inflation?

Let us consider now the question of whether de Sitter spacetime may be a self-consistent solution of the effective string equations (5.7) with the string sources included. The strings in cosmological universes like de Sitter spacetime have the equation of state (4.4)-(4.6). Since $e^\Psi = e^\Phi\,R^{1-D}$:

$$\bar{\rho} = e^\Psi \left(u\,R + \frac{d}{R} + s\right) \tag{5.19}$$

$$\bar{p} = \frac{e^\Psi}{D-1}\left(\frac{d}{R} - u\,R\right) \tag{5.20}$$

In the absence of dilaton potential and cosmological constant term, the string sources do not generate de Sitter spacetime as discussed in sec. III.A. We see that for $U = c = 0$, and $R = e^{HT}$, eqs.(5.11) yields to a contradiction (unless $D = 0$) for the value of Ψ , required to be $-HT +$ constant.

A self-consistent solution describing asymptotically de Sitter spacetime self-sustained by the string equation of state (5.19)-(5.20) is given by

$$\begin{aligned} R = e^{HT} \quad &, \quad H = \text{constant} > 0 \;, \\ 2U - c = D\,H^2 &= \text{constant} \\ \Psi_\pm = \mp HT &\pm i\pi + \log\frac{(D-1)\,H^2}{\rho_\pm} \\ \rho_+ \equiv u \quad &, \quad \rho_- \equiv d \end{aligned} \tag{5.21}$$

The branch Ψ_+ describes the solution for $R \to \infty$ ($T \to +\infty$), while the branch Ψ_- corresponds to $R \to 0$ ($T \to -\infty$). De Sitter spacetime with lorentzian signature self-sustained by the strings necessarily requires a constant imaginary piece $\pm i\pi$ in the dilaton field. This makes $e^\Psi < 0$ telling us that the gravitational constant $G \sim e^\Psi < 0$ here describes antigravity.

Is interesting to notice that in the euclidean signature case, i. e. (+++...++), the Ansatz $\dot{H} = 0$, $2U - c =$constant, yields a constant curvature geometry with a real dilaton, but which is of Anti-de Sitter type. This solution is obtained from eqs.(5.20)-(5.21) through the transformation

$$\hat{X}^0 = iT \quad , \quad \hat{H} = -iH \quad , \quad X^i = X^i \quad , \quad \Psi = \Psi \tag{5.22}$$

which maps the Lorentzian de Sitter metric into the positive definite one

$$d\hat{s}^2 = (d\hat{X}^0)^2 + e^{\hat{H}\hat{X}^0}\,(d\vec{X})^2. \tag{5.23}$$

The equations of motion (5.11) within the constant curvature Ansatz ($\dot{\hat{H}} = \ddot{\Psi} = 0$) are mapped onto the equations

$$\begin{aligned} (D-1)\,\hat{H}^2 - \left.\frac{\partial U}{\partial \Psi}\right|_R &= \bar{\rho} \\ \hat{H}\,\frac{d\Psi}{d\hat{X}^0} + \frac{R}{D-1}\left.\frac{\partial U}{\partial R}\right|_\Psi &= \bar{p} \\ -(\frac{d\Psi}{d\hat{X}^0})^2 + (D-1)\,\hat{H}^2 - 2\,\bar{\rho} - 2\,U + c &= 0\,, \end{aligned} \tag{5.24}$$

with the solution

$$\begin{aligned} R &= e^{\hat{H}\hat{X}^0} \quad , \quad \hat{H} = \text{constant} > 0 \quad , \\ c - 2\,U &= D\,\hat{H}^2 = \text{constant} \\ \Psi_\pm &= \mp\hat{H}\hat{X}^0 + \log\frac{(D-1)\,\hat{H}^2}{\rho_\pm} \\ \rho_+ &\equiv u \quad , \quad \rho_- \equiv d \end{aligned} \tag{5.25}$$

Both solutions (5.25) and (5.21) are mapped one into another through the transformation (5.22).

It could be recalled that in the context of (point particle) field theory, de Sitter spacetime (as well as anti-de Sitter) emerges as an exact selfconsistent solution of the semiclassical Einstein equations with the back reaction included [31] - [32]. (Semiclassical in this context, means that matter fields including the graviton are quantized to the one-loop level and coupled to the (c-number) gravity background through the expectation value of the energy-momentum tensor T_A^B . This expectation value is given by the trace anomaly: $< T_A^A >= \bar{\gamma}\ R^2$). On the other hand, the α' expansion of the effective string action admits anti-de Sitter spacetime (but not de Sitter) as a solution when the quadratic curvature corrections (in terms of the Gauss-Bonnet term) to the Einstein action are included [33]. It appears that the corrections to the anti-de Sitter constant curvature are qualitatively similar in the both cases, with α' playing the rôle of the trace anomaly parameter $\bar{\gamma}$ [32].

The fact that de Sitter inflation with true gravity $G \sim e^\Psi > 0$ does not emerge as a solution of the effective string equations does not mean that string theory excludes inflation.

What means is that the effective string equations are not enough to get inflation. The effective string action is a low energy field theory approximation to string theory containing only the *massless* string modes (*massless* background fields).

The vacuum energy scales to start inflation (physical or true vacuum) are typically of the order of the Planck mass [21] - [22] where the effective string action approximation breaks down. One must consider the massive string modes (which are absent from the effective string action) in order to properly get the cosmological condensate yielding de Sitter inflation. We do not have at present the solution of such problem.

TABLE 4. Asymptotic solution of the string effective equations (including the dilaton).

EFFECTIVE STRING EQUATIONS SOLUTIONS IN COSMOLOGY

Effective String equations	$R(X^0) \to 0$ behaviour	$R(X^0) \to \infty$ behaviour
$X^0 \to 0$	$\sim (X^0)^{+1/\sqrt{D-1}}$	$\sim (X^0)^{-1/\sqrt{D-1}}$
$X^0 \to \infty$	$\sim (X^0)^{-1/\sqrt{D-1}}$	$\sim (X^0)^{+1/\sqrt{D-1}}$

VI. MULTI-STRINGS AND SOLITON METHODS IN DE SITTER UNIVERSE

Among the cosmological backgrounds, de Sitter spacetime occupies a special place. This is, in one hand relevant for inflation and on the other hand string propagation turns to be specially interesting there [2] - [8]. String unstability, in the sense that the string proper length grows indefinitely is particularly present in de Sitter. The string dynamics in de Sitter universe is described by a generalized sinh-Gordon model with a potential unbounded from below [4]. The sinh-Gordon function $\alpha(\sigma, \tau)$ having a clear physical meaning : $H^{-1}e^{\alpha(\sigma,\tau)/2}$ determines the string proper length. Moreover the classical string equations of motion (plus the string constraints) turn to be integrable in de Sitter universe [4,5]. More precisely, they are equivalent to a non-linear sigma model on the grassmannian $SO(D,1)/O(D)$ with periodic boundary conditions (for closed strings). This sigma model has an associated linear system [34] and using it, one can show the presence of an infinite number of conserved quantities [35]. In addition, the string constraints imply a zero energy-momentum tensor and these constraints are compatible with the integrability.

The so-called dressing method [34] in soliton theory allows to construct solutions of non-linear classically integrable models using the associated linear system. In ref. [6] we systematically construct string solutions in three dimensional de Sitter spacetime. We start from a given exactly known solution of the string equations of motion and constraints in de

Sitter [5] and then we "dress" it. The string solutions reported there indeed apply to cosmic strings in de Sitter spacetime as well.

The invariant interval in D-dimensional de Sitter space-time is given by

$$ds^2 = dT^2 - \exp[2HT] \sum_{i=1}^{D-1} (dX^i)^2. \tag{6.1}$$

Here T is the so called cosmic time. In terms of the conformal time η,

$$\eta \equiv -\frac{\exp[-HT]}{H} \quad , -\infty < \eta \leq 0 \, ,$$

the line element becomes

$$ds^2 = \frac{1}{H^2\eta^2}[d\eta^2 - \sum_{i=1}^{D-1} (dX^i)^2] \, .$$

The de Sitter spacetime can be considered as a D-dimensional hyperboloid embedded in a D+1 dimensional flat Minkowski spacetime with coordinates $(q^0, ..., q^D)$:

$$ds^2 = \frac{1}{H^2}[-(dq^0)^2 + \sum_{i=1}^{D}(dq^i)^2] \tag{6.2}$$

where

$$\begin{aligned} q^0 &= \sinh HT + \frac{H^2}{2} \exp[HT] \sum_{i=1}^{D-1} (X^i)^2 \, , \\ q^1 &= \cosh HT - \frac{H^2}{2} \exp[HT] \sum_{i=1}^{D-1} (X^i)^2 \, , \\ q^{i+1} &= H \exp[HT] \, X^i \quad , \quad 1 \leq i \leq D-1, \ -\infty < T, \ X^i < +\infty. \end{aligned} \tag{6.3}$$

The complete de Sitter manifold is the hyperboloid

$$-(q^0)^2 + \sum_{i=1}^{D} (q^i)^2 = 1.$$

The coordinates (T, X^i) and (η, X^i) cover only the half of the de Sitter manifold $q^0 + q^1 > 0$.

We will consider a string propagating in this D-dimensional space-time.The string equations of motion (2.7) in the metric (6.2) take the form:

$$\partial_{+-} q + (\partial_+ q.\partial_- q) \, q = 0 \quad \text{with} \quad q.q = 1, \tag{6.4}$$

where . stands for the Lorentzian scalar product $a.b \equiv -a_0 b_0 + \sum_{i=1}^{D} a_i b_i$, $x_\pm \equiv \frac{1}{2}(\tau \pm \sigma)$ and $\partial_\pm q = \frac{\partial q}{\partial x_\pm}$. The string constraints (2.8) become for de Sitter universe

$$T_{\pm\pm} = \frac{\partial q}{\partial x_\pm} . \frac{\partial q}{\partial x_\pm} = 0. \tag{6.5}$$

Eqs.(6.4) describe a non compact O(D,1) non-linear sigma model in two dimensions. In addition, the (two dimensional) energy-momentum tensor is required to vanish by the constraints eqs.(6.5) . This system of non-linear partial differential equations can be reduced by choosing an appropriate basis for the string coordinates in the $(D+1)$-dimensional Minkowski space time $(q^0, ..., q^D)$ to a noncompact Toda model [4].

These equations can be rewritten in the form of a chiral field model on the Grassmanian $G_D = SO(D,1)/O(D)$. Indeed, any element $\mathbf{g} \in G_D$ can be parametrized with a real vector $q\rangle$ of the unit pseudolength

$$\mathbf{g} = 1 - 2|q\rangle\langle q|J, \quad \langle q|J|q\rangle = 1. \tag{6.6}$$

In terms of $\mathbf{g}$, the string equations (6.4) have the following form

$$2\,\mathbf{g}_{\xi\eta} - \mathbf{g}_\xi\,\mathbf{g}\,\mathbf{g}_\eta - \mathbf{g}_\eta\,\mathbf{g}\,\mathbf{g}_\xi = 0\ , \tag{6.7}$$

and the conformal constraints (6.5) become

$$\mathrm{tr}\,\mathbf{g}_\xi^2 = 0, \quad \mathrm{tr}\,\mathbf{g}_\eta^2 = 0\ . \tag{6.8}$$

The fact that $\mathbf{g} \in G_D$ implies that $\mathbf{g}$ is a real matrix with the following properties:

$$\mathbf{g} = J\mathbf{g}^{\mathrm{t}}J, \quad \mathbf{g}^2 = I, \quad tr\,\mathbf{g} = D-1\ , \quad \mathbf{g} \in SL(D+1,R). \tag{6.9}$$

These conditions are equivalent to the existence of the representation (6.6). Equation (6.7) is the compatibility condition for the following overdetermined linear system:

$$\Psi_\xi = \frac{U}{1-\lambda}\Psi, \quad \Psi_\eta = \frac{V}{1+\lambda}\Psi, \tag{6.10}$$

where

$$U = \mathbf{g}_\xi\,\mathbf{g}, \quad V = \mathbf{g}_\eta\,\mathbf{g}\ . \tag{6.11}$$

Or, in terms of the vector $q\rangle$

$$U = 2\,q_\xi\rangle\langle q\,J - 2\,q\rangle\langle q_\xi\,J,$$

$$V = 2\,q_\eta\rangle\langle q\,J - 2\,q\rangle\langle q_\eta\,J.$$

Eq. (6.6) can be easily inverted yielding q in terms of the matrix g:

$$q_0 = \sqrt{\frac{g_{00}-1}{2}}\ , \quad q_i = \sqrt{\frac{1-g_{ii}}{2}}\ \ 1 \le i \le D \text{ (no sum over } i) \tag{6.12}$$

The use of overdetermined linear systems to solve non-linear partial differential equations associated to them goes back to refs. [36]. (See refs. [37] - [38] for further references).

In order to fix the freedom in the definition of Ψ we shall identify

$$\Psi(\lambda = 0) = \mathbf{g}. \tag{6.13}$$

This condition is compatible with the above equations since the matrix function Ψ at the point $\lambda = 0$ satisfies the same equations as $\mathbf{g}$. Thus the problem of constructing exact solutions of the string equations is reduced to finding compatible solutions of the linear equations (6.10) such that $\mathbf{g} = \Psi(\lambda = 0)$ satisfies the constraints eqs.(6.8) and (6.9).

We concentrate below on the linear system (6.10) since this is the main tool to derive new string solutions in de Sitter spacetime.

In ref. [6] the dressing method was applied as follows. We started from the exact ring-shaped string solution $q_{(0)}$ [5] and we find the explicit solution $\Psi^{(0)}(\lambda)$ of the associated linear system, where λ stands for the spectral parameter. Then, we propose a new solution $\Psi(\lambda)$ that differs from $\Psi^{(0)}(\lambda)$ by a matrix rational in λ. Notice that $\Psi(\lambda = 0)$ provides in general a new string solution.

We then show that this rational matrix must have at least **four** poles, $\lambda_0, 1/\lambda_0, \lambda_0^*, 1/\lambda_0^*$, as a consequence of the symmetries of the problem. The residues of these poles are shown to be one-dimensional projectors. We then prove that these projectors are formed by vectors which can all be expressed in terms of an arbitrary complex constant vector $|x_0\rangle$ and the complex parameter λ_0. This result holds for arbitrary starting solutions $q_{(0)}$.

Since we consider closed strings, we impose a 2π-periodicity on the string variable σ . This restricts λ_0 to take discrete values that we succeed to express in terms of Pythagorean numbers. In summary, our solutions depend on two arbitrary complex numbers contained in $|x_0\rangle$ and two integers n and m . The counting of degrees of freedom is analogous to $2+1$ Minkowski spacetime except that left and right modes are here mixed up in a non-linear and precise way.

The vector $|x_0\rangle$ somehow indicates the polarization of the string. The integers (n, m) determine the string winding. They fix the way in which the string winds around the origin in the spatial dimensions (here S^2). Our starting solution $q_{(0)}(\sigma, \tau)$ is a stable string winded $n^2 + m^2$ times around the origin in de Sitter space.

The matrix multiplications involved in the computation of the final solution were done with the help of the computer program of symbolic calculation "Mathematica". The resulting solution $q(\sigma, \tau) = (q^0, q^1, q^2, q^3)$ is a complicated combination of trigonometric functions of σ and hyperbolic functions of τ. That is, these string solitonic solutions do not oscillate in time. This is a typical feature of string unstability [5] - [13] - [39]. The new feature here is that strings (even stable solutions) do not oscillate neither for $\tau \to 0$, nor for $\tau \to \pm\infty$.

We plot in figs. 1-7 the solutions for significative values of $|x_0\rangle$ and (m, n) in terms of the comoving coordinates (T, X^1, X^2)

$$T = \frac{1}{H}\log(q^0 + q^1) \ , \quad X^1 = \frac{1}{H}\frac{q^2}{q^0 + q^1} \ , \quad X^2 = \frac{1}{H}\frac{q^3}{q^0 + q^1} \tag{6.14}$$

The first feature to point out is that our solitonic solutions describe **multiple** (here five or three) strings, as it can be seen from the fact that for a given time T we find several different values for τ. That is, τ is a **multivalued** function of T for any fixed σ (fig.1-2). Each branch of τ as a function of T corresponds to a different string. This is a entirely new feature for strings in curved spacetime, with no analogue in flat spacetime where the time coordinate can always be chosen proportional to τ. In flat spacetime, multiple string solutions are described by multiple world-sheets. Here, we have a **single** world-sheet describing several independent and simultaneous strings as a consequence of the coupling with the spacetime

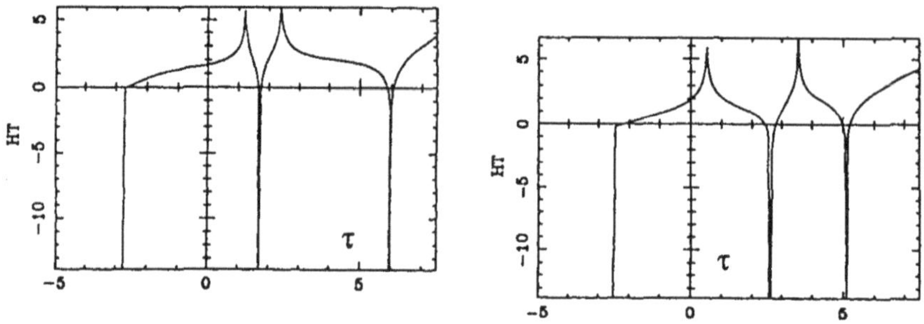

Figure 1: Plot of the function $HT(\tau)$, for two values of σ, for $n = 4, |x^0 >= (1 + i, .6 + .4i, .3 + .5i, .77 + .79i)$. The function $\tau(T)$ is multivalued, revealing the presence of five strings.

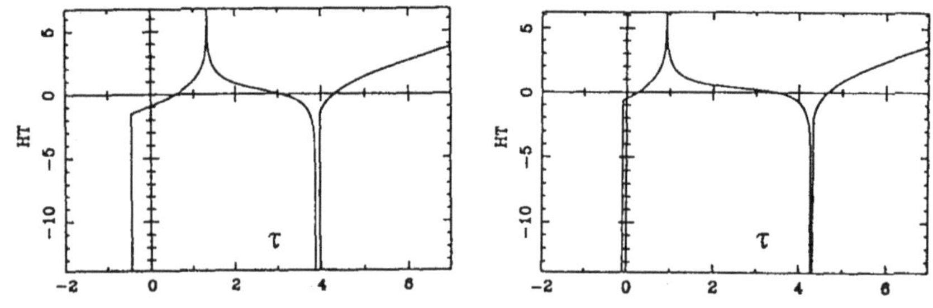

Figure 2: Same as fig.1, for $n = 4, |x^0 >= (1, -1, i, 1)$. Because of a degeneracy, there are now only three strings.

geometry. Notice that we consider *free* strings. (Interactions among the strings as splitting or merging are not considered). Five is the generic number of strings in our dressed solutions. The value five can be related to the fact that we are dressing a one-string solution ($q_{(0)}$) with *four* poles. Each pole adds here an unstable string.

In order to describe the real physical evolution, we eliminated numerically $\tau = \tau(\sigma, T)$ from the solution and expressed the spatial comoving coordinates X^1 and X^2 in terms of T and σ.

We plot $\tau(\sigma, T)$ as a function of σ for different fixed values of T in fig.3-4. It is a sinusoidal-type function. Besides the customary closed string period 2π, another period appears which varies on τ. For small τ , $\tau = \tau(\sigma, T)$ has a convoluted shape while for larger τ (here $\tau \leq 5$), it becomes a regular sinusoid. These behaviours reflect very clearly in the evolution of the spatial coordinates and shape of the string.

The evolution of the five (and three) strings simultaneously described by our solution as a function of T, for positive T is shown in figs. 5-7. One string is stable (the 5th one). The other four are unstable. For the stable string, (X^1, X^2) contracts in time precisely as e^{-HT}, thus keeping the proper amplitude $(e^{HT}X^1, e^{HT}X^2)$ and proper size constant. For this stable string $(X^1, X^2) \leq \frac{1}{H}$. ($1/H$ = the horizon radius). For the other (unstable) strings, (X^1, X^2) become very fast constant in time, the proper size expanding as the universe itself like e^{HT} . For these strings $(X^1, X^2) \geq \frac{1}{H}$. These exact solutions display remarkably the asymptotic string behaviour found in refs. [4,13].

In terms of the sinh-Gordon description, this means that for the strings outside the horizon the sinh-Gordon function $\alpha(\sigma, \tau)$ is the same as the cosmic time T up to a function of σ. More precisely,

$$\alpha(\sigma,\tau) \overset{T>>\frac{1}{H}}{=} 2H\,T(\sigma,\tau) + \log\left\{2H^2\left[(A^1(\sigma)')^2 + (A^2(\sigma)')^2\right]\right\} + O(e^{-2HT}). \qquad (6.15)$$

Here $A^1(\sigma)$ and $A^2(\sigma)$ are the X^1 and X^2 coordinates outside the horizon. For $T \to +\infty$ these strings are at the absolute *minimum* $\alpha = +\infty$ of the sinh-Gordon potential with infinite size. The string inside the horizon (stable string) corresponds to the *maximum* of the potential, $\alpha = 0$. $\alpha = 0$ is the only value in which the string can stay without being pushed down by the potential to $\alpha = \pm\infty$ and this also explains why only one stable string appears (is not possible to put more than one string at the maximum of the potential without falling down). These features are *generically* exhibited by our one-soliton multistring solutions, independently of the particular initial state of the string (fixed by $|x^0>$ and (n, m)). For particular values of $|x^0>$, the solution describes three strings, with symmetric shapes from $T = 0$, for instance like a rosette or a circle with festoons (fig. 5-7).

The string solutions presented here trivially embedd on D-dimensional de Sitter space-time ($D \geq 3$). It must be noticed that they exhibit the essential physics of strings in D-dimensional de Sitter universe. Moreover, the construction method used here works in any number of dimensions.

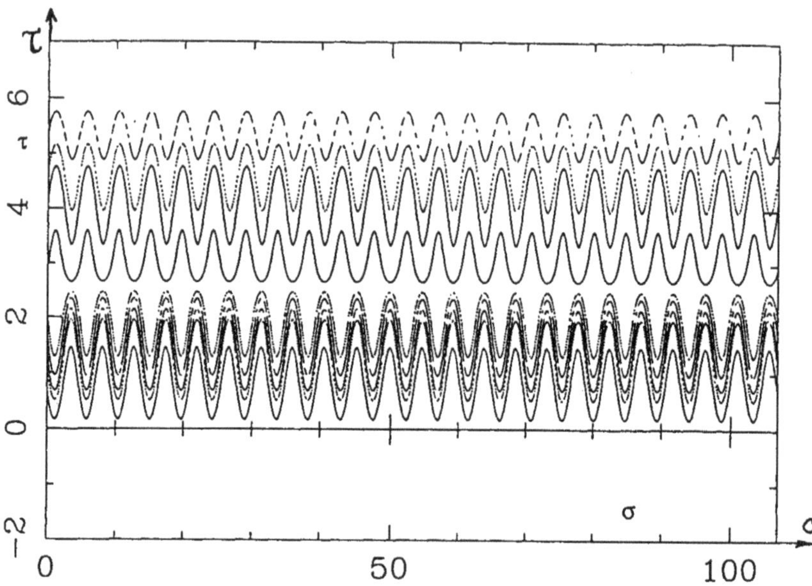

Figure 3: $\tau = \tau(\sigma, T)$ for fixed T for $n = 4, |x^0 >= (1, -1, i, 1)$. Three values of HT are displayed, corresponding to HT=0 (full line), 1 (dots), and 2 (dashed line). For each HT, three curves are plotted, which correspond to the three strings. They are ordered with τ increasing.

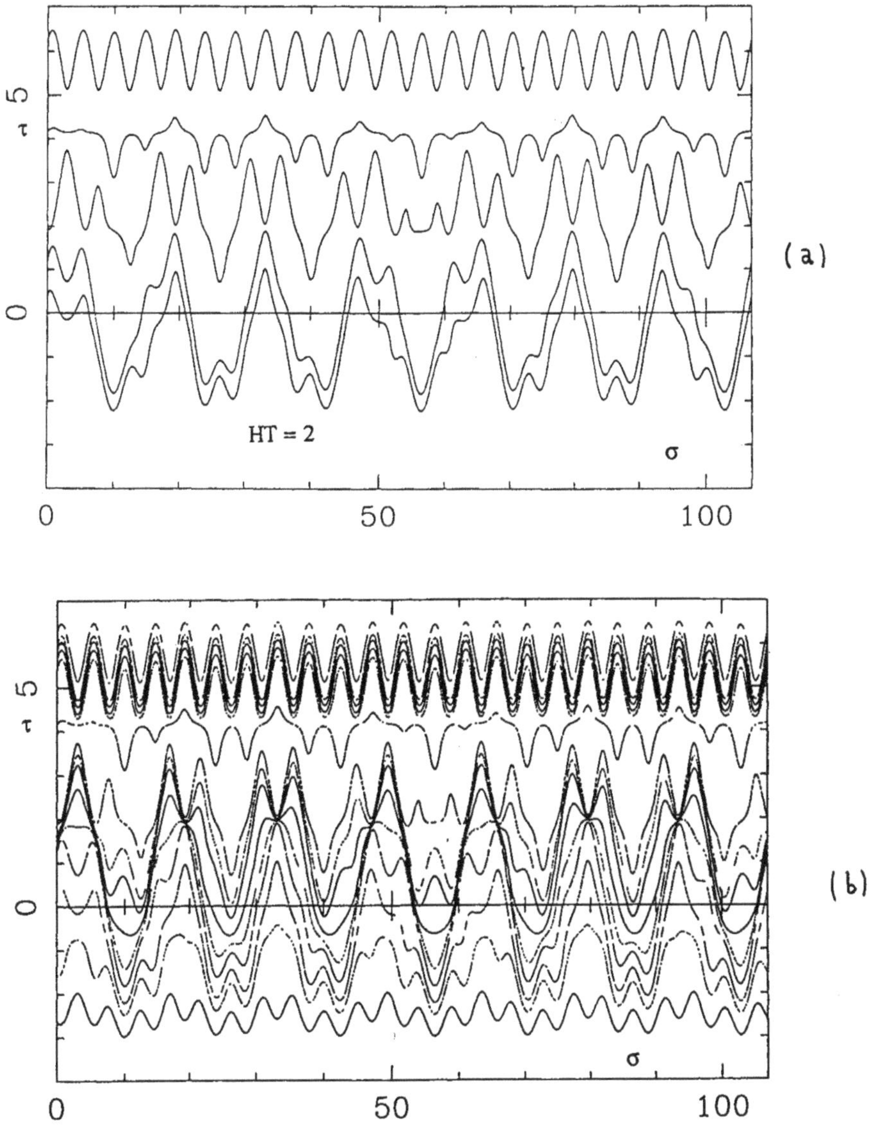

Figure 4: Same as fig. 3 for $n = 4, x^0 >= (1 + i, .6 + .4i, .3 + .5i, .77 + .79i)$. a) The five curves corresponding to the five strings at HT=2. b) The five curves for three values of HT: HT=0 (full line), 1 (dots), and 2 (dashed line).

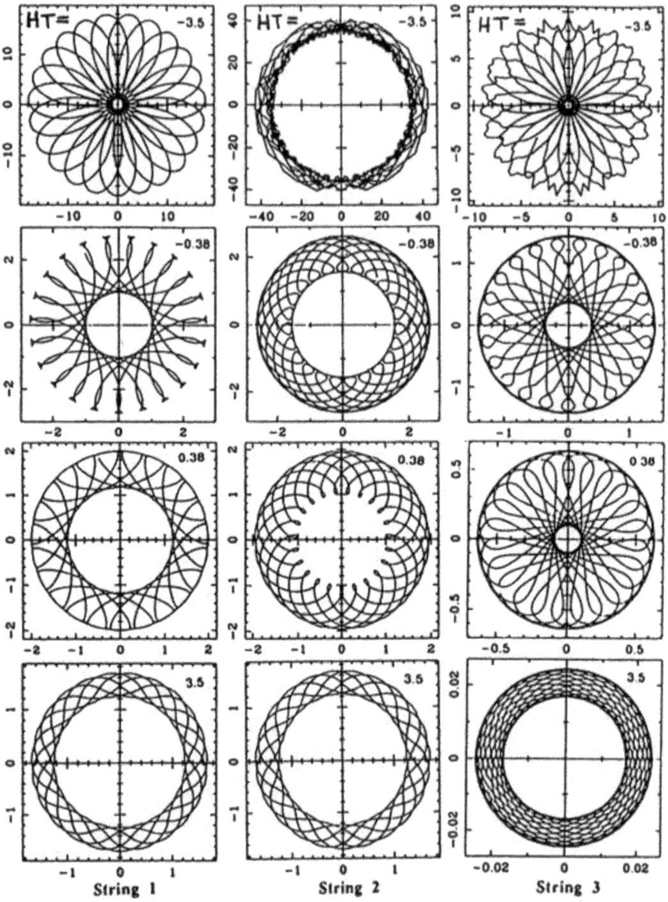

Figure 5: Evolution of the three strings, for $n = 4, |x^0> = (1, -1, i, 1)$. The comoving size of string (1) stays constant for $HT < -3$, then decreases around $HT = 0$, and stays constant again after $HT = 1$. The invariant size of string (2) is constant for negative HT, then grows as the expansion factor for $HT > 1$, and becomes identical to string (1). The string (3) has a constant comoving size for $HT < -3$, then collapses as e^{-HT} for positive HT.

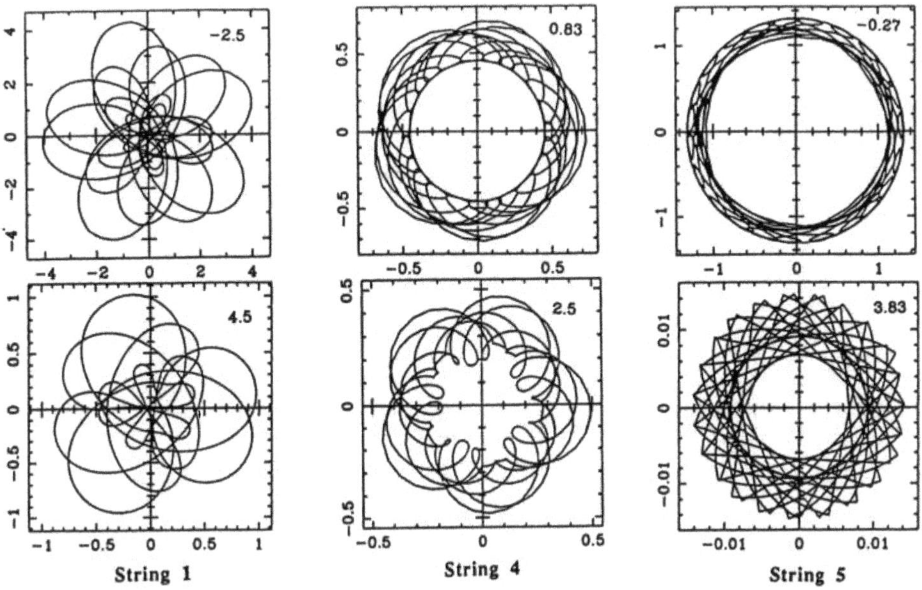

Figure 6: Evolution of three of the five strings for $n = 4, |x^0 >= (1 + i, .6 + .4i, .3 + .5i, .77 + .79i)$.

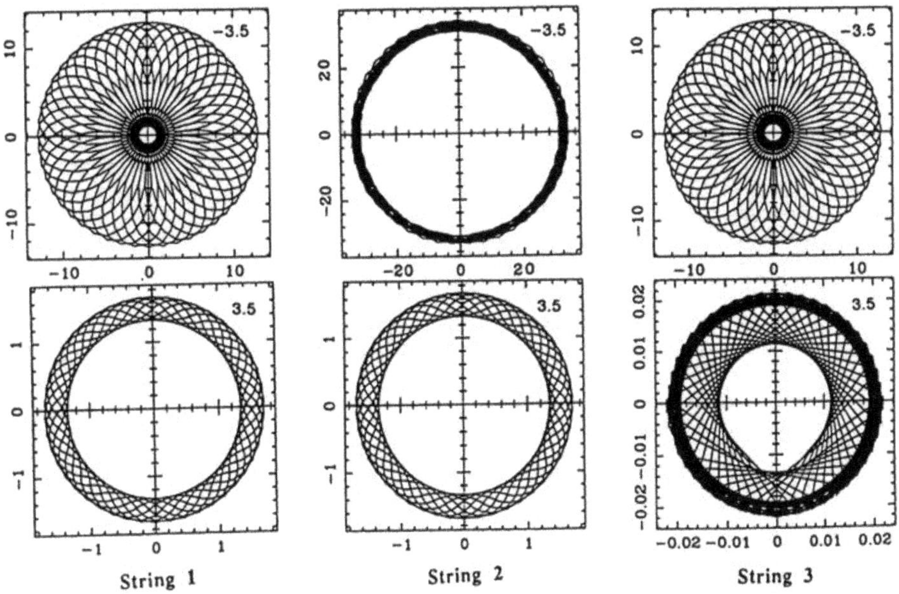

Figure 7: Evolution of the three strings for the degenerate case $n = 6, |x^0 >= (1, -1, i, 1)$.

REFERENCES

[1] H.J. de Vega and N. Sánchez in "String Quantum Gravity and the Physics at the Planck Scale", Proceedings of the Erice Workshop held in June 1992. Edited by N. Sánchez, World Scientific, 1993. Pages 73-185, and references given therein.

[2] H. J. de Vega and N. Sánchez, Phys. Lett. **B 197**, 320 (1987).

[3] H. J. de Vega and N. Sánchez, Phys. Rev. **D50**, 7202 (1994).

[4] H. J. de Vega and N. Sánchez, Phys. Rev. **D47**, 3394 (1993).

[5] H. J. de Vega, A. V. Mikhailov and N. Sánchez, Teor. Mat. Fiz. **94** (1993) 232.

[6] F. Combes, H. J. de Vega, A. V. Mikhailov and N. Sánchez, Phys. Rev. **D50**, 2754 (1994).

[7] H. J. de Vega, A. L. Larsen and N. Sánchez, Nucl. Phys. **B 427**, 643 (1994).

[8] I. Krichever, Funct. Anal. and Appl. **28**, 21 (1994), [Funkts. Anal. Prilozhen. **28**, 26 (1994)].

[9] H. J. de Vega and I. L. Egusquiza, Phys. Rev. **D49**, 763 (1994).

[10] A. L. Larsen and N. Sánchez, Phys. Rev. **D50**, 7493 (1994).

[11] S. Hawking and G. F. R. Ellis, 'The large scale structure of the spacetime', Cambridge Univ. Press, 1973.

[12] N. Sánchez and G. Veneziano, Nucl. Phys. **B333**, 253 (1990).

[13] M. Gasperini, N. Sánchez and G. Veneziano, Int. J. Mod. Phys. **A 6**, 3853 (1991) and Nucl. Phys. **B364**, 365 (1991).

[14] H. J. de Vega and N. Sánchez, Int. J. Mod. Phys. **A 7**, 3043 (1992).

[15] G. F. R. Ellis, Banff Lectures 1990, *in* Gravitation, eds. R. Mann and P. Wesson , World Scientific 1991.

[16] See for a review, T. W. B. Kibble, Erice Lectures at the Chalonge School in Astrofundamental Physics, N. Sánchez editor, World Scientific, 1992.

[17] A. Vilenkin, Phys. Rev. **D 24**, 2082 (1981) and Phys. Rep. **121**, 263 (1985). N. Turok and P. Bhattacharjee, Phys. Rev. **D 29**, 1557 (1984).

[18] J. D. Barrow, Nucl. Phys. **B 310**, 743 (1988).

[19] R. Myers, Phys. Lett. **B199**, 371 (1987). M. Mueller, Nucl. Phys. **B337**, 37 (1990). See for a review: A.A. Tseytlin in the Proceedings of the Erice School "String Quantum Gravity and Physics at the Planck Energy Scale", 21-28 June 1992, Edited by N. Sánchez, World Scientific, 1993.

[20] S. Weinberg, 'Gravitation and Cosmology', J. Wiley, 1972.

[21] E. W. Kolb and M. S. Turner, 'The Early Universe', Addison-Wesley, 1990.

[22] A. D. Linde, 'Particle Physics and Inflationary Cosmology', Harwood (1990).

[23] J. Polchinsky, Phys. Lett. **B209**, 252 (1988). J. Day and J. Polchinsky, Phys. Lett. **B220**, 387 (1989).

[24] D. Mitchell, N. Turok, R. Wilkinson and P. Jetzer, Nucl. Phys. **B 315**, 1 (1989). D. Mitchell, B. Sundborg and N. Turok, Nucl. Phys. **B 335**, 621 (1990).

[25] See for example, I. Antoniadis, C. Bachas, J. Ellis and D. V. Nanopoulos, Nucl. Phys. **B 328**, 117 (1989) and Phys. Lett. **B 257**, 278 (1991). B. A. Campbell, A. Linde and K. A. Olive, Nucl. Phys. **B 355**, 146 (1991).

B. A. Campbell, N. Kaloper and K. A. Olive, Phys. Lett. **B 277**, 265 (1992).
A.A. Tseytlin, Mod. Phys. Lett **A 6**, 1721 (1991).
A.A. Tseytlin and C. Vafa, Nucl. Phys. **B 372**, 443 (1992).
R. Brustein and P. J. Steinhardt, Phys. Lett. **B 302**, 196 (1993).
M. Gasperini and G. Veneziano, Mod. Phys. Lett. **A 8**, 370 (1993),
Phys. Lett. **B 277**, 256 (1992) and Astroparticle Physics **1**, 317 (1993).
E. Raiten, Nucl. Phys. **B 416**, 881 (1994).
R. Brustein and G. Veneziano, Phys. Lett. **B 329**, 429 (1994).
V. A. Kostelecký and M. J. Perry, Nucl. Phys. **B 414**, 174 (1994).
See in addition ref. [19].

[26] See for example:
M. Green, J. Schwarz, E. Witten, *Superstring Theory.*
Cambridge University Press. 1987.

[27] I.S. Gradshteyn and I.M. Ryzhik, Table of Integrals Series and Products,
(Academic Press, New York, fourth edition, 1965).

[28] See for example,
S. Weinberg, 'Gravitation and Cosmology', J. Wiley, 1972.

[29] F. Müller-Holstein, Class. Quant. Grav. **3**, 665 (1986).
K. G. Akdeniz et al. Mod. Phys. Lett. **A 6**, 1543 (1991) and
Phys. Lett. **B 321**, 329 (1994).

[30] M. V. Fischetti, J. B. Hartle and B. L. Hu, Phys. Rev. **D 20**, 1757 (1979).

[31] S. Wada and T. Azuma, Phys. Lett. **B 132**, 313 (1983).
V. Sahni and L. A. Kofman, Phys. Lett. **A 117**, 275 (1986).

[32] M. A. Castagnino, J. P. Paz and N. Sánchez, Phys. Lett. **B 193**, 13 (1987).

[33] D. G. Boulware and S. Deser, Phys. Rev. Lett. **55**, 2656 (1985).

[34] V. E. Zakharov and A. V. Mikhailov, JETP, **75**, 1953 (1978).

[35] H. J. de Vega, Phys. Lett. **B 87**, 233 (1979).

[36] C. S. Gardner, J. M. Greene, M. D. Kruskal and R. M. Miura,
Phys. Rev. Lett. **19**, 1095 (1967).
P. D. Lax, Comm. Pure and Appl. Math. **21** 467 (1968).

[37] M. J. Ablowitz and H. Segur, "Solitons and the Inverse scattering transformation"
SIAM Philadelphia 1981.
V.E. Zakharov, S.V. Manakov, S.P. Novikov and L.P. Pitaevsky,
"Soliton Theory; The Inverse Method", Nauka, Moscow, 1980.

[38] A. C. Scott, F. Y. F. Chu and D. W. MacLaughlin, Proc. IEEE, **61**, 1443 (1973).
G. L. Lamb, Elements of Soliton Theory, J. Wiley, NY (1980).

[39] H. J. de Vega and N. Sánchez, Nucl. Phys. **B309**, 552 and 577 (1988).
C. Loustó and N. Sánchez, Phys. Rev. **D47**, 4498 (1993).

Strings and Multi-Strings in Black Hole and Cosmological Spacetimes*

A.L. Larsen and N. Sánchez
Observatoire de Paris, DEMIRM. Laboratoire Associé au CNRS UA 336, Observatoire de Paris et École Normale Supérieure. 61, Avenue de l'Observatoire, 75014 Paris, France.

Abstract

Recent results on classical and quantum strings in a variety of black hole and cosmological spacetimes, in various dimensions, are presented. The curved backgrounds under consideration include the $2+1$ black hole anti de Sitter spacetime and its dual, the black string, the ordinary $D \geq 4$ black holes with or without a cosmological constant, the de Sitter and anti de Sitter spacetimes and static Robertson-Walker spacetimes. Exact solutions to the string equations of motion and constraints, representing circular strings, stationary open strings and dynamical straight strings, are obtained in these backgrounds and their physical properties (length, energy, pressure) are described. The existence of *multi-string* solutions, describing finitely or infinitely many strings, is shown to be a general feature of spacetimes with a positive or negative cosmological constant. Generic approximative solutions are obtained using the string perturbation series approach, and the question of the stability of the solutions is addressed.

Furthermore, using a canonical quantization procedure, we find the string mass spectrum in de Sitter and anti de Sitter spacetimes. New features as compared to the string spectrum in flat Minkowski spacetime appear, for instance the *fine-structure effect* at all levels beyond the graviton in both de Sitter and anti de Sitter spacetimes, and the *non-existence* of a Hagedorn temperature in anti de Sitter spacetime. We discuss the physical implications of these results. Finally, we consider the effect of spatial curvature on the string dynamics in Robertson-Walker spacetimes.

1 Introduction

The classical and quantum propagation of strings in curved spacetimes has attracted a great deal of interest in recent years. The main complication, as compared to the case of flat Minkowski spacetime, is related to the non-linearity of the equations of motion. It makes it possible to obtain the complete analytic solution only in a very few special cases like conical spacetime [1] and plane-wave/shock-wave backgrounds [2]. There are however also very general results concerning integrability and solvability for maximally symmetric spacetimes [3, 4] and gauged WZW models [5, 6]. These are the exceptional cases; generally the string equations of motion in curved spacetimes are not integrable and even if they are, it is usually an extremely difficult task to actually separate the equations, integrate them and finally write down the complete solution in closed form. Fortunately, there are several different ways to "attack" a system of coupled non-linear partial differential equations.

The systematic study of string dynamics in curved spacetimes and its associated physical phenomena was started in Refs.[7, 8]. Besides numerical methods, which will not be discussed here, approximative [7-10] and exact [11-14] methods for solving the string equations of motion and constraints in curved spacetimes, have been developed. Classical and quantum string dynamics have been investigated in black hole backgrounds [15-18], cosmological spacetimes [7, 11-14, 17-21], cosmic string spacetime [1], gravitational wave backgrounds [2], supergravity backgrounds (which are necessary for fermionic strings) [22], and near spacetime singularities [23]. Physical phenomena like the Hawking-Unruh effect in string theory [8, 24], horizon string stretching [8, 24], particle transmutation [15, 25], string scattering [1, 2, 15], mass spectrum and critical dimension [1, 7, 15, 27], string instability [7, 11-13, 16, 21] and multi-string solutions [11-13, 28] have been found.

In a generic D-dimensional curved spacetime with metric $g_{\mu\nu}$ and coordinates x^μ, ($\mu = 0, 1, ..., D-1$), the string equations of motion and constraints are:

$$\ddot{x}^\mu - x''^\mu + \Gamma^\mu_{\rho\sigma}(\dot{x}^\rho\dot{x}^\sigma - x'^\rho x'^\sigma) = 0, \tag{1.1}$$

$$g_{\mu\nu}\dot{x}^\mu x'^\nu = g_{\mu\nu}(\dot{x}^\mu\dot{x}^\nu + x'^\mu x'^\nu) = 0, \tag{1.2}$$

where dot and prime stand for derivative with respect to the world-sheet coordinates τ and σ, respectively and $\Gamma^\mu_{\rho\sigma}$ are the Christoffel symbols with respect to the metric $g_{\mu\nu}$. In the following we present recent results [26-29] on solutions to Eqs.(1.1)-(1.2) in a variety of curved spacetimes from cosmology, gravitation and string theory. We obtain explicit (exact and/or approximate) mathematical solutions, discuss the corresponding physical properties, we quantize the solutions in different ways (canonical quantization, semi-classical quantization) and find the physical content: the mass spectrum.

The presentation is organized as follows: In Section 2, we discuss the classical string dynamics in the 2 + 1 black hole anti de Sitter (BH-AdS) spacetime, recently found by Bañados et al [30]. We compare with the string dynamics in ordinary cosmological and black hole spacetimes. This clarifies the geometry (as seen by a string) of the BH-AdS spacetime. In Section 3, generalizing results from Section 2, we derive the quantum string mass spectrum in ordinary D-dimensional anti de Sitter spacetime. We discuss, in particular, the sectors of low and very high mass states. New physical phenomena arise like the fine-structure effect at all levels beyond the graviton and the non-existence of a Hagedorn critical temperature. The results are compared with corresponding results obtained in Minkowski and de Sitter spacetimes. Sections 4 and 5 are devoted to the investigation of the more general underlying structure of solutions to Eqs.(1.1)-(1.2). In Section 4, we find new classes of exact string and multi-string solutions in cosmological and black hole spacetimes, while in Section 5, we consider the effects of a non-zero spatial curvature on the classical and quantum string dynamics in Friedmann-Robertson-Walker (FRW) universes.

2 Classical String Dynamics in 2+1 BH-AdS

Anti de Sitter (AdS) spacetime is often considered to be a spacetime of minor importance in a cosmological context. However, first of all, it *is* a FRW universe and as such should not be neglected, secondly; it serves as a simple and convenient spacetime for comparison and understanding of results obtained in (say) Minkowski or de Sitter spacetimes and third, it has a tendency to show up (in disguise) as a solution in various models of dilaton-gravity and string theory. An example of the latter is represented by the $2+1$ BH-AdS spacetime of Bañados et al [30]. This spacetime background has arised much interest recently [31-35]. It describes a two-parameter family (mass M and angular momentum J) of black holes in 2+1 dimensional general relativity with metric:

$$ds^2 = (M - \frac{r^2}{l^2})dt^2 + (\frac{r^2}{l^2} - M + \frac{J^2}{4r^2})^{-1}dr^2 - Jdtd\phi + r^2d\phi^2. \qquad (2.1)$$

It has two horizons $r_\pm = \sqrt{\frac{Ml^2}{2} \pm \frac{l}{2}\sqrt{M^2l^2 - J^2}}$ and a static limit $r_{\text{erg}} = \sqrt{M}l$, defining an ergosphere, as for ordinary Kerr black holes. The spacetime is not asymptotically flat; it approaches anti de Sitter spacetime asymptotically with cosmological constant $\Lambda = -1/l^2$. The curvature is constant $R_{\mu\nu} = -(2/l^2)g_{\mu\nu}$ everywhere, except probably at $r = 0$, where it has at most a delta-function singularity. Notice the weak nature of the singularity at $r = 0$ in 2+1 dimensions as compared with the power law divergence of curvature scalars in $D > 3$ (We will not discuss here the geometry near $r = 0$. For a discusion, see Refs.[33, 35]). The spacetime, Eq.(2.1), is also a solution of the low energy effective action of string theory with zero dilaton field $\Phi = 0$, anti-symmetric tensor field $H_{\mu\nu\rho} = (2/l^2)\epsilon_{\mu\nu\rho}$ (i.e. $B_{\phi t} = r^2/l^2$) and $k = l^2$ [31]. Moreover, it yields an exact solution of string theory in 2+1 dimensions, obtained by gauging the WZWN sigma model of the group

$SL(2,R) \times R$ at level k [31, 32] (for non-compact groups, k does not need to be an integer, so the central charge $c = 3k/(k-2) = 26$ when k=52/23). This solution is the black string background [36]:

$$d\tilde{s}^2 = -(1-\tfrac{\mathcal{M}}{\tilde{r}})d\tilde{t}^2 + (1-\tfrac{Q^2}{\mathcal{M}\tilde{r}})d\tilde{x}^2 + (1-\tfrac{\mathcal{M}}{\tilde{r}})^{-1}(1-\tfrac{Q^2}{\mathcal{M}\tilde{r}})^{-1}\tfrac{l^2 d\tilde{r}^2}{4\tilde{r}^2},$$

$$\tilde{B}_{\tilde{x}\tilde{t}} = \tfrac{Q}{r}, \quad \tilde{\Phi} = -\tfrac{1}{2}\log \tilde{r}l, \tag{2.2}$$

which is related by duality [31, 37] to the 2+1 BH-AdS spacetime, Eq.(2.1). It has two horizons $\tilde{r}_\pm = r_\pm$, the same as the metric, Eq.(2.1), while the static limit is $\tilde{r}_{\rm erg} = J/(2\sqrt{M})$.

We first investigate the string propagation in the $2+1$ BH-AdS background by considering the perturbation series around the exact center of mass of the string:

$$x^\mu(\tau,\sigma) = q^\mu(\tau) + \eta^\mu(\tau,\sigma) + \xi^\mu(\tau,\sigma) + \ldots \tag{2.3}$$

The original method of Refs.[7, 8] can be conveniently formulated in covariant form. It is useful to introduce $D-1$ normal vectors n^μ_R $(R = 1,..,D-1)$, (which can be chosen to be covariantly constant by gauge fixing), and consider comoving perturbations δx_R, i.e. those seen by an observer travelling with the center of mass, thus $\eta^\mu = \delta x^R n^\mu_R$. After Fourier expansion, $\delta x^R(\tau,\sigma) = \sum_n C^R_n(\tau)e^{-in\sigma}$, the first order perturbations satisfy the matrix Schrödinger-type equation in τ:

$$\ddot{C}_{nR} + (n^2\delta_{RS} - R_{\mu\rho\sigma\nu}n^\mu_R n^\nu_S \dot{q}^\rho \dot{q}^\sigma)C^S_n = 0. \tag{2.4}$$

Second order perturbations ξ^μ and constraints are similarly covariantly treated, ξ^μ also satisfying Schrödinger-type equations with source terms, see Ref.[26].

For our purposes here it is enough to consider the non-rotating $(J = 0)$ 2+1 BH-AdS background and a radially infalling string. We have solved completely the c.m. motion $q^\mu(\tau)$ and the first and second order perturbations $\eta^\mu(\tau,\sigma)$ and $\xi^\mu(\tau,\sigma)$ in this background. Eqs.(2.4) become:

$$\ddot{C}_{nR} + (n^2 + \frac{m^2}{l^2})C_{nR} = 0; \quad R = \perp, \parallel \tag{2.5}$$

The first order perturbations are independent of the black hole mass, only the anti de Sitter (AdS) part emerges. All oscillation frequencies $\omega_n = \sqrt{n^2 + m^2/l^2}$ are real; there are no unstable modes in this case. The perturbations:

$$\delta x_R(\tau,\sigma) = \sum_n [A_{nR}e^{-i(n\sigma+\omega_n\tau)} + \tilde{A}_{nR}e^{-i(n\sigma-\omega_n\tau)}] \tag{2.6}$$

are completely finite and regular. This is also true for the second order perturbations which are bounded everywhere, even for $r \to 0$ $(\tau \to 0)$.

We have also computed the conformal generators L_n, (see Ref.[26]), and the string mass:

$$m^2 = 2\sum_n (2n^2 + \frac{m^2}{l^2})[A_{n\|}\tilde{A}_{-n\|} + A_{n\perp}\tilde{A}_{-n\perp}]. \tag{2.7}$$

The mass formula is modified (by the term m^2/l^2) with respect to the usual flat spacetime expression. This is due to the asymptotic (here AdS) behaviour of the spacetime. In ordinary $D \geq 4$ black hole spacetimes (without cosmological constant), which are asymptotically flat, the mass spectrum is the same as in flat Minkowski spacetime [15].

We compare with the string perturbations in the ordinary ($D \geq 4$) black hole anti de Sitter spacetime. In this case Eqs.(2.4) become:

$$\ddot{C}_{nS\perp} + (n^2 + m^2H^2 + \frac{Mm^2}{r^3})C_{nS\perp} = 0, \quad S = 1,2 \tag{2.8}$$

$$\ddot{C}_{n\|} + (n^2 + m^2H^2 - \frac{2Mm^2}{r^3})C_{n\|} = 0. \tag{2.9}$$

The transverse $\perp$-perturbations are oscillating with real frequencies and are bounded even for $r \to 0$. For longitudinal $\|$-perturbations, however, imaginary frequencies arise and instabilities develop. The ($\mid n \mid = 1$)-instability sets in at:

$$r_{\text{inst.}} = (\frac{2Mm^2}{1 + m^2H^2})^{1/3} \tag{2.10}$$

Lower modes become unstable even outside the horizon, while higher modes develop instabilities at smaller r and eventually only for $r \approx 0$. For $r \to 0$ (which implies $\tau \to \tau_0$) we find $r(\tau) \approx (3m\sqrt{M/2})^{2/3}(\tau_0 - \tau)^{2/3}$ and:

$$\ddot{C}_{nS\perp} + \frac{2}{9(\tau - \tau_0)^2}C_{nS\perp} = 0, \quad S = 1,2 \tag{2.11}$$

$$\ddot{C}_{n\|} - \frac{4}{9(\tau - \tau_0)^2}C_{n\|} = 0. \tag{2.12}$$

For $\tau \to \tau_0$ the $\|$-perturbations blow up while the string ends trapped into the $r = 0$ singularity. We see the important difference between the string evolution in the 2+1 BH-AdS background and the ordinary 3+1 (or higher dimensional) black hole anti de Sitter spacetime.

We also compare with the string propagation in the 2+1 black string background, Eq.(2.2) (with $J = 0$). In this case, Eqs.(2.4) become:

$$\ddot{C}_{n\perp} + n^2C_{n\perp} = 0, \tag{2.13}$$

$$\ddot{C}_{n\|} + (n^2 - \frac{2m^2M}{lr})C_{n\|} = 0. \tag{2.14}$$

The $\perp$-modes are stable, while $C_{n||}$ develop instabilities. For $r \to 0$ (which implies $\tau \to \tau_0$) we find $r(\tau) \approx \frac{m^2 M}{l}(\tau_0 - \tau)^2$ and:

$$\ddot{C}_{n||} - \frac{2}{(\tau_0 - \tau)^2} C_{n||} = 0, \tag{2.15}$$

with similar conclusions as for the ordinary 3+1 (or higher dimensional) black hole anti de Sitter spacetime.

In order to extract more information about the string evolution in these backgrounds, in particular exact properties, we consider the circular string ansatz:

$$t = t(\tau), \quad r = r(\tau), \quad \phi = \sigma + f(\tau), \tag{2.16}$$

in the equatorial plane ($\theta = \pi/2$) of the stationary axially symmetric backgrounds:

$$ds^2 = g_{tt}(r)dt^2 + g_{rr}(r)dr^2 + 2g_{t\phi}(r)dtd\phi + g_{\phi\phi}(r)d\phi^2. \tag{2.17}$$

This includes all the cases of interest here: The 2+1 BH-AdS spacetime, the black string, as well as the equatorial plane of ordinary Einstein black holes. The string dynamics, determined by Eqs.(1.1)-(1.2), is then reduced to a system of second order ordinary differential equations and constraints, also described as a Hamiltonian system:

$$\dot{r}^2 + V(r) = 0; \qquad V(r) = g^{rr}(g_{\phi\phi} + E^2 g^{tt}), \tag{2.18}$$

$$\dot{t} = -Eg^{tt}, \qquad \dot{f} = -Eg^{t\phi}; \quad E = -P_t = \text{const.}, \tag{2.19}$$

which in all backgrounds considered here are solved in terms of either elementary or elliptic functions. The dynamics of the circular strings takes place at the r-axis in the $(r, V(r))$-diagram and from the properties of the potential $V(r)$ (minima, zeroes, asymptotic behaviour for large r and the value $V(0)$), general knowledge about the string motion can be obtained. On the other hand, the line element of the circular string turns out to be:

$$ds^2 = g_{\phi\phi}(d\sigma^2 - d\tau^2), \quad \text{i.e.} \quad S(\tau) = \sqrt{g_{\phi\phi}(r(\tau))}, \tag{2.20}$$

$S(\tau)$ being the invariant string size. For all the static black hole AdS spacetimes (2+1 and higher dimensional): $S(\tau) = r(\tau)$.

For the rotating 2+1 BH-AdS spacetime:

$$V(r) = r^2(\frac{r^2}{l^2} - M) + \frac{J^2}{4} - E^2, \tag{2.21}$$

(see Fig.1). $V(r)$ has a global minimum $V_{\text{min}} < 0$ between the two horizons r_+, r_- (for $Ml^2 \geq J^2$, otherwise there are no horizons). The vanishing of $V(r)$ at $r = r_{01,2}$ (see Ref.[26]) determines three different types of solutions: (i) For $J^2 > 4E^2$, there are two positive zeroes $r_{01} < r_{02}$, the string never comes outside the static limit, never falls into $r = 0$ neither (there is a barrier between $r = r_{01}$ and $r = 0$). The mathematical solution oscillates between r_{01} and r_{02} with $0 < r_{01} < r_- < r_+ < r_{02} < r_{erg}$. It may be interpreted as a string travelling between the different universes described by the maximal analytic extension of the manifold. (ii) For $J^2 < 4E^2$, there is only one positive zero r_0 outside the static limit and there is no barrier preventing the string from collapsing into $r = 0$. The string starts at $\tau = 0$ with maximal size $S_{\max}^{(ii)}$ outside the static limit, it then contracts through the ergosphere and the two horizons and eventually collapses into a point $r = 0$. For $J \neq 0$, it may be still possible to continue this solution into another universe like in the case (i). (iii) $J^2 = 4E^2$ is the limiting case where the maximal string size equals the static limit: $S_{\max}^{(iii)} = l\sqrt{M}$. In this case $V(0) = 0$, thus the string contracts through the two horizons and eventually collapses into a point $r = 0$.

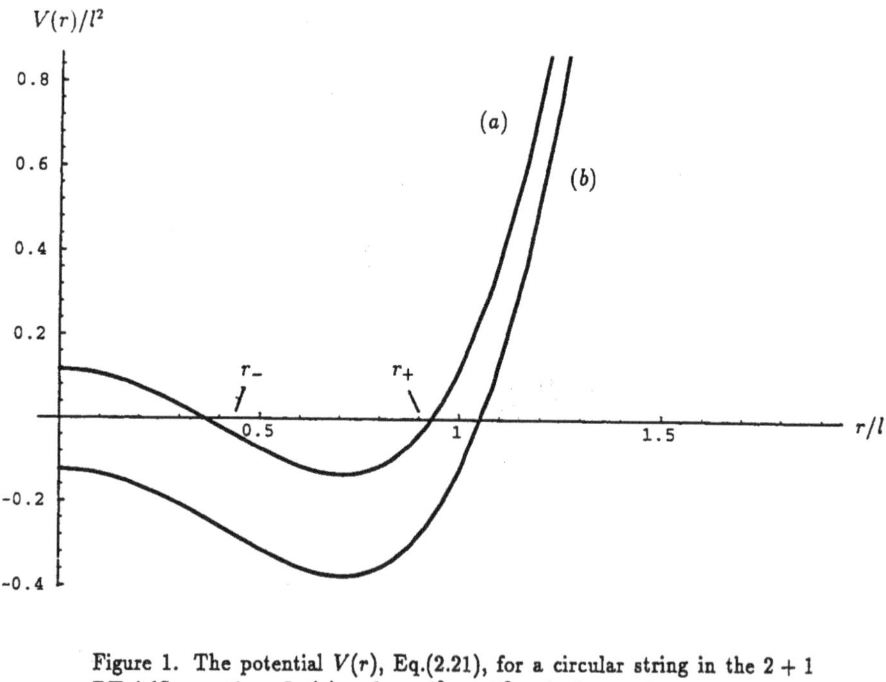

Figure 1. The potential $V(r)$, Eq.(2.21), for a circular string in the $2+1$ BH-AdS spacetime. In (a) we have $J^2 > 4E^2$ and a barrier between the inner horizon and $r = 0$, while (b) represents a case where $J^2 < 4E^2$ and a string will always fall into $r = 0$. The static limit is $r_{erg} = l$.

The exact general solution in the three cases (i)-(iii) is given by:

$$r(\tau) = \left| r_m - \frac{1}{c_1\wp(\tau-\tau_0)+c_2} \right|, \tag{2.22}$$

where:

$$r_m = S_{\text{max}} = \sqrt{\frac{Ml^2}{2}}\sqrt{1+\sqrt{1-\frac{4V(0)}{Ml^2}}}, \quad V(0) = \frac{J^2}{4} - E^2. \tag{2.23}$$

c_1, c_2 are constants in terms of (l, M, r_m), given in Ref.[26], and $\wp$ is the Weierstrass elliptic $\wp$-function with invariants (g_2, g_3), discriminant Δ and roots (e_1, e_2, e_3), also given in Ref.[26]. The three cases (i)-(iii) correspond to the cases $\Delta > 0$, $\Delta < 0$ and $\Delta = 0$, respectively. Notice that $S^{(ii)}_{\text{max}} > S^{(iii)}_{\text{max}} = l\sqrt{M} > S^{(i)}_{\text{max}}$. In the case (i), $r(\tau)$ can be written in terms of the Jacobian elliptic function $\text{sn}[\tau^*, k]$, $\tau^* = \sqrt{e_1 - e_3}\,\tau$, $k = \sqrt{(e_2-e_3)/(e_1-e_3)}$. It follows that the solution (i) oscillates between the two zeroes r_{01} and r_{02} of $V(r)$, with period 2ω, where ω is the real semi-period of the Weierstrass function: $\omega = K(k)/\sqrt{e_1 - e_3}$, in terms of the complete elliptic integral of the first kind $K(k)$. We have:

$$r(0) = r_m, \quad r(\omega) = \sqrt{Ml^2 - r_m^2}, \quad r(2\omega) = r_m, \ldots \tag{2.24}$$

In the case (ii) ($\Delta < 0$) two roots (e_1, e_3) become complex, the string collapses into a point $r = 0$ and we have:

$$r(0) = r_m, \quad r(\frac{\omega_2}{2}) = 0, \quad r(\omega_2) = r_m, \ldots \tag{2.25}$$

where ω_2 is the real semi-period of the Weierstrass function for this case. In the case (iii) ($\Delta = 0$) the elliptic functions reduce to hyperbolic functions:

$$r(\tau) = \frac{\sqrt{M}l}{\cosh(\sqrt{M}\tau)}, \tag{2.26}$$

so that:

$$r(-\infty) = 0, \quad r(0) = r_m = \sqrt{M}l, \quad r(+\infty) = 0. \tag{2.27}$$

Here, the string starts as a point, grows until $r = r_m$ (at $\tau = 0$), and then it contracts until it collapses again ($r = 0$) at $\tau = +\infty$. In this case the string makes only one oscillation between $r = 0$ and $r = r_m$.

Notice that for the static background ($J = 0$), the only allowed motion is (ii), i.e. $r_m > r_{\text{hor}} = \sqrt{M}l$ (there is no ergosphere and only one horizon in this case), with:

$$r_m = \sqrt{\frac{Ml^2}{2}}\sqrt{1+\sqrt{1+\frac{4E^2}{Ml^2}}}. \tag{2.28}$$

For $J = 0$, the string collapses into $r = 0$ and stops there. The Penrose diagram of the 2+1 BH-AdS spacetime for $J = 0$ is very similar to the Penrose diagram of the ordinary ($D \geq 4$) Schwarzschild spacetime, so the string motion outwards from $r = 0$ is unphysical because of the causal structure. The coordinate time $t(\tau)$ can be expressed in terms of the incomplete elliptic integral of the third kind Π, see Ref.[26]. The string has its maximal size r_m at $\tau = 0$, passes the horizon at $\tau = \tau_{\text{hor}}$ (expressed in terms of an incomplete elliptic integral of the first kind) and falls into $r = 0$ for $\tau = \omega_2/2$, ω_2 being the real semi-period of the Weierstrass function. That is, we have:

$$r(0) = r_m, \quad r(\tau_{\text{hor}}) = \sqrt{M}l, \quad r(\omega_2/2) = 0,$$
$$t(0) = 0, \quad t(\tau_{\text{hor}}) = \infty, \tag{2.29}$$

and $t(\omega_2/2)$ can be expressed in terms of the Jacobian zeta function Z, see Ref.[26].

We also study the circular strings in the ordinary $D \geq 4$ spacetimes. In the generic 3+1 Kerr anti de Sitter (or Kerr de Sitter) spacetime, the potential $V(r)$ is however quite complicated, see Ref.[26]. The general circular string solution involves higher genus elliptic functions and it is not necessary to go into details here. We will compare with the non-rotating cases, only.

It is instructive to recall [38] the circular string in Minkowski spacetime (Min), for which $V(r) = r^2 - E^2$ (Fig.2a), the string oscillates between its maximal size $r_m = E$, and $r = 0$ with the solution $r(\tau) = r_m \mid \cos\tau \mid$.

In the Schwarzschild black hole (S) $V(r) = r^2 - 2Mr - E^2$ (Fig.2b), the solution is remarkably simple: $r(\tau) = M + \sqrt{M^2 + E^2}\cos\tau$. The mathematical solution oscillates between $r_m = M + \sqrt{M^2 + E^2}$ and $M - \sqrt{M^2 + E^2} < 0$, but because of the causal structure and the curvature singularity the motion can not be continued after the string has collapsed into $r = 0$.

For anti de Sitter spacetime (AdS), we find: $V(r) = r^2(1 + H^2r^2) - E^2$ (Fig.2c). The string oscillates between $r_m = \frac{1}{\sqrt{2}H}\sqrt{-1 + \sqrt{1 + 4H^2E^2}}$ and $r = 0$ with the solution:

$$r(\tau) = r_m \mid \text{cn}[(1 + 4H^2E^2)^{1/4}, k] \mid, \tag{2.30}$$

which is periodic with period 2ω :

$$\omega = \frac{K(k)}{(1 + 4H^2E^2)^{1/4}}, \quad k = \sqrt{\frac{\sqrt{1 + 4H^2E^2} - 1}{2\sqrt{1 + 4H^2E^2}}}. \tag{2.31}$$

For Schwarzschild anti de Sitter spacetime (S-AdS), we find $V(r) = r^2(1 + H^2r^2) - 2Mr - E^2$ (Fig.2d) and:

$$r(\tau) = r_m - \frac{1}{d_1\wp(\tau) + d_2}, \quad r(0) = r_m \tag{2.32}$$

d_1, d_2 are constants given in terms of (M, H, r_m), see Ref.[26] (r_m is the root of the equation $V(r) = 0$, which has in this case exactly one positive solution). The invariants, the discriminant and the roots are also given explicitly in

Ref.[26]. The string starts with $r = r_m$ at $\tau = 0$, it then contracts and eventually collapses into the $r = 0$ singularity. The existence of elliptic function solutions for the string motion is characteristic of the presence of a cosmological constant. For $\Lambda = 0 = -3H^2$ the circular string motion reduces to simple trigonometric functions. From Fig.2 and our analysis we see that the circular string motion is qualitatively very similar in all these backgrounds (Min, S, AdS, S-AdS): the string has a maximal *bounded* size and then it contracts towards $r = 0$. There are however physical and quantitative differences: in Minkowski and pure anti de Sitter spacetimes, the string truly oscillates between r_m and $r = 0$, while in the black hole cases (S, S-AdS), there is only one half oscillation, the string motion stops at $r = 0$. This also holds for the 2+1 BH-AdS spacetime with $J = 0$ (Fig.1b). Notice also that in all these cases, $V(0) = -E^2 < 0$ and $V(r) \sim r^\alpha$; $(\alpha = 2, 4)$ for $r >> E$.

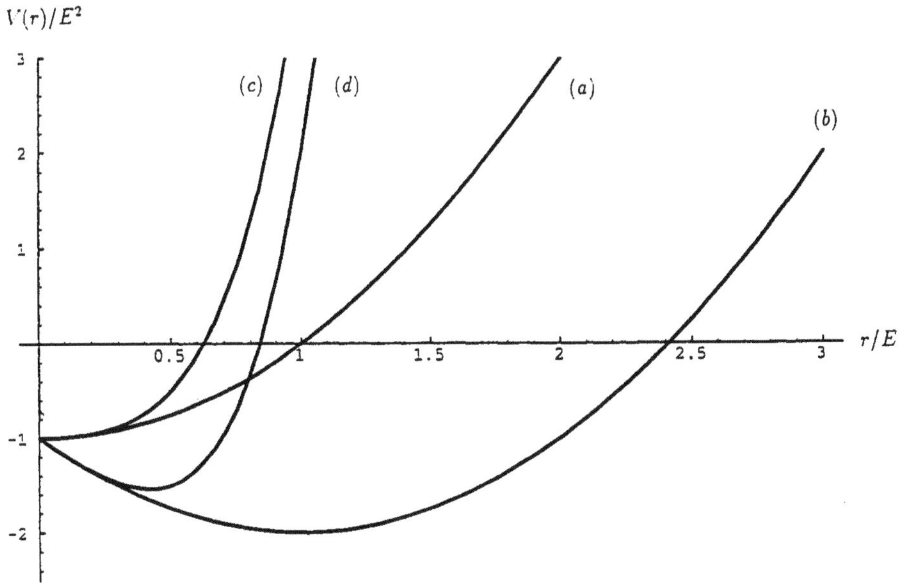

Figure 2. The potential $V(r)$, Eq.(2.18), for a circular string in the equatorial plane of the four $3 + 1$ dimensional spacetimes: (a) Minkowski (Min), (b) Schwarzschild black hole (S), (c) anti de Sitter (AdS), and (d) Schwarzschild anti de Sitter space (S-AdS).

The similarity can be pushed one step further by considering small perturbations around the circular strings. We find [18]:

$$\ddot{C}_n + (n^2 + \frac{r}{2}\frac{da(r)}{dr} + \frac{r^2}{2}\frac{d^2a(r)}{dr^2} - \frac{2E^2}{r^2})C_n = 0, \tag{2.33}$$

determining the Fourier components of the comoving perturbations. For the spacetimes of interest here, $a(r) = 1 - 2M/r + H^2r^2$ (Min, S, AdS, S-AdS), or $a(r) = r^2/l^2 - 1$ (2+1 BH-AdS), the comoving perturbations are regular except near $r = 0$, where we find (for all cases) $r(\tau) \approx -E(\tau - \tau_0)$ and:

$$\ddot{C}_n - \frac{2}{(\tau - \tau_0)^2}C_n = 0. \tag{2.34}$$

It follows that not only the unperturbed circular strings, but also the comoving perturbations around them behave in a similar way in all these non-rotating backgrounds (2+1 and higher dimensional). This should be contrasted with the string perturbations around the center of mass, which behave differently in these backgrounds. It must be noticed that for rotating ($J \neq 0$) spacetimes, the circular string behaviour is qualitatively different from the non-rotating ($J = 0$) spacetimes. For large J, both in the 2+1 BH-AdS as well as in the 3+1 ordinary Kerr-AdS spacetimes, non-collapsing circular string solutions exist. The potential $V(r) \to +\infty$ for $r \to 0$ and no collapse into $r = 0$ is possible.

The dynamics of circular strings in curved spacetimes is determined by the string tension, which tends to contract the string, and by the local gravity (which may be attractive or repulsive). In all the previous non-rotating backgrounds, the local gravity is attractive (i.e. $da(r)/dr > 0$), and it acts together with the string tension in the sense of contraction. But in spacetimes with regions in which repulsion (i.e. $da(r)/dr < 0$) dominates, the strings can expand with unbounded radius (unstable strings [11, 13]). It may also happen that the string tension and the local gravity be of the same order, i.e. the two opposite effects can balance, and the string is stationary. De Sitter spacetime provides an example in which all such type of solutions exist [11, 13]. In de Sitter spacetime, $V(r)$ is unbounded from below for $r \to \infty$ ($V(r) \sim -r^4$) and unbounded expanding circular strings are present. In addition, an interesting new feature appears in the presence of a positive cosmological constant: the existence of *multi-string* solutions [9-11]. The world-sheet time τ turns out to be a multi (finite or infinite) valued function of the physical time. That is, one single world-sheet where $-\infty \leq \tau \leq +\infty$, can describe many (even infinitely many [13]) different and independent strings (in flat spacetime, one single world-sheet describes only one string). In the S, AdS and S-AdS spacetimes, the multi-string feature for circular strings is absent.

The main conclusions of this section are given in Tables 1, 2. Further details can be found in Ref.[26]

Table 1. String motion described by the string perturbation series approach in the 2+1 BH-AdS, ordinary black hole-AdS, de Sitter (dS) and black string backgrounds.

String Perturbation Series Approach
$x^\mu(\tau,\sigma) = q^\mu(\tau) + \eta^\mu(\tau,\sigma) + \xi^\mu(\tau,\sigma) + \ldots$ $\eta^\mu = n^\mu_R \delta x^R, \quad \delta x^R(\tau,\sigma) = \sum_n C^R_n(\tau) e^{-in\sigma}; \ R = 1, \ldots, (D-1)$ $\ddot{C}_{nR} + (n^2 \delta_{RS} - R_{\mu\rho\sigma\nu} n^\mu_R n^\nu_S \dot{q}^\rho \dot{q}^\sigma) C^S_n = 0$

2+1 Black Hole-ADS	Ordinary ($D \geq 4$) Black Hole-ADS
$\ddot{C}_{n\perp} + (n^2 + \frac{m^2}{l^2}) C_{n\perp} = 0$ $\ddot{C}_{n\parallel} + (n^2 + \frac{m^2}{l^2}) C_{n\parallel} = 0$	$\ddot{C}_{n\perp} + (n^2 + m^2 H^2 + \frac{Mm^2}{r^3}) C_{n\perp} = 0$ $\ddot{C}_{n\parallel} + (n^2 + m^2 H^2 - \frac{2Mm^2}{r^3}) C_{n\parallel} = 0$
No instability, $\omega_n = \sqrt{n^2 + \frac{m^2}{l^2}}$	Instability, $r_{\text{inst}} = (\frac{2Mm^2}{1+m^2H^2})^{1/3}$
δx^R, ξ^μ bounded everywhere. $\delta x_R = \sum_n [A_{nR} e^{-i(n\sigma + \omega_n \tau)} + \tilde{A}_{nR} e^{-i(n\sigma - \omega_n \tau)}]$	Unbounded perturbations $C_{n\parallel}$ for $r(\tau) \approx (3m\sqrt{M/2})^{2/3} (\tau_0 - \tau)^{2/3} \rightarrow 0$: $\ddot{C}_{n\parallel} - \frac{4}{9(\tau_0 - \tau)^2} C_{n\parallel} = 0$
$m^2 = 2 \sum_n (2n^2 + \frac{m^2}{l^2}) \sum_R A_{nR} \tilde{A}_{-nR}$	$m^2(H = 0) = 4 \sum_n n^2 \sum_R A_{nR} \tilde{A}_{-nR}$

Ordinary ($D \geq 4$) de Sitter	Black String
$\ddot{C}_{nR} + (n^2 - m^2 H^2) C_{nR} = 0$ Unbounded modes for $\mid n \mid < mH$	$\ddot{C}_{n\perp} + n^2 C_{n\perp} = 0$ $\ddot{C}_{n\parallel} + (n^2 - \frac{2Mm^2}{lr}) C_{n\parallel} = 0$
Instability, $\omega_n = \sqrt{n^2 - m^2 H^2}$	Instability, $r_{\text{inst}} = \frac{2Mm^2}{l}$
$\delta x_R = \sum_n [A_{nR} e^{-i(n\sigma + \omega_n \tau)} + \tilde{A}_{nR} e^{-i(n\sigma - \omega_n \tau)}]$	Unbounded perturbations $C_{n\parallel}$ for $r(\tau) \approx \frac{Mm^2}{l} (\tau_0 - \tau)^2 \rightarrow 0$: $\ddot{C}_{n\parallel} - \frac{2}{(\tau_0 - \tau)^2} C_{n\parallel} = 0$
$m^2 = 2 \sum_n (2n^2 - m^2 H^2) \sum_R A_{nR} \tilde{A}_{-nR}$	$m^2 = 4 \sum_n n^2 \sum_R A_{nR} \tilde{A}_{-nR}$

Table 2. Circular exact string solutions in the indicated backgrounds. $S(\tau)$ is the invariant string size.

Exact Circular Strings

$$ds^2 = g_{tt}(r)dt^2 + g_{rr}(r)dr^2 + 2g_{t\phi}(r)dtd\phi + g_{\phi\phi}d\phi^2$$
$$t = t(\tau), \quad r = r(\tau), \quad \phi = \sigma + f(\tau), \quad (\theta = \pi/2)$$
$$\dot r^2 + V(r) = 0; \quad V(r) = g^{rr}(E^2 g^{tt} + g_{\phi\phi}), \quad \dot t = -Eg^{tt}, \quad \dot f = -Eg^{t\phi}$$
$$ds^2 = g_{\phi\phi}(-d\tau^2 + d\sigma^2), \quad \text{i.e.} \quad S(\tau) = \sqrt{g_{\phi\phi}(r(\tau))}$$

2+1 Black Hole ADS	Ordinary ($D \geq 4$) Black Hole ADS
$V(r) = r^2(r^2/l^2 - M) + J^2/4 - E^2$ $V(0) = J^2/4 - E^2, \quad V(r \to \infty) \propto r^4 \to \infty$ $S(\tau) = r(\tau) = r_m - (c_1\wp(\tau - \tau_0) + c_2)^{-1}$ $S_{\max} \doteq r_m = \sqrt{\frac{Ml^2}{2}}\sqrt{1 + \sqrt{1 - \frac{4V(0)}{Ml^2}}}$ For $J^2 > 4E^2$, $S_{\min} > 0$, no collapse. For $J^2 \leq 4E^2$, $S_{\min} = 0$, collapse. No unbounded string size. No multi-string solutions.	$V(r) = r^2(1 + H^2r^2) - 2Mr - E^2$ $V(0) = -E^2, \quad V(r \to \infty) \propto r^4 \to \infty$ $S(\tau) = r(\tau) = r_m - (c_1\wp(\tau - \tau_0) + c_2)^{-1}$ $V(r_m) \equiv 0; \quad S_{\max} = r_m = r_m(M, H, E)$ String contracts from $r(0) = r_m$ until it collapses into the $r = 0$ singularity. No unbounded string size. No multi-string solutions.
Ordinary ($D \geq 4$) Black Hole DS	**Black String**
$V(r) = r^2(1 - H^2r^2) - 2Mr - E^2$ $V(0) = -E^2, \quad V(r \to \infty) \propto -r^4 \to -\infty$ Contracting, expanding or stationary solutions, depending on the balance between the string tension and the local gravity. $M = 0: \; S^2(\tau) = H^{-2}\wp(\tau - \tau_0) + H^{-2}/3$ Contracting, expanding or oscillating. Unbounded string size. Multi-string solutions.	$\tilde V(r) = J^2/(4r^2) - M/r^2 + 1/l^2 - E^2$ $\tilde V(0) = \infty \; (J \neq 0), \; \tilde V(0) = -\infty \; (J = 0)$ $\tilde S(\tau) = 1/r(\tau), \quad \tilde V(\infty) = 1/l^2 - E^2$ For $J \neq 0$, all solutions are bounded (finite $\tilde S$). For $J = 0$ unbounded size solutions exist as well. $J = 0, El = 1: \; \tilde S(\tau) = (\mid 2\sqrt{M}\tau \mid)^{-1/2}$ $\tilde S(-\infty) = 0, \; \tilde S(0) = \infty, \; \tilde S(\infty) = 0$ Unbounded string size. Multi-string solutions.

3 Quantum String Spectrum in Ordinary AdS

In the first part of the previous section, the classical string dynamics was studied in the $2+1$ BH-AdS spacetime. In this section we go further in the investigation of the physical properties of strings in AdS spacetime, by performing string quantization. Since the 2+1 BH-AdS spacetime is locally AdS, the results of Section 2 can be extended to classical strings in D-dimensional AdS spacetime as well. In AdS spacetime, the string motion is oscillatory in time and is *stable*; all fluctuations around the string center of mass are well behaved and bounded. Local gravity of AdS spacetime is always negative and string instabilities do not develop. The string perturbation series approach, considering fluctuations around the center of mass, is particularly appropriate in AdS spacetime, the natural dimensionless expansion parameter being $\lambda = \alpha'/l^2 > 0$, where α' is the string tension and the 'Hubble constant' $H = 1/l$. The negative cosmological constant of AdS spacetime is related to the 'Hubble constant' H by $\Lambda = -(D-1)(D-2)H^2/2$.

All the spatial ($\mu = 1, ..., D-1$) modes in D-dimensional AdS, oscillate with frequency $\omega_n = \sqrt{n^2 + m^2\alpha'\lambda}$, which are real for all n (m being the string mass). In this section, we perform a canonical quantization procedure. From the conformal generators L_n, $\tilde{L}_n$ and the constraints $L_0 = \tilde{L}_0 = 0$, imposed at the quantum level, we obtain the mass formula (the quantum version of Eq.(2.7) written in units where α' appears explicitly):

$$m^2\alpha' = (D-1)\sum_{n>0}\Omega_n(\lambda) + \sum_{n>0}\Omega_n(\lambda)\sum_{R=1}^{D-1}[(\alpha_n^R)^\dagger\alpha_n^R + (\tilde{\alpha}_n^R)^\dagger\tilde{\alpha}_n^R], \qquad (3.1)$$

where:

$$\Omega_n(\lambda) = \frac{2n^2 + m^2\alpha'\lambda}{\sqrt{n^2 + m^2\alpha'\lambda}}, \qquad (3.2)$$

and we have applied symmetric ordering of the operators. The operators α_n^R, $\tilde{\alpha}_n^R$ satisfy:

$$[\alpha_n^R, (\alpha_n^R)^\dagger] = [\tilde{\alpha}_n^R, (\tilde{\alpha}_n^R)^\dagger] = 1, \qquad \text{for all } n > 0. \qquad (3.3)$$

For $\lambda << 1$, which is clearly fulfilled in most interesting cases, we have found the lower mass states $m^2\alpha'\lambda << 1$ and the quantum mass spectrum [27]. Physical states are characterized by the eigenvalue of the number operator:

$$N = \frac{1}{2}\sum_{n>0} n \sum_{R=1}^{D-1}[(\alpha_n^R)^\dagger\alpha_n^R + (\tilde{\alpha}_n^R)^\dagger\tilde{\alpha}_n^R], \qquad (3.4)$$

and the ground state is defined by:

$$\alpha_n^R \mid 0 >= \tilde{\alpha}_n^R \mid 0 >= 0, \qquad \text{for all } n > 0. \qquad (3.5)$$

We find that $m^2\alpha' = 0$ is an *exact* solution of the mass formula in $D = 25$ and that there is a graviton at $D = 25$, which indicates, as in de Sitter space [7], that the critical dimension of AdS is 25 (although it should be stressed that the question whether these spacetimes are solutions to the full β-function equations remains open). As in Minkowski spacetime, the ground state is a tachyon. Remarkably enough, for $N \geq 2$ we find that a generic feature of all excited states beyond the graviton, is the presence of a *fine structure* effect: for a given eigenvalue $N \geq 2$, the corresponding states have different masses. For the lower mass states the expectation value of the mass operator in the corresponding states (generically labelled $\mid j >$) turns out to have the form (see Ref.[27]):

$$< j \mid m^2\alpha' \mid j >_{\text{AdS}} = a_j + b_j\lambda^2 + c_j\lambda^3 + O(\lambda^4). \tag{3.6}$$

The collective index "j" generically labels the state $\mid j >$ and the coefficients a_j, b_j, c_j are all well computed numbers, different for each state, their precise values are given in Ref.[27]. The corrections to the mass in Minkowski spacetime appear to order λ^2. Therefore, the leading Regge trajectory for the lower mass states is:

$$J = 2 + \frac{1}{2}m^2\alpha' + O(\lambda^2). \tag{3.7}$$

In Minkowski spacetime the mass and number operator of the string are related by $m^2\alpha' = -4 + 4N$. In AdS (as well as in de Sitter (dS)), there is no such simple relation between the mass and the number operators; the splitting of levels increases considerably for very large N. The *fine structure* effect we find here is also present in de Sitter space. Up to order λ^2, the lower mass states in dS and in AdS are the same, the differences appear to the order λ^3. The lower mass states in de Sitter spacetime are given by Eq.(3.6) but with the λ^3-term getting an opposite sign $(-c_j\lambda^3)$.

For the very high mass spectrum, we find more drastic effects. States with very large eigenvalue N, namely $N >> 1/\lambda$, have masses:

$$< j \mid m^2\alpha' \mid j >_{\text{AdS}} \approx d_j\lambda N^2 \tag{3.8}$$

and angular momentum:

$$J^2 \approx \frac{1}{\lambda}m^2\alpha', \tag{3.9}$$

where d_j are well computed numbers different for each state, their precise values are given in Ref.[27]. Since $\lambda = \alpha'/l^2$, we see from Eq.(3.8) that the masses of the high mass states are *independent* of α'. In Minkowski spacetime, very large N states all have the same mass $m^2\alpha' \approx 4N$, but here in AdS the masses of the high mass states with the same eigenvalue N are *different* by factors d_j. In addition, because of the fine structure effect, states with different N can get mixed up. For high mass states, the level spacing *grows* with N (instead of being constant as in Minkowski spacetime). As a consequence, the density of states $\rho(m)$ as a function of mass grows like $\text{Exp}[(m/\sqrt{\mid \Lambda \mid})^{1/2}]$ (instead of $\text{Exp}[m\sqrt{\alpha'}]$ as in Minkowski spacetime), and *independently* of α'. The partition function for a gas of strings at a temperature β^{-1} in AdS spacetime is well defined for all finite temperatures β^{-1}, discarding the existence of the Hagedorn temperature.

For the low mass states ($m^2\alpha'\lambda << 1$) in anti de Sitter spacetime, our results can be written as [27]:

$$< j \mid m^2(\alpha', l) \mid j >= \frac{4(N-1)}{\alpha'} + \frac{1}{l^2}\sum_{n=0}^{\infty} a_{jn}(\alpha'/l^2)^n, \qquad (3.10)$$

where "j" is a collective index labelling the state $\mid j >$. It is now important to notice that $a_{j0} = 0$ for *all* the low mass states [27], i.e. there is no "constant" term on the right hand side of Eq.(3.10). A non-zero a_{j0}-term would give rise to a α'-independent contribution to the string mass. Its absense, on the other hand, means that the first term on the right hand side of Eq.(3.10) is super-dominant (since, in all cases, $\alpha'/l^2 = \lambda << 1$) and that the string scale is therefore set by $1/\alpha'$.

For the high mass states ($m^2\alpha'\lambda >> 1$) we found instead [27]:

$$< j \mid m^2(\alpha', l) \mid j > \approx \frac{d_j}{l^2} N^2, \qquad \text{for } N >> l^2/\alpha' \qquad (3.11)$$

where the number d_j depends on the state. The right hand side of Eq.(3.11) is exactly like a non-zero dominant a_{j0}-term in Eq.(3.10). For the high mass states the scale is therefore set by $1/l^2$ which is equal to the absolute value of the cosmological constant Λ (up to a geometrical factor) and *independent* of α'. This suggests that for $\lambda << 1$, the masses of *all* string states can be represented by a formula of the form (3.10). For the low mass states $a_{j0} = 0$, while for the high mass states a_{j0} becomes a large positive number.

In the black string background, Eq.(2.2), we have calculated explicitly the first and second order string fluctuations around the center of mass [27]. We then determined the world-sheet energy-momentum tensor and we derived the mass formula in the asymptotic region. The mass spectrum is equal to the mass spectrum in flat Minkowski spacetime. Therefore, for a gas of strings at temperature β^{-1} in the asymptotic region of the black string background, the partition function goes like:

$$Z(\beta) \sim \int^{\infty} dm\; e^{-m(\beta-\sqrt{\alpha'})}, \qquad (3.12)$$

which is only defined for $\beta > \sqrt{\alpha'}$, i.e. there is a Hagedorn temperature:

$$T_{\rm Hg} = (\alpha')^{-1/2}. \qquad (3.13)$$

In higher dimensional ($D \geq 4$) black hole spacetimes the next step now would be to set up a scattering formalism, where a string from an asymptotic in-state interacts with the gravitational field of the black hole and reappears in an asymptotic out-state [15]. However, this is not possible in the black string background. In the uncharged black string background under consideration here, *all* null and timelike geodesics incoming from spatial infinity pass through the horizon and fall into the physical singularity [36]. No "angular momentum", as in the case of scattering off the ordinary Schwarzschild black hole, can prevent a point particle from falling into the singularity. The string solutions considered here are based on perturbations around the string center of mass which follows, at least approximately, a point particle geodesic. Therefore, a string incoming from spatial infinity inevitably falls into the singularity in the black string background.

Table 3. Characteristic features of the quantum string mass spectrum in anti de Sitter (AdS) and de Sitter (dS) spacetimes.

Quantum Strings	
Anti de Sitter spacetime (AdS)	de Sitter spacetime (dS)
Classical motion is stable and oscillatory in time with real frequencies $\omega_n = \sqrt{n^2 + m^2\alpha'^2 H^2}$	Classical motion is unstable with frequencies $\omega_n = \sqrt{n^2 - m^2\alpha'^2 H^2}$ Unbounded string size and energy for large de Sitter radius, $R \to \infty$.
The mass formula is well defined for all m. There is an infinite number of states with arbitrary high masses. $m^2\alpha' = 0$ is an exact solution at $D = 25$.	Real mass solutions only for $m < 1/(\alpha' H)$. Finite number of states, $N_{\max} \approx 0.15/(\alpha' H^2)$. $m^2\alpha' = 0$ is an exact solution at $D = 25$.
The coupling to the gravitational background produces a Fine structure effect at all levels in the mass spectrum. The number of levels considerably increases with respect to flat space.	Fine structure effect appears in low mass spectrum. Is similar to AdS; the differences appear to order $(\alpha' H^2)^3$
For the high mass states: $< m^2 > \sim \mid \Lambda \mid N^2, \quad J^2 \sim m^2/\mid \Lambda \mid$ Both are *independent* of α' ! The level spacing grows with N. $\rho(m) \sim \mathrm{Exp}[m/\sqrt{\mid \Lambda \mid})^{1/2}]$, high m. No Hagedorn temperature exists.	The similar region of high mass states does not exist in the de Sitter spacetime.

4 New Classes of Exact String and Multi-String Solutions in Curved Spacetimes

In this section we return to the string equations of motion and constraints, Eqs.(1.1)-(1.2), to look for new classes of exact solutions. In most spacetimes, quite general families of exact solutions can be found by making an appropriate ansatz, which exploits the symmetries of the underlying curved spacetime. In axially symmetric spacetimes, a convenient ansatz corresponds to circular strings, as we saw in Section 2. Such an ansatz effectively decouples the dependence on the spatial world-sheet coordinate σ, and the string equations of motion and constraints reduce to non-linear coupled ordinary differential equations, Eqs.(2.18)-(2.19). In this section we will make instead an ansatz which effectively decouples the dependence on the temporal world-sheet coordinate τ. This ansatz, which we call the "stationary string ansatz" is dual to the "circular string ansatz" in the sense that it corresponds to a formal interchange of the world-sheet coordinates (τ, σ), as well as of the azimuthal angle ϕ and the stationary time t in the target space:

$$\tau \leftrightarrow \sigma, \qquad t \leftrightarrow \phi. \tag{4.1}$$

The stationary string ansatz will describe stationary strings when t (and τ) are timelike, for instance in anti de Sitter spacetime (in static coordinates) and outside the horizon of a Schwarzschild black hole. On the other hand, if t (and τ) are spacelike, for instance inside the horizon of a Schwarzschild black hole or outside the horizon of de Sitter spacetime (in static coordinates), the stationary string ansatz will describe dynamical propagating strings. Considering for simplicity a static line element in the form:

$$ds^2 = -a(r)dt^2 + \frac{dr^2}{a(r)} + r^2(d\theta^2 + \sin^2\theta d\phi^2), \tag{4.2}$$

the stationary string ansatz reads explicitly:

$$t = \tau, \qquad r = r(\sigma), \qquad \phi = \phi(\sigma), \qquad \theta = \pi/2. \tag{4.3}$$

The string equations of motion and constraints, Eqs.(1.1)-(1.2), reduce to two separated first order ordinary differential equations:

$$\phi' = \frac{L}{r^2}, \qquad r'^2 + V(r) = 0; \quad V(r) = -a(r)[a(r) - \frac{L^2}{r^2}], \tag{4.4}$$

where L is an integration constant. The qualitative features of the possible string configurations can be read off directly from the shape of the potential $V(r)$. Thereafter, the detailed analysis of the quantitative features can be performed by explicitly solving the (integrable) system of equations (4.4). The induced line element on the world-sheet is given by:

$$ds^2 = a(r)(-d\tau^2 + d\sigma^2). \tag{4.5}$$

Thus, if $a(\tau)$ is negative, the world-sheet coordinate τ becomes spacelike while σ becomes timelike and the stationary string ansatz (4.3) describes a dynamical string. If $a(\tau)$ is positive, the ansatz describes a stationary string.

In this section we solve explicitly Eqs.(4.4) and we analyze in detail the solutions and their physical interpretation in Minkowski, de Sitter, anti de Sitter, Schwarzschild and $2+1$ black hole anti de Sitter spacetimes. In all these cases, the solutions are expressed in terms of elliptic (or elementary) functions. We furthermore analyze the physical properties, energy and pressure, of these solutions. The energy and pressure densities of the strings can be obtained from the $3+1$ dimensional spacetime energy-momentum tensor:

$$\sqrt{-g}T^{\mu\nu} = \frac{1}{2\pi\alpha'}\int d\tau d\sigma(\dot{X}^\mu\dot{X}^\nu - X'^\mu X'^\nu)\delta^{(4)}(X - X(\tau,\sigma)). \qquad (4.6)$$

After integration over a spatial volume that completely encloses the string [39], the energy-momentum tensor takes the form of a fluid:

$$T^\mu{}_\nu = \text{diag.}(-E, P_1, P_2, P_3). \qquad (4.7)$$

In Minkowski spacetime (Min), the potential defined in Eqs.(4.4) is given by:

$$V_{\text{M}}(r) = \frac{L^2}{r^2} - 1, \qquad (4.8)$$

and the solution of Eqs.(4.4) describes one infinitely long straight string with "impact-parameter" L. The equation of state, relating energy and pressure densities, takes the well-known form Ref.[28]:

$$dE = -dP_2 = \text{const.}, \quad P_1 = P_3 = 0. \qquad (4.9)$$

In anti de Sitter spacetime (AdS), the potential is given by:

$$V_{\text{AdS}}(r) = -(1 + H^2r^2)[1 + H^2r^2 - \frac{L^2}{r^2}]. \qquad (4.10)$$

The radial coordinate $r(\sigma)$ is periodic with finite period T_σ, which can be expressed in terms of a complete elliptic integral, see Ref.[28]. For $\sigma \in [0, T_\sigma]$, the solution describes an infinitely long stationary string in the wedge $\phi \in\ ]0, \Delta\phi[$, where:

$$\Delta\phi = 2k\sqrt{\frac{1-2k^2}{1-k^2}}[\Pi(1-k^2,\ k) - K(k)] \ \in\ ]0,\ \pi[\qquad (4.11)$$

The elliptic modulus k, which is a function of HL [28], parametrizes the solutions, $k \in\ ]0, 1/\sqrt{2}[$. The azimuthal angle is generally not a periodic function of σ, thus when the spacelike world-sheet coordinate σ runs through the range $]-\infty, +\infty[$, the solution describes an *infinite number* of infinitely long stationary open strings. The general solution is therefore a *multi-string* solution. Until now multi-string solutions were only found in de Sitter spacetime [11-13]. Our results show that multi-string solutions are a general feature of spacetimes with a cosmological constant (positive or negative). The solution in anti de Sitter spacetime describes a *finite* number of strings if the following relation holds:

$$N\Delta\phi = 2\pi M. \tag{4.12}$$

Here N and M are integers, determining the number of strings and the winding in azimuthal angle, respectively, for the multi-string solution, see Fig.3. The equation of state for a full multi-string solution takes the form ($P_3 = 0$):

$$dP_1 = dP_2 = -\frac{1}{2}dE, \quad \text{for } r \to \infty \tag{4.13}$$

corresponding to extremely unstable strings [21].

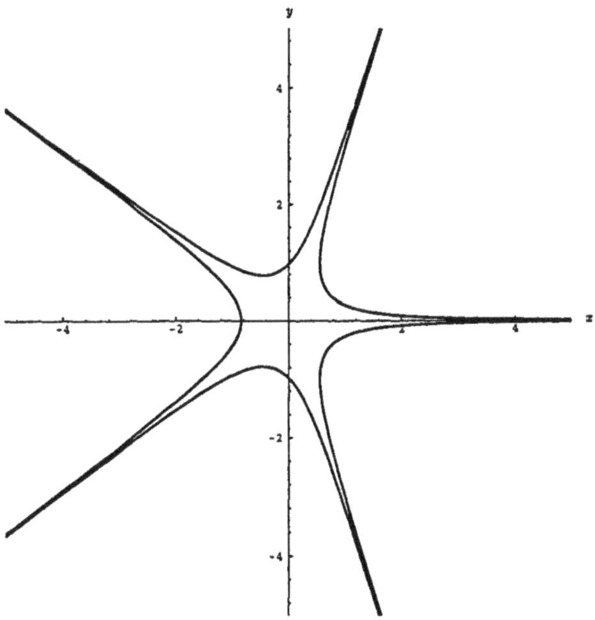

Figure 3. The $(N, M) = (5, 1)$ multi-string solution in anti de Sitter spacetime. The (N, M) multi-string solutions describe N stationary strings with M windings in the Azimuthal angle ϕ, in anti de Sitter spacetime.

In de Sitter spacetime (dS), the potential is given by:

$$V_{\rm dS}(r) = -(1 - H^2r^2)[1 - H^2r^2 - \frac{L^2}{r^2}]. \tag{4.14}$$

In this case we have to distinguish between solutions inside the horizon (where r is timelike) and solutions outside the horizon (where r is spacelike). Inside the horizon, the generic solution describes one infinitely long open stationary string winding around $r = 0$. For special values of the constants of motion, corresponding to a relation, which formally takes the same form as Eq.(4.12), the solution describes a closed string of finite length $l = N\pi/H$. The integer N in this case determines the number of "leaves", see Fig.4. The energy is positive and finite and grows with N. The pressure turns out to vanish identically, thus the equation of state corresponds to cold matter.

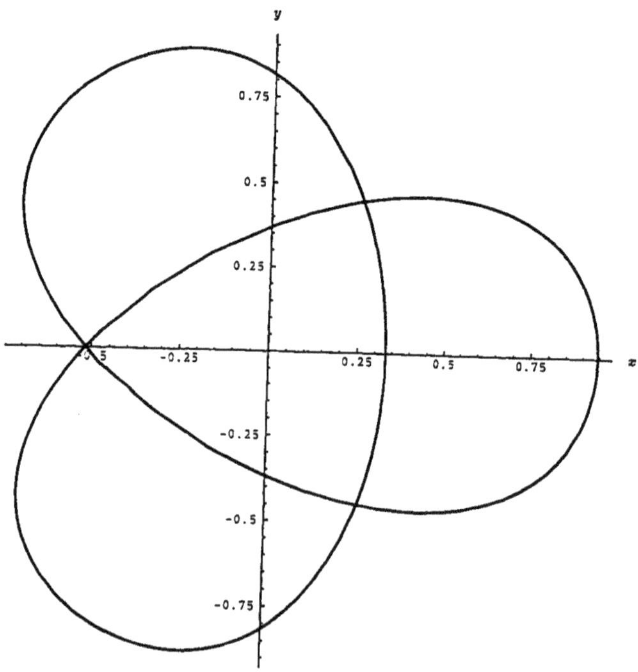

Figure 4. The $(N, M) = (3, 2)$ stationary string solution inside the horizon of de Sitter spacetime. Besides the circular string, this is the simplest stationary closed string configuration in de Sitter spacetime.

Outside the horizon, the world-sheet coordinate τ becomes spacelike while σ becomes timelike, thus we define:

$$\tilde{\tau} \equiv \sigma, \qquad \tilde{\sigma} \equiv \tau, \tag{4.15}$$

and the string solution is conveniently expressed in hyperboloid coordinates. The radial coordinate $r(\tilde{\tau})$ is periodic with a finite period $T_{\tilde{\tau}}$. For $\tilde{\tau} \in [0, T_{\tilde{\tau}}]$, the solution describes a straight string incoming non-radially from spatial infinity, scattering at the horizon and escaping towards infinity again, Fig.5.

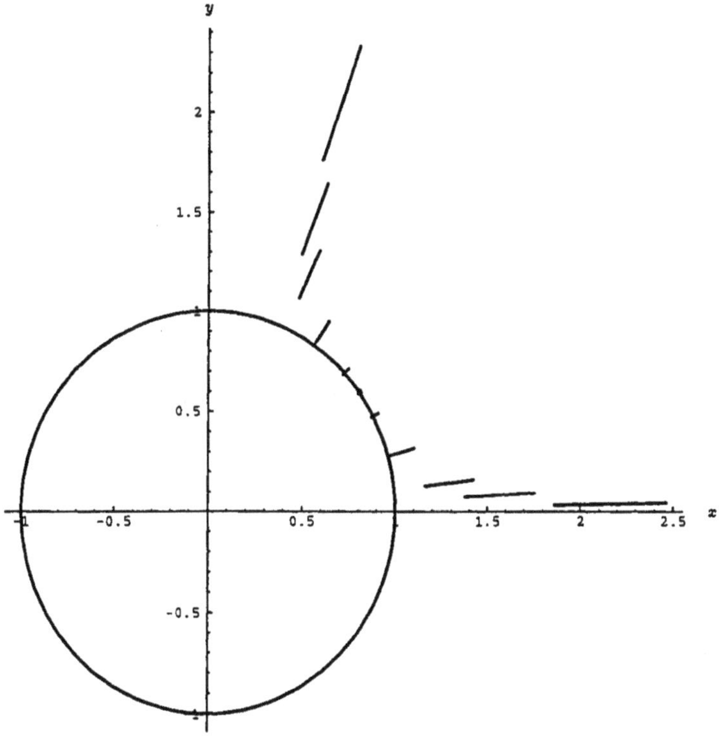

Figure 5. Schematic representation of the time evolution of the $(N, M) = (5, 1)$ dynamical multi-string solution, outside the horizon of de Sitter space-time. Only one of the 5 strings is shown; the others are obtained by rotating the figure by the angles $2\pi/5$, $4\pi/5$, $6\pi/5$ and $8\pi/5$. During the "scattering" at the horizon, the string collapses to a point and re-expands.

The string length is zero at the horizon and grows indefinetely in the asymptotic regions. As in the case of anti de Sitter spacetime, the azimuthal angle is generally not a periodic function, thus when the timelike world-sheet coordinate $\tilde{\tau}$ runs through the range $]-\infty, +\infty[$, the solution describes an *infinite* number of dynamical straight strings scattering at the horizon at different angles. The general solution is therefore a multi-string solution. In particular, a multi-string solution describing a *finite* number of strings is obtained if a relation of the form (4.12) is fulfilled. It turns out that the solution describes *at least* three strings. The energy and pressures of a full multi-string solution have also been computed. In the asymptotic region they fulfill an equation of state corresponding to extremely unstable strings, i.e. like Eq.(4.13).

In the Schwarzschild black hole background (S), the potential is given by:

$$V_S(r) = -(1 - 2m/r)[1 - 2m/r - \frac{L^2}{r^2}]. \tag{4.16}$$

No multi-string solutions are found in this case. Outside the horizon the solution of Eqs.(4.4) describes one infinitely long stationary open string. This solution was already derived in Ref.[40], and we shall not discuss it here. Inside the horizon, where τ becomes spacelike while σ becomes timelike, we make the redefinitions (4.15) and the solution is conveniently expressed in terms of Kruskal coordinates, see Ref.[28]. The solution describes one straight string infalling *non-radially* towards the singularity, see Fig.6. At the horizon, the string length is zero and it grows *indefinitely* when the string approaches the spacetime singularity. Thereafter, the solution can not be continued, see Ref.[28].

In the $2+1$ black hole anti de Sitter spacetime (BH-AdS) [30], the potential is given by:

$$V_{\text{BH-AdS}}(r) = -(\frac{r^2}{l^2} - 1)[\frac{r^2}{l^2} - 1 - \frac{L^2}{r^2}]. \tag{4.17}$$

Outside the horizon, the solutions "interpolate" between the solutions found in anti de Sitter spacetime and outside the horizon of the Schwarzschild black hole. The solutions thus describe infinitely long stationary open strings. As in anti de Sitter spacetime, the general solution is a *multi-string* describing *infinitely* many strings. In particular, for certain values of the constants of motion, corresponding to the condition of the form (4.12), the solution describes a finite number of strings. In the simplest version of the $2+1$ BH-AdS background ($M = 1,\ J = 0$), it turns out that the solution describes *at least* seven strings, see Ref.[28]. Inside the horizon, we make the redefinitions (4.15) and the solution is conveniently expressed in terms of Kruskal-like coordinates, see Ref.[28]. The solution is similar to the solution found inside the horizon of the Schwarzschild black hole, but there is one important difference at $r = 0$. As in the Schwarzschild black hole background, the solution describes one straight string infalling non-radially towards $r = 0$, and beyond this point the solution can not be continued because of the global structure of the spacetime. At the horizon the string size is zero and during the fall towards $r = 0$, the string size *grows* but stays *finite*. This should be compared with the straight string inside the horizon of the Schwarzschild black hole, where the string size grows indefinetely. The physical reason for this difference is that the point $r = 0$ is not a strong curvature singularity in the 2 + 1 BH-AdS spacetime.

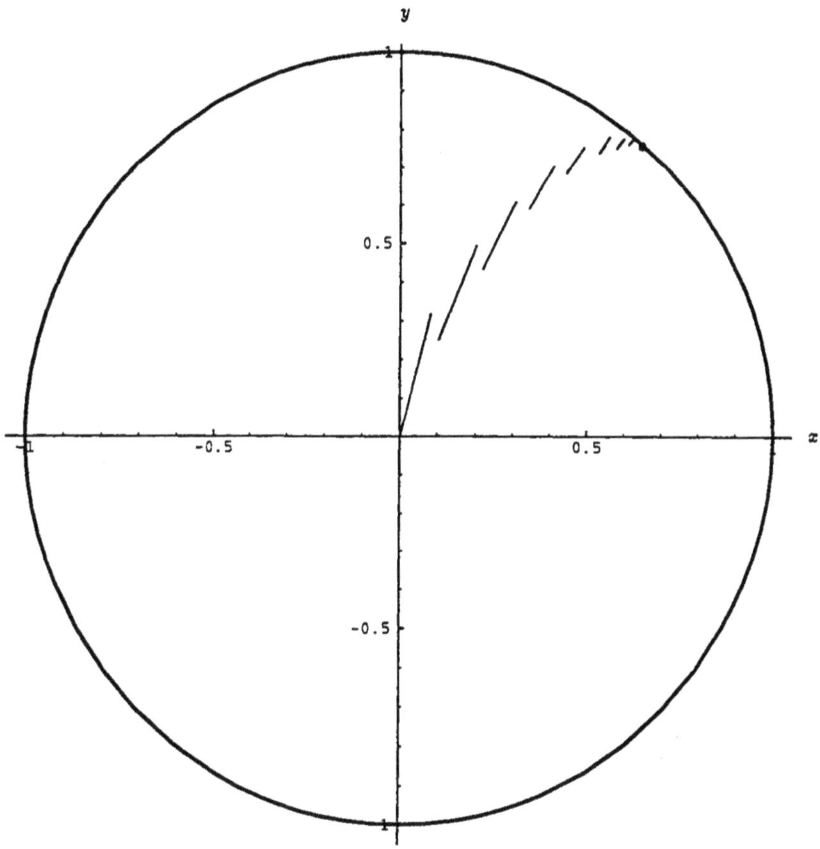

Figure 6. The dynamical straight string inside the horizon of the Schwarzschild black hole. The string infalls non-radially towards the singularity with indefinetely growing length.

We have considered also the Schwarzschild de Sitter and Schwarzschild anti de Sitter spacetimes. These spacetimes contain all the features of the spacetimes already discussed: singularities, horizons, positive or negative cosmological constants. All the various types of string solutions found in the other spacetimes (open, closed, straight, finitely and infinitely long, multi-strings) are therefore present in the different regions of the Schwarzschild-de Sitter and Schwarzschild-anti de Sitter spacetimes. The details are given in Ref.[28].

We close this section with a few remarks on the stability of the solutions. Generally, the question of stability must be addressed by considering small perturbations around the exact solutions. In Ref.[10], a covariant formalism describing physical perturbations propagating along an arbitrary string configuration embedded in an arbitrary curved spacetime, was developed. The resulting equations determining the evolution of the perturbations are however very complicated in the general case, although partial (analytical) results have been obtained in special cases for de Sitter [18, 43] and Schwarzschild black hole [10, 43, 44] spacetimes. The exact solutions found in this section fall essentially into two classes: dynamical and stationary. The dynamical string solutions outside the horizon of de Sitter (or S-dS) and inside the horizon of Schwarzschild (or S-dS, S-AdS) spacetimes, are already unstable at the zeroth order approximation (i.e. without including small perturbations), in the sense that their physical length grows indefinetely. For the stationary string solutions the situation is more delicate. The existence of the stationary configurations is based on an exact balance between the string tension and the local attractive or repulsive gravity. For that reason, it can be expected that the configurations are actually unstable for certain modes of perturbation, especially in strong curvature regions.

The main conclusions of this section are given in Table 4. Further details can be found in Ref.[28].

Table 4. Short summary of the features of the string solutions found in this section. In anti de Sitter spacetime and outside the horizon of de Sitter and $2+1$ BH-AdS spacetimes, the solutions describe a *finite* number of strings provided a condition of the form $N\Delta\phi = 2\pi M$ is fulfilled, where $\Delta\phi$ is the angle betwen the "arms" of the string and (N, M) are integers.

Line element:	$ds^2 = -a(r)dt^2 + \frac{dr^2}{a(r)} + r^2(d\theta^2 + \sin^2\theta d\phi^2)$
Ansatz:	$t = \tau, \quad r = r(\sigma) \quad \phi = \phi(\sigma), \quad \theta = \pi/2$
String solution:	$\phi' = \frac{L}{r^2} \quad r'^2 + V(r) = 0; \quad V(r) = -a(r)[a(r) - \frac{L^2}{r^2}]$
String length element:	$dl = \sqrt{a(r(\sigma))}\, d\sigma$

MINKOWSKI, $a(r) = 1, \quad V(r) = \frac{L^2}{r^2} - 1$

The solution describes one infinitely long stationary straight string.

$\frac{dE}{dl} = -\frac{dP_2}{dl} = \text{const.}, \qquad P_1 = P_3 = 0$

ANTI DE SITTER, $a(r) = 1 + H^2r^2, \quad V(r) = -(1 + H^2r^2)[1 + H^2r^2 - \frac{L^2}{r^2}]$

The solution describes a finite or infinite number of infinitely long stationary strings.

$\frac{dE}{dl} = -2\frac{dP_1}{dl} = -2\frac{dP_2}{dl}, \qquad P_3 = 0, \qquad$ (asymptotically for $r \to \infty$)

DE SITTER, $a(r) = 1 - H^2r^2, \quad V(r) = -(1 - H^2r^2)[1 - H^2r^2 - \frac{L^2}{r^2}]$

Inside the horizon, the solution describes one finitely or infinitely long stationary string winding around $r = 0$.

$P_1 = P_2 = P_3 = 0$, as cold matter, $\qquad E$ expressed in terms of elliptic functions.

Outside the horizon, the solution describes a finite or infinite number of dynamical straight strings "scattering at the horizon". The string length vanishes at the horizon, but stretches indefinetely at spatial infinity.

$\frac{dE}{dl} = -2\frac{dP_1}{dl} = -2\frac{dP_2}{dl}, \qquad P_3 = 0, \qquad$ (asymptotically for $r \to \infty$)

SCHWARZSCHILD, $a(r) = 1 - 2m/r, \quad V(r) = -(1 - 2m/r)[1 - 2m/r - \frac{L^2}{r^2}]$

Outside the horizon, the solution describes one infinitely long stationary string.

Inside the horizon, the solution describes one dynamical straight string. The string size is zero at the horizon, and grows *indefinetely* as the string falls towards $r = 0$.

2+1 BLACK HOLE-ADS, $a(r) = \frac{r^2}{l^2} - 1, \quad V(r) = -(\frac{r^2}{l^2} - 1)[\frac{r^2}{l^2} - 1 - \frac{L^2}{r^2}]$

Outside the horizon, the solution describes a finite or infinite number of infinitely long stationary string.

Inside the horizon, the solution describes one dynamical straight string. The string size is zero at the horizon, and grows *finitely* as the string falls towards $r = 0$.

5 Spatial Curvature Effects on the String Dynamics

The propagation of strings in Friedmann-Robertson-Walker (FRW) cosmologies has been investigated using both exact and approximative methods, see for example Refs.[7, 11-14, 17-20] (as well as numerical methods, which shall not be discussed here). Except for anti de Sitter spacetime, which has negative spatial curvature, the cosmologies that have been considered until now, have all been spatially flat. In this section we will consider the physical effects of a non-zero (positive or negative) curvature index on the classical and quantum strings. The non-vanishing components of the Riemann tensor for the generic D-dimensional FRW line element, in comoving coordinates:

$$ds^2 = -dt^2 + a^2(t)\frac{d\vec{x}d\vec{x}}{(1+\frac{K}{4}\vec{x}\vec{x})^2}, \tag{5.1}$$

are given by:

$$R_{itit} = \frac{-aa_{tt}}{(1+\frac{K}{4}\vec{x}\vec{x})^2}, \qquad R_{ijij} = \frac{a^2(K+a_t^2)}{(1+\frac{K}{4}\vec{x}\vec{x})^4}; \quad i \neq j \tag{5.2}$$

where $a = a(t)$ is the scale factor and K is the curvature index. Clearly, a non-zero curvature index introduces a non-zero spacetime curvature; the exceptional case provided by $K = -a_t^2 =$ const., corresponds to the Milne-Universe. From Eqs.(5.2), it is also seen that the curvature index has to compete with the first derivative of the scale factor. The effects of the curvature index are therefore most conveniently discussed in the family of FRW-universes with constant scale factor, the so-called static Robertson-Walker spacetimes. This is the point of view we take in the present section.

We consider both the closed ($K > 0$) and the hyperbolic ($K < 0$) static Robertson-Walker spacetimes, and all our results are compared with the already known results in the flat ($K = 0$) Minkowski spacetime. We determine the evolution of circular strings, derive the corresponding equations of state (using Eq.(4.6)), discuss the question of strings as self-consistent solutions to the Einstein equations [20], and we perform a semi-classical quantization. We also find all the stationary string configurations in these spacetimes.

The radius of a classical circular string in the spacetime (5.1), for $a = 1$, is determined by:

$$\dot{r}^2 + V(r) = 0; \qquad V(r) = (1-Kr^2)(r^2 - b\alpha'^2), \tag{5.3}$$

where b is an integration constant ($b\alpha'^2 = E$, in the notation of Eqs.(2.18)-(2.19)). This equation is solved in terms of elliptic functions and all solutions describe oscillating strings (Fig.7 shows the potential $V(r)$ for $K > 0$, $K = 0$, $K < 0$).

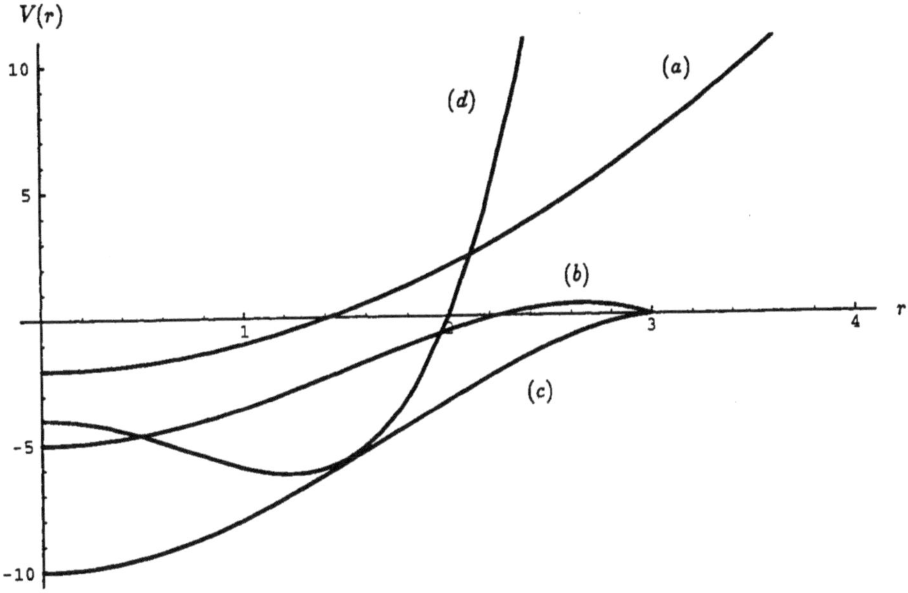

Figure 7. The potential $V(r)$ introduced in Eqs.(5.3) for a circular string in the static Robertson-Walker spacetimes: (a) flat ($K = 0$), (b) closed ($K > 0$ *and* $bK\alpha'^2 \leq 1$), (c) closed ($K > 0$ *and* $bK\alpha'^2 > 1$), (d) hyperbolic ($K < 0$).

For $K > 0$, when the spatial section is a hypersphere, the string either oscillates on one hemisphere or on the full hypersphere. The energy is always positive while the average pressure can be positive, negative or zero, depending on the precise value of an elliptic modulus; the equation of state is given explicitly in Table 5, in the different cases. Interestingly enough, we find that the circular strings provide a self-consistent solution to the Einstein equations with a selected value of the curvature index. Self-consistent solutions to the Einstein equations with string sources have been found previously in the form of power law expanding universes [20].

Table 5. Classical circular strings in the static Robertson-Walker spacetimes. Notice that a self-consistent solution to the Einstein equations, with the string back-reaction included, can be obtained only for $K > 0$.

CLASSICAL CIRCULAR STRING SOLUTIONS

Spacetime: $ds^2 = -dt^2 + \frac{dr^2}{1-Kr^2} + r^2 d\phi^2$

String solution: $\phi = \sigma, \; t(\tau) = \sqrt{b}\alpha'\tau, \; \dot{r}^2 + V(r) = 0; \; V(r) = (1 - Kr^2)(r^2 - b\alpha'^2)$

String length, energy and pressure: $l(\tau) = 2\pi|r(\tau)|, \; T^{\mu}{}_{\nu}(\text{string}) = \text{diag.}(-\rho, P, P)$

MINKOWSKI ($K = 0$):

$r(\tau) = \sqrt{b}\alpha' \cos\tau$

$\rho = \sqrt{b}, \quad P = -\frac{\sqrt{b}}{2}\cos 2\tau, \quad < P >= 0,$ as cold matter.

Is not a self-consistent solution to the Einstein equations.

CLOSED ($K > 0$):

$$r(\tau) = \begin{cases} \sqrt{b}\alpha' \mathrm{sn}[\tau, k_-] & \text{for } \sqrt{bK}\alpha' \leq 1, \quad k_- \equiv \sqrt{bK}\alpha' \in [0,1] \\ \frac{1}{\sqrt{K}}\mathrm{sn}[\tau/k_+, k_+] & \text{for } \sqrt{bK}\alpha' > 1, \quad k_+ \equiv 1/(\sqrt{bK}\alpha') \in]0,1[\end{cases}$$

$$\rho = \sqrt{b}, \quad P = \begin{cases} \frac{k_-}{2\sqrt{K}\alpha'}(1 - 2\mathrm{sn}^2[\tau, k_-]) & \text{for } \sqrt{bK}\alpha' \leq 1 \\ \frac{1}{2k_+\sqrt{K}\alpha'}(1 - 2k_+^2 \mathrm{sn}^2[\tau/k_+, k_+]) & \text{for } \sqrt{bK}\alpha' > 1 \end{cases}$$

$$< P >= \begin{cases} \frac{1}{2\sqrt{K}\alpha'}[\frac{k_-^2 - 2}{k_-} + \frac{2}{k_-}\frac{E(k_-)}{K(k_-)}] & \text{for } \sqrt{bK}\alpha' \leq 1, \quad \text{negative.} \\ \frac{1}{2k_+\sqrt{K}\alpha'}[\frac{2E(k_+)}{K(k_+)} - 1] & \text{for } \sqrt{bK}\alpha' > 1, \quad \text{positive or negative.} \end{cases}$$

A self-consistent solution to the Einstein equations is obtained for $k_+ = 0.9089...$

HYPERBOLIC ($K < 0$):

$r(\tau) = \frac{k}{\sqrt{-K}}\mathrm{sd}[\mu\tau, k], \quad \mu \equiv \sqrt{1 - Kb\alpha'^2}, \quad k \equiv \sqrt{-Kb\alpha'^2}/\mu \in [0,1[$

$\rho = \sqrt{b}, \quad P = \frac{k - 2k(1-k^2)\mathrm{sd}^2[\mu\tau,k]}{2\sqrt{1-k^2}\sqrt{-K}\alpha'}$

$< P >= \frac{1}{2\sqrt{1-k^2}\sqrt{-K}\alpha'}[-k + \frac{2}{k}(1 - \frac{E(k)}{K(k)})],$ positive.

Is not a self-consistent solution to the Einstein equations.

We have semi-classically quantized the circular string solutions of Eq.(5.3). We used an approach developed in field theory by Dashen et. al. [41], based on the stationary phase approximation of the partition function. The result of the stationary phase integration is expressed in terms of the function:

$$W(m) \equiv S_{\text{cl}}(T(m)) + m\, T(m), \tag{5.4}$$

where S_{cl} is the action of the classical solution, m is the mass and the period $T(m)$ is implicitly given by:

$$\frac{dS_{\text{cl}}}{dT} = -m. \tag{5.5}$$

The bound state quantization condition then becomes [41]:

$$W(m) = 2\pi n, \quad n \in N_0 \tag{5.6}$$

for n 'large'. Parametric plots of $K\alpha' W$ as a function of $K\alpha'^2 m^2$ in the different cases, are shown in Fig.8. The strings oscillating on one hemisphere give rise to a finite number N_- of states with the following mass-formula Ref.[29]:

$$m_-^2 \alpha' \approx \pi\, n, \qquad N_- \approx \frac{4}{\pi K \alpha'}. \tag{5.7}$$

As in flat Minkowski spacetime, the scale of these string states is set by α'. The strings oscillating on the full hypersphere give rise to an infinity of more and more massive states with the asymptotic mass-formula Ref.[29]:

$$m_+^2 \approx K\, n^2. \tag{5.8}$$

The masses of these states are independent of α', the scale is set by the curvature index K. Notice also that the level spacing grows with n. A similar result was found recently for strings in anti de Sitter spacetime [42].

For $K < 0$, when the spatial section is a hyperboloid, both the energy and the average pressure of the oscillating strings are positive. The equation of state is given in Table 5. In this case, the strings can not provide a self-consistent solution to the Einstein equations. After semi-classical quantization, we find an infinity of more and more massive states. The mass-formula is given by (see Ref.[29]):

$$\sqrt{-Km^2\alpha'^2}\, \log \sqrt{-Km^2\alpha'^2} \approx -\frac{\pi}{2} K\alpha'\, n \tag{5.9}$$

Notice that the level spacing grows faster than in Minkowski spacetime but slower than in the closed static Robertson-Walker spacetime.

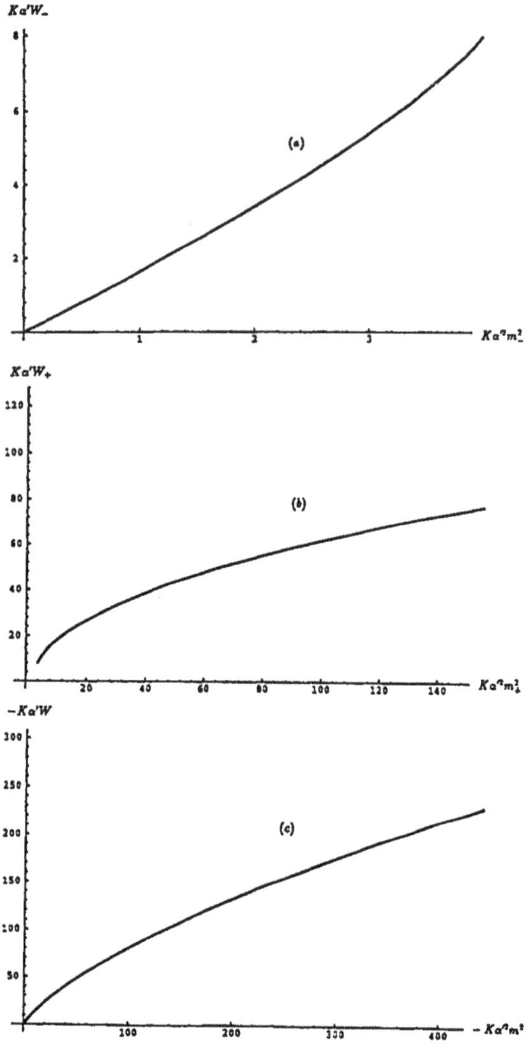

Figure 8. Parametric plot of $K\alpha'W$ as a function of $K\alpha'^2m^2$ in the three cases: (a) $K > 0$, $bK\alpha'^2 \leq 1$ (strings oscillating on one hemisphere, only finitely many states), (b) $K > 0$, $bK\alpha'^2 > 1$ (strings oscillating on the full hypersphere, infinitely many states), (c) $K < 0$ (strings oscillating on the hyperboloid, infinitely many states).

Table 6. Semi-classical quantization of the circular strings in the static Robertson-Walker spacetimes. Notice in particular the different behaviour of the high mass spectrum of strings in the three cases.

SEMICLASSICAL QUANTIZATION OF CIRCULAR STRINGS

Stationary phase: $W(m) = S^{\text{cl}}(T(m)) + m\,T(m)$

Stationary phase point: $\frac{\partial S^{\text{cl}}}{\partial T} = -m,$ determines the period T

Quantization condition: $W(m) = 2\pi\,n, \quad n \in N_0$

MINKOWSKI ($K = 0$):

Mass: $m = 2\sqrt{b}$

Stationary phase: $W = 2\pi\sqrt{b}\alpha'$

Quantization condition: $\alpha' m^2 = 4\,n, \quad n \in N_0$

There are infinitely many states; "$m^2 \propto n$"

CLOSED ($K > 0$):

Mass: $m = \begin{cases} \frac{2k_-}{\sqrt{K\alpha'}} & \text{for } \sqrt{bK}\alpha' \leq 1, \quad k_- \equiv \sqrt{bK}\alpha' \in [0,1] \\ \frac{2}{\sqrt{K\alpha'}k_+} & \text{for } \sqrt{bK}\alpha' > 1, \quad k_+ \equiv 1/(\sqrt{bK}\alpha') \in]0,1[\end{cases}$

Stationary phase: $W = \begin{cases} \frac{8}{K\alpha'}[E(k_-) - (1 - k_-^2)K(k_-)] & \text{for } \sqrt{bK}\alpha' \leq 1 \\ \frac{8}{K\alpha'}\frac{E(k_+)}{k_+} & \text{for } \sqrt{bK}\alpha' > 1 \end{cases}$

Quantization condition: $\begin{cases} m_-^2\alpha' \approx \pi\,n & \text{for } \sqrt{bK}\alpha' \leq 1, \quad \text{on one hemisphere.} \\ m_+^2 \approx K\,n^2 & \text{for } \sqrt{bK}\alpha' > 1, \quad \text{on full hypersphere.} \end{cases}$

Altogether there are infinitely many states; "$m \propto n$", for high states

---

HYPERBOLIC ($K < 0$):

Mass: $m = \frac{2}{\sqrt{-K\alpha'}}\frac{k}{\sqrt{1-k^2}}, \quad k \equiv \sqrt{\frac{-Kb\alpha'^2}{1-Kb\alpha'^2}}$

Stationary phase: $W = \frac{8}{K\alpha'}\frac{E(k)-K(k)}{\sqrt{1-k^2}}$

Quantization condition: $\sqrt{-Km^2\alpha'^2}\log\sqrt{-Km^2\alpha'^2} \approx -\frac{\pi}{2}K\alpha'\,n$

There are infinitely many states; "$m \log m \propto n$", for high states

On the other hand, the stationary strings are determined by Eqs.(4.4):

$$\phi' = \frac{L}{r^2}, \qquad r'^2 + V(r) = 0; \quad V(r) = (1 - Kr^2)(\frac{L^2}{r^2} - 1), \qquad (5.10)$$

where L is an integration constant (Fig.9 shows the potential $V(r)$ for $K > 0$, $K = 0$, $K < 0$).

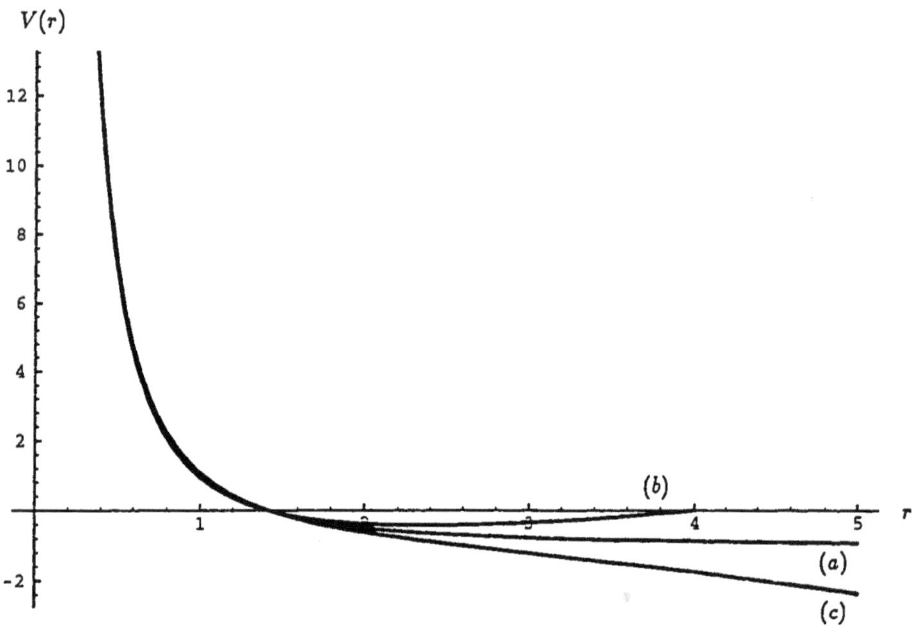

Figure 9. The potential $V(r)$ introduced in Eqs.(5.10) for a stationary string in the static Robertson-Walker spacetimes: (a) flat ($K = 0$), (b) closed ($K > 0$ *and* $KL^2 \leq 1$), (c) hyperbolic ($K < 0$).

For $K > 0$, all the stationary string solutions describe circular strings winding around the hypersphere. The equation of state is of the extremely unstable string type [21]. For $K < 0$, the stationary strings are represented by infinitely long open configurations with an angle between the two "arms" given by:

$$\Delta\phi = \pi - 2\arctan(\sqrt{-K}L). \qquad (5.11)$$

The energy density is positive while the pressure densities are negative. No simple equation of state is found for these solutions.

We also computed the first and second order fluctuations around a static string center of mass, using the string perturbation series approach [7] and its covariant versions [9, 26]. Up to second order, the mass-formula for arbitrary values of the curvature index (positive or negative) is identical to the well-known flat spacetime mass-formula; all dependence on K cancels out.

Table 7. Stationary strings in the static Robertson-Walker spacetimes. Notice that the pressure densities are always negative in all three cases.

STATIONARY STRING SOLUTIONS

Spacetime: $ds^2 = -dt^2 + \frac{dr^2}{1-Kr^2} + r^2 d\phi^2$

String solution: $t = \tau$, $\phi' = \frac{L}{r^2}$, $r'^2 + V(r) = 0$; $V(r) = (1 - Kr^2)(\frac{L^2}{r^2} - 1)$

String length, energy and pressure: $dl = d\sigma$, $T^{\mu}{}_{\nu}(\text{string}) = \text{diag.}(-\rho, P_i)$

MINKOWSKI ($K = 0$):

$r(\sigma) = \sqrt{\sigma^2 + L^2}$, $\phi(\sigma) = \arctan(\sigma/L)$, straight strings.

$\frac{d\rho}{d\sigma} = \frac{1}{2\pi\alpha'}$, $\frac{dP_y}{d\sigma} = -\frac{1}{2\pi\alpha'}$

The equation of state corresponds to extremely unstable strings in 1+1 dimensions.

CLOSED ($K > 0$):

$r^2(\sigma) = \frac{1}{K} \frac{KL^2 + \tan^2(\sqrt{K}\sigma)}{1 + \tan^2(\sqrt{K}\sigma)}$, $\phi(\sigma) = \arctan[\frac{\tan(\sqrt{K}\sigma)}{\sqrt{K}L}]$, circular strings.

$\rho = \frac{1}{\sqrt{K}\alpha'}$, $P_x = P_y = -\frac{1}{2\sqrt{K}\alpha'}$

The equation of state corresponds to extremely unstable strings in 2+1 dimensions.

HYPERBOLIC ($K < 0$):

$r^2(\sigma) = \frac{1}{K} \frac{KL^2 - \tanh^2(\sqrt{-K}\sigma)}{1 - \tanh^2(\sqrt{-K}\sigma)}$, $\phi(\sigma) = \pm\left\{\frac{\pi}{2} - \arctan[\sqrt{-K}L \tanh^{-1}(\sqrt{-K}\sigma)]\right\}$

Infinitely long open strings, $\Delta\phi = \pi - 2\arctan(\sqrt{-K}L) \in]0, \pi]$

$\frac{d\rho}{d\sigma} = \frac{1}{2\pi\alpha'}$, $\frac{dP_x}{d\sigma} = \frac{8KL^2}{\pi\alpha'(1-KL^2)^2} e^{-\sqrt{-K}|\sigma|}$, $\frac{dP_y}{d\sigma} = \frac{-8}{\pi\alpha'(1-KL^2)^3} e^{-\sqrt{-K}|\sigma|}$

The pressure densities are negative but asymptotically go to zero exponentially. No simple equation of state in the general case.

6 Conclusion

We first studied the string propagation in the 2+1 BH-AdS background. We found the first and second order perturbations around the string center of mass as well as the mass formula, and compared with the ordinary black hole AdS spacetime and with the black string background as well. These results were then generalized to ordinary D-dimensional anti de Sitter spacetime. The classical string motion in anti de Sitter spacetime is stable in the sense that it is oscillatory in time with real frequencies and the string size and energy are bounded. Quantum mechanically, this reflects in the mass operator, which is well defined for any value of the wave number n, and arbitrary high mass states (and therefore an infinite number of states) can be constructed. This is to be contrasted with de Sitter spacetime, where string instabilities develop, in the sense that the string size and energy become unbounded for large de Sitter radius. For low mass states (the stable regime), the mass operator in de Sitter spacetime is given by Eq.(3.1) but with

$$\Omega_n(\lambda)_{\rm dS} = \frac{2n^2 - m^2\alpha'\lambda}{\sqrt{n^2 - m^2\alpha'\lambda}}.$$

Real mass solutions can be defined only up to some *maximal mass* of the order $m^2\alpha' \approx 1/\lambda$ [7]. For $\lambda << 1$, real mass solutions can be defined only for $N \leq N_{\rm max} \sim 0.15/\lambda$ (where N is the eigenvalue of the number operator) and therefore there exists a *finite* number of states only. These features of strings in de Sitter spacetime have been recently confirmed within a different (semi-classical) quantization approach based on *exact* circular string solutions [42].

The presence of a cosmological constant Λ (positive or negative) increases considerably the number of levels of different eigenvalue of the mass operator (there is a splitting of levels) with respect to flat spacetime. That is, a non-zero cosmological constant *decreases* (although does not remove) the degeneracy of the string mass states, introducing a *fine structure effect*. For the low mass states the level spacing is approximately constant (up to corrections of the order λ^2). For the high mass states, the changes are more drastic and they depend crucially on the sign of Λ. A value $\Lambda < 0$ causes the *growing* of the level spacing linearly with N instead of being constant as in Minkowski space. Consequently, the density of states $\rho(m)$ grows with the exponential of $\sqrt{m}$ (instead of m as in Minkowski space) discarding the existence of a Hagedorn temperature in AdS spacetime, and the possibility of a phase transition. In addition, another important feature of the high mass string spectrum in AdS spacetime is that it becomes independent of α'. The string scale for the high mass states is given by $\mid \Lambda \mid$, instead of $1/\alpha'$ for the low mass states.

We have found new classes of exact string solutions obtained either by the circular string ansatz, Eq.(2.16), or by the stationary string ansatz, Eq.(4.3), in a variety of curved backgrounds including Schwarzschild, de Sitter and anti de Sitter spacetimes. Many different types of solutions have been found: oscillating circular strings, closed stationary strings, infinitely long stationary strings, dynamical straight strings and multi-string solutions describing finitely or infinitely many stationary or dynamical strings. In all the cases we have obtained the exact solutions in terms of either elementary or elliptic functions. Furthermore, we have analyzed the physical properties (length, energy, pressure) of the string solutions.

Finally, we solved the equations of motion and constraints for circular strings in static Robertson-Walker spacetimes. We computed the equations of state and found that there exists a self-consistent solution to the Einstein equations in the case of positive spatial curvature. The solutions were quantized semi-classically using the stationary phase approximation method, and the resulting spectra were analyzed and discussed. We have also found all stationary string configurations in these spacetimes and we computed the corresponding physical quantities, string length, energy and pressure.

Acknowledgements:
A.L. Larsen is supported by the Danish Natural Science Research Council under grant No. 11-1231-1SE

References

[1] H.J. de Vega and N. Sánchez, Phys. Rev. D42 (1990) 3969.
H.J. de Vega, M. Ramón-Medrano and N. Sánchez, Nucl. Phys. B374 (1992) 405.

[2] H.J. de Vega and N. Sánchez, Phys. Lett. B244 (1990) 215, Phys. Rev. Lett. 65C (1990) 1517, IJMP A7 (1992) 3043, Nucl. Phys. B317 (1989) 706 and ibid 731.
D. Amati and C. Klimcik, Phys. Lett. B210 (1988) 92.
M. Costa and H.J. de Vega, Ann. Phys. 211 (1991) 223 and ibid 235.
C. Loustó and N. Sánchez, Phys. Rev. D46 (1992) 4520.

[3] V.E. Zakharov and A.V. Mikhailov, JETP 47 (1979) 1017.
H. Eichenherr, in "Integrable Quantum Field Theories", ed. J. Hietarinta and C. Montonen (Springer, Berlin, 1982).

[4] H.J. de Vega and N. Sánchez, Phys. Rev. D47 (1993) 3394.

[5] I. Bars and K. Sfetsos, Mod. Phys. Lett. A7 (1992) 1091.

[6] H.J. de Vega, J.R. Mittelbrunn, M.R. Medrano and N. Sánchez, Phys. Lett. B232 (1994) 133 and "The Two-Dimensional Stringy Black Hole: a new Approach and a Pathology", PAR-LPTHE 93/14.

[7] H.J. de Vega and N. Sánchez, Phys. Lett. B197 (1987) 320.

[8] H.J. de Vega and N. Sánchez, Nucl. Phys. B299 (1988) 818.

[9] P.F. Mende, in "String Quantum Gravity and Physics at the Planck Energy Scale". Proceedings of the Erice Workshop held in June 1992, ed. N. Sánchez (World Scientific, 1993).

[10] A.L. Larsen and V.P. Frolov, Nucl. Phys. B414 (1994) 129.

[11] H.J. de Vega, A.V. Mikhailov and N. Sánchez, Teor. Mat. Fiz. 94 (1993) 232, Mod. Phys. Lett. A29 (1994) 2745.

[12] F. Combes, H.J. de Vega, A.V. Mikhailov and N. Sánchez, Phys. Rev. D50 (1994) 2754.

[13] H.J. de Vega, A.L. Larsen and N. Sánchez, Nucl. Phys. B427 (1994) 643.

[14] I. Krichever, "Two-Dimensional Algebraic-Geometrical Operators with Self-Consistent Potentials", Landau Institute Preprint, March, 1994, to appear in Funct. Analisis and Appl.

[15] H.J. de Vega and N. Sánchez, Nucl. Phys. B309 (1988) 552 and ibid, 577.

[16] C.O. Loustó and N. Sánchez, Phys. Rev. D47 (1993) 4498.

[17] H.J. de Vega and I.L. Egusquiza, Phys. Rev. D50 (1994) 763.

[18] A.L. Larsen, Phys. Rev. D50 (1994) 2623.

[19] A.L. Larsen and M. Axenides, Phys. Lett. B318 (1993) 47.

[20] H.J. de Vega and N. Sánchez, Phys. Rev. D50 (1994) 7202.

[21] N. Sánchez and G. Veneziano, Nucl. Phys. B333 (1990) 253.
M. Gasperini, N. Sánchez and G. Veneziano, IJMP A6 (1991) 3853, Nucl. Phys. B364 (1991) 365.

[22] H.J. de Vega, M. Ramón-Medrano and N. Sánchez, Nucl. Phys. B374 (1992) 425.

[23] H.J. de Vega and N. Sánchez, Phys. Rev. D45 (1992) 2783.
M. Ramón-Medrano and N. Sánchez, Class. Quantum Grav. 10 (1993) 2007.

[24] N. Sánchez, Phys. Lett. B195 (1987) 160.

[25] H.J. de Vega, M. Ramón-Medrano and N. Sánchez, Nucl. Phys. B351 (1991) 277.

[26] A.L. Larsen and N. Sánchez, Phys. Rev. D50 (1994) 7493.

[27] A.L. Larsen and N. Sánchez, "Mass Spectrum of Strings in Anti de Sitter Spacetime", DEMIRM-Paris-94048, to appear in Phys. Rev. D.

[28] A.L. Larsen and N. Sánchez, "New Classes of Exact Multi-String Solutions in Curved Spacetimes", DEMIRM-Paris-95003, to appear in Phys. Rev. D.

[29] A.L. Larsen and N. Sánchez, "The Effect of Spatial Curvature on the Classical and Quantum Strings", DEMIRM-Paris-95004.

[30] M. Bañados, C. Teitelboim and J. Zanelli, Phys. Rev. Lett. 69 (1992) 1849.

[31] G.T. Horowitz and D.L. Welch, Phys. Rev. Lett. 71 (1993) 328.

[32] N. Kaloper, Phys. Rev. D48 (1993) 2598.

[33] M. Bañados, M. Henneaux, C. Teitelboim and J. Zanelli, Phys. Rev. D48 (1993) 1506.

[34] C. Farina, J. Gamboa and A.J. Seguí-Santonja, Class. Quantum Grav. 10 (1993) L193.
N. Cruz, C. Martínez and L. Peña, Class. Quantum Grav. 11 (1994) 2731.

[35] A. Achúcarro and M.E. Ortiz, Phys. Rev. D48 (1993) 3600.
D. Cangemi, M. Leblanc and R.B. Mann, Phys. Rev. D48 (1993) 3606.

[36] J. Horne and G.T. Horowitz, Nucl. Phys. B368 (1992) 444.

[37] T. Buscher, Phys. Lett. B201 (1988) 466, B194 (1987) 59.

[38] A. Vilenkin, Phys. Rev. D24 (1981) 2082.

[39] H.J. de Vega and N. Sánchez, Int. J. Mod. Phys. A7 (1992) 3043.

[40] V.P. Frolov, V.D. Skarzhinsky, A.I. Zelnikov and O. Heinrich, Phys. Lett. B224 (1989) 255.

[41] R. Dashen, B. Hasslacher and A. Neveu, Phys. Rev. D11 (1975) 3424.

[42] H.J. de Vega, A.L. Larsen and N. Sánchez, "Semi-Classical Quantization of Circular Strings in de Sitter and anti de Sitter Spacetimes", DEMIRM-Paris-94049, submitted to Phys. Rev. D.

[43] A.L. Larsen, "Stable and Unstable Circular Strings in Inflationary Universes", NORDITA-94/14-P, to appear in Phys. Rev. D.

[44] A.L. Larsen, Nucl. Phys. B412 (1994) 372.

BLACK HOLES AND EXACT SOLUTIONS IN STRING THEORY

A.A. TSEYTLIN*†
Theoretical Physics Group, Blackett Laboratory
Imperial College, London SW7 2BZ, U.K.

Abstract

We review some recent results about exact classical solutions in string theory. In particular, we consider four dimensional extremal electric black holes which are related via dimensional reduction to the exact five dimensional fundamental string solutions. We also comment on the issue of non-trivial α' corrections to non-extremal black holes.

1. Introduction

String theory is the only presently known model of a unified theory which includes gravity and at the same time is consistent at the quantum level. This, in principle, enables one to study its quantum-gravitational implications for cosmology, black hole physics, space-time singularities, etc.

The basic objects of string theory are (closed) strings which, when first-quantised, have a finite number of massless (graviton or metric $G_{\mu\nu}$, antisymmetric tensor $B_{\mu\nu}$, dilaton ϕ, vectors, other scalars and spinors) and an infinite number of massive ($M^2 \sim 1/\alpha' \sim M^2_{Planck}$) particle-like excitations. Expanded near the trivial flat space vacuum the theory is parametrised by two constants: $\alpha' \sim L^2_{Planck}$ and dimensionless string coupling $g = \exp \phi_0$.The low-energy (large scale) classical string dynamics is described by an effective action [1,2]

$$S = \int d^D x \sqrt{G} e^{-2\phi} [R + 4(\partial\phi)^2 + ... + \alpha' R^2 + ...] \ ,$$

* e-mail address: tseytlin@ic.ac.uk
† On leave from Lebedev Physics Institute, Moscow, Russia.

which starts with the Einstein term and contains contributions of all orders in α'. In general, the dilaton ϕ plays the role of the (logarithm of the) effective string coupling [3,2].

One of the important problems in string theory is to find exact black hole solutions in four dimensions. While the vacuum Schwarzschild solution of Einstein equations is also a leading-order solution in string theory, it is modified by α' corrections. It is straightforward to find other solutions to the leading order string effective equations describing charged dilatonic black holes [4,5,6], as well as to compute the first nontrivial α' corrections to the Schwarzschild solution [7].

To be able to study the Planck-scale structure of the theory one should go beyond the solutions of the leading-order equations, i.e. one should look for solutions which are exact in α'. Though the explicit form of the full string effective action is not known, one of the remarkable properties of string theory is that it is possible to by-pass this complication by using the equivalence [8,9] between extrema of the effective action and 2-dimensional conformal theories (CFT's). Knowing the corresponding CFT is also crucial for an adequate "stringy" interpretation of a solution.

Exact black hole solutions have been constructed in two [10] and three [11] dimensions starting with (gauged) WZW models (or coset CFT's). In addition, using the result that the extremal limit of a black five-brane [12] is an exact superstring solution [13], one can obtain an exact five dimensional extreme black hole (which has a magnetic charge associated with the antisymmetric tensor field). It has recently been shown that one can also dimensionally reduce the exact solution of [13] down to four dimensions [14] to find an extreme magnetically charged black hole [15].[1]

Below we shall discuss the exact string solution [21] describing an extremal four dimensional electrically charged black hole.[2] This solution can be considered [21] as four-dimensional "image" of an exact five-dimensional solution describing the field of a fundamental string [22,23] which winds around a compact dimension.

In Section 2 we shall review the structure of the string effective action and suggest a possibility that string theory is "asymptotically free", i.e. that the effective string coupling goes to zero at small distances and thus there is a sense in studying classical solutions.

---

[1] These exact five and four dimensional black hole solutions are asymptotically flat. In addition, there are coset CFT or gauged WZW constructions of just the 'throat' regions of some four dimensional extreme magnetically charged dilatonic black holes [16,17] (the fact that the form of the throat solution is unchanged by α' corrections was noticed earlier [18]). Also, one can use gauged WZW models to construct axisymmetric black-hole type solutions [19,20].

[2] The electrically charged solutions are qualitatively different from their magnetic analogues. In particular, the extremal magnetically charged solutions found in four and five dimensions are products of a time-like line and a euclidean solution. The extremal electrically charged solution, on the other hand, has a nontrivial time-like direction.

In Section 3 we shall consider the leading-order terms in the string effective action and present the corresponding dilatonic black hole solution [4,5,6]. We shall also comment on how α' corrections can possibly modify non-extremal black hole solutions using the analogy with the Born-Infeld effective action of the open string theory. In Section 4 we shall briefly discuss two classes of exact classical string solutions: "plane fronted waves" and "fundamental string" backgrounds. Their close relation to extremal electric black hole solutions (implying the exactness of the latter) will be explained in Section 5. Section 6 contains concluding remarks.

2. String effective action and "asymptotic freedom"

In general, the low-energy effective action in string theory is given by a sum of classical, quantum (string loop) and non-perturbative contributions

$$S = S_c + S_l + S_n \ . \tag{2.1}$$

Several leading terms in the (bosonic string) classical effective action are [1,2,9,24][3]

$$S_c = \int d^D x \sqrt{G}\, e^{-2\phi}\{c_0 + [\, R + 4\nabla^2\phi - 4(\nabla\phi)^2 - \frac{1}{12}(H_{\mu\nu\lambda})^2\,] \tag{2.2}$$

$$+\frac{1}{4}\alpha' \, [\, R_{\mu\nu\lambda\kappa}R^{\mu\nu\lambda\kappa} - \frac{1}{2}R^{\mu\nu\kappa\lambda}H^{\rho}{}_{\mu\nu}H_{\kappa\lambda\rho}$$

$$+\frac{1}{24}H_{\mu\nu\lambda}H^{\nu}{}_{\rho\alpha}H^{\rho\sigma\lambda}H_{\sigma}{}^{\mu\alpha} - \frac{1}{8}(H_{\mu\alpha\beta}H_{\nu}{}^{\alpha\beta})^2\,] + O(\alpha'^3)\} \ ,$$

where $c_0 = -2(D-26)/3\alpha'$. The loop contribution is ultraviolet-finite (with $\sqrt{\alpha'}$ playing the role of an effective cutoff). For example, in the superstring theory S_l has the following structure

$$S_l = \sum_{n=0}^{\infty} \int d^D x \sqrt{G}\, e^{2n\phi}(\frac{1}{\alpha'}R^4 + ... + \text{non}-\text{local terms}) \ . \tag{2.3}$$

In field theory non-perturbative terms are usually expected to involve factors like $\exp(-1/g^2)$ where g is a coupling constant. Since in string theory the latter is related to the dilaton, the non-perturbative contributions will generate non-trivial terms in S (dilaton potential, etc.)

$$S_n = \int d^D x \sqrt{G}\, V(\phi) + ... \ , \quad V = \frac{c_1}{g^2(\phi)} \exp[-\frac{c_2}{g^2(\phi)}] + ... \ , \quad g(\phi) = e^{\phi} \ . \tag{2.4}$$

---

[3] The form of such action is not unique: a class of actions related by field redefinitions which are local, covariant, background-independent, power series in α' (depending on dilaton only through its derivatives not to mix different orders of string loop perturbation theory) will correspond to the same string S-matrix [25,26].

The non-perturbative corrections should be responsible for supersymmetry breaking and generation of a mass for the dilaton. Since the dilaton has gravitational-type couplings it should get a mass to avoid contradiction with observations (for a discussion and references see e.g. [27,28]).

The existence of a (perturbatively) massless scalar dilaton (which also determines the strength of effective string coupling) is one of the universal and most important consequences of string theory. While dilaton should be effectively massive at macroscopic distances or late stages of evolution of the Universe, it may be massless at small scales or early times. In fact, one can contemplate a possibility of *"asymptotic freedom"* in string theory: it may be that e^{ϕ} is very small at small scales/times so that string theory is effectively classical in early Universe (or at small distances), i.e. both perturbative and non-perturbative string corrections are negligible there. Dilaton may grow at "intermediate" scales/times eventually making non-perturbative corrections to become relevant and to generate a potential (mass) for the dilaton. It seems that the proper dilaton boundary condition for, e.g., black hole or cosmological solutions in string theory should be $\phi \to -\infty$ at $r \to 0$ or $t \to 0$, that is small string coupling at small scales.[4]

This "asymptotic freedom" suggestion implies that massless dilaton should play an important role in early Universe (for example, the issue of primordial black holes should be addressed on the basis of classical string theory with a massless dilaton). Assuming that string coupling is small at small distances, in what follows we shall ignore both perturbative and non-perturbative contributions to the string effective action and discuss the *classical* black hole solutions in string theory. The condition that string coupling should vanish at small scales will, in fact, be fulfilled for exact solutions we shall describe.

3. Leading-order dilatonic black hole solutions and α' corrections

The leading-order terms in (2.2) may contain also the Yang-Mills gauge field term which may be of heterotic or Kaluza-Klein origin. Here we shall assume the second possibility. If one considers a bosonic string in a space with some compact isometric directions one can get extra vector fields from "non-diagonal" components of the metric and the antisymmetric tensor.

For example, if we start with the leading-order term in the $D = 5$ bosonic string action

$$S_5 = \int d^5x\sqrt{G}\, e^{-2\phi}\ \{\ R\ + 4(\partial_\mu\phi)^2 - \frac{1}{12}(H_{\mu\nu\lambda})^2\ + O(\alpha')\}\ , \tag{3.1}$$

[4] There are, in fact, classical string solutions (in particular, the exact fundamental string solutions discussed below) that have such behaviour (see also [5,22,29,30,31,21]). For a review of cosmological solutions see [27].

and assume that all the fields are independent of $x^5 = y$, we obtain the four dimensional reduced action

$$S_4 = \int d^4x\sqrt{\hat{G}}\, \mathrm{e}^{-2\phi+\sigma} \{ \hat{R} + 4(\partial_\mu\phi)^2 - 4\partial_\mu\phi\partial^\mu\sigma \tag{3.2}$$

$$-\frac{1}{12}(\hat{H}_{\mu\nu\lambda})^2 - \frac{1}{4}\mathrm{e}^{2\sigma}(\mathcal{F}_{\mu\nu})^2 - \frac{1}{4}\mathrm{e}^{-2\sigma}(\mathcal{B}_{\mu\nu})^2 + O(\alpha')\} ,$$

where we have defined

$$G_{55} \equiv \mathrm{e}^{2\sigma}, \quad \mathcal{F}_{\mu\nu} = 2\partial_{[\mu}\mathcal{A}_{\nu]} , \quad \mathcal{B}_{\mu\nu} = 2\partial_{[\mu}\mathcal{B}_{\nu]} , \quad \mathcal{A}_\mu \equiv G^{55}G_{\mu 5} , \quad \mathcal{B}_\mu \equiv B_{\mu 5} , \tag{3.3}$$

$$\hat{G}_{\mu\nu} \equiv G_{\mu\nu} - G_{55}\mathcal{A}_\mu\mathcal{A}_\nu , \qquad \hat{H}_{\lambda\mu\nu} = 3\partial_{[\lambda}B_{\mu\nu]} - 3\mathcal{A}_{[\lambda}\mathcal{B}_{\mu\nu]} .$$

This action can be considered also, for example, as a part of $N = 4, D = 4$ supergravity action obtained by dimensional reduction from $D = 10$ supergravity. Setting $\varphi = 2\phi - \sigma$ the action (3.2) becomes [32]

$$S_4 = \int d^4x\sqrt{\hat{G}}\, \mathrm{e}^{-\varphi} \{ \hat{R} + (\partial_\mu\varphi)^2 - (\partial_\mu\sigma)^2 \tag{3.4}$$

$$-\frac{1}{12}(\hat{H}_{\mu\nu\lambda})^2 - \frac{1}{4}\mathrm{e}^{2\sigma}(\mathcal{F}_{\mu\nu})^2 - \frac{1}{4}\mathrm{e}^{-2\sigma}(\mathcal{B}_{\mu\nu})^2 + O(\alpha')\} .$$

In the Einstein frame (where there is no dilaton factor in front of R) (3.4) takes the form

$$S_4 = \int d^4x\sqrt{\hat{G}_E}\, \{ \hat{R}_E - \frac{1}{2}(\partial_\mu\varphi)^2 - (\partial_\mu\sigma)^2 \tag{3.5}$$

$$-\frac{1}{12}\mathrm{e}^{-2\varphi}(\hat{H}_{\mu\nu\lambda})^2 - \frac{1}{4}\mathrm{e}^{-\varphi+2\sigma}(\mathcal{F}_{\mu\nu})^2 - \frac{1}{4}\mathrm{e}^{-\varphi-2\sigma}(\mathcal{B}_{\mu\nu})^2 + O(\alpha')\} .$$

Thus, in general, the four dimensional theory contains two scalars, two vectors, and the antisymmetric tensor, as well as the metric. In certain special cases, the nontrivial part of the action (3.5) can be expressed in terms of only one scalar and one vector, so that it becomes

$$S_4 = \int d^4x\sqrt{\hat{G}_E}\, \{ \hat{R}_E - \frac{1}{2}(\partial_\mu\psi)^2 - \frac{1}{4}\mathrm{e}^{-a\psi}(\mathcal{F}_{\mu\nu})^2 + O(\alpha')\} . \tag{3.6}$$

For example, if one sets $\phi = 0$ and $H_{\mu\nu\lambda} = 0$ in the $D = 5$ action, or equivalently $\varphi = -\sigma$, $\hat{H}_{\mu\nu\lambda} = 0 = \mathcal{B}_{\mu\nu}$ directly in (3.5), one obtains (3.6) with $\psi = -a\sigma$ and $a = \sqrt{3}$. This is, of course, the standard Kaluza-Klein reduction of the Einstein action. Another possibility which we shall mostly consider below is to set $\sigma = 0$ ($G_{55} = 1$), $\hat{H}_{\mu\nu\lambda} = 0$ and either the two vector fields proportional to each other, or let one of them vanish. This case corresponds to (3.6) with $\psi = 2\phi$ and $a = 1$.

This action admits, of course, the standard Schwarzschild solution ($R_{\mu\nu} = 0, \psi = 0, \mathcal{F}_{\mu\nu} = 0$) but not the Reissner-Nordstrom one (because of non-trivial coupling to the gauge field the dilaton cannot remain constant). The dilaton-vector coupling plays a crucial role in modifying the properties of black hole - type solutions (this coupling becomes irrelevant only after the dilaton is "frozen" at a minimum of a non-perturbative potential).

The charged dilatonic black hole solutions (which are string analogues of Reissner-Nordstrom one) were found in [4,5,6] (for a review see [29]). The electrically charged solution has the folowing form ($\psi = 2\phi, a = 1$)

$$ds^2 = e^{2\phi} ds_E^2 \ , \quad ds_E^2 = -(1 - \frac{2M}{r})dt^2 + (1 - \frac{2M}{r})^{-1} dr^2 + r^2(1 - \frac{2Q^2}{Mr})d\Omega_2^2 \ , \qquad (3.7)$$

$$\mathcal{A}_t = -\frac{Q}{r} \ , \quad e^{2\phi} = 1 - \frac{2Q^2}{Mr} \ .$$

This metric has the same causal structure as Schwarzschild (e.g. there is no inner horizon) and singularity at $r = 0$.

Taking the extremal limit $Q = M$ and redefining $r \to r + 2M$ one finds that the solution simplifies to

$$ds^2 = -F^2(r)dt^2 + dr^2 + r^2 d\Omega_2^2 \ , \qquad (3.8)$$

$$\mathcal{A}_t = e^{2\phi} = F(r) = (1 + \frac{2M}{r})^{-1} \ .$$

Note that the spatial part of the metric (3.8) is flat (this is one of the reasons why this solution has an exact generalisation to all orders in α'). Also, $e^{\phi} \to 0$ at $r \to 0$ so that the string coupling becomes weak near the singularity (it turns out that this remarkable behaviour is not modified by α'-corrections).

There are various indications that extreme black holes play a special role in string theory, being gravitational analogues of solitons [4,5,33,29,34]: they cannot be created from non-extremal ones by a classical process, their entropy is zero, solutions can be linearly superposed, are supersymmetric and stable when embedded into supergravity theory [35], etc. A solitonic interpretation of extreme electric dilatonic black holes is strengthened by the results of [21] discussed below: (i) extreme electric dilatonic $D = 4$ black hole solution can be represented as a dimensional reduction of the $D = 5$ fundamental string solution solution of [22,23] which itself has a solitonic interpretation and is an exact classical string solution [31,32], so that (ii) extreme electric black hole solution is not modified by α' corrections. The absence of α' corrections is related to the extremality property of the solution, i.e. to a special balance between the metric and the gauge field (or the metric and the antisymmetric tensor in the case of $D = 5$ fundamental string solution).

At the same time, the situation with non-extremal black holes is much more complicated: they are modified by α' corrections in a non-trivial way.[5] For example, starting

[5] As was already mentioned above, there is a field redefinition ambiguity in the effective action. When saying that a solution is not modified by α' corrections we understand that this is true modulo a local field redefinition (i.e. in a specific "scheme"): a solution in a generic scheme will be related to the leading-order one by a *local* field redefinition. A solution is said to receive non-trivial α' corrections if there does not exist a scheme where the exact solution is equal to the leading-order one.

with the Schwarzschild solution (with $\phi = \text{const}$) and solving the effective equations corresponding to S_c (2.2) (assuming the central charge deficit c_0 is cancelled by some internal degrees of freedom)

$$R_{\mu\nu} + 2D_\mu D_\nu \phi + \frac{1}{2}\alpha' R_{\mu\lambda\rho\kappa} R_\nu{}^{\lambda\rho\kappa} + O(\alpha'^2) = 0 \ , \tag{3.9}$$

$$-D^2\phi + 2(\partial\phi)^2 + \frac{1}{8}\alpha' R_{\mu\lambda\rho\kappa} R^{\mu\lambda\rho\kappa} + O(\alpha'^2) = 0 \ , \tag{3.10}$$

one finds that to the leading order in α' the metric and dilaton get $O(1/r^3)$ corrections [7]. Thus the dilaton is no longer constant. Interestingly, e^ϕ does decrease towards the center of black hole. Since the Hawking-Penrose singularity condition ($R_{\mu\nu} l^\mu l^\nu \geq 0, \ l^2 \leq 0$) may not be satisfied for the full α'-dependent effective equations, one may expect that α'-corrections, once summed up, may eliminate the $r = 0$ singularity by modifying the metric $G_{00} = -1 + \frac{2M}{r}$, for example, in the following way (see also [36])

$$G_{00} = -1 + \frac{2M}{r} \exp(-c\frac{\alpha' M}{r^3}) \ . \tag{3.11}$$

An indication that higher order α' terms in the effective action may eliminate the black hole singularity comes from analogy with the open string theory. There the derivative-independent part of the effective action for abelian gauge field can be found exactly and is given by the Born-Infeld action [37] (for derivative corrections see [38])

$$S = \int d^D x \sqrt{\det\, (\delta_{\mu\nu} + 2\pi\alpha' \mathcal{F}_{\mu\nu})} + O(\alpha'^5 \mathcal{F}^2 \partial\mathcal{F}\partial\mathcal{F}) \ . \tag{3.12}$$

While in the Maxwell theory the field of a point-like charge is singular at the origin and its energy is infinite, in the Born-Infeld theory the field is regular at $r = 0$ where the electric field takes its maximal value and its total energy is finite [39]. From the point of view of the distribution of the electric field ($\rho_{eff} = divE/4\pi$) the source is no longer point-like but has an effective radius $r_0 \sim \sqrt{\alpha'}$ (for example, in $D = 4$ $E_r = \mathcal{F}_{rt} = Q/\sqrt{r^4 + r_0^4}$, $r_0^2 = 2\pi\alpha' Q$). Since both open and closed string theories are non-local with a characteristic scale of $\sqrt{\alpha'}$, one may expect that the Schwarzschild singularity is smeared in a similar way. Unfortunately, an analogue of the Born-Infeld action is hard to derive in the closed string case, i.e. there are technical difficulties on the way to verification of this conjecture.

4. Exact string solutions: "plane waves" and "fundamental strings"

Backgrounds with null Killing vectors seem to play a special role in string theory. It was observed in [40] that the "plane-wave" spaces (u, v are "light-cone" coordinates: $u, v = y \pm t$; $i = 1, ..., D-2$)

$$ds^2 = dudv + K(x,u)du^2 + dx^i dx_i \ , \tag{4.1}$$

represent exact classical string solutions if K is subject to the leading-order equation $R_{\mu\nu} = 0$, i.e. $\partial_i \partial^i K = 0$ (for example, $K = f(u)(x_1^2 - x_2^2)$). There exist generalisations of this exact solution to the presence of antisymmetric tensor and dilaton (see [41,42,31,32] and refs. there). A simple and adequate way to specify a string solution is to write down the expression for the string action in the corresponding background. For example, a generalisation of (4.1) is represented by the folowing conformal sigma model ($\mathcal{R}$ is 1/4 of the curvature of the world-sheet metric)

$$L_{PW} = (G_{\mu\nu} + B_{\mu\nu})(x)\ \partial x^\mu \bar{\partial} x^\nu + \alpha' \mathcal{R} \phi(x) \tag{4.2}$$

$$= \partial u \bar{\partial} v + K(x,u) \partial u \bar{\partial} u + 2A_i(x,u) \partial u \bar{\partial} x^i + \ \partial x^i \bar{\partial} x_i + \alpha' \mathcal{R} \phi(u) \ ,$$

where the functions K, A_i, ϕ satisfy

$$\partial^i \partial_i K = 2\partial^i \partial_u A_i + 4\partial_u^2 \phi \ , \quad \partial_i F^{ij} = 0 \ . \tag{4.3}$$

The $SO(D-3)$ invariant "plane wave" (4.1) with $K = K(r)$, $r^2 = x^i x_i$, i.e.

$$K = 1 + \frac{2M}{r^{D-4}} \ , \quad D > 4 \ ; \quad K = \ -2M \ln \frac{r}{r_0} \ , \quad D = 4 \ , \tag{4.4}$$

is related (by space-time duality in the y direction [43,29]) to the following "fundamental string" (FS) solution [22]

$$ds^2 = F(r) dudv + dr^2 + r^2 d\Omega_{D-3}^2 \ , \tag{4.5}$$

$$B_{uv} = \frac{1}{2} F(r) \ , \qquad e^{2\phi} = F(r) \ , \quad F \equiv K^{-1} \ , \tag{4.6}$$

i.e. [31]

$$L_{FS} = F(r) \partial u \bar{\partial} v + \partial x^i \bar{\partial} x_i + \alpha' \mathcal{R} \phi(r) \ . \tag{4.7}$$

This background represents a "solitonic" string-like solution of the "vacuum" effective equations following from (3.1). At the same time, it can be considered also as a solution describing a field outside an infinite straight fundamental string source along y direction [22]. The string source provides the δ-function in the equation for F: $\partial^i\partial_i F^{-1} \sim \delta(r)$.[6]

Fundamental string solution is an extremal limit of a class of leading-order "black string" solutions with axionic charge [12] (which can be obtained by starting with a direct product of the Schwarzschild solution and a spatial direction and making a boost and duality transformation to generate an axionic charge, i.e. a non-trivial $B_{\mu\nu}$ [29]). FS is thus similar to the extremal black holes (with the role of the electric charge being played by the axionic one).

In contrast to non-extremal solutions (e.g. the neutral $D = 4$ "cosmic string" field with a conical singularity) which are modified by α' corrections, the FS is exact (in a specially chosen scheme) [31]. This is due to the special balance between the metric and the antisymmetric tensor (and dilaton) resulting in a very simple form of the corresponding string action (4.7) which, in particular, has conserved chiral currents (or additional infinite-dimensional symmetries) and is conformally invariant. This balance is also responsible for the fact that the equation for F^{-1} is linear, i.e. that FS solutions with parallel sources can be linearly superposed (this is also the reason why the force on a static test string which is parallel to the string source vanishes [22,46]).

The FS solution has a curvature singularity at $r = 0$ (and no horizon) and is stable and supersymmetric when embedded into a supergravity theory [22,23]. The effective string coupling e^{ϕ} decreases to zero at $r = 0$ in agreement with the "asymptotic freedom" assumption, suggesting that the classical approximation can be trusted at small r. In any case, the $r = 0$ curvature singularity does not necessarily mean a singularity of the string theory since a propagation of test quantum strings in this background may not be pathological.[7]

There are generalisations of the FS solution (e.g. including momentum in the direction of the string and Kaluza-Klein type charges). Like (4.5), they are leading-order solutions [23,48,49] which turn out not to be modified by α' corrections [32]. A particular generalised fundamental string solution which we shall consider below is [23,47] (cf. (4.7))

$$L'_{FS} = F(r)\partial u \bar{\partial} v + \partial u \bar{\partial} u + \partial x^i \bar{\partial} x_i + \alpha' \mathcal{R}\phi(r) \ , \tag{4.8}$$

[6] As discussed in [44,45], there exists a (wormhole-type) resummation of the string perturbation theory in which the semiclassical approximation is governed by the sum of the string action in a background of massless fields and the classical effective action for the latter. FS can then be considered as a solution corresponding to such combined action. This suggests that the need for a δ-function source at $r = 0$ is related to a non-perturbative solitonic nature of the FS solution (note that the string action and the effective action are of different orders in string coupling).

[7] In fact, scattering of test strings in this background appears to be similar to their scattering in flat space [46,47].

where F and ϕ are the same as in (4.4), (4.6). Assuming that the y-direction is periodic, (4.8) describes a field outside a closed fundamental string winding around the y-circle.[8] All the main properties of the FS solution are true also for (4.8). A novel feature is that the corresponding metric has a regular ergosphere [21].

5. Extremal electric $D = 4$ black hole as dimensional reduction of exact $D = 5$ fundamental string solution

String theory naturally leads to spaces of dimension $D > 4$. If the scale of "extra" internal dimensions is small, non-trivial higher-dimensional solutions may look like $D = 4$ ones. To get extremal black holes we need gauge fields. They may appear as in Kaluza-Klein model if internal dimensions are compactified on a torus. From the low energy field theory point of view we have to start with the full (containing terms of all orders in α') massless string effective action in, say, five dimensions. Assuming the fifth direction y is periodic, we can expand the metric, antisymmetric tensor and dilaton in Fourier series in y and explicitly integrate over y. The result will be a $D = 4$ effective action containing massless fields as well as an infinite tower of massive modes with masses proportional to a compactification scale. Any exact solution of the $D = 5$ theory which does not depend on y can then be directly interpreted as a solution of the equations of the $D = 4$ "compactified" theory with all massive modes set equal to zero (but all "massless" α'-terms included). In string theory there is a simpler alternative – to perform the dimensional reduction at the more fundamental level of the string action itself [32]. That amounts to a separation of the action into gauge-invariant $D = 4$ part and "internal" coordinate part, with the latter one been coupled to gauge fields originating from (5μ)-components of the metric and the antisymmetric tensor (cf. (3.3)).

For example, consider the five dimensional plane fronted wave (4.1),(4.4)

$$ds_5^2 = dudv + K(r)du^2 + dr^2 + r^2 d\Omega_3^2 \ , \qquad K(r) = 1 + \frac{2M}{r} \ . \tag{5.1}$$

To reduce to four dimensions let us take u to be an internal coordinate y and $v/2$ to be the time coordinate t in four dimensions. Rewriting (5.1) in the form

$$ds_5^2 = -K^{-1}dt^2 + dr^2 + r^2 d\Omega_2^2 + K(dy + K^{-1}dt)^2 \ , \tag{5.2}$$

we conclude that the four dimensional metric, abelian gauge field and the scalar $\sigma = \frac{1}{2}\ln G_{55}$ are

$$ds^2 = -K^{-1}(r)dt^2 + dr^2 + r^2 d\Omega_2^2 \ , \quad A_t = K^{-1}(r) \ , \quad \sigma = \frac{1}{2}\ln K(r) \ . \tag{5.3}$$

[8] In general, the string source which is needed to support this solution (the vanishing of β-functions in $uv-$ and $uu-$ directions) should be represented by $u = f(\tau + \sigma)$, $v = h(\tau - \sigma)$, i.e. in addition to winding there may be a momentum flow along the string.

One thus finds the extremal electrically charged "Kaluza-Klein" black hole which is a solution corresponding to the action (3.6) with $a = \sqrt{3}$ and $\psi = \sigma$ [4]. The $D = 5$ solution (5.1) can be viewed as a field of a string source along u direction boosted to the speed of light [4]. At the same time, its "image" in the four dimensional space-time orthogonal to u-axis is the extremal electric black hole.[9] Since (5.1) is an exact string solution we conclude that (5.3) is also not changed by α' corrections.

Starting with the $D = 5$ generalised FS solution (4.8) and dimensionally reducing to $D = 4$ along u direction one finds [21,32] the following $D = 4$ background[10]

$$ds^2 = -F^2(r)dt^2 + dx_i dx^i \ , \tag{5.4}$$

$$\mathcal{A}_t = -\mathcal{B}_t = F(r) \ , \qquad e^{2\phi} = F(r) \ .$$

Since the original $D = 5$ background had a non-trivial antisymmetric tensor we got $D = 4$ solution with two vector fields (cf. (3.3)). The latter, however, are proportional to each other so (5.4) can be identified with the solution corresponding to (3.6) (with $a = 1, \psi = 2\phi$) which describes the extremal electrically charged dilatonic black hole (3.8). The relation via dimensional reduction to the exact $D = 5$ generalised FS solution implies that this extremal dilatonic black hole is also an exact string solution [21].

6. Conclusions

It may be that string theory is "asymptotically free", i.e. the string coupling is small at small distances so that the dilaton is massless and the theory is nearly classical there. Then the dilaton should play an important role in predictions of string theory about early Universe, primordial black holes, etc. The gravitational-type coupling of the dilaton to gauge fields implies, in particular, that properties of the charged black hole solutions are modified compared to the pure Einstein-Maxwell theory.

There exists a large class of classical string solutions with null Killing vectors which do not receive α' corrections, i.e. are exact. Particular examples are plane-wave type backgrounds and solitonic fundamental string solutions. The effective string coupling (exponential of the dilaton) does indeed decrease at small distances for the latter solutions.

It turns out that these solutions are related by dimensional reduction to extremal electric dilatonic black holes. In particular, the $D = 5$ solution describing the field of a

---

[9] The same $D = 4$ black hole background is found by dimensionally reducing the $D = 5$ FS solution along $y = \frac{1}{2}(u + v)$ direction [50]. This is not surprising since the two (plane wave and FS) $D = 5$ backgrounds are related by duality.

[10] Dimensional reduction along $u + v$ direction gives equivalent solution [32].

source represented by a closed fundamental string winding around a compact 5-th dimension appears to be equivalent to the $D = 4$ extremal dilatonic black hole with an electric charge of Kaluza-Klein origin. As a result, this extremal black hole remains string solution to all orders in α'.

There are various indications that extremal black holes form a class of "solitonic" string solutions which are very different from non-extremal ones. It appears that non-extremal black holes (in particular, the Schwarzschild one) *are* modified by α' corrections in an essential way. A difficult unsolved problem is to find an exact classical string solution which generalises the Schwarzschild solution to all orders in α'. It is usually conjectured that it is quantum corrections that should be responsible for elimination of black hole singularities. However, in non-local theories like string theory singularities may be absent already at the classical level. By analogy with the open string case, one may expect that the modification of the Schwarzschild black hole may be similar to the one of the Coulomb point-charge solution in the Born-Infeld effective theory: the singularity at the origin disappears being "regularised" by the string non-locality scale $\sqrt{\alpha'}$.

7. Acknowledgement

I would like to thank G. Horowitz for useful discussions and collaboration and acknowledge also the support of PPARC.

8. References

[1] J. Scherk, *Nucl. Phys.* **B31** (1971) 222; J. Scherk and J. Schwarz, *Nucl. Phys.* **B81** (1974) 118; T. Yoneya, *Progr. Theor. Phys.* **51** (1974) 1907.
[2] E. Fradkin and A. Tseytlin, *Phys. Lett.* **B158** (1985) 316; *Nucl. Phys.* **B261** (1985) 1.
[3] E. Witten, *Phys. Lett.* **B149** (1984) 351.
[4] G. Gibbons, *Nucl. Phys.* **B207** (1982) 337.
[5] G. Gibbons and K. Maeda, *Nucl. Phys.* **B298** (1988) 741.
[6] D. Garfinkle, G. Horowitz and A. Strominger, *Phys. Rev.* **D43** (1991) 3140; **D45** (1992) 3888(E).
[7] C. Callan, R. Myers and M. Perry, *Nucl. Phys.* **B311** (1988) 673; R. Myers, *Nucl. Phys.* **B289** (1987) 701.
[8] C. Lovelace, *Phys. Lett.* **B135** (1984) 75; *Nucl. Phys.* **B273** (1986) 413.
[9] C. Callan, D. Friedan, E. Martinec and M. Perry, *Nucl. Phys.* **B262** (1985) 593.
[10] E. Witten, *Phys. Rev.* **D44** (1991) 314.
[11] G. Horowitz and D. Welch, *Phys. Rev. Lett.* **71** (1993) 328; N. Kaloper, *Phys. Rev.* **D48** (1993) 2598.

[12] G. Horowitz and A. Strominger, *Nucl. Phys.* **B360** (1991) 197.

[13] C. Callan, J. Harvey and A. Strominger, *Nucl. Phys.* **B359** (1991) 611.

[14] R. Khuri, *Nucl. Phys.* **B387** (1992) 315; *Phys. Lett.* **B294** (1992) 325; J. Gauntlett, J. Harvey and J. Liu, *Nucl. Phys.* **B409** (1993) 363.

[15] M. Duff, R. Khuri, R. Minasyan and J. Rahmfeld, *Nucl. Phys.* **B418** (1994) 195.

[16] S. Giddings, J. Polchinski and A. Strominger, *Phys. Rev.* **D48** (1993) 5784.

[17] C. Jonson and R. Myers, "Taub-NUT dyons in heterotic string theory", IASSNS-HEP-94/50, hep-th/9406069.

[18] M. Gürses, *Phys. Rev.* **D46** (1992) 2522.

[19] P. Horava, *Phys. Lett.* **B278** (1992) 101.

[20] D. Gershon, *Phys. Rev.* **D49** (1994) 999.

[21] G. Horowitz and A. Tseytlin, "Extremal black holes as exact string solutions", preprint Imperial/TP/93-94/51, UCSBTH-94-24, hep-th/9408040.

[22] A. Dabholkar, G. Gibbons, J. Harvey and F. Ruiz Ruiz, *Nucl. Phys.* **B340** (1990) 33.

[23] D. Waldram, *Phys. Rev.* **D47** (1993) 2528.

[24] R. Metsaev and A. Tseytlin, *Nucl. Phys.* **B293** (1987) 385.

[25] A. Tseytlin, *Phys. Lett.* **B176** (1986) 92; *Nucl. Phys.* **B276** (1986) 391.

[26] D. Gross and E. Witten, *Nucl. Phys.* **B277** (1986) 1.

[27] A. Tseytlin, "String cosmology and dilaton", in: *String Quantum Gravity and Physics at the Planck scale, Proceedings of the 1992 Erice workshop*, ed. N. Sánchez (World Scientific, Singapore, 1993) p.202; hep-th/9206067; D1, 223, 1992, hep-th/9203033.

[28] T. Damour and A. Polyakov, *Nucl. Phys.* **B423** (1994) 532.

[29] G. Horowitz, in: *String Theory and Quantum Gravity '92, Proc. of the 1992 Trieste Spring School*, ed. J. Harvey et al. (World scientific, Singapore, 1993); hep-th/9210119.

[30] M. Cvetič and A. Tseytlin, *Nucl. Phys.* **B416** (1994) 137.

[31] G. Horowitz and A. Tseytlin, "On exact solutions and singularities in string theory", preprint Imperial/TP/93-94/38; hep-th/9406067.

[32] G. Horowitz and A. Tseytlin, "A new class of exact solutions in string theory", preprint Imperial/TP/93-94/54, UCSBTH-94-31, hep-th/9409021.

[33] C. Holzhey and F. Wilczek, *Nucl. Phys.* **B380** (1992) 447.

[34] G. Gibbons, talks at LMS Symposium "Quantum concepts in time and space", Durham, July 1994 and International Congress of Mathematical Physics, Paris, July 1994.

[35] R. Kallosh, A. Linde, T. Ortín, A. Peet and A. Van Proeyen, *Phys. Rev.* **D46** (1992) 5278.

[36] B. Harms and Y. Leblanc, "Conjectures on nonlocal effects in string black holes", UAHEP-935, hep-th/9307042.

[37] E. Fradkin and A. Tseytlin, *Phys. Lett.* **B163** (1985) 123.

[38] O. Andreev and A. Tseytlin, *Nucl. Phys.* **B311** (1988) 205; *Mod. Phys. Lett.* **A3** (1988) 1349.

[39] M. Born, *Proc. Roy. Soc.* **A 143** (1934) 410; M. Born and L. Infeld, *Proc. Roy. Soc.* **A 144** (1934) 425.

[40] R. Güven, *Phys. Lett.* **B191** (1987) 275; D. Amati and C. Klimčík, *Phys. Lett.* **B219** (1989) 443; G. Horowitz and A. Steif, *Phys. Rev. Lett.* **64** (1990) 260.

[41] A. Tseytlin, *Nucl. Phys.* **B390** (1993) 153.

[42] E. Bergshoeff, R. Kallosh and T. Ortín, *Phys. Rev.* **D47** (1993) 5444; E. Bergshoeff, I. Entrop and R. Kallosh, *Phys. Rev.* **D49** (1994) 6663.

[43] J. Horne, G. Horowitz and A. Steif, *Phys. Rev. Lett.* **68** (1992) 568.

[44] A. Tseytlin, *Phys. Lett.* **B251** (1990) 530.

[45] A. Tseytlin, "Exact string solutions and duality", to appear in: *Proceedings of the 2nd Journée Cosmologie, Observatoire de Paris, June 2-4, 1994*, ed. H. De Vega and N. Sánchez (World Scientific, Singapore); hep-th/9407099.

[46] C. Callan and R. Khuri, *Phys. Lett.* **B261** (1991) 363; R. Khuri, *Nucl. Phys.* **B403** (1993) 335.

[47] J. Gauntlett, J. Harvey, M. Robinson and D. Waldram, *Nucl. Phys.* **B411** (1994) 461.

[48] A. Sen, *Nucl. Phys.* **B388** (1992) 457.

[49] D. Garfinkle, *Phys. Rev.* **D46** (1992) 4286.

[50] J. Gauntlett, talk presented at the conference "Quantum Aspects of Black Holes", Santa Barbara, June 1993.

ENTROPY OF BLACK HOLES

VALERI FROLOV

CIAR Cosmology Program; Theoretical Physics Institute,
*University of Alberta, Edmonton, Canada T6G 2J1**

Abstract

The problem of the dynamical origin of the entropy of a black hole is discussed. Black hole's internal degrees of freedom are identified with states of all physical fields (including the gravitational one) propagating in the black hole's interior. The expression for the density matrix $\hat{\rho}^H$ of a black-hole is given and the statistical-mechanical entropy of a black hole S^{SM}, defined as $S^{SM} = -\mathrm{Tr}(\hat{\rho}^H \ln \hat{\rho}^H)$, is calculated. For the natural choice of the cut-off parameter, connected with quantum fluctuations of the horizon, S^{SM} is of the same order of magnitude as the Bekenstein-Hawking entropy S^{BH}. Different fields (including the gravitational one) give independent contribution to S^{SM}. Since the geometry of a black hole depends on the mass parameter M, which in the state of thermal equilibrium in its turn is determined by the temperature T, the operations of differentiation with respect to the temperature T and tracing over the states of a black hole of mass M do not commute. Hence the stardard statistical-mechanical proof of equality of S^{SM} and the thermodynamical entropy S^{TD} is not valid for black holes. As the result for any physical field the leading (divergent) terms of one-loop contributions to S^{BH} (which is the standard thermodynamical entropy S^{TD} determined for a black hole) vanish and the Bekenstein-Hawking entropy S^{BH} is defined by the tree-level expression. This property allows to give a simple explanation of the universality of S^{BH}.

1 Black-Hole Entropy Problem

According to the thermodynamical analogy in black hole physics, the entropy of a black hole in the Einstein theory of gravity equals $S^{BH} = A_H/(4l_P^2)$, where A_H is the area of a black hole surface and $l_P = (\hbar G/c^3)^{1/2}$ is the Planck length [1, 2]. In black hole physics the Bekenstein-Hawking entropy S^{BH} plays essentially the same role an

*E-mail address: frolov@phys.ualberta.ca

in the usual thermodynamics. In particular it allows one to estimate what part of the internal energy of a black hole can be transformed into work. Four laws of black hole physics which form the basis in the thermodynamical analogy were formulated in [3]. The generalized second law [1, 2, 4] (see also [5, 6, 7, 8] and references therein) implies that when a black hole is a part of the thermodynamical system the total entropy (i.e. the sum of the entropy of a black hole and the entropy of the surrounding matter) does not decrease. The success of the thermodynamical analogy in black hole physics allows one to hope that this analogy may be is even deeper and it is possible to develop statistical-mechanical foundation of black hole thermodynamics.

Thermodynamical and statistical-mechanical definitions of the entropy are logically different. *Thermodynamical entropy* S^{TD} is defined by the response of the free energy F of the system on the change of its temperature:

$$dF = -S^{TD} dT. \tag{1.1}$$

(This definition applied to a black hole determines its Bekenstein-Hawking entropy.)

Statistical-mechanical entropy S^{SM} is defined as

$$S^{SM} = -\mathrm{Tr}(\hat{\rho} \ln \hat{\rho}), \tag{1.2}$$

where $\hat{\rho}$ is the density matrix describing the internal state of the system under consideration. It is also possible to introduce the *informational entropy* S^I by counting different possibilities to prepare a system in a final state with given macroscopical parameters from different initial states

$$S^I = -\sum_n p_n \ln p_n, \tag{1.3}$$

with p_n being the probabilities of different initial states. In standard case all three definitions give the same answer.

Is the analogy between black holes thermodynamics and the 'standard' thermodynamics complete? Do there exist internal degrees of freedom of a black hole which are responsible for its entropy? Is it possible to apply the statistical-mechanical and informational definitions of the entropy to black holes and how are they related with the Bekenstein-Hawking entropy? These are the questions which are to be answered.

Historically first attempts of the statistical-mechanical foundation of the entropy of a black hole were connected with the informational approach [2, 9]. According to this approach the black hole entropy is interpreted as "the logarithm of the number of quantum mechanically distinct ways that the hole could have been made"[9, 5]. The so defined informational entropy of a black hole is simply related with the amount of information lost by stretching the horizon, and as was shown by Thorne and Zurek it is equal to the Bekenstein-Hawking entropy [9, 5].

The dynamical origin of the entropy of a black hole and the relation between the statistical-mechanical and Bekenstein-Hawking entropy have remained unclear. In the present talk I describe some new results obtained in this direction.

2 Euclidean Approach

The Euclidean approach provides the simplest way to obtain the expressions temperature and entropy of a black hole. The calculations in the framework of the Euclidean approach initiated by Gibbons and Hawking [10, 11] relate the entropy of a black hole with the tree-level contribution of the gravitational action, namely the action of the Euclidean black hole instanton. This approach is based on the following simple observation. The Wick's rotation of time $t \to -i\tau$ transforms the Schwarzschild metric

$$ds^2 = -Fdt^2 + F^{-1}dr^2 + r^2 d\omega^2, \tag{2.1}$$

$$d\omega^2 = d\theta^2 + \sin^2\theta d\varphi^2, \qquad F = 1 - 2M/r, \tag{2.2}$$

into its Euclidean version

$$ds^2 = Fd\tau^2 + F^{-1}dr^2 + r^2 d\omega^2. \tag{2.3}$$

The latter is regular at the horizon $r = r_+ = 2M$ (where $F = 0$) only provided τ is periodic with the period $2\pi/\kappa$ (κ is the surface gravity of a black hole, for the Schwarzschild black hole $\kappa = 1/4M$). A propagator for a field ϕ in the Schwarzschild black hole background obtained by analytical continuation from the regular Euclidean black hole space is automatically periodic in imaginary time it with the period $\beta = 2\pi/\kappa$. This property implies in particular that a black hole is in thermal equlibrium with surrounding thermal radiation of the field ϕ, provided the temperature of radiation, measured at infinity, is $T_{BH} = \beta^{-1} = \kappa/2\pi$.

The partition function Z for the field ϕ ($Z \equiv \exp(-\beta F)$, F is the free energy) can be expressed either as

$$Z = \mathrm{Tr}\, e^{-\beta\hat{H}} = \sum \exp(-\beta E_n), \tag{2.4}$$

where E_n is the energy (eigenvalue of the Hamiltonian $\hat{H}$ for the field ϕ), or as the Euclidean path integral

$$Z = \int D[\phi] \exp(-I_E[\phi]), \tag{2.5}$$

where the integration is taken over all fields ϕ that are real on the Euclidean section and are periodic in the imaginary time coordinate τ with period β. Here $D[\varphi]$ is the measure of the space of fields ϕ and I_E is the Euclidean action of the field configuration $\phi(x)$[1]. In more general case when there are several fields, ϕ must be understood as the collective variable describing all the fields. In particular one needs also to include integration over the gravitational perturbations. The natural generalization is the integration in Eq.(2.5) over all the gravitational fields configurations (not only perturbations) obeying the periodicity conditions. The corresponding action I_E includes

[1]Strictly speaking the path integral (2.5) depends upon renormalization. Allen [14] has shown that $\ln Z$ for Z defined by Eq.(2.4) and Eq.(2.5) can differ by a term linear in β which does not contribute to the entropy.

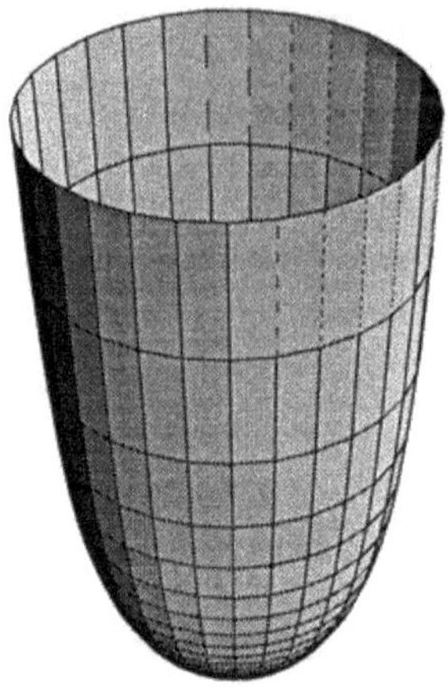

Figure 1: The imbedding diagram for the two-dimensional $\tau - r$ section of the Hawking-Gibbons instanton.

the Euclidean Einstein action. The state of the system is determined by the choice of the boundary conditions on the metrics that one integrates over. For the canonical ensemble for the gravitational fields inside a sperical box of radius r_B at temperature T one must integrate over all the metrics inside r_B which are periodically identified in the imaginary time direction with period $\beta = T^{-1}$. Such a partition function must describe in particular the thermal ensemble of black holes.

By using the stationary-phase approximation one gets

$$\beta F \equiv -\ln Z = I_E[\phi_0] - \ln Z_1 + \ldots . \tag{2.6}$$

Here ϕ_0 is the (generally speaking, complex) solution of classical field equations for action $I[\phi]$ obeying the required periodicity and boundary conditions. Besides the tree-level contribution $I_E[\phi_0]$, the expression (2.6) includes also one-loop corrections $\ln Z_1$, connected with the contributions of the fields perturbations on the background ϕ_0, as well as higher order terms in loops expansion, denoted by $(\ldots)$. For the above described canonical ensemble of gravitational fields the leading tree-level contribution is given by the Euclidean gravitational action for the Euclidean black hole solution (the so called Hawking-Gibbons instanton). The imbedding diagram for the two-dimensional $\tau - r$ section of the Hawking-Gibbons instanton is shown in Fig.1.

The tree-level contribution of the black hole to the free energy of the system is [12, 13]

$$F_0 = r_B\left(1 - \sqrt{1 - r_+/r_B}\right) - \pi r_+^2 \beta^{-1}. \tag{2.7}$$

Here $r_+ = 2M$ is the gravitational radius of a black hole of mass M, which for a given temperature β^{-1} at the boundary r_B is defined by the relation $\beta = 4\pi r_+(1-r_+/r_B)^{1/2}$. According to its definition the Bekenstein-Hawking (thermodynamical) entropy of a black hole S^{BH} is defined by the response of the free energy of a system including a black hole on the change of the temperature. One can easily verify that

$$S^{BH} \equiv S^{TD} = -\frac{dF_0}{dT} \equiv \beta^2 \frac{dF_0}{d\beta} = \frac{A_H}{4l_P^2}. \tag{2.8}$$

(It is assumed that r_+ in F_0 is expessed in terms of β and r_B before differentiation with respect to β.) One-loop and higher-loops contributions in Eq.(2.6) describe corrections to the total entropy of the system including a black hole due to the presence of radiation in its exterior. In particular for one-loop contribution of a field ϕ one has the following expression $\ln Z_1 = \frac{1}{2}\mathrm{Tr}\ln \hat{D}$, where $\hat{D}$ is the field operator for the field ϕ inside the box r_B. The one-loop contribution contains divergences and required the renormalization. Since the Euclidean black hole background is regular the corresponding contribution $\ln Z_1$ after its renormilazation is finite. Analogous calculations using spacetime complexification allows one to obtained the thermodynamical entropy for a rotating charged black hole.

3 Dynamical Degrees of Freedom

In the Euclidean approach the entropy of a black hole is defined by a tree-level contribution to the partition function and it has pure topological origin. It remains unclear whether there exist any real dynamical degrees of freedom which are responsible for it. The problem of the dynamical origin of the black hole entropy was intensively discussed recently. The basic idea which was proposed is to relate the dynamical degrees of freedom of a black hole with its quantum excitations. This idea has different realizations[15, 16, 17, 18, 19, 20][2]. Here I discuss the recent proposal [16, 20] to identify the dynamical degrees of freedom of a black hole with the states of all fields (including the gravitational one) which are propagating inside the black hole.

In order to specify the corresponding internal states consider at first the process of quantum particles creation by the gravitational field of a black hole. This process can be described as the effect of parametric excitations of zero-point fluctuations propagating in the time-dependent gravitational field of a black hole. Fig.2 illustrates such a process. Particles (shown by waved lines) are created in pairs. The creation of a particle outside the event horizon is necessarily accompined by a creation of another particle inside the horizon. The latter particle has negative total energy[3].

[2]For recent review of the problem of the dynamical origin of the entropy of a black hole, see [21].

[3]It should be reminded that any isolated black hole at late time after its formation is stationary. The energy is defined with respect to the corresponding Killing vector. For a static black hole the Killing vector is spacelike inside the event horizon, which makes it possible the existence of the negative energy states.

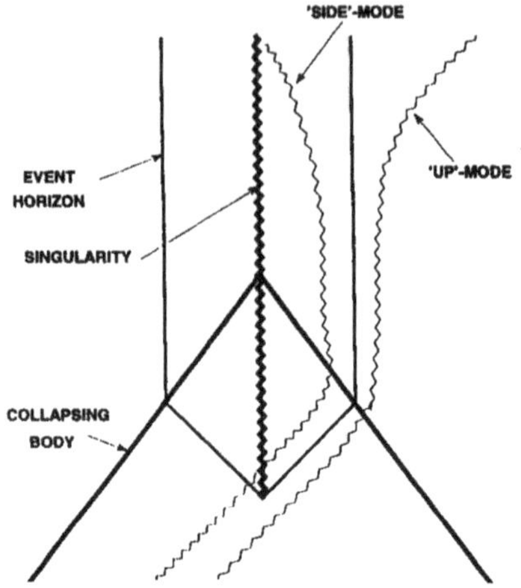

Figure 2: Spacetime of a spherically symmetric black hole formed by gravitational collapse of matter.

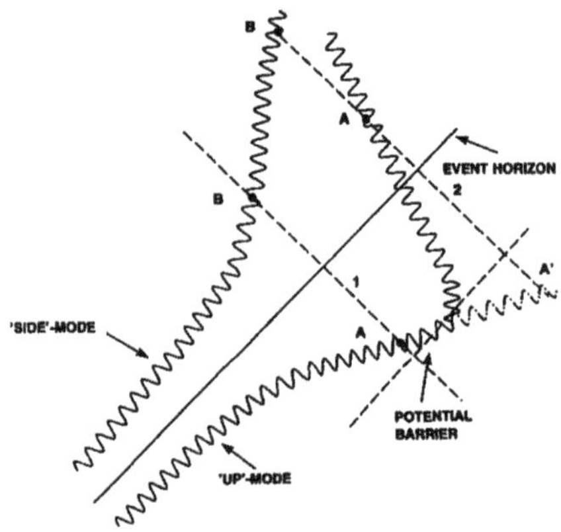

Figure 3: A pair created by a black hole consists of two particles: external (A) and internal (B). The state of the particle A is described by an 'up'-mode, while the state of the particle B is described by a 'side'-mode. Almost all the particles A are reflected by the gravitational potential barrier and fall down inside a black hole. Those particles which succeed to penetrate the gravitational potential barrier form the Hawking radiation (A'). At the moment of time 1, when the components of the pair areseparated by the event horizon, the pair contribute to the entropy of a black hole. At later time 2, when both of the particles are located inside the event horizon they are described by pure state and do not contribute to the entropy of the black hole.

As the result of the Hawking process of pairs creation these modes with negative energies are permanently excited and as we shall see later their state is described by a thermal density matrix. As for the particles created outside a black hole (external particles), only very small their number can penetrate the potential barrier and reach infinity. Namely those particles form the Hawking radiation of a black hole. All other external particles are reflected by the potential barier and fall down into the black hole (see Fig.3). During the time when they are still outside the horizon, the corresponding internal modes (which are described by a thermal density matrix) give the contribution to the black hole entropy.

In order to make the definition of the black hole entropy more concrete we assume that there exists a stationary black hole and denote by $\hat{\rho}^{\rm init}$ the density matrix describing in the Heisenberg representation the initial state of quantum fields propagating in its background. One may consider e.g. the in-vacuum state for a black hole evaporating in the vacuum, or the Hartle-Hawking state for a black hole in equilibrium with thermal radiation. For an exterior observer the system under consideration consists of two parts: a black hole and radiation outside of it. The state of radiation outside the black hole is described by the density matrix which is obtained from $\hat{\rho}^{\rm init}$ by averaging it over the states which are located inside the black hole and are invisible in its exterior

$$\hat{\rho}^{\rm rad} = {\rm Tr}^{\rm inv}\hat{\rho}^{\rm init}. \tag{3.1}$$

For an isolated black hole this density matrix $\hat{\rho}^{\rm rad}$ in particular describes its Hawking radiation at infinity. For a black hole in thermal equilibrium with radiation inside a cavity the density matrix $\hat{\rho}^{\rm rad}$ describes the state of thermal radiation.

Analogously we define the density matrix describing the state of a black hole as

$$\hat{\rho}^{H} = {\rm Tr}^{\rm vis}\hat{\rho}^{\rm init}. \tag{3.2}$$

The trace-operators ${\rm Tr}^{\rm vis}$ and ${\rm Tr}^{\rm inv}$ in these relations mean that the trace is taken over the states located either outside ('visible') or inside ('invisible') the event horizon, correspondingly. We define the entropy of a black hole as

$$S^{H} = -{\rm Tr}^{\rm inv}(\hat{\rho}^{H}\ln\hat{\rho}^{H}). \tag{3.3}$$

The proposed definition of the entropy of a black hole is similar to the definition of the entropy of a usual black body (Fig.4). The definition is invariant in the following sense. Independent changes of vacuum definitions for 'visible' and 'invisible' states do not change the value of S^{H}. Bogolubov's transformations describing an independent changes of the vacuum states inside and outside the black hole can be represented by the unitary operator $\hat{U} = \hat{U}^{\rm vis}\otimes\hat{U}^{\rm inv}$, where $\hat{U}^{\rm vis}$ and $\hat{U}^{\rm inv}$ are unitary operators in the Hilbert spaces of 'visible' and 'invisible' particles, correspondingly. The above used trace operators are invariant under such transformations.

In order to define the states one usually use modes expansion. The modes are characterized by a complete set of quantum numbers. Due to the symmetry properties

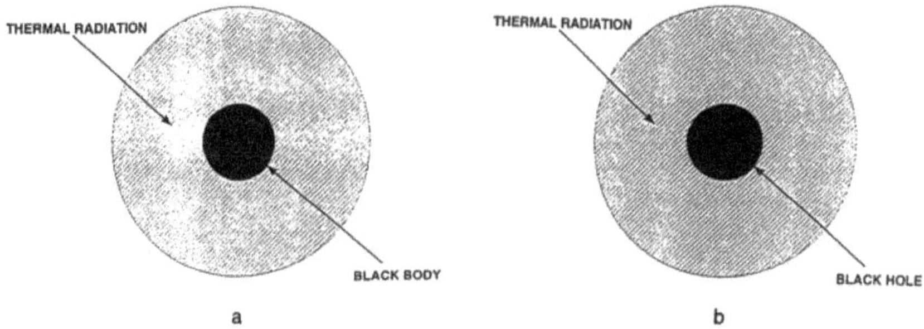

Figure 4: Black-hole entropy definition. Consider a thermal radiation inside a cavity which is in thermal equilibrium either with a black body (a) or with a black hole (b). For both systems their equilibrium state is described by some total density matrix. The density matrix for a black body can be obtained from the total density matrix by tracing it with respect to the degrees of freedom of the radiation outside the body. In a similar manner the black hole density matrix, which is used to determine its entropy, is defined by tracing the total density matrix of the system with respect to the states of the fields (radiation) in the black hole exterior .

one can choose such a subset J of quantum numbers connected with conservation laws (such as orbital and azimuthal angular momenta, helicity and so on) that guarantees the factorization of the density matrices. In the absence of mutual interaction of different fields the subset J necessarily includes also the parameters identifying the type of the field (e.g. mass, spin, and charge). The factorization in particular means that

$$\hat{\rho}^{\text{init}} = \otimes_J \hat{\rho}_J^{\text{init}}, \tag{3.4}$$

where $\hat{\rho}_J^{\text{init}}$ is acting in the Hilbert space $\mathcal{H}_J$ of states with the chosen quantum numbers J, while the complete Hilbert space is $\mathcal{H} = \otimes_J \mathcal{H}_J$. The factorization also means that the separation into 'visible' and 'invisible' states can be done independently in each subset of modes with a fixed J so that

$$\hat{\rho}^H = \otimes_J \hat{\rho}_J^H \qquad \hat{\rho}_J^H = \text{Tr}_J^{\text{vis}} \hat{\rho}_J^{\text{init}}, \tag{3.5}$$

where all the operators with subscript J are acting in the Hilbert space $\mathcal{H}_J$[4].

[4]Instead of non-normalizable monochromatic waves $f_{\omega lm}$ with fixed frequency ω it is convenient to consider normalized wave packets constructed from the monochromatic waves

$$f_{jnlm} = \delta^{-1/2}\kappa^{-1} \int_{j\delta}^{(j+1)\delta} \exp(2\pi i n \kappa^{-1}\omega/\delta) f_{\omega lm}\, d\omega,$$

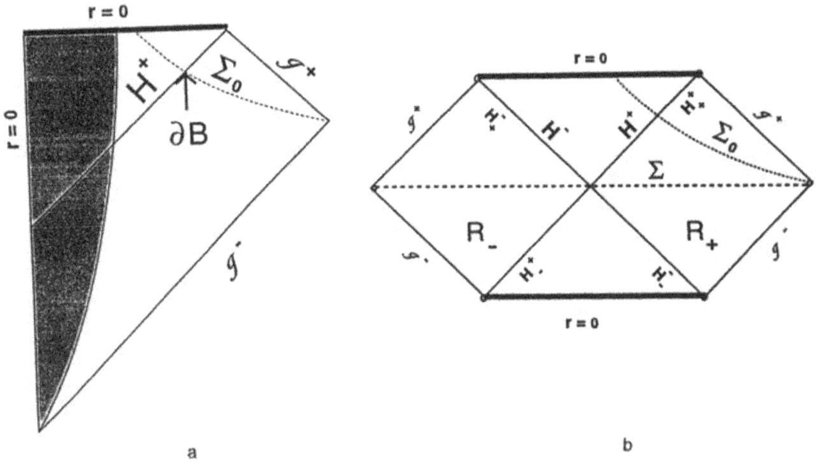

Figure 5: Penrose diagrams for a Schwarzschild-Kruskal black hole formed by gravitational collapse of matter (a) and for its 'eternal version' (b).

We illustrate the main steps of the calculations of the contribution of a given field to the density matrix of a black hole for a spherically-symmetric black hole. (The presence of charge and rotation does not create problems.) Moreover for simplicity we assume that a black hole is surrounded by a spherical mirror-like boundary of size $r_B < 3M$, so that the black hole is in thermal equilibrium with the radiation inside the cavity and the state of the system is described by the Hartle-Hawking vacuum state. The easiest way to introduce this state as well as to give definition of the 'up'- and 'side'-modes we use to describe the states of 'visible' and 'invisible' particles one can use the following useful trick proposed by Hawking in his original paper on the black hole evaporation [22]. At late time after the formation of a black hole the spacetime with the high accuracy is described by a stationary (in our case static) metric. For a given black hole we define its 'eternal version' as a spacetime of an eternal black hole which has the same global parameters (mass, charge, angular momentum) as the given black hole (see Fig.5). Modes of the field ϕ propagating at late-time in an 'original' black hole can be traced back in time in the spacetime of its 'eternal version' up to the initial global Cauchy surface Σ described by the equation $t = 0$, where t is the global time parameter defined by the Killing vector. 'Up'-modes are defined as positive (in time t) solutions vanishing in the left wedge R_-, and 'side'-modes are defined as negative (in time t) solutions vanishing in the right wedge R_+. In the

where $0 < \delta \ll 1$. In what follows we shall be working with these type wavepackets and assume that the collective index $J = (j, n, l, m)$. We denote by λ the collective intex $\lambda = (\omega, l, m)$.

presence of mirror-like boundaries B and B' they are to satisfy the corresponding boundary conditions at B and B'[5].

By using the linear combinations of the operators of creation and annihilation for 'up'- and 'side'-particles one can construct the operators which annihilate the Hartle-Hawking vacuum $|HH\rangle$. For our choice of the initial state $\hat{\rho}^{\text{init}} = |HH\rangle\langle HH|$. This density matrix can be evidently expressed as the function of the 'up' and 'side' creation and annihilation operators. In order to separate the states into 'visible' and 'invisible' we consider a spacelike or null surface Σ_0 which cross the event horizon at late time after black hole formation (Fig.5a). An 'up'-particle is 'visible' if it crosses the surface Σ_0 of a chosen moment of time outside the horizon. For this mode J, the density matrix $\hat{\rho}^H_J$, describing the state of the corresponding 'side'-particle, is obtained by tracing $\hat{\rho}^{\text{init}}_J$ over the Hilbert space of 'up'-particles with index J. As the result one obtains the thermal density matrix of the form [16]

$$\hat{\rho}^H_J = \rho^0_J \exp\left[-(\omega_J/T_H)\hat{\alpha}^{*SIDE}_J \hat{\alpha}^{SIDE}_J\right]. \tag{3.6}$$

In other words Hawking radiation of a black hole is accompined by thermal excitation of 'side' modes (black hole's internal degrees of freedom). For the 'up'-modes J crossing Σ_0 inside the horizon, both 'up' and 'side' particles are 'invisible'. We denote by $\hat{\rho}^\bullet_J$, the density matrix for such a pair . The density matrix $\hat{\rho}^\bullet_J$ evidently describes the pure state, so that such a pair does not contribute to the entropy of a black hole. For the total density matrix $\hat{\rho}^H$ of a black hole we have the following representation

$$\hat{\rho}^H = \prod_J{}' \hat{\rho}^H_J \prod_J{}'' \hat{\rho}^\bullet_J. \tag{3.7}$$

Here the prime indicates that the product includes only those states for which the 'up'-modes are 'visible', and double prime indicates that the product includes states for which the 'up'-modes are 'invisible'.

The expession (3.6) was obtained for the special choice of the initial state. For another choice of the initial state this expression must be modified. What is important that the modifications practically do not influent the expression for $\hat{\rho}^H_J$ for late-time modes J. A 'side'-mode propagating inside a black hole close to the horizon at late-time being traced back in time reaches $\mathcal{J}^-$ with a huge blue shift proportional to $\exp \kappa t$, where t is time past after the formation of the black hole. It means that in order to change the distribution for the mode J (e.g. to add an addinitional quantum in this state) one needs to send from $\mathcal{J}^-$ the excitation which has exponentiall large energy. At sufficiently late time this energy for given frequency is much larger than the mass of the black hole. For this reason the density matrix for 'side'-modes at late time will have the universal form (3.6).

[5] For more details concerning the definition of 'up' and 'side' modes and their relations with other standard modes ('in', 'out', and 'down') defined in a black hole's exterior, see [20]

4 Statistical-Mechanical Entropy

By using the density matrix of a black hole $\hat{\rho}^H$ one can calculate the corresponding statistical-mechanical entropy $S^{SM} = -\mathrm{Tr}(\hat{\rho}^H \ln \hat{\rho}^H)$. The main contribution to the entropy of a black hole is given by 'side' modes of fields located in the very close vicinity of the horizon. Contributions of different fields enter S^{SM} additively. The factorization property (3.5) allows to write

$$S^{SM} = \sum_J S_J^{SM}, \qquad S_J^{SM} = -\mathrm{Tr}_J^{\mathrm{inv}}(\hat{\rho}_J^H \ln \hat{\rho}_J^H), \tag{4.1}$$

where all the operators with subscript J are acting in the Hilbert space $\mathcal{H}_J$. In accordance with this decomposition one has

$$S^{SM} = \sum_J{}' s(\beta\omega_J), \tag{4.2}$$

where

$$s(\beta\omega) = \frac{\beta\omega}{e^{\beta\omega} - 1} - \ln(1 - e^{-\beta\omega}), \tag{4.3}$$

is the entropy of a single oscillator with the frequency ω at temperature $T = 1/\beta$ and the prime in sum indicates that the summation is taken over the modes for which the corresponding 'up'-particle is 'visible' at the chosen moment of time Σ_0. The terms $\hat{\rho}_J^{\bullet}$ correponding to a pure state of a pair inside a black hole do not contribute to S^{SM}.

Returning from wave packets J to monochromatic waves λ one obtains the following result for the contribution of a chosen field to S^{SM}

$$S^{SM} = \int d\boldsymbol{x} \sum_\lambda \mu_\lambda(\boldsymbol{x}) s(\beta\omega_\lambda). \tag{4.4}$$

Here

$$\mu_\lambda(\boldsymbol{x}) = g^{\tau\tau} g^{1/2} [R_\lambda(\boldsymbol{x})]^2 \tag{4.5}$$

is a phase space density of quantum modes and $R_\lambda(\boldsymbol{x})$ are spatial harmonics corresponding to the mode with a collective quantum number $\lambda = (\omega, l, m)$.

The so defined S^{SM} contains a divergence, connected with the integration over the space regions near the horizon and is of the form [16]

$$S^{SM} \approx \frac{\alpha}{\varepsilon} \tag{4.6}$$

where $\varepsilon = (l/r_+)^2$ and l is the proper distance cut-off parameter. For a conformal massless scalar field the dimensionless parameter $\alpha = 1/90$. For a fixed value l of the cut-off parameter, the value of S^{SM} does not depend on the particular choice of

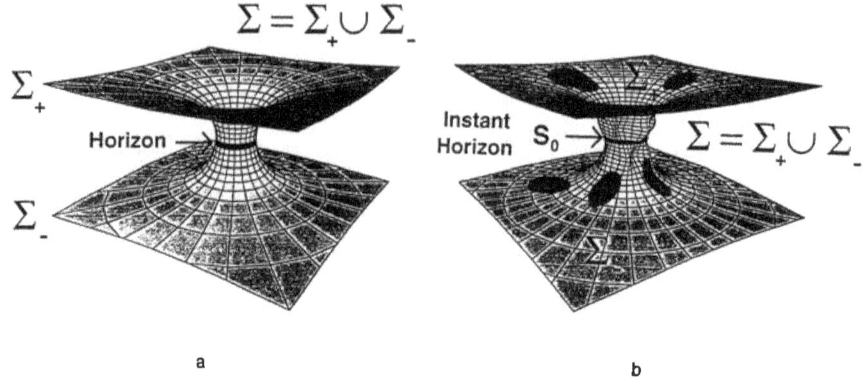

Figure 6: The Einstein-Rosen bridge for the Schwarzschild-Kruskal spacetime (a) and its deformation (b). The deformation arises due to the gravitational perturbations and presence of other fisical fields (schematically shown by black spots).

the surface Σ_0, which was introduced to define 'visible' and 'invisible' particles. One may expect that quantum fluctuations of the horizon may provide natural cut-off and make S^{SM} finite. Simple estimations of the cut-off parameter show that $l \approx l_P$ and $S^{SM} \approx S^{BH}$ [16][6].

5 No-Boundary Wave Function

Another approach to the problem of dynamical degrees of freedom of a black hole was proposed in Ref.[20]. Its basic idea is the following. The study of propagation of perturbations of physical fields in the spacetime of a real black hole (Fig.5a) can be reduced to the analogous problem for its 'eternal version' (Fig.5b). In particular one can trace back in time the perturbations in the space of the 'eternal version' until they cross the section Σ decribed by the equation $t = 0$. This section is known as the Einstein-Rosen bridge (Fig.6a). The number and properties of the physical fields depends on the particular model. In any case the gravitational field must be included. The initial data at Σ for the gravitational perturbations at Σ can be

[6]The loop expansion can be formulated as the expansion in powers of $\hbar$, and hence S^{SM}, which is one-loop quantity, must contain extra $\hbar$ with respect to the tree-level contribution S^{BH}. Nevertheless the calculations show that S^{SM} and S^{BH} are of the same order of magnitude. It happens because the cut-off parameter needed to make S^{SM} finite is also dependent on $\hbar$: $\varepsilon \sim \hbar$.

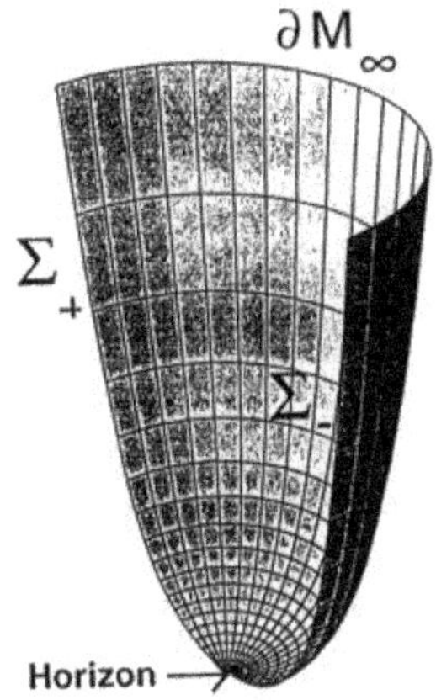

Figure 7: The imbedding diagram for the two-dimensional $\tau - r$ section of the half of the Hawking-Gibbons instanton.

related with small deformations of the geometry of the Einstein-Rosen bridge. The space of physical configurations of a system including a black hole can be related to the space of 'deformations' of the Einstein-Rosen bridge of the eternal black hole and possible configurations of other (besides the gravitational) fields on it, which obey the constrains and preserve asymptotic flatness (Fig.6b). In a spacetime of an 'eternal version' of a black hole, perturbations with initial data located on the inner part of the Einstein-Rosen bridge (Σ_-) are propagating to the future remaining entirely inside the horizon, and hence the corresponding perturbations in a 'physical' black hole also always remain under the horizon. That is why these data should be identified with internal degrees of freedom of a black hole. This construction allows natural generalization to the case when the deformations of the Einstein-Rosen bridge are not small. A quantum state of a black hole can be described by a wavefunction defined as a functional on the configuration space of deformations of the Einstein-Rosen bridge. In this representation deformations of the external (Σ_+) and internal (Σ_-) parts of the Einstein-Rosen bridge naturally represent degrees of freedom of matter outside the black hole and black hole's internal degrees of freedom.

The no-boundary ansatz (analogous to Hartle-Hawking ansatz in quantum cosmology) singles out a state which plays the role of a ground state of the system [20]. In order to describe this anzatz for a black hole we consider a half of the Hawking-Gibbons instanton shown in Fig.1, i.e. the space of the Euclidean black hole with the metric (2.3) and $\tau \in (0, \pi/\kappa)$ (Fig.7). Besides the boundary at infinity $\partial \boldsymbol{M}_\infty$ this Euclidean space $\boldsymbol{M}$ possesses only one boundary $\partial \boldsymbol{M}$ which is isometric to the

Einstein-Rosen bridge.

The no-boundary wavefunction of a black hole [20] is defined as a path integral

$$\boldsymbol{\Psi}({}^3g(\boldsymbol{x}),\varphi(\boldsymbol{x})) = \int \mathcal{D}\,{}^4g\;\mathcal{D}\phi\; \mathrm{e}^{-\boldsymbol{I}[{}^4g,\,\phi]} \tag{5.1}$$

of the exponentiated gravitational action $\boldsymbol{I}[{}^4g,\phi]$ over Euclidean 4-geometries and matter-field configurations on those spacetime histories of physical fields $\phi = \phi(x)$ on $\boldsymbol{M}$ that generate the Euclidean 4-geometries asymptotically flat at the infinity $\partial\boldsymbol{M}_\infty$ of spacetime and are subject to the conditions $({}^3g(\boldsymbol{x}),\,\varphi(\boldsymbol{x}))$, $\boldsymbol{x}\in\partial\boldsymbol{M}$, – the collection of 3-geometry and boundary matter fields on $\partial\boldsymbol{M}$, which are just the argument of the wavefunction (5.1). $\boldsymbol{I}[{}^4g,\phi]$ is the Lagrangian gravitational action in terms of these fields. The integration measure $\mathcal{D}\,{}^4g\;\mathcal{D}\phi$ involves the local functional measure the structure of which is not very important for our purposes.

By its construction the no-boundary wavefunction of a black hole is symmetric with respect to the transposition of the interior and exterior parts of the Einstein-Rosen bridge. We call this property *duality*. For a 'real' black hole formed in the gravitational collapse, this exact symmetry is broken. Nevertheless, since there is a close relation between physics of a 'real' black hole and its 'eternal version', the duality of the above type plays an important role and allows one, for example, to explain why the approach based on identifying the dynamical degrees of freedom of a black hole with its external modes gives formally the same answer for the dynamical entropy of a black hole as our approach.

In the semiclassical approximation the no-boundary wavefunction Eq.(5.1) takes the form

$$\Psi(\varphi, M) = P\,\mathrm{e}^{-2\pi M^2 - \boldsymbol{I}_2[\phi(\varphi)]}\,, \tag{5.2}$$

where $\boldsymbol{I}_2[\phi(\varphi)]$ is a quadratic term of the action in the linearized physical fields, and $\phi(\varphi)$ is a solution of the corresponding field equations on $\boldsymbol{M}$ matching the boundary conditions φ on $\partial\boldsymbol{M}$. Denote by $\varphi_{\lambda,\pm}$ the coefficients in the decomposition of the field $\varphi(\boldsymbol{x})$ on $\Sigma_\pm$ (an external and internal parts of the Einstein-Rosen bridge $\Sigma = \partial\boldsymbol{M}$) in the basis of spatial harmonics $\varphi_\pm(\boldsymbol{x}) = \sum_\lambda \varphi_{\lambda,\pm} R_\lambda(\boldsymbol{x})$. Then the quadratic form $\boldsymbol{I}_2[\phi(\varphi)]$ is

$$\boldsymbol{I}_2(\varphi_+,\varphi_-) = \frac{1}{2}\sum_\lambda \left\{ \frac{\omega_\lambda \cosh(\beta_0\,\omega_\lambda/2)}{\sinh(\beta_0\,\omega_\lambda/2)}\,(\varphi^2_{\lambda,+} + \varphi^2_{\lambda,-}) - \frac{2\,\omega_\lambda}{\sinh(\beta_0\,\omega_\lambda/2)}\,\varphi_{\lambda,+}\varphi_{\lambda,-} \right\}. \tag{5.3}$$

For study the fields contribution to the statistical-mechanical entropy in the one-loop approximation it is sufficient to fix mass M of a black hole as a parameter in the wave function, and consider only the part describing fields perturbations. It is possible to show that for a fixed mass parameter M the part $\exp -\boldsymbol{I}_2(\varphi_+,\varphi_-)$ of the no-boundary wave function (5.2) with $\boldsymbol{I}_2(\varphi_+,\varphi_-)$ given by (5.3) describes the Hartle-Hawking vacuum state of the corresponding perturbations ϕ.

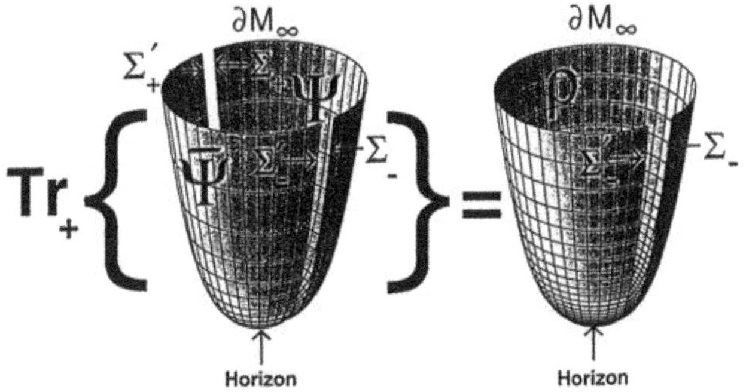

Figure 8: The wavefunction $\Psi(\varphi_+, \varphi_-)$ is defined as a path integral over the physical degrees of freedom in the τ-foliation of the Euclidean spacetime on a half-instanton. The density matrix $\rho(\varphi'_-, \varphi_-) = \mathrm{Tr}_+ |\Psi >< \Psi|$ is proportional to the analogous functional integral, but on the whole instanton, since the proposed wave function can be interpreted as an amplitude of an Euclidean evolution from the initial state φ_- to the final state φ_+ during the time interval $\beta/2$. Then $\overline{\Psi}(\varphi_+, \varphi'_-)$ implies the evolution from φ_+ to φ'_- and the density matrix is the kernel of the evolution from φ_- to φ'_- during the Euclidean time interval $0 < \tau < \beta$. The arguments of the density matrix $\varphi'_-(\boldsymbol{x}')$ and $\varphi_-(\boldsymbol{x})$ are the values of the fields on different sides $\boldsymbol{\Sigma}_-{}'$ and $\boldsymbol{\Sigma}_-$ of the cut in the instanton.

By tracing over the external variables one obtains the density matrix of a black hole and can calculate its statistical-mechanical entropy [20] (see Fig.8). The result coincides with (4.6). The reason why the calculations based on a 'real' black hole and on its 'eternal version' give the same result for the statistical-mechanical entropy of a black hole is the following. As it was already mentioned the late-time occupation of 'side'-modes does not depend of the particular choice of the initial state, so that to compare the results of the calculations one can choose the Hartle-Hawking state for the 'real' black hole, which is virtually the same as for the no-boundary wave function. On the other hand the main contribution to the entropy is connected with modes, propagating in the very close vicinity of the horizon, which are highly blue-shifted and for which the geometric-optics approximation works extremely well when one propagates these modes to the Einstein-Rosen bridge Σ in the eternal version of a black hole. For these reasons the counting of modes, contributing to the entropy for a 'real' black hole and for its 'eternal version' give the same answer.

6 Why the Entropy is $A/4$?

In the above approach (as well as in other dynamical approaches) different fields give independent contributions to S^{SM}. Even if we assume that the cut-off parameter depends on the number of fields in Nature it is virtually impossible to exclude the dependence of the expression for S^{SM} on the fields properties. This behavior of S^{SM} differs this quantity from the universally defined Bekenstein-Hawking entropy S^{BH}. This was considered by many as a puzzle. In Ref.[23] a simple solution to this puzzle was proposed: the standard relation $S^{TD} = S^{SM}$ is simply not valid for black holes. In order to explain the reason why the standard proof of this relation in the statistical mechanics is not directly applicable to black holes we reproduce here its main steps.

For the canonical ensemble one has

$$\exp(-\beta F) = \mathrm{Tr}(\exp(-\beta \hat{H}), \qquad \hat{\rho} = \exp[\beta(F - \hat{H})]. \tag{6.1}$$

The thermodynamical entropy S^{TD} is defined as

$$S^{TD} \equiv \beta^2 \frac{dF}{d\beta} = -\beta F - \beta \exp(\beta F) \frac{d[\exp(-\beta F)]}{d\beta}. \tag{6.2}$$

The first term in the right-hand-side of Eq.(6.2) can be written in the form $-\mathrm{Tr}(\beta F \hat{\rho})$. One can prove that $S^{TD} = -\mathrm{Tr}(\hat{\rho} \ln \hat{\rho})$ if one shows that the second term in the right-hand-side of Eq.(6.2) can be written in the form $\mathrm{Tr}(\beta \hat{H} \hat{\rho})$. It can be done if and only if the differentiation with respect to β commutes with Tr-operation. In the case of a black hole, where the number and properties of the internal states depend on the mass of a black hole, which in turn depends on the equilibrium temperature, this commutation property is not valid.

In order to demonstrate this one can write the contribution of a chosen field to the free energy of a black hole in the form, which is similar to (4.4)

$$F = \int d\boldsymbol{x} \sum_{\lambda} \mu_{\lambda}(\boldsymbol{x}) f(\beta \omega_{\lambda}) + \ldots, \tag{6.3}$$

$$f(\beta\omega) = \beta^{-1} \ln(1 - e^{-\beta\omega}), \tag{6.4}$$

where $(\ldots)$ denotes the terms independent of β which do not contribute to S^{SM}. After summation over l, m and integration over the spatial volume one gets

$$F = \int d\omega N(\omega | M, \varepsilon) f(\beta\omega) + \ldots, \tag{6.5}$$

where $N(\omega|M, \varepsilon)$ is the density of number of states. For the thermal equilibrium the mass M of a black hole is related with temperature ($\beta \equiv T^{-1} = 8\pi M$). That is why for the calculation of the response dF on the change of the temperature dT one needs to take into account additional dependence of N on T. This additional dependence of N on T reflects the fact that operations d/dT and Tr do not commute for our system. As the result S^{TD} differs from S^{SM} and one has[7] $S^{TD} = S^{SM} + \Delta S$.

[7]The independent of β terms denoted by dots in Eq.(6.3) also contribute to ΔS. It is interesting to note that this relation can be used to give a simple explanation of the entropy 'renormalization' procedure, which was postulated by Thorne and Zurek [9]

One can relate S^{SM} with the variation of the free energy $F(T, M)$ with respect to temperature for fixed value of the black hole mass M. The additional term ΔS is related with the variation of $F(T, M)$ with respect to M for fixed temperature T. One can obtain S^{TD} directly by varying with respect to T the expression $F(T, M(T))$, where the mass of a black hole is expressed in terms of the equilibrium temperature T. The latter expression $F(T, M(T))$ remains finite for $\varepsilon = 0$. In other words for each of the fields the additional term ΔS exactly cancels its leading (divergent near the horizon) contribution to the black hole statistical-mechanical entropy [23]. As the result of this cancellation one-loop corrections to the thermodynamical entropy of a black hole (describing the contribution to this quantity of the internal dynamical degrees of freedom of a black hole) are small. This explains the universality (independence on number and properties of fields) of the expression for the thermodynamical (Bekenstein-Hawking) entropy of a black hole.

7 Acknowledgements

This work was supported by the Natural Sciences and Engineering Research Council of Canada.

References

[1] J. D.Bekenstein, Nuov.Cim.Lett. **4** (1972) 737.

[2] J. D. Bekenstein, Phys.Rev. **D7** (1973) 2333.

[3] J. M. Bardeen, B. Carter, and S. W. Hawking, Comm.Math.Phys. **31**, (1973) 181.

[4] J. D. Bekenstein, Phys.Rev. **D9** (1974) 3292.

[5] K. S. Thorne, W. H. Zurek, and R. H. Price, in *Black Holes: The Membrane Paradigm* , edited by K. S. Thorne, R. H. Price, and D. A. MacDonald (Yale University Press, New Haven, 1986), p.280.

[6] I. Novikov and V. Frolov, *Physics of Black Holes.* (Kluwer Academic Publ., Dordrech-Boston-London), 1989.

[7] R. W. Wald, In:*Black Hole Physics* (Eds.V. DeSabbata and Z. Zhang), (Kluwer Academic Publ., Dordrech-Boston-London), 1992).

[8] V. Frolov and D. N. Page, Phys.Rev.Lett. **71** (1993) 3902.

[9] W. H. Zurek and K. S. Thorne, Phys.Rev.Lett. **54** (1985) 2171.

[10] G. W. Gibbons and S. W. Hawking, Phys.Rev. **D15** (1976) 2752.

[11] S. W. Hawking, In: *General Relativity: An Einstein Centenary Survey.* (eds. S. W. Hawking and W. Israel), Cambridge Univ.Press, Cambridge, 1979.

[12] J. W. York, Phys.Rev. **D33** (1986) 2092.

[13] H. W. Braden, J. D. Brown, B. F. Whiting, and J. W. Jork, Phys.Rev. **D42** (1990) 3376.

[14] B. Allen, Phys.Rev. **D33** (1986) 3640.
[15] G. 't Hooft, Nucl.Phys. **B256** (1985) 727.
[16] V. Frolov and I. Novikov, Phys.Rev. **D48** (1993) 4545.
[17] S. Carlip and C. Teitelboim, Preprint gr-qc/9312002 (1993).
[18] L. Susskind and J. Uglum, Phys.Rev. **D50** (1994) 2700.
[19] D. Garfinkle, S. B. Giddings, and A. Strominger, Phys.Rev. **D49** (1994) 958.
[20] A. I. Barvinsky, V. P. Frolov, and A. I. Zelnikov, preprint Thy 13-94, gr-qc/9404036 (1994), (to appear in Phys.Rev.D).
[21] J. D. Bekenstein, preprint gr-qc/9409015 (1994).
[22] S. W. Hawking, Comm.Math.Phys. **43**, (1975) 199.
[23] V. Frolov, preprint Thy 22-94, gr-qc/9406037 (1994).

CMU-94-31
DOR-ER/40682-85
10-1-94

QUANTUM FIELD THEORY IN A NON-EQUILIBRIUM SETTING: FORMALISM AND APPLICATIONS

D. Boyanovsky[(a)], H.J. de Vega[(b)] and R. Holman[(c)]

(a) Department of Physics and Astronomy, University of Pittsburgh, Pittsburgh, PA. 15260, U.S.A.
(b) Laboratoire de Physique Theorique et Hautes Energies[*] University Pierre et Marie Curie, Tour 16, 1er etage, 4, Place Jussieu 75252 Paris, Cedex 05, France
(c) Department of Physics, Carnegie Mellon University, Pittsburgh, PA. 15213, U. S. A.

Abstract

Applications of field theory to cosmology will typically require an understanding of quantum field theory in extreme, non-equilibrium settings. This can be seen for example in the production of topological defects via the Kibble mechanism, during the slow roll period of inflationary models, and, most especially, during the reheating phase in the post inflationary universe. I describe the formalism we have developed to address these problems and discuss some new results we have obtained with it.

1 Introduction

Some of the most interesting aspects of the confluence between particle physics and cosmology have to do with non-equilibrium phenomena. The inflationary Universe [1], at least in its new[2], and chaotic incarnations[3], requires that the so-called inflaton, whose energy density drives the supraluminal expansion, be so weakly coupled so as to have no chance to be in thermal equilibrium with its surroundings. Other examples of non-equilibrium behavior are the formation of topological defects via the Kibble mechanism[4] as well as the decay or thermal activation of a false vacuum state in quantum field theory.

Despite the fact that the above mentioned processes are inherently non-equilibrium phenomena, most of the work concerning them uses an equilibrium based formalism. In the case of inflation, the effective potential has been the tool most often used used to describe the evolution of the inflaton during its slow roll period, as well as during the reheating phase, while in the case of thermal activation of metastable states, the Matsubara imaginary time formalism is used to arrive at the activation rate (see for example [5]).

Clearly, non-equilibrium problems should be attacked with non-equilibrium tools. Such tools have been devised quite some time ago by Schwinger and independently by Keldysh[6]. They involve considering the time evolution of the density matrices describing the system.

By means of this formalism, known as the *Closed Time Path* (CTP) formalism, we should be able to treat the actual out of equilibrium evolution of any system (at least in principle!).

What I'd like to describe in this article is some work we have done on the application of the CTP formalism to the problem of understanding how dissipation occurs in quantum field theories[7]. This problem has a direct bearing on the problem of reheating in inflationary models, a problem that has recently received renewed attention, partly due to the fact that previous treatments of this problem have been shown to be inadequate in various ways[10].

We start in the next section by giving an overview of the CTP formalism. We show here why evolution in real time gives rise to the closed time contour and its concomitant doubling of fields. Section 3 then develops the treatment of dissipation in quantum field theories. We pay particular attention to the case in which a continous symmetry is spontaneously broken. In this case the preferred mode of energy dissipation is to create Goldstone modes. This gives rise to one of the more interesting results of our analysis, namely that the actual endpoint of the evolution of the zero mode of the radial component (for example in the $O(N)$ model), is very different than what one would expect just from an examination of the effective potential.

In section 4 we discuss some further applications of our work such as the "slow roll" period and reheating phases of inflationary models.

2 The Closed Time Path Formalism

To understand the origins of the CTP formalism, it is easiest to start from the evolution of the density matrix of the system. For simplicity, we first consider the case of one dimendional quantum mechanics; the generalization to field theory is straightforward. We have the relation, coming from the quantum Liouville equation:

$$\rho(t) = U(t,t_0)\rho(t_0)U^\dagger(t,t_0), \tag{1}$$

where $U(t,t_0) \equiv \exp(-iH_{\rm evol}(t-t_0))$ is the time evolution operator, and $H_{\rm evol}$ is the *evolution* Hamiltonian; we make the distinction in the Hamiltonians here since will consider the case in which for $t < t_0$ the system is described by an equilibrium density matrix with a given Hamiltonian H_0, while for $t > t_0$ the evolution is described by another.

Now the time evolution operator has a path integral representation[9]:

$$\langle x|U(t,t_0)|x'\rangle = \int \mathcal{D}x(\cdot)\ \exp(\frac{i}{\hbar}S[x(\cdot)]), \tag{2}$$

where the paths in the integral are those satisfying $x(t) = x, x(t_0) = x'$. Similarly, we have:

$$\langle x|U^\dagger(t,t_0)|x'\rangle = \int \mathcal{D}y(\cdot)\ \exp(-\frac{i}{\hbar}S[y(\cdot)]), \tag{3}$$

with $y(t_0) = x$, $y(t) = x'$. If we assume that the initial density matrix is that for a thermally distributed system with inverse temperature β, then $\rho(t_0) \propto \exp(-\beta H(t_0))$. The temperature dependence of the Hamiltonian is just that discussed above.

We can then write:

$$\begin{aligned}\langle x|\rho(t)|x'\rangle &= \int dx_1 dx_2 \langle x|U(t,t_0)|x_1\rangle\ \langle x_1|\rho(t_0)|x_2\rangle\ \langle x_2|U^\dagger(t,t_0)|x'\rangle \\ &= \int dx_1 dx_2 \langle x|U(t,t_0)|x_1\rangle\ \langle x_1|U(t_0-i\beta,t_0)|x_2\rangle\ \langle x_2|U^\dagger(t,t_0)|x'\rangle \qquad (4)\end{aligned}$$

From this result we can construct the trace of the (unnormalized) density matrix:

$$\begin{aligned}\mathrm{Tr}(\rho(t)) &= \int dx \langle x|\rho(t)|x\rangle = \\ &\int dx\ dx_1\ dx_2 \langle x_1|U(t_0-i\beta,t_0)|x_2\rangle\ \langle x_2|U(t_0,t)|x\rangle\ \langle x|U(t,t_0)|x_1\rangle. \qquad (5)\end{aligned}$$

The time evolution above corresponds to a time contour that starts at t_0 runs to t, goes back to t_0 and then finally runs down along the imaginary axis to $t_0 - i\beta$. The Matsubara formalism corresponds to just considering this last segment; it is the other two segments that allow us to address questions involving the time evolution of the system.

Since all of the important physical observables are averages taken with the above density matrix, you can easily convince yourselves that the relevant generating functional for this contour is (returning to the field theoretic case now):

$$\begin{aligned}Z[J^+,J^-,J^\beta] &= \int D\Phi D\Phi_1 D\Phi_2 \int D\Phi^+ D\Phi^- D\Phi^\beta e^{i\int_T^{T'}\{\mathcal{L}[\Phi^+,\mathcal{L}^+]-\mathcal{L}[\Phi^-,\mathcal{L}^-]\}} \times \\ &\exp\left(i\int_T^{T-i\beta_i} \mathcal{L}[\Phi^\beta, J^\beta]\right), \qquad (6)\end{aligned}$$

with the boundary conditions $\Phi^+(T) = \Phi^\beta(T-i\beta_i) = \Phi$, $\Phi^+(T') = \Phi^-(T') = \Phi_2$, $\Phi^-(T) = \Phi^\beta(T) = \Phi_1$. Here we have introduced sources J^+, J^-, J^β along the three segments of the contour, so as to allow for the construction of correlation functions and ensemble averages via the usual functional differentiation techniques. We can write the generating functional as:

$$\begin{aligned}Z[J^+,J^-,J^\beta] = \exp\left\{i\int_T^{T'} dt \left[\mathcal{L}_{int}(-i\delta/\delta J^+) - \mathcal{L}_{int}(i\delta/\delta J^-)\right]\right\} \times \\ \exp\left\{i\int_T^{T-i\beta_i} dt\mathcal{L}_{int}(-i\delta/\delta J^\beta)\right\}\exp\left\{\frac{i}{2}\int_c dt_1 \int_c dt_2 J_c(t_1)J_c(t_2)G_c(t_1,t_2)\right\}, \qquad (7)\end{aligned}$$

where J_c stands for the currents evaluated along the contour and the G_c's are the Green's functions along the contour[11, 12]; we have suppressed the spatial arguments.

By use of the Riemann-Lesbegue lemma, it can be shown that correlation functions involving fields on either the $\pm$ contours (running along the real axis, modulo the $i\epsilon$ prescription necessary to define the Green's functions) and the pure imaginary contour must vanish. Thus, in order to calculate real time correlation functions, we can write the generating functional in its final form[11]:

$$
\begin{aligned}
Z[J^+,J^-] &= \exp\left\{i\int_T^{T'} dt\left[\mathcal{L}_{int}(-i\delta/\delta J^+)-\mathcal{L}_{int}(i\delta/\delta J^-)\right]\right\}\times \\
&\exp\left\{\frac{i}{2}\int_T^{T'} dt_1\int_T^{T'} dt_2 J_a(t_1)J_b(t_2)G_{ab}(t_1,t_2)\right\}, \qquad (8)
\end{aligned}
$$

with $a, b = +, -$.

The Green's functions are to be found along the contour and are given by:

$$
\begin{aligned}
\langle T\Phi(\vec{r}_1,t_1)\Phi(\vec{r}_2,t_2)\rangle &= -iG^{++}(\vec{r}_1,t_1;\vec{r}_2,t_2) \qquad (9)\\
\langle \tilde{T}\Phi(\vec{r}_1,t_1)\Phi(\vec{r}_2,t_2)\rangle &= -iG^{--}(\vec{r}_1,t_1;\vec{r}_2,t_2) \qquad (10)\\
\langle \Phi(\vec{r}_1,t_1)\Phi(\vec{r}_2,t_2)\rangle &= iG^{+-}(\vec{r}_1,t_1;\vec{r}_2,t_2), \qquad (11)
\end{aligned}
$$

with T and $\tilde{T}$ the time ordering and anti-time ordering symbols. The Green's functions above are written in terms of the homogeneous solutions to the quadratic form of the Lagrangian (free fields) $G^>$, $G^<$ as

$$
\begin{aligned}
G^{++}(\vec{r}_1,t_1;\vec{r}_2,t_2) &= G^>(\vec{r}_1,t_1;\vec{r}_2,t_2)\Theta(t_1-t_2)+G^<(\vec{r}_1,t_1;\vec{r}_2,t_2)\Theta(t_2-t_1) \qquad (12)\\
G^{--}(\vec{r}_1,t_1;\vec{r}_2,t_2) &= G^>(\vec{r}_1,t_1;\vec{r}_2,t_2)\Theta(t_2-t_1)+G^<(\vec{r}_1,t_1;\vec{r}_2,t_2)\Theta(t_1-t_2) \qquad (13)\\
G^{+-}(\vec{r}_1,t_1;\vec{r}_2,t_2) &= -G^<(\vec{r}_1,t_1;\vec{r}_2,t_2) \qquad (14)\\
G^{-+}(\vec{r}_1,t_1;\vec{r}_2,t_2) &= -G^>(\vec{r}_1,t_1;\vec{r}_2,t_2) = -G^<(\vec{r}_2,t_2;\vec{r}_1,t_1) \qquad (15)\\
G^>(\vec{r}_1,t_1;\vec{r}_2,t_2) &= i\langle\Phi(\vec{r}_1,t_1)\Phi(\vec{r}_2,t_2)\rangle \qquad (16)\\
G^<(\vec{r}_1,T;\vec{r}_2,t_2) &= G^>(\vec{r}_1,T-i\beta_i;\vec{r}_2,t_2). \qquad (17)
\end{aligned}
$$

The condition (17) is recognized as the periodicity condition in imaginary time and is a result of considering an equilibrium situation for $t < t_0$. The functions $G^>$, $G^<$, which are the homogeneous solutions of the quadratic form, with appropriate boundary conditions, will be constructed explicitly below.

The Feynman rules are the same as in regular perturbation theory, with the proviso that there are *two interaction vertices* corresponding to the $\pm$ branches, with opposite signs (arising from the complex conjugation of the unitary time evolution operator for evolution backwards in time).

Having the techniques for computing non-equilibrium quantities in hand, we turn the the question of how to first define an order parameter for the system, and then how to arrive at its quantum corrected equation of motion.

To define the order parameter $\phi(t)$, we first integrate $\Phi(\vec{x},t)$ over all space to project the zero mode, and then coarse-grain over all fluctuations by taking the expectation value of this integrated quantity with respect to the state specified by $\rho(t)$. Since we can take the state to be translationally invariant, this is equivalent to setting:

$$\phi(t) = \langle \Phi(\vec{x}) \rangle \equiv \mathrm{Tr}(\rho(t)\Phi(\vec{x},t)). \tag{18}$$

To get the equation of motion for the order parameter $\phi(t)$ we can use the tadpole method[13].This entails expanding the field $\Phi(\vec{x},t)$ in terms of the order parameter and flutuations about it: $\Phi(\vec{x},t) = \phi(t) + \psi(\vec{x},t)$. We then insert this decomposition into the non-equilibrium generating functional and demand that $\langle \psi^{\pm}(\vec{x},t) = 0$. Suppose that we start with the standard quartic scalar field theory with Lagrangian:

$$\mathcal{L} = \frac{1}{2}\partial_\mu \Phi \partial^\mu \Phi - \frac{1}{2}m^2\Phi^2 - \frac{\lambda}{4!}\Phi^4. \tag{19}$$

Then the expectation value $\langle \psi^{\pm}(\vec{x},t)$ is computed using perturbation theory with the "non-equilibrium" Lagrangian:

$$\mathcal{L}[\phi+\psi^+]-\mathcal{L}[\phi+\psi^-] = \left\{\frac{\delta\mathcal{L}}{\delta\phi}\psi^+ + \mathcal{L}_0[\psi^+] - \lambda\left(\frac{\phi^2(\psi^+)^2}{4} + \frac{\phi(\psi^+)^3}{6} + \frac{(\psi^+)^4}{4!}\right)\right\} - \left\{\psi^+ \to \psi^-\right\}. \tag{20}$$

This formalism can be generalized to the case where continous symmetries, both local and global, exist.

In the next section we apply the above formalism to the description of dissipative processes in scalar field theories.

3 Dissipation in Field Theories

The equations of motion for a scalar field are second order in time, and thus would appear to be time- reversal invariant. However, it is hard to see how a time reversal invariant theory would lead to reheating in inflationary models, for example. Recall that at the end of inflation, the energy densities of matter and radiation have been redshifted to nearly zero, so that the universe is quite cold. In order to restart the standard Big-Bang cosmology, a large amount of entropy must be dumped after inflation is completed. The standard technique for doing this is to note that the zero mode of the inflaton is oscillating about the position of the true vacuum. These oscillations contain energy which, if it could be transferred to modes that could thermalize quickly (compared to the Hubble parameter), could reheat the Universe.

The standard lore[14] has it that if the inflaton is coupled to other, lighter, fields, then this coupling will induce a *dissipative* term that will allow for the conversion of coherent inflaton oscillation energy into thermal energy. The equation of motion will then become:

$$\ddot{\phi}(t) + \Gamma\dot{\phi}(t) + V'(\phi(t)) = 0, \tag{21}$$

where Γ is the decay width of the inflaton into the lighter modes.

The claim is then that the couplings of the inflaton give rise to dissipative dynamics! Whereas on the face of it this might seem reasonable, nowhere is there a derivation from first principles as to how this might actually occur.

What our formalism is able to do is to actually track the exchange of energy between the zero mode and the other modes of the theory. In our latest work[7], we consider a self-coupled scalar field, at zero temperature (as it would be at the end of inflation) and ask how the non-zero momentum modes backreact on the zero mode and affect its evolution. What we find is quite different than the above equation of motion. We summarize our results below and refer the reader to our paper for the details.

3.1 Single Scalar Field

In the case of a single scalar field, we can construct the equation of motion within the amplitude expansion, i.e. assuming $\phi(t) << m$. Doing this yields:

$$\begin{aligned}\ddot{\phi}(t)+m^2\phi(t) \;&+\; \frac{\lambda}{6}\phi^3(t)+\frac{\lambda}{2}\phi(t)\int\frac{d^3k}{(2\pi)^3}\frac{1}{2\omega_k}-\frac{\lambda^2}{4}\phi^3(t)\int\frac{d^3k}{(2\pi)^3}\frac{1}{4\omega_k^3}+\\ &\frac{\lambda^2}{4}\phi(t)\int_{-\infty}^{t}dt'\phi(t')\dot{\phi}(t')\int\frac{d^3k}{(2\pi)^3}\frac{\cos[2\omega_k(t-t')]}{2\omega_k^3}=0. \end{aligned} \tag{22}$$

The above equation shows both mass and coupling constant renormalization explicitly. It also shows a term proportional to $\dot{\phi}(t)$. However, this term is (i) *nonlinear* in the amplitude, (ii) convolved with a *non-Markovian* kernel:

$$\mathcal{K}(t-t')\equiv\int\frac{d^3k}{(2\pi)^3}\frac{\cos[2\omega_k(t-t')]}{2\omega_k^3} \tag{23}$$

In the zero temperature situation we are dealing with here, there is *no* limit in which $\mathcal{K}(t-t')\simeq D\delta(t-t')$. This can be seen by trying to integrate both sides with respect to t; there is an infrared divergence that makes D be infinite.

In the case in which the reflection symmetry is spontaneously broken, the equation of motion can be linearized about the minimum. However, the non-trivial kernel still persists.

We can also go beyond the amplitude expansion and do a full one-loop calculation. This has to be done numerically, but what we find is clear. If one performs a consistent expansion in the coupling λ, then ***perturbation theory must break down at late times!*** This failure of perturbation theory to describe dissipation is clearly understood by means of a very elementary but yet illuminating example. Consider a damped harmonic oscillator

$$\ddot{q}+\Gamma\dot{q}+q=0, \tag{24}$$

with $\Gamma\sim\mathcal{O}(\lambda)$ where λ is a small perturbative coupling. One can attempt to solve eq.(24) in a perturbative expansion in Γ. That is, set $q(t)=q_0(t)+\Gamma q_1(t)+\cdots$. The solution for $q_1(t)$ may then be found by the Laplace transform:

$$q_1(t)\propto\Gamma t\cos(t) \tag{25}$$

which clearly exhibits resonant behavior. If eq.(24) had been obtained as an effective equation of motion in a *perturbative expansion* in λ, this would be the consistent manner to solve this equation. However, we would be led to conclude that perturbation theory breaks down at long times. The correct solution is

$$q(t) = e^{-\frac{\Gamma}{2}t}\cos[\omega(\Gamma^2)t] \approx \cos(t) - \frac{\Gamma}{2}t\cos(t) + \mathcal{O}(\Gamma^2). \tag{26}$$

We see that the first order correction in Γ is correctly reproduced by perturbation theory, but in order to find appreciable damping, we must wait a time $\sim \mathcal{O}(1/\Gamma)$ at which perturbation theory becomes unreliable.

In order to properly describe dissipation and damping one must *resum* the perturbative expansion. One could in principle keep the first order correction and exponentiate it in an *ad-hoc* manner with the hope that this would be the correct resummation. Although ultimately this *may be* the correct procedure, it is by no means warranted in a field-theoretic perturbative expansion, since in field theory, dissipation is related to particle production and open multiparticle channels, both very subtle and non-linear mechanisms.

A way to incorporate some of the required nonperturbative features is via the Hartree approximation. This allows us to deal with a quadratic Hamiltonian, but incorporates the nonlinearities via a consistency condition.

The Hartree approximation scheme is implemented by use of the following factorizations for both the $\pm$ components of the fluctuation field:

$$\begin{aligned}
\psi^3(\vec{x},t) &\rightarrow 3\langle\psi^2(\vec{x},t)\rangle\psi(\vec{x},t) \\
\psi^4(\vec{x},t) &\rightarrow 6\langle\psi^2(\vec{x},t)\rangle\psi^2(\vec{x},t) - 3\langle\psi^2(\vec{x},t)\rangle^2.
\end{aligned} \tag{27}$$

Translational invariance requires that $\langle(\psi^{\pm}(\vec{x},t))^2(\vec{x},t)\rangle$ be a function of time alone. Since this is an equal time correlation function, we have that:

$$\langle(\psi^+(\vec{x},t))^2\rangle = \langle(\psi^-(\vec{x},t))^2\rangle \stackrel{\text{def}}{=} \langle\psi^2(t)\rangle. \tag{28}$$

Using this factorization, we can rewrite the CTP path integral and compute the tadpoles $\langle\psi^{\pm}(\vec{x},t)\rangle$ to arrive at the following equations of motion:

$$\ddot{\phi} + m^2\phi + \frac{\lambda}{6}\phi^3 + \frac{\lambda}{2}\phi\langle\psi^2(t)\rangle = 0 \tag{29}$$

$$\langle\psi^2(t)\rangle = \int\frac{d^3k}{(2\pi)^3}\left[-iG_k^<(t,t)\right] = \int\frac{d^3k}{(2\pi)^3}\frac{\left|U_k^+(t)\right|^2}{2\omega_k(0)} \tag{30}$$

$$\left[\frac{d^2}{dt^2} + \omega_k^2(t)\right]U_k^+(t) = 0 \quad \omega_k^2(t) = \vec{k}^2 + \mathcal{M}^2(t)$$

$$\mathcal{M}^2(t) = V''(\phi) + \frac{\lambda}{2}\langle\psi^2(t)\rangle = m^2 + \frac{\lambda}{2}\phi^2(t) + \frac{\lambda}{2}\langle\psi^2(t)\rangle \tag{31}$$

The initial conditions for the mode functions are

$$U_k^+(0) = 1 \quad , \qquad \dot{U}_k^+(0) = -i\omega_k(0). \tag{32}$$

This approximation is equivalent to summing the so-called "daisy" diagrams.

There are some subtleties in the Hartree approximation having to do with how the theory is renormalized. However, the renormalization program can be performed at which point we can use numerical methods to follow the evolution of the zero mode and of the quantum fluctuations $\langle\psi^2(t)\rangle$. We can also compute the particle number, where particles are defined using the in modes at $t = 0$ and the time dependence of the particle number arises from the fact that the density matrix is evolving in time. Thus the "vacuum" at any time t is different than that at $t = 0$, and are related via a Bogoliubov transformation.

We show the results of a sample run in Figs. 1-3. Here we have chosen the parameters $g \equiv \lambda/8\pi^2 = 0.1$, $\eta(\tau) \equiv \phi(\tau)/M = 5$, and $\Lambda/M = 100$; M is the renormalized mass of the field, τ is time measured in units of M, while Λ is the momentum cutoff required in order to calculate $\Sigma(\tau) \equiv \langle\psi^2(t)\rangle$ numerically.

The features to note are that initially the zero mode is heavily damped. During this damping phase, the quantum fluctuations grow, and particles are produced. Eventually the quantity $g\Sigma$ grows to be $\mathcal{O}(1)$, at which point the channels that were open for energy transfer out of the zero mode shut off, leaving the zero mode oscillating. We do *not* see evidence for exponential damping at late times, contrary to what the $\Gamma\dot{\phi}$ approximation would have predicted.

The case of broken symmetry for a single scalar field in the Hartree approximation is much the same, except for the possibility that some momentum modes are unstable due to the presence of a negative mass^2 in the mode equations. Eventually, the quantum fluctuations grow to shut off this instability, and then the behavior is much the same as in the unbroken symmetry case.

Some of these results, such as the form of the kernel were obtained by others, most notably Paz[8]. However, only the short time behavior was analyzed there.

3.2 The O(N) Model in the Large N Approximation

Whereas the Hartree approximation did encode some of the non perturbative features of dissipation, for a single scalar field, this approximation is not controllable. That is, there is no small parameter that shows that the diagrams discarded by this approximation were in fact smaller than the ones kept.

The Hartree approximation *can* be made controlled in the large N approximation to the $O(N)$ model. Furthermore this model is interesting in that we can look at the case in which a *continous* symmetry is spontaneously broken giving rise to Goldstone bosons.

The Lagrangian density is the following:

$$\mathcal{L} = \frac{1}{2}\partial_\mu\vec{\phi}\cdot\partial^\mu\vec{\phi} - V(\sigma,\vec{\pi})$$

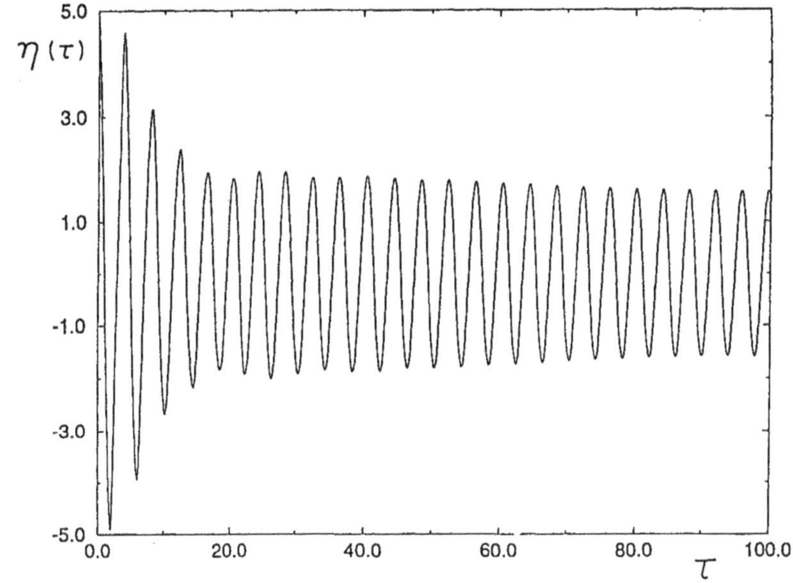

Fig. 1: $\eta(\tau)$ vs τ in the Hartree approximation, unbroken symmetry case. $g = 0.1$; $\eta(0) = 5$; $\Lambda/M_R = 100$.

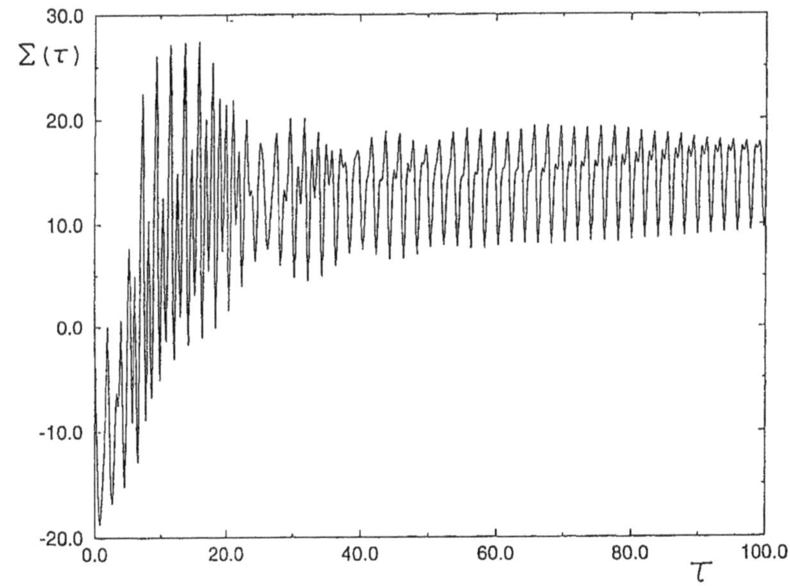

Fig. 2: $\Sigma(\tau)$ vs τ for the same case as in fig.(1).

$$V(\sigma, \vec{\pi}) = \frac{1}{2} m^2 \vec{\phi} \cdot \vec{\phi} + \frac{\lambda}{8N} (\vec{\phi} \cdot \vec{\phi})^2 \tag{33}$$

for λ fixed in the large N limit. Here $\vec{\phi}$ is an $O(N)$ vector, $\vec{\phi} = (\sigma, \vec{\pi})$ and $\vec{\pi}$ represents the $N-1$ pions.

We will assume that there is symmetry breaking in this theory, so that $m^2 < 0$.

To implement the Hartree factorization here, we must be careful so as to arrive at an approximation scheme that is consistent with the Ward Identities for the theory. This is not hard to do and we again refer the reader to our paper. The resulting equations are:

$$\sigma(\vec{x}, t) = \sigma_0(t) + \chi(\vec{x}, t), \tag{34}$$

$$\chi^4 \rightarrow 6\langle \chi^2 \rangle \chi^2 + constant \tag{35}$$

$$\chi^3 \rightarrow 3\langle \chi^2 \rangle \chi \tag{36}$$

$$(\vec{\pi} \cdot \vec{\pi})^2 \rightarrow \left(2 + \frac{4}{N-1}\right) \langle \vec{\pi}^2 \rangle \vec{\pi}^2 + constant \tag{37}$$

$$\vec{\pi}^2 \chi^2 \rightarrow \langle \vec{\pi}^2 \rangle \chi^2 + \vec{\pi}^2 \langle \chi^2 \rangle \tag{38}$$

$$\vec{\pi}^2 \chi \rightarrow \langle \vec{\pi}^2 \rangle \chi, \tag{39}$$

where by "constant" we mean the operator independent expectation values of the composite operators which will not enter into the time evolution equation of the order parameter. To obtain a large N limit, we define

$$\langle \vec{\pi}^2 \rangle = N \langle \psi^2 \rangle \tag{40}$$

$$\sigma_0(t) = \phi(t) \sqrt{N} \tag{41}$$

with

$$\langle \psi^2 \rangle \approx \mathcal{O}(1) \; ; \; \langle \chi^2 \rangle \approx \mathcal{O}(1) \; ; \; \phi \approx \mathcal{O}(1). \tag{42}$$

We will approximate further by neglecting the $\mathcal{O}(\frac{1}{N})$ terms in the formal large N limit.

The tadpole method then yields the following equations:

$$\ddot{\phi}(t) + \phi(t) \left[m^2 + \frac{\lambda}{2} \phi^2(t) + \frac{\lambda}{2} \langle \psi^2(t) \rangle \right] = 0 \; ; \; \langle \psi^2(t) \rangle = \int \frac{d^3k}{(2\pi)^3} \frac{\left| U_k^+(t) \right|^2}{2\omega_{\pi k}^2(0)} \tag{43}$$

$$\left[\frac{d^2}{dt^2} + \omega_{\pi k}^2(t) \right] U_k^+(t) = 0 \; ; \; \omega_{\pi k}^2(t) = \vec{k}^2 + \mathcal{M}_\pi^2(t), \tag{44}$$

with

$$\mathcal{M}_\pi^2(t) = m^2 + \frac{\lambda}{2} \phi^2(t) + \frac{\lambda}{2} \langle \psi^2(t) \rangle \tag{45}$$

$$\mathcal{M}_\chi^2(t) = m^2 + \frac{3\lambda}{2} \phi^2(t) + \frac{\lambda}{2} \langle \psi^2(t) \rangle. \tag{46}$$

We now analyze these equations numerically. We will also calculate the number of Goldstone bosons produced by the zero mode. Clearly this will be the preferred method of energy dissipation since these are *massless* modes, so that large amounts of energy can be channeled into the production of large numbers of soft Goldstone particles.

What we find is the following. As expected, the spectrum of Goldstone particles produced is strongly peaked about $k = 0$ at late times. We also find that the damping of the zero mode of the radial field coincides with the growth of the quantum fluctuations and the explosive growth in the number of particles.

We also find some very unexpected results. In Figs.4-6 we have the behavior of the zero mode, quantum fluctuations and particle number for the case in which $\eta(0) = 10^{-4}, g = 10^{-12}$, so that the field begins very close to the top of the potential hill. This initial condition corresponds to a "slow-roll" scenario. The fluctuations must grow for a long time before $g\Sigma(\tau)$ becomes $\mathcal{O}(1)$, during which the field expectation value evolves classically reaching the classical turning point at $\eta_{tp} = \sqrt{2}$, and then bounces back. But by the time it gets near the origin again, the fluctuations have grown dramatically absorbing most of the energy of the field expectation value and completely damping its motion. In this case *almost all* the initial potential energy has been converted into particles. This is a remarkable result. The conclusion of this analysis is that the strong dissipation by Goldstone bosons dramatically changes the dynamics of the phase transition. For slow-roll initial conditions the scalar field relaxes to a final value which is very close to the origin. This is the minimum of the effective action, rather than the minimum of the tree-level effective potential. Thus dissipative effects by Goldstone bosons introduce a very strong dynamical correction of the effective action leading to a very shallow effective potential (the effective action for constant field configuration). The condition for this situation to happen is that the period of the classical trajectory is of the same order of magnitude as the time scale of growth for the fluctuations.

The weak coupling estimate for this dynamical non-equilibrium time scale is $\tau_s \approx \ln(1/g)/2$, which is obtained by requiring that $g\Sigma(\tau) \approx 1$. For weak coupling the mode functions grow as $U_q^+(\tau) \approx e^\tau$ for $q^2 < 1$, and $\Sigma(\tau) \approx e^{2\tau}$. The numerical analysis confirms this time scale for weak coupling. For weakly coupled theories, this non-equilibrium time scale is much larger than the static correlation length (in units in which $c = 1$) and the only relevant time scale for the dynamics.

What we are seeing here is that the system will search for minima of the effective action that are *consistent* with the value of the energy of the system. This energy is conserved, so depending on where the field starts and how large the coupling constant is, the energy will redistributed amongst the modes in different ways.

This is a striking effect of the dynamical evolution of the system, and one which can have interesting effects in cosmology. For example, the $O(N)$ model can describe topological defects. From a cosmolgical perspective, the important quantity is the vacuum expectation value of the field which breaks the symmetry. What we have seen here is that this quanitity can be very different than what the effective *potential* would have led one to expect.

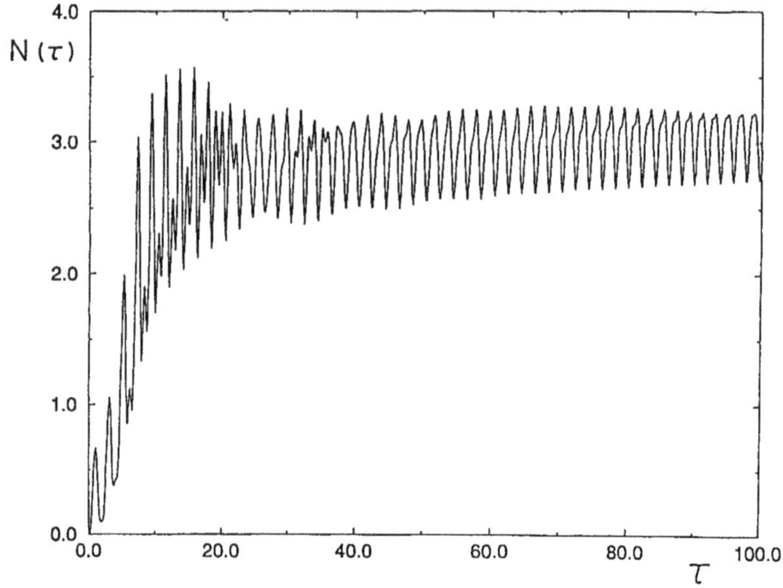

Fig. 3: $N(\tau)$ vs τ for the same case as in fig.(1).

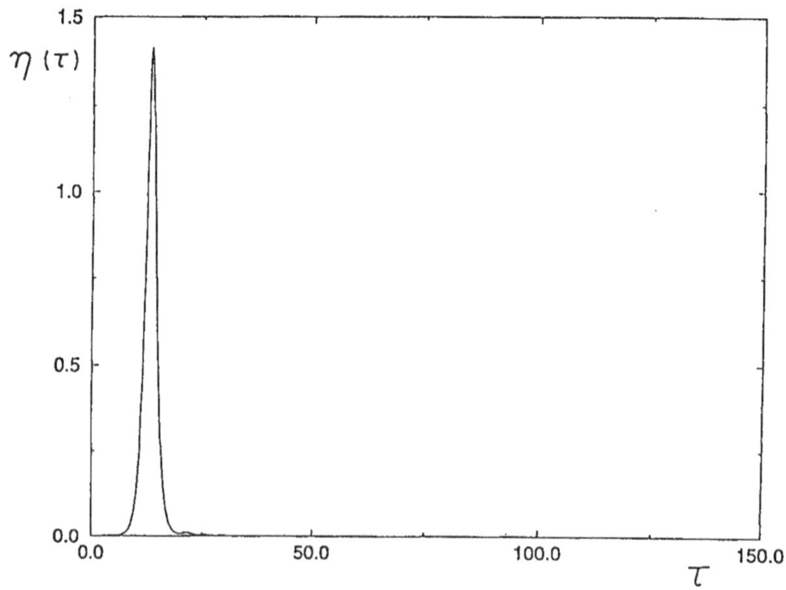

Fig. 4: $\eta(\tau)$ vs τ in the large N approximation in the O(N) model, broken symmetry case. $g = 10^{-12}$; $\eta(0) = 10^{-5}$; $\Lambda/\mu_R = 100$.

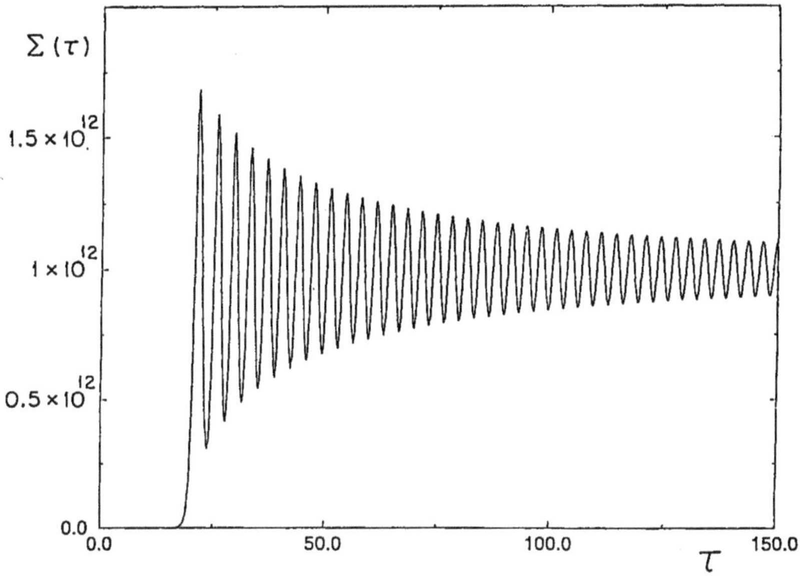

Fig. 5: $\Sigma(\tau)$ vs τ for the same case as in fig.(4).

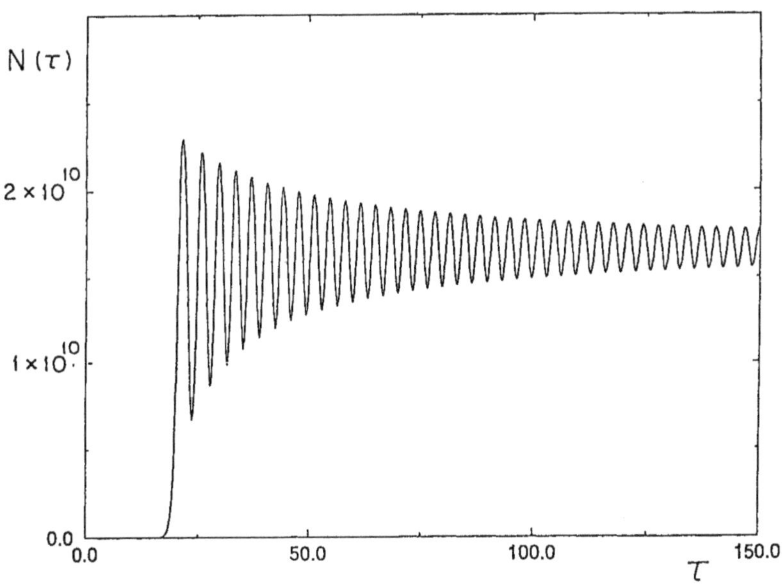

Fig. 6: $N(\tau)$ vs τ for the same case as in fig.(4).

4 Further Thoughts and Conclusions

What we believe we have in hand is a formalism that allows us to understand the dynamical, non-equilibrium evolution of a field theory. Applied to the problem of dissipation, we see that it is the back reaction of the produced particles that gives rise to the damping of the zero mode oscillations. We can make use of this to understand how this can be consistent with time reversal invariance as well as why a formalism such as the CTP one, which yields "in-in" correlation functions as opposed to the "in-out" ones of the usual path integral is needed.

The dynamics is completely determined by the set of equations of motion for the field expectation value and the mode functions for the fluctuations described above. These equations are solved by providing non equilibrium initial conditions on the field expectation value, its derivative and the mode functions and their derivatives at the initial time $t = 0$. The problem is then evolved in time by solving the coupled *second order* differential equations. We emphasize the fact that the equations are second order in time because these are time reversal invariant. Now consider evolving this set of equations up to a positive time t_0, at which we stop the integration and find the value of the field expectation value, its derivative, the value of *all* the mode functions and their derivatives. Because this is a system of differential equations which is second order in time, we can take these values at t_0 as initial conditions at this time and evolve *backwards* in time reaching the initial values at $t = 0$. Notice that doing this involves beginning at a time t_0 in an excited state with (generally) a large number of particles. The conditions at this particular time are such that the energy stored in this excited state is focused in the back reaction to the field expectation value that acquires this energy and whose amplitude will begin to grow.

What other problems can we address with our technology?

We have not yet solved the problem that got us into thinking about non-equilibrium in the first place: the reheating problem. To do this we must couple the inflaton to other fields and compute the growth in the particle number of those fields. We then would have to understand how those particles then reach an equilibrium distribution via thermalization. We are investigating this problem, and have found some interesting facets to it. In particular, in the case that the other fields are also scalars, there can be mixing in the mass matrix of the modes of the inflaton and those of the other field. This mixing depends on the zero modes of both fields and is time dependent. This leads to some complications in implementing our formalism that we have not yet sorted out.

The slow-roll portion of inflationary evolution is another regime in which fields are not in equilibrium. We are currently tackling the problem of understanding the effect of quantum fluctuations on this evolution by making use of the semi-classical Einstein equations:

$$R_{\mu\nu} - \frac{1}{2}g_{\mu\nu}R = 8\pi G\langle T_{\mu\nu}\rangle. \tag{47}$$

The expectation value is taken with the state described by a density matrix as developed above.

Finally, we are trying to understand the behavior of theories with gauge fields, with an eye towards being able to follow the dynamics of formation of topological defects.

5 Acknowledgements

I would like to thank Norma Sanchez for inviting to this wonderful School. I would also like to thank my collaborators D. Boyanovsky, H. de Vega, A. Singh and D-S. Lee for the great times had while working on this. This work was supported in part by the DOE contract # DE-FG02-91-ER40682.

References

[*] Laboratoire Associé au CNRS, UA280.

[1] A.H.Guth, Phys. Rev D23, 347 (1981).

[2] A. Albrecht and P.J. Steinhardt, Phys. Rev. Lett. 48, 1220 (1982). A.D. Linde, Phys. Lett. 108B, 389 (1982).

[3] A.D. Linde, Phys. Lett. 162B, 281 (1985).

[4] T.W.B. Kibble, J. Phys. A9, 1387, (1982).

[5] A.D. Linde "Particle Physics and Inflationary Cosmology", Harwood Academic Publishers (1990).

[6] J. Schwinger, J. Math. Phys. 2, 407 (1961); P. M. Bakshi and K. T. Mahanthappa, J. of Math. Phys. 4, 1 (1963); ibid. 12; L. V. Keldysh, Sov. Phys. JETP 20, 1018 (1965).

[7] D. Boyanovsky, H. J. de Vega, R. Holman, A. Singh, and D-S.Lee, "Dissipation via Particle Production in Scalar Field Theories", CMU prerint CMU-HEP-94-23, (submitted to Phys. Rev. D) (1994).

[8] J.P. Paz, Phys. Rev. D.41, 1054 (1990); Phys.Rev. D42, 529 (1990).

[9] R.P. Feynman and A.R. Hibbs "Quantum Mechanics and Path Integrals", McGraw-Hill, New York (1965).

[10] L.Koffman, A.D. Linde, A.A. Starobinski, Stanford University preprint "Reheating After Inflation" SU-ITP-94-13 and hep/th 9405187 (1994); Y. Shtanov, J. Traschen and R. Brandenberger, Brown University preprint "Universe Reheating After Inflation" BROWN-HET-957 (1994).

[11] E. Calzetta and B. L. Hu, Phys. Rev. D35, 495; *ibid* D37, 2838 (1988).

[12] R. D. Jordan, Phys. Rev. D33, 444 (1986).

[13] S. Weinberg, Phys. Rev. D9, 3357 (1974).

[14] A. Albrecht, P. Steinhardt, M.S. Turner, and F. Wilczek, Phys. Rev. Lett. 48, 1437 (1982); L. Abbott, E. Farhi and M. Wise, Phys. Lett. 117B, 29 (1982); M. Morikawa and M. Sasaki, Prog. Theor. Phys. 72, 782 (1984); M. Morikawa, Phys. Rev. D33, 3607 (1986); A. Ringwald, Ann. of Physics (N.Y.) 177, 129, (1987).

ASTROPHYSICAL AND COSMOLOGICAL CONSEQUENCES OF A GAUGE THEORY OF GRAVITY

ANTHONY LASENBY, CHRIS DORAN and STEPHEN GULL
MRAO, Cavendish Laboratory, Madingley Road,
Cambridge CB3 0HE, UK.

Abstract

A new gauge-theory description of gravity is presented, employing gauge fields in a flat background spacetime. These fields ensure that all physical relations are independent of the position and orientation of the matter fields in this background. The language of 'geometric algebra' best expresses the physical and mathematical content of the theory and is employed throughout. A method of working directly with the physical fields is developed and applied to the case of a radially-symmetric time-varying perfect fluid. A gauge is found in which the physics reduces to a set of Newtonian equations. The insistence on finding global solutions alters the physical picture of the horizon around a black hole, and enables one to discuss the properties of field lines inside the horizon created by a point charge held at rest outside it. Some applications to cosmology are discussed, and a study of the Dirac equation in a cosmological background reveals that the only models consistent with homogeneity are spatially flat.

1 Introduction

The most successful theories of physics created to date are underpinned by the principle of local gauge invariance. This principle is based on the belief that global symmetry transformations, in which the transformation is applied simultaneously at all points in spacetime, are incompatible with the requirements of locality. Instead, any global symmetry must be replaced by a local equivalent in which the symmetry holds independently from point to point. Local symmetry is guaranteed by the existence of fields with certain transformation properties — gauge fields — and these fields are responsible for inter-particle forces. But what of gravity: can general relativity be formulated as a gauge theory? This question has troubled physicists for many years [1, 2, 3, 4]. It was Kibble [2] who, elaborating on work by Utiyama [1], was the first to recover aspects of general relativity (GR) from a gauging argument. Kibble used the 10-component Poincaré group of passive coordinate transformations (consisting of four translations and six rotations) as the global symmetry group. Kibble's use of passive transformations was criticised by Hehl *et al.* [5], who reproduced Kibble's derivation from the standpoint of active transformations of the matter fields. In both cases the authors arrived at a more general version of GR known as a 'spin-torsion' theory, and it is now generally accepted that torsion is an inevitable feature of a gauge theory based on the Poincaré group.

There are a number of conceptual difficulties associated with the approaches adopted by Kibble and Hehl *et al.* For example, Hehl *et al.* state that '*coordinates and frames are regarded as fixed once and for all, while the matter fields are replaced by fields that have been rotated or translated*'. It follows that their derivation can only affect the properties of the matter fields, and not the properties of spacetime itself. Yet, once the gauge fields have been introduced, the authors identify these fields as determining the curvature and torsion of a Riemann-Cartan spacetime. Despite being untouched by the derivation of the gravitational fields, spacetime itself is supposed to have become an active participant in physics! Here we propose an alternative gauge theory of gravity which we believe is free from such conceptual difficulties. Our starting point is the parameterisation of events by *vectors* in a 4-dimensional spacetime. This spacetime is 'flat', though it might more appropriately be referred to as 'structureless'. All matter fields are now parameterised by a vector variable x, but in so doing are we not returning to some Newtonian notion of an absolute space? The answer is no — but only if we insist that all physical laws consist of local relationships between fields, and not between a field and its spacetime position. The location of the matter fields in the background spacetime is then irrelevant. We therefore aim to construct a theory where the relationships between fields are invariant under local changes of the position and orientation of fields in our background space. The result is a gauge theory in which spacetime itself does not play an active role in physics.

Besides the conceptual clarity afforded by the theory outlined here, it contains some considerable practical benefits. In particular, once one chooses a gauge by prescribing how matter fields are parameterised by vectors, one is then free to utilise all the advantages of working in a flat space. These advantages are best exploited using the language of 'geometric algebra' [6] and its spacetime version 'spacetime algebra' [7]. It is well known that many problems in two and three dimensions can be handled most efficiently by representing points with vectors and working in a coordinate-free manner. Spacetime algebra generalises these techniques so that they are applicable to problems in spacetime. Naturally associated with the representation of points by vectors is the vector derivative — the derivative with respect to vector position. The vector derivative plays a central role in the spacetime algebra form of both the Maxwell and Dirac theories. Furthermore, in both cases the derivative appears in a first-order form for which an inverse function can be found (in the form of a Green's function). The resultant first-order propagator theory is highly efficient computationally [6, 8]. The theory developed here enables these same techniques to be employed in gravitational problems, and the gauge structure ensures that all physical predictions are independent of the means by which points were parameterised with vectors. In addition to these 'flatspace' techniques, we have developed a new technique which deals directly with the physically-observable objects. We call this the '*intrinsic*' method. The method consists of manipulating abstract operators satisfying certain bracket relations, and is remarkably powerful in many applications.

This paper is intended to offer an outline review of our theory. The full formal development is contained in [9]. We begin with an introduction to geometric algebra and its application to spacetime physics. We then turn to a discussion of the gauge fields that must be introduced to construct a theory in which all relations between physical fields are independent of their position and orientation in the background space. Having determined the necessary fields, we discuss the appropriate field equations for them. These equations are then studied for the case of a radially-symmetric perfect fluid. We find that a particular choice of time variable leads to some dramatic simplifications and results in a set of equations which are almost entirely Newtonian! A number of applications of this system are then discussed. Our employment of a background vector space forces us to rethink the physics of horizons and their formation during a collapse process. Not surprisingly, since we work with a background flat spacetime, the possibility of wormholes to new universes does not arise! The example of electromagnetism reveals the importance of finding global solutions to the field equations, which is a crucial point where our theory departs from GR. We then turn to applications in cosmology. The 'intrinsic' method leads quickly to a pair of first-order equations which work directly with the Hubble velocity. The Friedmann equations are easily recovered and a novel treatment of particle horizons is given. Finally, studying the Dirac equation in a cosmological background, we conclude that the only cosmological models consistent with spatial homogeneity are spatially flat. We end with a summary of the main conclusions.

2 Geometric Algebra

The following comprises a brief introduction to the ideas and conventions of geometric algebra. Further details can be found in the series of papers [8, 10, 11, 12], in the books by Hestenes [7, 13] and Hestenes & Sobczyk [6] and in the introductory articles by Hestenes [14, 15] and Vold [16, 17]. The applications to a gauge theory of gravity were first discussed in [18] and [19].

A 'geometric algebra' is a graded vector space with an associative product that is distributive over addition. The grade-0 elements of this space are real scalars and commute with all higher-grade elements (known as 'multivectors'). The grade-1 elements are vectors and are usually given lower-case Roman symbols (a, b). The geometric product is distinguished by the property that the square of any vector is a scalar. It then follows from the identity

$$(a+b)^2 = (a+b)(a+b) = a^2 + (ab+ba) + b^2 \tag{2.1}$$

that the symmetrised product of two vectors is a scalar. We use this to decompose the geometric product of two vectors into a scalar term

$$a\cdot b \equiv \tfrac{1}{2}(ab+ba) \tag{2.2}$$

and a grade-2 term, called a *bivector*,

$$a\wedge b \equiv \tfrac{1}{2}(ab-ba). \tag{2.3}$$

Hence we now have

$$ab = a\cdot b + a\wedge b \tag{2.4}$$

for the full geometric product. The significant feature of the geometric product is that it mixes two different types of object: scalars and bivectors. This is not problematic — the addition implied by (2.4) is that used when, for example, a real number is added to an imaginary number to form a complex number. Indeed, it was Clifford's desire to extend the complex field that led him to propose the geometric product.

Forming further geometric products of vectors produces higher-grade multivectors. Multivectors in which all elements have the same grade are termed *homogeneous* and are usually written as A_r to show that A contains only a grade-r component. Multivectors inherit an associative product, and the geometric product of a grade-r multivector A_r with a grade-s multivector B_s decomposes into

$$A_rB_s = \langle AB\rangle_{r+s} + \langle AB\rangle_{r+s-2}\ldots + \langle AB\rangle_{|r-s|}. \tag{2.5}$$

The symbol $\langle M\rangle_r$ denotes the projection onto the grade-r component of M. The projection onto the grade-0 (scalar) component of M is written $\langle M\rangle$. The scalar part of a product of multivectors satisfies the cyclic reordering property

$$\langle A\ldots BC\rangle = \langle CA\ldots B\rangle. \tag{2.6}$$

The '$\cdot$' and '$\wedge$' symbols are retained for the lowest-grade and highest-grade terms of the series (2.5), so that

$$A_r\cdot B_s \equiv \langle AB\rangle_{|s-r|} \tag{2.7}$$
$$A_r\wedge B_s \equiv \langle AB\rangle_{s+r}, \tag{2.8}$$

which are called the interior and exterior products respectively. We also define the commutator product

$$A \times B \equiv \tfrac{1}{2}(AB - BA). \tag{2.9}$$

The associativity of the geometric product ensures that the commutator product satisfies the Jacobi identity

$$A \times (B \times C) + B \times (C \times A) + C \times (A \times B) = 0. \tag{2.10}$$

Finally, we introduce an operator ordering convention. *In the absence of brackets, inner, outer and scalar products take precedence over geometric products.* Thus $a \cdot b\, c$ means $(a \cdot b)\, c$, not $a \cdot (bc)$. This convention helps to eliminate unwieldy numbers of brackets. Summation convention and natural units ($\hbar = c = \epsilon_0 = G = 1$) are employed throughout.

2.1 The Spacetime Algebra

Of central importance to many applications is the geometric algebra of spacetime, the *spacetime algebra* [7]. To describe the spacetime algebra (STA) it is helpful to introduce a set of four orthonormal basis vectors $\{\gamma_\mu\}$, $\mu = 0 \ldots 3$, satisfying

$$\gamma_\mu \cdot \gamma_\nu = \eta_{\mu\nu} = \mathrm{diag}(+ - - -). \tag{2.11}$$

The vectors $\{\gamma_\mu\}$ satisfy the same algebraic relations as Dirac's γ-matrices, but they now form a set of four independent basis vectors for spacetime, not four components of a single vector in an internal 'spin-space'. The relation between Dirac's matrix algebra and the STA is described in more detail elsewhere [11].

The full STA is spanned by the quantities

$$1, \quad \{\gamma_\mu\}, \quad \{\sigma_k,\, i\sigma_k\}, \quad \{i\gamma_\mu\}, \quad i, \tag{2.12}$$

where

$$\sigma_k \equiv \gamma_k \gamma_0, \tag{2.13}$$

and

$$i \equiv \gamma_0\gamma_1\gamma_2\gamma_3 = \sigma_1\sigma_2\sigma_3. \tag{2.14}$$

The $\{\sigma_k\}$ form an orthonormal frame of spatial vectors in the space relative to the γ_0 direction. The algebraic properties of the $\{\sigma_k\}$ are the same as those of the Pauli spin matrices, though again they are to be interpreted geometrically and not as components of a vector in spin-space. The highest-grade element (or 'pseudoscalar') is denoted by i. The symbol i is used because the square of i is -1, but the pseudoscalar must not be confused with the unit scalar imaginary employed in quantum mechanics. Since we are working with a space of even dimension, i *anti*commutes with odd-grade elements, and only commutes with even-grade elements.

The split of the six spacetime bivectors into relative vectors $\{\sigma_k\}$ and relative bivectors $\{i\sigma_k\}$ is a frame-dependent operation — different observers determine different relative spaces. This fact is best illustrated with the Faraday bivector F. The 'spacetime split' [7, 20] of F into the γ_0-system is made by separating F into parts which anticommute and commute with γ_0. Thus

$$F = \boldsymbol{E} + i\boldsymbol{B} \tag{2.15}$$

where

$$\begin{aligned} \boldsymbol{E} &= \tfrac{1}{2}(F - \gamma_0 F \gamma_0) &(2.16)\\ i\boldsymbol{B} &= \tfrac{1}{2}(F + \gamma_0 F \gamma_0). &(2.17)\end{aligned}$$

Both $\boldsymbol{E}$ and $\boldsymbol{B}$ are spatial vectors in the γ_0-frame, and $i\boldsymbol{B}$ is a spatial bivector. Equation (2.15) decomposes F into separate electric and magnetic fields, and the explicit appearance of γ_0 in the formulae for $\boldsymbol{E}$ and $\boldsymbol{B}$ shows how this split is observer-dependent. Where required, relative (or spatial) vectors in the γ_0-system are written in bold type to record the fact that in the STA they are actually bivectors. This distinguishes them from spacetime vectors, which are left in normal type. No problems arise for the $\{\sigma_k\}$, which are unambiguously spacetime bivectors, so these are also left in normal type.

2.2 Geometric Calculus

Throughout this paper we employ the derivative with respect to a vector argument a. This is written as ∂_a and is defined in terms of its directional derivatives $b \cdot \partial_a$, where

$$b\cdot\partial_a F(a) \equiv \lim_{\tau \mapsto 0} \frac{F(a+\tau b) - F(a)}{\tau}. \tag{2.18}$$

Then, in terms of an arbitrary vector basis $\{e_k\}$ and reciprocal basis $\{e^k\}$ ($e_j\cdot e^k = \delta_j^k$) the full derivative is defined by

$$\partial_a \equiv e^j e_j\cdot\partial_a. \tag{2.19}$$

This definition shows how the derivative ∂_a inherits the vector properties of its argument a, as well as a calculus from equation (2.18). The vector derivative is employed frequently to manipulate linear functions in a frame-independent manner.

The derivative with respect to spacetime position x is of special significance. It is called the *vector derivative*, and is given the symbol

$$\nabla = \nabla_x \equiv \partial_x. \tag{2.20}$$

In terms of a set of Cartesian coordinates $\{x^\mu\}$, the vector derivative can be written

$$\nabla = \gamma^\mu \partial_\mu, \tag{2.21}$$

where $\partial_\mu \equiv \partial_{x^\mu}$. Hence, just as the γ-matrices are replaced by vectors in spacetime, objects such as $x^\mu \gamma_\mu$ and $\nabla = \gamma^\mu \partial_\mu$ become frame-free vectors. The usefulness of the geometric product for the vector derivative is illustrated by electromagnetism. In tensor notation, Maxwell's equations take the form

$$\partial_\mu F^{\mu\nu} = J^\nu \qquad \partial_{[\alpha} F_{\mu\nu]} = 0, \tag{2.22}$$

which have the STA equivalents

$$\nabla \cdot F = J \qquad \nabla \wedge F = 0. \tag{2.23}$$

But now we can utilise the geometric product to combine these into the single equation

$$\nabla F = J. \tag{2.24}$$

The significance of the ∇ operator is that it possesses an inverse, so a first-order propagator theory can be developed for it [6, 8]. This is not possible for the individual $\nabla \cdot$ and $\nabla \wedge$ operators.

The vector derivative acts on objects to its immediate right unless brackets are present. So, in the expression ∇AB the ∇ acts only on A, but in the expression $\nabla(AB)$ the ∇ acts on both A and B. If the ∇ is intended to only act on B then this is written as $\dot{\nabla} A \dot{B}$, where the overdot denotes the multivector on which the derivative acts. For example, Leibniz' rule can be written in the form

$$\nabla(AB) = \dot{\nabla} \dot{A} B + \dot{\nabla} A \dot{B}. \tag{2.25}$$

2.3 Linear Algebra

Geometric algebra offers many advantages over tensor calculus when used for developing the theory of linear functions [6, 21, 15]. A linear function mapping vectors to vectors is written with an underbar $\underline{f}(a)$. The adjoint function is written with an overbar, $\overline{f}(a)$, so that

$$a \cdot \underline{f}(b) = \overline{f}(a) \cdot b, \tag{2.26}$$

and hence

$$\overline{f}(a) = \partial_b \langle \underline{f}(b) a \rangle. \tag{2.27}$$

As explained, the use of the derivative operator ∂_b keeps all expressions free from requiring an explicit coordinate frame. Of course, the ∂_b and b vectors can be replaced by the sum over a set of frame vectors and their reciprocals, if desired.

Linear functions extend to act on multivectors via

$$\underline{f}(a \wedge b \wedge \ldots \wedge c) \equiv \underline{f}(a) \wedge \underline{f}(b) \ldots \wedge \underline{f}(c), \tag{2.28}$$

so that $\underline{f}$ is now a grade-preserving linear function mapping multivectors to multivectors. In particular, since the pseudoscalar of a space, I say, is unique up to a scale factor, we can define

$$\det \underline{f} = \underline{f}(I)I^{-1}. \tag{2.29}$$

Viewed as linear functions over the entire geometric algebra, $\underline{f}$ and $\overline{f}$ are related by the fundamental formulae

$$\begin{aligned} A_r \cdot \overline{f}(B_s) &= \overline{f}[\underline{f}(A_r) \cdot B_s] \qquad r \leq s \\ \underline{f}(A_r) \cdot B_s &= \underline{f}[A_r \cdot \overline{f}(B_s)] \qquad r \geq s, \end{aligned} \tag{2.30}$$

which are derived in [6, Chapter 3]. The formulae for the inverse functions are found as special cases of (2.30),

$$\begin{aligned} \underline{f}^{-1}(A) &= \det(f)^{-1}\overline{f}(AI)I^{-1} \\ \overline{f}^{-1}(A) &= \det(f)^{-1}I^{-1}\underline{f}(IA). \end{aligned} \tag{2.31}$$

Of particular importance here is the geometric algebra description of rotations. A vector a is rotated to a new vector a' by

$$a' = Ra\tilde{R} \tag{2.32}$$

where R is a *rotor*. In the STA, rotors are elements of the even subalgebra satisfying

$$R\tilde{R} = 1 \tag{2.33}$$

and any rotor can be written as $\pm\exp\{B\}$, where B is a bivector. The usefulness of rotors in the description of rotations lies in the ease with which they extend to act on multivectors. So, for example,

$$(Ra\tilde{R})\wedge(Rb\tilde{R}) = \tfrac{1}{2}(Rab\tilde{R} - Rba\tilde{R}) = Ra\wedge b\tilde{R} \tag{2.34}$$

and the generalisation to an arbitrary multivector is simply

$$M \mapsto RM\tilde{R}. \tag{2.35}$$

Furthermore, this formula is equally valid for boosts as well as spatial rotations. No other mathematical language affords such a simple description of Lorentz transformations.

3 The Gravitational Gauge Fields

In this Section we identify the dynamical variables which describe gravitational interactions. Our aim is to achieve a theory in which all reference to spacetime position is removed, and all that remains are the intrinsic relations between fields at a point (these fields will include the gravitational field). The 'position' in spacetime of an object will then be defined *intrinsically* by the values of the fields at that position, and not by a set of arbitrary coordinates. At the same time, we do not want to lose the advantages of representing spacetime positions by vectors. To satisfy both of these demands, we must require that the actual position vector of a field be irrelevant. That is, all equations should be unchanged in form if the fields are moved from one spacetime position to another. Furthermore, these changes of position must be local in character if the physics at a point is to be genuinely free of the position vector of the point. In Section 2.2 we saw how to combine Maxwell's equations into the single equation $\nabla F = J$. The STA form of the Dirac equation [11, 22] also employs the same differential operator ∇ — the vector derivative. We must therefore focus attention on this operator to see how to make our field equations local.

We start by considering a scalar field $\phi(x)$ and form its vector derivative $\nabla\phi(x)$. Suppose now that $\phi(x)$ is moved around to form a new field $\phi'(x)$,

$$\phi'(x) \equiv \phi(x'), \tag{3.1}$$

where

$$x' = f(x) \tag{3.2}$$

and $f(x)$ is an arbitrary (differentiable) map between spacetime position vectors. The map $f(x)$ should not be thought of as a map between manifolds, or as moving points around. The function $f(x)$ is just a rule for relating one position vector to another within a single vector space. If we now consider ∇ acting on the new scalar field ϕ' we form the quantity $\nabla\phi[f(x)]$. To evaluate this we return to the definition of the vector derivative and construct

$$\begin{aligned} a\cdot\nabla\phi[f(x)] &= \lim_{\epsilon\to 0}\frac{1}{\epsilon}\left(\phi f(x+\epsilon a) - \phi f(x)\right) \\ &= \lim_{\epsilon\to 0}\frac{1}{\epsilon}\Big(\phi[f(x)+\epsilon\underline{f}(a)] - \phi f(x)\Big) \\ &= \underline{f}(a)\cdot\nabla_{x'}\phi(x'), \end{aligned} \tag{3.3}$$

where $\underline{f}(a) = \underline{f}_x(a) = a\cdot\nabla f(x)$ and the subscript on $\nabla_{x'}$ records that the derivative is now with respect to the new vector position variable x'. It follows that

$$\nabla_x = \overline{f}(\nabla_{x'}) \tag{3.4}$$

and hence that $\nabla\phi'(x) = \overline{f}[\nabla_{x'}\phi(x')]$.

In physics we frequently need to relate the gradient of some scalar to a vector field. In three dimensions, for example, a static electric field can be written as the gradient of the scalar potential ϕ. We wish to move to a situation where all relations are unaffected by arbitrary displacements. To achieve this we must introduce a gauge field which assembles with the vector derivative to form an object which just changes its position dependence under local translations. We construct such an object by replacing ∇ with a new derivative $\overline{h}(\nabla)$, where $\overline{h}(a)$ has an arbitrary position dependence and is a linear function of a. If we wish to make the position dependence explicit we will write $\overline{h}_x(a)$ or $\overline{h}(a,x)$ (recalling that $\overline{h}(a,x)$ is linear on a and non-linear on x). Under local translations the gauge field $\overline{h}(a)$ is defined to transform to the new field $\overline{h}'(a)$, where

$$\overline{h}'(a,x) \equiv \overline{h}(\overline{f}^{-1}(a), f(x)) = \overline{h}_{x'}\overline{f}^{-1}(a) \tag{3.5}$$

so that

$$\overline{h}_x(\nabla_x) \mapsto \overline{h}_{x'}\overline{f}^{-1}(\nabla_x) = \overline{h}_{x'}(\nabla_{x'}). \tag{3.6}$$

This transformation law ensures that the vector $\overline{h}(\nabla\phi(x))$ now translates as

$$\overline{h}(\nabla\phi(x)) \mapsto \overline{h}_{x'}[\nabla_{x'}\phi(x')] \tag{3.7}$$

and hence just changes its vector position under *arbitrary* local translations. This is the type of behaviour we are after, since ultimately all that should matter are the relations between fields at a point, and the actual position vector of that point (and its relation to nearby points) should be irrelevant. The same intrinsic relations should therefore hold if the fields are moved arbitrarily from one vector position to another. The introduction of the $\overline{h}$-field ensures that derivatives can also be moved around arbitrarily. The $\overline{h}$-field is not a connection in the conventional Yang-Mills sense. The coupling to derivatives is different, as is the transformation law (3.5). This is unsurprising, since the group of arbitrary translations $x \mapsto f(x)$ is infinite-dimensional (if we were considering maps between manifolds then this would form the group of diffeomorphisms). There is no doubt, however, that the $\overline{h}$-field derived from this group embodies the idea of replacing a global symmetry by a local one, so clearly deserves to be called a gauge field.

Henceforth, we refer to any quantity that just changes its position dependence under local translations as behaving *covariantly* under translations. The $\overline{h}$-field enables us to form derivatives of covariant objects which are also covariant under translations. When we come to calculate with this theory, we fix a gauge by choosing a labeling of spacetime points with vectors. In this way we are still free to exploit all the advantages of representing points with vectors. Of course, all the physical predictions of the theory will remain independent of the actual gauge choice.

Having arrived at an operator which transforms covariantly under local translations, we must now consider rotations (*i.e.* Lorentz transformations). Returning to Maxwell's equations in the form $\nabla F = J$, we see that if F is a solution with current J, then

$$F' = RF(x')\tilde{R}, \quad x' = \tilde{R}xR \tag{3.8}$$

is a rotated solution with current $J' = RJ(x')\tilde{R}$. When the ∇ operator is replaced by $\overline{h}(\nabla)$ we have already taken care of covariance under the displacement from x

to x', so the rotation of spacetime position is no longer necessary. All that remains of the transformation is the rotation of F at a point. The reason that this is a symmetry is as follows. Suppose we have two covariant multivector fields A and B satisfying the equation $A(x) = B(x)$. Then the rotated fields $A'(x) = RA(x)\tilde{R}$ and $B'(x) = RB(x)\tilde{R}$ also satisfy the same equation. Hence the intrinsic physical relation between A and B is independent of how we choose to represent their directions in spacetime. This is as it must be if we are to eliminate all spacetime-dependence from the final relations. For the case of the Maxwell equations, the fields F and J are now related by $\overline{h}(\nabla)F = J$. If we rotate the F and J fields, this equation will only remain satisfied if we also rotate the $\overline{h}(a)$ field. This makes sense, because the quantity $\overline{h}(\nabla\phi)$ is also covariant under translations, so we should be free to form relations between this and other covariant objects. We are therefore led to the transformation law for the $\overline{h}(a)$ field under rotations at a point:

$$\overline{h}(a) \mapsto R\overline{h}(a)\tilde{R}, \tag{3.9}$$

which ensures that the equation $\overline{h}(\nabla)F = J$ is now covariant under a global rotation of $\overline{h}(a)$, F and J.

We have now achieved a complete decoupling of rotations and translations, a situation which is not reached in derivations based solely on infinitesimal transformations [2, 5]. But we are still only part way to achieving our goal. The $A(x)$ and $B(x)$ fields above can be rotated by different amounts at each point in space without changing the intrinsic content of the equation $A(x) = B(x)$. Hence we must demand that the equations are invariant under local rotations, where the rotor R is an arbitrary function of position. To see how to achieve this we replace $\overline{h}(\nabla)$ by $\overline{h}(\partial_a)a\cdot\nabla$ and focus attention on the $a\cdot\nabla$ operator. Acting on a multivector $A(x)$ which has been subjected to a position-dependent rotation we find that

$$a\cdot\nabla(RA\tilde{R}) = Ra\cdot\nabla A\tilde{R} + a\cdot\nabla RA\tilde{R} + RAa\cdot\nabla\tilde{R}. \tag{3.10}$$

Since $R\tilde{R} = 1$ for a rotation we have

$$a\cdot\nabla R\tilde{R} = -Ra\cdot\nabla\tilde{R}, \tag{3.11}$$

hence $a\cdot\nabla R\tilde{R}$ is equal to minus its reverse and so must be a bivector (an element of the Lie algebra of the rotation group). We can therefore write

$$a\cdot\nabla(RA\tilde{R}) = Ra\cdot\nabla A\tilde{R} + 2(a\cdot\nabla R\tilde{R})\times(RA\tilde{R}). \tag{3.12}$$

To construct a covariant derivative we must therefore add a 'connection' term to $a\cdot\nabla$ to construct the operator

$$\mathcal{D}_a = a\cdot\nabla + \Omega(a)\times . \tag{3.13}$$

Here $\Omega(a) = \Omega(a,x)$ is a bivector-valued linear function of a with an arbitrary x-dependence. The operation of commuting a multivector with a bivector is grade-preserving. So, even though it is not a scalar operator, $\mathcal{D}_a$ preserves the grade of the multivector on which it acts.

Under local rotations we demand that $\Omega(a)$ transforms as

$$\Omega(a) \mapsto \Omega'(a) = R\Omega(a)\tilde{R} - 2a\cdot\nabla R\tilde{R}. \tag{3.14}$$

Since $\Omega(a)$ is now an arbitrary function of position, however, it cannot in general be transformed away by the application of a rotor. The equations (3.13) and (3.14) ensure that under under a local rotation

$$\mathcal{D}_a{}'(RA\tilde{R}) = R\mathcal{D}_a A\tilde{R}, \tag{3.15}$$

as required for a covariant derivative. Furthermore, the operator (3.13) is a *derivation* since it satisfies Leibniz' rule

$$\mathcal{D}_a(AB) = (\mathcal{D}_a A)B + A(\mathcal{D}_a B), \tag{3.16}$$

as follows from the identity

$$\Omega(a)\times(AB) = (\Omega(a)\times A)B + A(\Omega(a)\times B). \tag{3.17}$$

Under local translations $\Omega(a)$ must transform in the same way as $a\cdot\nabla R\tilde{R}$, hence

$$\Omega_x(a) \mapsto \Omega_{x'}\underline{f}(a) = \Omega(\underline{f}(a), f(x)), \tag{3.18}$$

where the subscript is again used to label position dependence. Since we wish to deal with quantities that are covariant under translations, we define the bivectors

$$\omega(a) = \Omega\underline{h}(a), \tag{3.19}$$

so that $\omega(a,x) \mapsto \omega(a,x')$ under a local translation. In addition, we introduce the directional derivative L_a,

$$L_a \equiv a\cdot\overline{h}(\nabla) \tag{3.20}$$

and the covariant directional derivative $a\cdot\mathcal{D}$,

$$a\cdot\mathcal{D}A \equiv a\cdot\overline{h}(\nabla)A + \omega(a)\times A. \tag{3.21}$$

(Note that this notation implies that $\underline{h}^{-1}(a)\cdot\mathcal{D} = \mathcal{D}_a$.) From (3.21) we define the covariant extension of the vector derivative by

$$\mathcal{D}A = \partial_a a\cdot\mathcal{D}A. \tag{3.22}$$

As with the vector derivative, $\mathcal{D}$ inherits the algebraic properties of a vector. We can therefore write

$$\mathcal{D}A = \mathcal{D}\cdot A + \mathcal{D}\wedge A, \tag{3.23}$$

where

$$\mathcal{D}\cdot A \equiv \partial_a\cdot(a\cdot\mathcal{D}A) \tag{3.24}$$
$$\mathcal{D}\wedge A \equiv \partial_a\wedge(a\cdot\mathcal{D}A). \tag{3.25}$$

General considerations have led us to the introduction of two new gauge fields, the $\overline{h}(a,x)$ linear function and the $\Omega(a,x)$ bivector-valued linear function, both of which are arbitrary functions of position vector x. This gives a total of $4\times4+4\times6 = 40$ scalar degrees of freedom. The $\overline{h}(a)$ and $\Omega(a)$ fields are incorporated into the vector derivative to form the operator $\mathcal{D} = \overline{h}(\partial_a)\mathcal{D}_a$, which acts covariantly on multivector fields. Thus we can begin to construct equations whose intrinsic content is free from the manner in which we choose to represent spacetime positions with vectors.

4 The Field Equations

Having introduced the $\overline{h}$- and Ω-fields, we look to construct a covariant set of field equations. We start by defining the field-strength via

$$R(a\wedge b)\times A \equiv [\mathcal{D}_a, \mathcal{D}_b]A, \tag{4.1}$$

$$\Rightarrow R(a\wedge b) = a\cdot\nabla\Omega(b) - b\cdot\nabla\Omega(a) + \Omega(a)\times\Omega(b). \tag{4.2}$$

$R(a\wedge b)$ is a bivector-valued linear function of its bivector argument $a\wedge b$. We write this as $R(B)$, where B denotes an arbitrary bivector. As always, the position dependence can be made explicit by writing $R(B,x)$ or $R_x(B)$.

The definition (4.2) ensures that under local rotations $R(B)$ transforms as

$$R(B)\mapsto R'(B) = RR(B)\tilde{R}, \tag{4.3}$$

and under local translations $R(B)$ transforms as

$$R(B)\mapsto R'(B) = R(\underline{f}(B), f(x)). \tag{4.4}$$

A covariant quantity can therefore be constructed by defining

$$\mathcal{R}(B) \equiv R\underline{h}(B). \tag{4.5}$$

Under local translations and rotations $\mathcal{R}(B)$ has the following transformation laws:

$$\begin{aligned} \text{Translations:}\quad \mathcal{R}'(B,x) &= \mathcal{R}(B,x') \\ \text{Rotations:}\quad \mathcal{R}'(B,x) &= R\mathcal{R}(\tilde{R}BR)\tilde{R}. \end{aligned} \tag{4.6}$$

Any linear function with transformation laws of this type is referred to as a covariant tensor. We have begun to employ a notation which is very helpful for the theory developed here. Covariant quantities such as $\mathcal{R}(B)$ and $\mathcal{D}$ are written with calligraphic ('curly') symbols. This helps keep track of the true covariant quantities, so we can easily read off the 'intrinsic' physical relations between fields.

It is useful to extend (4.2) to a form that gives $\mathcal{R}(B)$ directly. This is achieved by writing

$$\mathcal{R}(a\wedge b) = L_a\omega(b) - L_b\omega(a) + \omega(a)\times\omega(b) - \omega(c), \tag{4.7}$$

where

$$c = \underline{h}^{-1}(L_a\underline{h}(b) - L_b\underline{h}(a)). \tag{4.8}$$

From $\mathcal{R}(B)$ we define the following contractions:

$$\begin{aligned} \text{Ricci Tensor:} \quad \mathcal{R}(b) &= \partial_a\cdot\mathcal{R}(a\wedge b) \\ \text{Ricci Scalar:} \quad \mathcal{R} &= \partial_a\cdot\mathcal{R}(a) \\ \text{Einstein Tensor:} \quad \mathcal{G}(a) &= \mathcal{R}(a) - \tfrac{1}{2}a\mathcal{R}. \end{aligned} \tag{4.9}$$

The use of the symbol $\mathcal{R}$ for all of the Riemann tensor, Ricci tensor and Ricci scalar does not pose a problem, since the argument always specifies which is required. Both $\mathcal{R}(a)$ and $\mathcal{G}(a)$ are covariant tensors, since they inherit the same transformation properties as $\mathcal{R}(B)$.

The natural way to proceed now is to construct an invariant action integral and vary this with respect to the $\overline{h}$- and ω-functions. This is the approach adopted in [9] and here we simply quote the necessary results. In the full gauge-theoretic treatment, gravity is introduced by minimally coupling the $\overline{h}$- and ω-fields to the Dirac action. The only action integral which results in a self-consistent theory is the Hilbert action (together with a possible cosmological term)

$$S = \int |d^4x| \det \underline{h}^{-1}(\tfrac{1}{2}\mathcal{R} + \Lambda - \kappa\mathcal{L}_m), \tag{4.10}$$

where $\mathcal{L}_m$ describes the matter content, Λ is the cosmological term and $\kappa = 8\pi G$. In the absence of spin, the equations obtained from varying this equation are

$$\mathcal{D}\wedge\overline{h}(a) = \overline{h}(\nabla\wedge a) \tag{4.11}$$

and

$$\mathcal{G}(a) - \Lambda a = \kappa\mathcal{T}(a), \tag{4.12}$$

where $\mathcal{T}(a)$ is the matter stress-energy tensor. The latter equation shows that our gauge theory is very close to standard GR. Indeed, one can recover the field equations of GR by introducing a set of coordinates $\{x^\mu\}$ with an associated coordinate frame $\{e_\mu\}$. A metric tensor is then defined by

$$g_{\mu\nu} = \underline{h}^{-1}(e_\mu)\cdot\underline{h}^{-1}(e_\nu). \tag{4.13}$$

But crucial differences do exist between the gauge theory developed here and conventional GR. One of these is immediately apparent from our use of a flat background, which is that the differential equations (4.11) and (4.12) have equivalent integral forms. This is particularly significant for the treatment of singularities in our theory [9]. Further differences will emerge when we consider some applications in Section 6.

The significance of equation (4.11) becomes clearer once one forms

$$\begin{aligned} \bar{h}(\dot{\nabla})\wedge\dot{\bar{h}}(c) &= -\partial_d\wedge(\omega(d)\cdot\bar{h}(c)) \\ \Rightarrow \langle b\wedge a\bar{h}(\dot{\nabla})\wedge\dot{\bar{h}}(c)\rangle &= -\langle b\wedge a\partial_d\wedge(\omega(d)\cdot\bar{h}(c))\rangle \\ \Rightarrow [\dot{L}_a\dot{\underline{h}}(b) - \dot{L}_b\dot{\underline{h}}(a)]\cdot c &= [a\cdot\omega(b) - b\cdot\omega(a)]\cdot\bar{h}(c) \end{aligned} \tag{4.14}$$

where, as always, the overdots determine the scope of a differential operator. The left-hand side of equation (4.14) contains a term that appears in the commutator of L_a and L_b, which suggests that we form

$$\begin{aligned} [L_a, L_b] &= [L_a\underline{h}(b) - L_b\underline{h}(a)]\cdot\nabla \\ &= [\dot{L}_a\dot{\underline{h}}(b) - \dot{L}_b\dot{\underline{h}}(a)]\cdot\nabla + (L_a b - L_b a)\cdot\bar{h}(\nabla) \\ &= [a\cdot\omega(b) - b\cdot\omega(a) + L_a b - L_b a]\cdot\bar{h}(\nabla). \end{aligned} \tag{4.15}$$

We can therefore write

$$[L_a, L_b] = L_c \tag{4.16}$$

where

$$c = a\cdot\omega(b) - b\cdot\omega(a) + L_a b - L_b a. \tag{4.17}$$

This 'bracket' structure summarises the content of (4.11). It also simplifies the definition of $\mathcal{R}(a\wedge b)$ (4.7) since the vector c defined by equation (4.8) can now be found from equation (4.17).

The general technique we use for studying the field equations is to let $\omega(a)$ contain a set of arbitrary functions, and then use (4.17) to find relations between them. In doing so we lose some of the information contained in the 'wedge' equation (4.11). This information is recovered by employing the symmetry properties of $\mathcal{R}(B)$ and the Bianchi identity. The symmetry of $\mathcal{R}(B)$ is summarised by the single equation

$$\partial_a\wedge\mathcal{R}(a\wedge b). \tag{4.18}$$

As $\mathcal{R}(B)$ maps bivectors to bivectors it has, at most, 36 degrees of freedom. Equation (4.18) gives a set of 16 scalar equations, reducing the number of degrees of freedom in $\mathcal{R}(B)$ to 20 — the expected number for the Riemann tensor. It is notable how easy this calculation is in geometric algebra! The Bianchi identity follows from a simple application of the Jacobi identity, and can be written

$$\partial_a\wedge(a\cdot\mathcal{D}\mathcal{R}(B) - \mathcal{R}(a\cdot\mathcal{D}B)) = 0. \tag{4.19}$$

In terms of the L_a and $\omega(a)$, equation (4.19) becomes

$$\partial_a\wedge[L_a\mathcal{R}(B) - \mathcal{R}(L_a B) + \omega(a)\times\mathcal{R}(B) - \mathcal{R}(\omega(a)\times B)] = 0. \tag{4.20}$$

The contracted Bianchi identity is

$$\partial_a \cdot [L_a \mathcal{G}(b) - \mathcal{G}(L_a b) + \omega(a) \times \mathcal{G}(b) - \mathcal{G}(\omega(a) \times b)] = 0, \tag{4.21}$$

and it follows from (4.12)) that the matter stress-energy tensor must satisfy the same equation. This is the covariant version of conservation of the stress-energy tensor.

The final relation we need before applying this formalism is to invert (4.11) to give $\omega(a)$ in terms of $\overline{h}(a)$. We first write equation (4.11) as

$$\overline{h}(\dot{\nabla}) \wedge \dot{\overline{h}} \, \overline{h}^{-1}(a) + \partial_b \wedge (\omega(b) \cdot a) = 0. \tag{4.22}$$

On defining

$$B(a) \equiv -\overline{h}(\dot{\nabla}) \wedge \dot{\overline{h}} \, \overline{h}^{-1}(a) = \overline{h}(\dot{\nabla} \wedge \dot{\overline{h}}^{-1}(a)) \tag{4.23}$$

equation (4) becomes

$$\partial_b \wedge (\omega(b) \cdot a) = B(a). \tag{4.24}$$

We solve this equation by first 'protracting' to give

$$\partial_a \wedge \partial_b \wedge (\omega(b) \cdot a) = 2\partial_b \wedge \omega(b) = \partial_b \wedge B(b) \tag{4.25}$$

and then dotting with a, leaving

$$\omega(a) - \partial_b \wedge (a \cdot \omega(b)) = \tfrac{1}{2} a \cdot (\partial_b \wedge B(b)). \tag{4.26}$$

Hence, using equation (4.24) again, we find that

$$\omega(a) = -B(a) + \tfrac{1}{2} a \cdot (\partial_b \wedge B(b)). \tag{4.27}$$

This completes the required set of equations.

5 Spherically-Symmetric Matter Distributions

We now apply the formalism of Section 4 to a time-dependent radially-symmetric perfect fluid. We temporarily drop the cosmological term from equation (4.12). It will be replaced when we come to consider cosmological applications. The first step is to introduce a set of polar coordinates. In terms of the fixed $\{\gamma_\mu\}$ frame we define:

$$\begin{aligned} t &\equiv x \cdot \gamma_0 \\ r &\equiv \sqrt{(x \wedge \gamma_0)^2} \\ \cos\theta &\equiv x \cdot \gamma^3 / r \\ \tan\phi &\equiv (x \cdot \gamma^2)/(x \cdot \gamma^1). \end{aligned} \tag{5.1}$$

The associated coordinate frame is

$$\begin{aligned} e_t &\equiv \gamma_0 \\ e_r &\equiv x\wedge\gamma_0\,\gamma_0/r \\ e_\theta &\equiv r\cos\theta(\cos\phi\,\gamma_1+\sin\phi\,\gamma_2)-r\sin\theta\,\gamma_3 \\ e_\phi &\equiv r\sin\theta(-\sin\phi\,\gamma_1+\cos\phi\,\gamma_2) \end{aligned} \tag{5.2}$$

and the dual-frame vectors are denoted as $\{e^t, e^r, e^\theta, e^\phi\}$. We will also frequently employ the unit vectors $\hat{\theta}$ and $\hat{\phi}$ defined by

$$\hat{\theta} \equiv e_\theta/r \tag{5.3}$$

$$\hat{\phi} \equiv e_\phi/(r\sin\theta). \tag{5.4}$$

Rotational symmetry is imposed on $\overline{h}(a)$ as follows. If B is a constant spatial bivector ($e_t\cdot B=0$) and

$$R = e^{B/2} \tag{5.5}$$

$$x' = \tilde{R}xR, \tag{5.6}$$

then rotating $\overline{h}(a)$ and translating it to the back-rotated position x' should leave $\overline{h}(a)$ unchanged. Hence rotational symmetry requires that

$$R\overline{h}_{x'}(\tilde{R}aR)\tilde{R} = \overline{h}(a). \tag{5.7}$$

This places a strong restriction on $\overline{h}(a)$, and the most general form that $\overline{h}(a)$ can take is given by

$$\begin{aligned} \overline{h}(e^t) &= f_1e^t+f_2e^r \\ \overline{h}(e^r) &= g_1e^r+g_2e^t \\ \overline{h}(e^\theta) &= \alpha e^\theta \\ \overline{h}(e^\phi) &= \alpha e^\phi, \end{aligned} \tag{5.8}$$

where f_1, f_2, g_1, g_2 and α are all functions of t and r only. Rotational symmetry alone does not fix all the gauge freedom in the $\overline{h}$-function because we are free to reparameterise t and r without altering the functional form of (5.8). Once we have identified the intrinsic quantities, however, a natural gauge choice will emerge in which $\overline{h}(a)$ is expressed solely in terms of physically-determined variables.

Conventional GR would now proceed by constructing the line element from (5.8). This has the general form

$$ds^2 = g_{00}(t,r)\,dt^2+g_{01}(t,r)\,dt\,dr+g_{11}(t,r)\,dr^2+g_{22}(t,r)(d\theta^2+\sin^2\theta\,d\phi^2). \tag{5.9}$$

From (5.9) the connection coefficients are then found, and the Riemann and Einstein tensors computed. This procedure results in a set of complicated non-linear second-order equations which are difficult to solve. Moreover, the gauge-transformation properties of the quantities involved are hard to recover. Here we follow a different and, to our knowledge, entirely new approach. We use equation (4.27) to find what terms must be present in $\omega(a)$ for the $\overline{h}$-field specified by (5.8), but leave the individual coefficients as arbitrary functions. The only terms from $\overline{h}(a)$ which are kept in $\omega(a)$ are those which are undifferentiated. These arise from frame derivatives and the algebra is usually simplified if these terms are left explicit in $\omega(a)$. Applying this procedure to (5.8) we are led to the following form for $\omega(a)$:

$$\begin{aligned} \omega(e_t) &= G e_r e_t \\ \omega(e_r) &= F e_r e_t \\ \omega(\hat{\theta}) &= S\hat{\theta} e_t + (T - \alpha/r) e_r \hat{\theta} \\ \omega(\hat{\phi}) &= S\hat{\phi} e_t + (T - \alpha/r) e_r \hat{\phi}, \end{aligned} \tag{5.10}$$

where G, F, S and T are functions of t and r only.

The bracket relations obtained from (5.10) are as follows:

$$\begin{aligned} [L_t, L_r] &= GL_t - FL_r & \qquad [L_r, L_{\hat{\theta}}] &= -TL_{\hat{\theta}} \\ [L_t, L_{\hat{\theta}}] &= -SL_{\hat{\theta}} & [L_r, L_{\hat{\phi}}] &= -TL_{\hat{\phi}} \\ [L_t, L_{\hat{\phi}}] &= -SL_{\hat{\phi}} & [L_\theta, L_\phi] &= 0. \end{aligned} \tag{5.11}$$

Here L_t, L_r, L_θ and L_ϕ are abbreviations for the associated coordinate frame derivatives, so $L_\theta = e_\theta \cdot \overline{h}(\nabla)$ *etc.* For the unit vectors we of course have $L_{\hat{\theta}} = \hat{\theta} \cdot \overline{h}(\nabla)$ and $L_{\hat{\phi}} = \hat{\phi} \cdot \overline{h}(\nabla)$. Use of the unit vectors $\hat{\theta}$ and $\hat{\phi}$ eliminates the need to calculate some coordinate derivatives which cancel out of the final results.

We now apply equations (4.7) and (4.17) to compute the Riemann tensor and find that

$$\begin{aligned} \mathcal{R}(e_r e_t) &= (L_r G - L_t F + G^2 - F^2) e_r e_t \\ \mathcal{R}(\hat{\theta} e_t) &= (-L_t S + GT - S^2)\hat{\theta} e_t + (-L_t T + SG - ST) e_r \hat{\theta} \\ \mathcal{R}(\hat{\phi} e_t) &= (-L_t S + GT - S^2)\hat{\phi} e_t + (-L_t T + SG - ST) e_r \hat{\phi} \\ \mathcal{R}(e_r \hat{\theta}) &= (L_r T - FS + T^2) e_r \hat{\theta} + (L_r S - FT + ST)\hat{\theta} e_t \\ \mathcal{R}(e_r \hat{\phi}) &= (L_r T - FS + T^2) e_r \hat{\phi} + (L_r S - FT + ST)\hat{\phi} e_t \\ \mathcal{R}(\hat{\theta}\hat{\phi}) &= (-S^2 + T^2 - (\alpha/r)^2)\hat{\theta}\hat{\phi}. \end{aligned} \tag{5.12}$$

In arriving at (5.12) the bracket structure is used to simplify the derivatives of α/r. Since $\alpha/r = L_{\hat{\theta}}\theta$, we have

$$L_t(\frac{\alpha}{r}) = L_t L_{\hat{\theta}}\theta = [L_t, L_{\hat{\theta}}]\theta = -S\frac{\alpha}{r} \tag{5.13}$$

and

$$L_r(\frac{\alpha}{r}) = L_r L_{\hat{\theta}}\theta = [L_r, L_{\hat{\theta}}]\theta = -T\frac{\alpha}{r}. \tag{5.14}$$

To proceed, we must now decide what form the matter stress-energy tensor should take. We assume that the matter is modelled by an ideal fluid so we can write

$$\mathcal{T}(a) = (\rho + p) a \cdot v v - p a, \tag{5.15}$$

where ρ is the energy density, p is the pressure and v is the covariant fluid velocity ($v^2 = 1$). Radial symmetry means that v can only be in the e_t and e_r directions, so v must be of the form $(\cosh u\, e_t + \sinh u\, e_r)$. But, in the $\overline{h}$-function defined by equation (5.8), we are free to perform an arbitrary radial boost as this does not change the functional form of $\overline{h}(a)$. It follows that we are free to apply any radial boost to v. We use this freedom to set $v = e_t$ so that the matter stress-energy tensor becomes

$$\mathcal{T}(a) = (\rho + p)a \cdot e_t e_t - pa. \tag{5.16}$$

Contracting (5.12) to form the Ricci and Einstein tenors and equating with (5.16) we obtain the following set of equations:

$$\begin{aligned} L_t S &= 2A + GT - S^2 - 4\pi p && (5.17)\\ L_t T &= S(G - T) && (5.18)\\ L_r S &= T(F - S) && (5.19)\\ L_r T &= -2A + FS - T^2 - 4\pi\rho && (5.20)\\ L_r G - L_t F &= F^2 - G^2 + 4A + 4\pi(\rho + p) && (5.21) \end{aligned}$$

where

$$A \equiv \tfrac{1}{4}(-S^2 + T^2 - (\alpha/r)^2). \tag{5.22}$$

Feeding these relations back into (5.12) we find that the Riemann tensor can be written concisely as

$$\mathcal{R}(B) = (A + \frac{2\pi}{3}\rho)(B + 3e_r e_t B e_r e_t) + 4\pi[(\rho + p)B \cdot e_t e_t - \frac{2}{3}\rho B]. \tag{5.23}$$

The first term in this expression for the Riemann tensor is the Weyl tensor, and the second is the term that contracts to give the Ricci tensor. This latter contribution to $\mathcal{R}(B)$ does not appear to have been given a name in the literature. It is perhaps best thought of as the 'source' term.

The remaining equations are contained in the Bianchi identities. It turns out that, in this case, the full Bianchi identities are satisfied using only the above equations and the contracted identity (4.21). The contracted Bianchi identity produces the pair of equations

$$\mathcal{D} \cdot (\rho e_t) + p\mathcal{D} \cdot e_t = 0 \tag{5.24}$$
$$(\rho + p)(e_t \cdot \mathcal{D} e_t) \wedge e_t - (\mathcal{D} p) \wedge e_t = 0, \tag{5.25}$$

which reduce to

$$\begin{aligned} L_t \rho &= -(F + 2S)(\rho + p) && (5.26)\\ L_r p &= -G(\rho + p). && (5.27) \end{aligned}$$

Equations (5.13), (5.14), (5.17)–(5.21), (5.26) and (5.27) and the bracket relation

$$[L_t, L_r] = GL_t - FL_r \tag{5.28}$$

form the complete set of intrinsic equations. The structure is closed, in that it is easily verified that the bracket relation (5.28) is consistent with the known derivatives. There is a remarkable (pseudo)duality between these equations, in that the structure is almost completely unchanged under the simultaneous interchange $F \leftrightarrow G$, $S \leftrightarrow T$, $\rho \leftrightarrow p$ and $L_t \leftrightarrow L_r$. The duality is not exact because of the α/r term (though this term can be viewed as an artefact of the r coordinate) and the matter terms. Elsewhere [9] we exhibit a similar duality in static rotating systems where, for the matter-free case, the duality is exact.

To simplify this structure we start by forming the derivatives of A. From equations (5.13), (5.14) and (5.17)–(5.21) we find that

$$\begin{aligned} L_tA + 3SA &= 2\pi Sp & (5.29)\\ L_rA + 3TA &= -2\pi T\rho. & (5.30) \end{aligned}$$

The above results for the derivatives of A, and equations (5.18) and (5.19), suggest that we should look for an integrating factor for the $L_t + S$ and $L_r + T$ operators. Such a function, X say, should have the properties that

$$\begin{aligned} L_tX &= SX & (5.31)\\ L_rX &= TX. & (5.32) \end{aligned}$$

A function with these properties can only be found if the derivatives are consistent with the bracket relation (5.28). We check this by forming

$$\begin{aligned} [L_t, L_r]X &= L_t(TX) - L_r(SX)\\ &= X(L_tT - L_rS)\\ &= X(SG - FT)\\ &= GL_tX - FL_rX, & (5.33) \end{aligned}$$

so the properties of X are consistent with (5.28) and we can assume the existence of such a function. Once a particular form of $\overline{h}(a)$ has been chosen X can be found from S and T by partial integration and self-consistency of this procedure is guaranteed by (5.33).

Now that X is available to us we can simplify many of the above equations. We start by defining

$$M \equiv -2X^3A \tag{5.34}$$

so that

$$\begin{aligned} L_tM &= -4\pi SX^3p\\ L_rM &= 4\pi TX^3\rho. \end{aligned} \tag{5.35}$$

We next find that

$$L_t(X\frac{\alpha}{r}) = 0, \qquad L_r(X\frac{\alpha}{r}) = 0 \tag{5.36}$$

from which it follows that

$$X\frac{\alpha}{r} = c, \tag{5.37}$$

where c is a constant. Equations (5.31) and (5.32) only define X up to a constant multiple, however, so we can absorb the constant c into X and write

$$\frac{\alpha}{r} = \frac{1}{X}. \tag{5.38}$$

It is now clear that X plays the role of an intrinsic distance variable. A natural gauge choice is therefore to set the distance scale so that

$$r = X \tag{5.39}$$

which implies that

$$\alpha = 1. \tag{5.40}$$

In making this gauge choice we ensure that the distance scale of the background vector space matches the scale defined physically by the gravitational fields.

From the form of $\overline{h}(a)$ (5.8) and equations (5.31) and (5.32) we now see that

$$g_1 = L_r r = Tr \tag{5.41}$$

$$g_2 = L_t r = Sr \tag{5.42}$$

and it follows that M is given by

$$M = \frac{r}{2}({g_2}^2 - {g_1}^2 + 1). \tag{5.43}$$

In addition, we now find that G and F satisfy

$$L_t g_1 = G g_2 \tag{5.44}$$

$$L_r g_2 = F g_1. \tag{5.45}$$

The 'Newtonian' Gauge

To complete the solution we need to make a gauge choice for the t coordinate. Unlike the distance variable X, no natural time coordinate has yet emerged. We must therefore look for some additional criteria to motivate a choice of time coordinate. Returning to the derivatives of M (5.35) we find that these formulae can be inverted to yield

$$\frac{\partial M}{\partial t} = \frac{-4\pi g_1 g_2 r^2(\rho + p)}{f_1 g_1 - f_2 g_2} \tag{5.46}$$

$$\frac{\partial M}{\partial r} = \frac{4\pi r^2(f_1 g_1 \rho + f_2 g_2 p)}{f_1 g_1 - f_2 g_2}. \tag{5.47}$$

The second equation reduces to a simple classical relation if we choose $f_2 = 0$ as we then obtain

$$\partial_r M = 4\pi r^2 \rho, \tag{5.48}$$

which says that $M(r,t)$ is determined by the amount of mass-energy in a sphere of radius r. There are other reasons for choosing the time variable such that $f_2 = 0$. For example, we can then use the bracket structure to solve for f_1. With $f_2 = 0$ we have

$$\begin{aligned} L_t &= f_1\partial_t + g_2\partial_r && (5.49)\\ L_r &= g_1\partial_r && (5.50) \end{aligned}$$

and the bracket relation (5.28) implies that

$$\begin{aligned} L_r f_1 &= -Gf_1 \\ \Rightarrow\ \partial_r f_1 &= \frac{G}{g_1} f_1 \\ \Rightarrow\ f_1 &= \epsilon(t)\exp\{-\int^r \frac{G}{g_1} dr\}. && (5.51) \end{aligned}$$

The function $\epsilon(t)$ can be absorbed by a further rescaling of t, so we are left with the simple result that

$$f_1 = \exp\{-\int^r \frac{G}{g_1} dr\}. \tag{5.52}$$

Another reason why $f_2 = 0$ is a natural gauge choice is seen when the pressure is zero. In this case equation (5.27) forces G to be zero, and equation (5.52) then sets $f_1 = 1$. A free-falling particle with $v = e_t$ (*i.e.* comoving with the fluid) then has

$$\dot{t}e_t + \dot{r}e_r = e_t + g_2 e_r, \tag{5.53}$$

where the dots denote differentiation with respect to the affine parameter. Since $\dot{t} = 1$ the time coordinate t matches the proper time of all observers comoving with the fluid. So, in the absence of pressure, we are able to recover a global 'Newtonian' time on which all observers can agree (provided all clocks are correlated initially, which is not hard to do). Furthermore, it is also clear from (5.53) that g_2 is the velocity of the particle. Hence equation (5.46), which reduces to

$$\frac{\partial M}{\partial t} = -4\pi g_2 r^2 \rho \tag{5.54}$$

in the absence of pressure, has a simple Newtonian interpretation — it equates the work with the rate of flow of energy density. The equation for M (5.43) in the form

$$\tfrac{1}{2}{g_2}^2 - \frac{M}{r} = \tfrac{1}{2}({g_1}^2 - 1) \tag{5.55}$$

The $\overline{h}$-function	The ω-function
$\overline{h}(e^t) = f_1 e^t$ $\overline{h}(e^r) = g_1 e^r + g_2 e^t$ $\overline{h}(e^\theta) = e^\theta$ $\overline{h}(e^\phi) = e^\phi$	$\omega(e_t) = G e_r e_t$ $\omega(e_r) = F e_r e_t$ $\omega(\hat{\theta}) = g_2/r\, \hat{\theta} e_t + (g_1 - 1)/r\, e_r \hat{\theta}$ $\omega(\hat{\phi}) = g_2/r\, \hat{\phi} e_t + (g_1 - 1)/r\, e_r \hat{\phi}$
Directional derivatives	$L_t = f_1 \partial_t + g_2 \partial_r$ $L_r = g_1 \partial_r$
Equations relating the $\overline{h}$- and ω-functions	$L_t g_1 = G g_2$ $L_r g_2 = F g_1$ $f_1 = \exp\{\int^r -G/g_1\, dr\}$
Definition of M	$M \equiv \frac{1}{2} r ({g_2}^2 - {g_1}^2 + 1)$
Remaining derivatives	$L_t g_2 = G g_1 - M/r^2 - 4\pi r p$ $L_r g_1 = F g_2 + M/r^2 - 4\pi r \rho$
Matter derivatives	$L_t M = -4\pi g_2 r^2 p \quad L_t \rho = -(2g_2/r + F)(\rho + p)$ $L_r M = 4\pi g_1 r^2 \rho \quad L_r p = -G(\rho + p)$
Riemann tensor	$\mathcal{R}(B) = 4\pi[(\rho + p) B \cdot e_t e_t - 2\rho/3\, B]$ $\quad - \frac{1}{2}(M/r^3 - 4\pi\rho/3)(B + 3 e_r e_t B e_r e_t)$
Fluid stress-energy tensor	$\mathcal{T}(a) = (\rho + p) a \cdot e_t e_t - p a$

Table 1: *Equations governing a radially-symmetric perfect fluid.*

is also now familiar from Newtonian physics — it is a Bernoulli equation for zero pressure and total (non-relativistic) energy $({g_1}^2 - 1)/2$. For these various reasons we refer to $f_2 = 0$ as defining the 'Newtonian' gauge. This is the gauge choice we employ for many of the applications to follow.

The equations of 'intrinsic fluid dynamics' in the Newtonian gauge are summarised in Table 1.

6 Applications

In the following sections we develop a number of applications of the equations summarised in Table 1. We start by treating the well-known case of a static, radially-symmetric star.

6.1 Static Matter Distributions

For a static matter distribution, ρ and p are functions of r only. It follows from the derivatives of M that

$$M(r) = \int_0^r 4\pi r'^2 \rho(r')\, dr' \tag{6.1}$$

and

$$L_t M = g_2 4\pi r^2 \rho = -g_2 4\pi r^2 p. \tag{6.2}$$

But, for any physical matter distribution, ρ and p must both be positive, in which case equation (6.2) can only be satisfied if

$$g_2 = 0 \tag{6.3}$$

$$\Rightarrow F = 0. \tag{6.4}$$

Since $g_2 = 0$ we see that g_1 is given simply in terms of M by

$${g_1}^2 = 1 - 2M/r, \tag{6.5}$$

which recovers contact with the standard line element for a static, radially-symmetric field.

The remaining equation of use is that for $L_t g_2$ which now gives

$$G g_1 = M/r^2 + 4\pi r p. \tag{6.6}$$

Equations (6.5) and (6.6) now combine with that for $L_r p$ to yield the famous Oppenheimer-Volkov equation

$$\frac{\partial p}{\partial r} = -\frac{(\rho + p)(M(r) + 4\pi r^3 p)}{r(r - 2M(r))}. \tag{6.7}$$

By this point we have successfully recovered all the usual equations governing a non-rotating star. The work involved in recovering these equations from the full time-dependent case is minimal, and the final form of $\overline{h}(a)$ is very simple (it is a diagonal function). What's more, the meaning of the t and r coordinates is clear, since they have been defined operationally. The final equations found here do not differ from those of GR, however, so no observable differences can be expected for the case of a stationary star.

6.2 Point-Source Solutions — Black Holes

The next solution of interest is obtained when the matter is concentrated at a single point ($r = 0$). For such a solution, $\rho = p = 0$ everywhere away from the source, and the matter equations reduce to

$$\left.\begin{array}{l} L_t M = 0 \\ L_r M = 0 \end{array}\right\} \Rightarrow M = \text{constant}. \tag{6.8}$$

Maintaining the symbol M for this constant we now find that the equations reduce to

$$L_t g_1 = G g_2 \tag{6.9}$$
$$L_r g_2 = F g_1 \tag{6.10}$$

and

$${g_1}^2 - {g_2}^2 = 1 - 2M/r. \tag{6.11}$$

There are no further equations which yield new information. In the vacuum, therefore, we have an under-determined system of equations and some additional gauge-fixing is needed to choose an explicit form of $\overline{h}(a)$. The reason for this is that in the vacuum region the Riemann tensor reduces to

$$\mathcal{R}(B) = -\frac{M}{2r^3}(B + 3e_r e_t B e_r e_t). \tag{6.12}$$

This tensor is now *invariant* under boosts in the $e_r e_t$ plane. In the presence of matter the source contribution to $\mathcal{R}(B)$ breaks this invariance since the e_t vector is transformed to a different vector under a radial boost. This new symmetry of $\mathcal{R}(B)$ appears as soon as ρ and p vanish and manifests itself as a new freedom in the choice of $\overline{h}$-function.

Given that this new freedom exists, we should look for a choice of g_1 and g_2 which simplifies the physics as far as possible. If we attempt to reproduce the Schwarzschild solution we would set $g_2 = 0$, but then we immediately run into difficulties with g_1, which is not defined for $r < 2M$. We must therefore look for an alternative gauge choice. We will see in the following section that for collapsing dust g_1 controls the energy of infalling matter at $r = \infty$. A sensible gauge choice is therefore to set

$$g_1 = 1 \tag{6.13}$$

so that

$$g_2 = -\sqrt{2M/r} \tag{6.14}$$
$$G = 0 \tag{6.15}$$
$$F = -M/(g_2 r^2) \tag{6.16}$$
$$f_1 = 1. \tag{6.17}$$

In this gauge the $\overline{h}$-function takes the remarkably simple form

$$\overline{h}(a) = a - \sqrt{2M/r}\, a\cdot e_r e_t, \tag{6.18}$$

which only differs from the identity through a single term. The geodesic equation for a radially-infalling particle (with unit mass) reduces to

$$\begin{aligned} \dot{r}^2 &= 2M/r + \sinh^2 u_0 & (6.19)\\ (1-2M/r)\dot{t} &= \cosh u_0 + \dot{r}\sqrt{2M/r}. & (6.20) \end{aligned}$$

The $\dot{r}$ equation shows immediately that $\ddot{r} = -M/r^2$ and the constant $\sinh^2 u_0$ can be identified with twice the initial kinetic energy of the infalling particle. At the horizon ($r = 2M$) $\dot{r} = -\cosh u_0$, so there is no pole in $\dot{t}$ (6.20). All particles cross the horizon and reach the singularity in a finite coordinate time. Some possible trajectories are illustrated in Figure 1.

In the case where the particle is dropped from rest at $r = \infty$ equations (6.19) and (6.20) reduce to

$$\dot{r} = -\sqrt{2M/r}, \quad \dot{t} = 1, \tag{6.21}$$

and we recover an entirely Newtonian description of the motion. The properties of a black hole are so simple in the gauge defined by (6.18) that it is astonishing that it is almost never seen in the literature. This is presumably because the line element associated with (6.18) does not look as natural as the $\overline{h}$-function itself, and hides the underlying simplicity of the system. In the Newtonian gauge one hardly needs to modify classical reasoning at all to understand the processes involved — all particles just cross the horizon and fall into the singularity in a finite coordinate time. And the horizon is located at $r = 2M$ precisely because we can apply Newtonian arguments! The only departures from Newtonian physics lie in relativistic corrections to the proper-time taken for infall, and in modifications to the equations for angular motion which lead to the familiar results for orbital precession.

Horizons and Birkhoff's Theorem

The bulk of the literature on black holes works with the Schwarzschild solution, which is obtained by setting $g_2 = 0$. In this case

$$g_1 = \sqrt{1-2M/r} \tag{6.22}$$

which is only defined for $r > 2M$. GR starts to invoke coordinate transformations at this point, usually resulting in the advanced or retarded Eddington-Finkelstein form of the solution. In our gauge theory, however, the picture is different; $g_2 = 0$ is simply *not* an allowed solution since it does not result in a globally-defined $\overline{h}$-function. It is over such questions regarding the global nature of fields that differences between

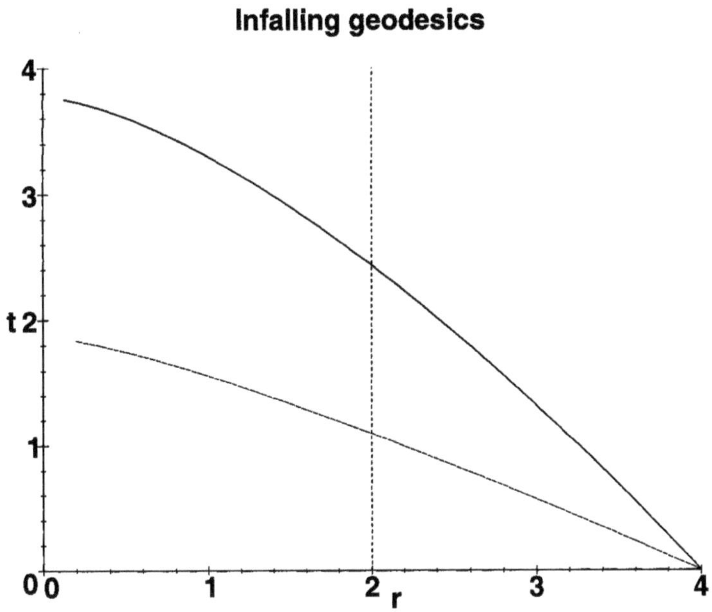

Figure 1: *Possible particle trajectories for radially-infalling particles in the Newtonian gauge. The upper curve is for a particle released with* $\dot{r} = -\sqrt{2M/r}$ *at* $r = 4M$*, while the lower curve is for a photon. The vertical dotted line indicates the horizon.*

our 'flatspace' gauge theory and GR start to emerge. Here the differences manifest themselves quite clearly. We are forced to a solution in which $g_2 \neq 0$, so $\overline{h}(a)$ is not diagonal and hence not time-reverse symmetric. Time reversal is achieved by combining the translation

$$f(x) = -e_t x e_t = x' \tag{6.23}$$

with the reflection

$$\overline{h}'(a) = -e_t \overline{h}(a) e_t. \tag{6.24}$$

Hence the time-reversed solution is given by

$$\overline{h}^*(a) = e_t \overline{h}_{x'}(e_t a e_t) e_t. \tag{6.25}$$

Applying this to the solution (6.18) we obtain

$$\overline{h}^*(a) = a + \sqrt{2M/r}\, a\cdot e_r e_t, \tag{6.26}$$

which has changed the sign of the off-diagonal term. The result is a solution in which particles inside the horizon are swept out. Once outside, the force is still attractive but particles cannot re-enter back through the horizon. Despite the fact that the Riemann tensor is unchanged by time-reversal, it is impossible to find an $\overline{h}$-function that is also time-reverse symmetric. In our theory the black hole has more memory about its formation than simply its mass M. It also remembers that it was formed in a particular time direction. The appearance of a horizon is thus associated with the onset time asymmetry, which is very satisfying from a physical viewpoint.

Another form of the Schwarzschild solution that runs into difficulties in our theory is that obtained by using Kruskal coordinates. These coordinates introduce a 'double-cover' of spacetime so that each value of r determines two distinct hypersurfaces. This is clearly not possible in our theory without some radical redefinition of how r is viewed as a function of spacetime position x. In GR the Kruskal-Szerkes solution is the maximal continuation of the Schwarzschild metric and is viewed as giving the complete description of a radially-symmetric black hole [23]. The fact that it is ruled out of our gauge theory means that our allowed solutions are not 'maximal' and forces us to address the issue of geodesic incompleteness. For the solution (6.18) geodesics exist which cannot be extended into the past for all values of their affine parameter. But, if we adopt the view that the black hole must have formed in the past from some collapse process, then there must have been a time before which the horizon did not exist. Any geodesic must therefore have come from a region in the past where no horizon was present, so there is no question of the geodesics being incomplete. A model of such a collapse process is discussed in the following section. We therefore arrive at a consistent picture in which the formation of a horizon retains information about the direction of time in which collapse occurred, and all geodesics from the past emanate from a period before the horizon formed. This picture is very different from GR, which is happy to deal with eternal, time-symmetric black holes. This shift to a picture with a fixed time direction is typical of the transition from a second-order theory to a first-order one [8].

It is finally worth commenting on how the above affects Birkhoff's theorem. One form of Birkhoff's theorem is that the gravitational fields outside any radially-symmetric distribution of matter are necessarily static. This is seen immediately from equation (6.8) which shows that the Riemann tensor is a function of r only. However, Birkhoff's theorem is frequently used to argue that the line element outside a radially-symmetric body can always be brought to the form [24]

$$ds^2 = (1 - 2M/r)\,dt^2 - (1 - 2M/r)^{-1}\,dr^2 - r^2(d\theta^2 + \sin^2\theta\,d\phi^2). \tag{6.27}$$

As we have seen, this ceases to be correct in our theory if a horizon is present. In this case $\overline{h}(a)$ is independent of t, but cannot be time-reverse symmetric.

6.3 Collapsing Dust

The simplest model for collapsing matter is one in which the pressure is set to zero so that the situation describes collapsing dust. In this case $G = 0$, $f_1 = 1$ and t is the time measured by freely-falling observers (from ∞). The equations of Table 1 reduce to

$$\begin{aligned} F &= \partial_r g_2 && (6.28)\\ M(r,t) &= \int_0^r 4\pi r'^2 \rho(r',t)\, dr', && (6.29) \end{aligned}$$

which define F and M on a time slice, together with the update equations

$$\begin{aligned} \partial_t g_2 + g_2 \partial_r g_2 &= -M/r^2 && (6.30)\\ \partial_t M + g_2 \partial_r M &= 0. && (6.31) \end{aligned}$$

As g_2 is the velocity of a particle comoving with the fluid, equations (6.30) and (6.31) provide a completely Newtonian description of the fluid. Equation (6.30) is the Euler equation with an inverse-square gravitational force, and (6.31) is the equation for conservation of mass. It is also worth noting that the L_t derivative now plays the role of the 'matter' or 'comoving' derivative for the fluid.

Equations (6.30) and (6.31) enable ρ and M to be propagated from one time-slice to the next so, given suitable initial conditions, one can propagate into the future. The system of equations is in fact soluble analytically [9], but here we just consider a numerical simulation. From an initial density profile $\rho(r, t_0)$ and velocity profile $g_2(r, t_0)$, equation (6.29) is used to obtain $M(r, t_0)$. Equations (6.30) and (6.31) are then used to update M and g_2. On any given time-slice, ρ and F can be recovered using equations (6.28) and (6.29), and g_1 can be recovered from g_2 and M. The results of such a simulation are displayed in Figure 2.

Any particle on a radial path has a covariant velocity of the form

$$v = \cosh u\, e_t + \sinh u\, e_r. \tag{6.32}$$

The underlying trajectory has $\dot{x} = \underline{h}(v)$, so the radial motion is determined by

$$\dot{r} = \cosh u\, g_2 + \sinh u\, g_1. \tag{6.33}$$

Since g_2 is negative for collapsing matter, the particle can only achieve an outward velocity if ${g_1}^2 > {g_2}^2$. A horizon therefore forms at the point where

$$2M(r,t)/r = 1. \tag{6.34}$$

The appearance of this horizon is illustrated in Figure 2. It is conventional to extend the horizon back in time along the past light-cone to the origin ($r = 0$), since any particle inside this surface could not have reached the point where $2M/r - 1$ first drops to zero, and hence is also trapped [25].

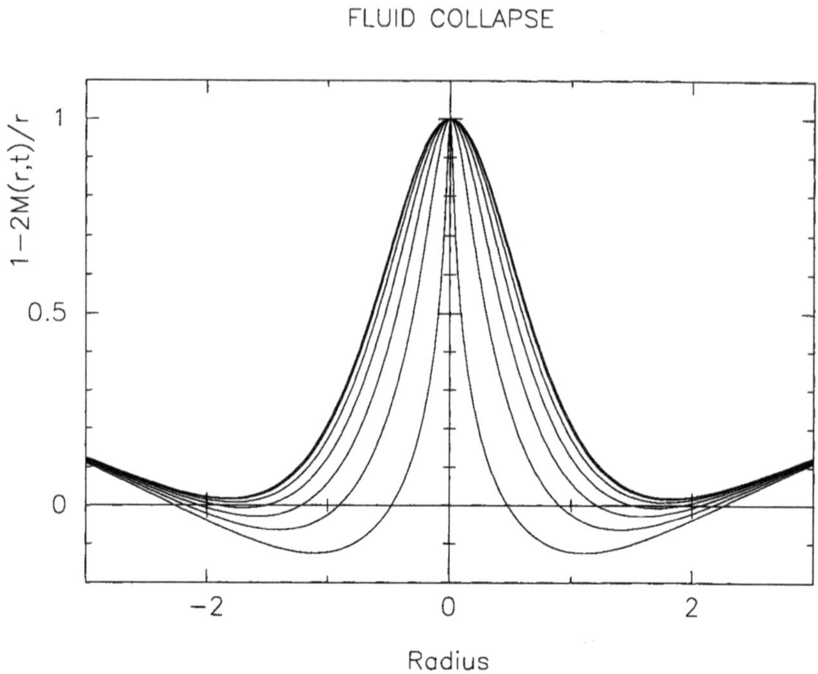

Figure 2: *Simulation of collapsing dust in the Newtonian gauge. Successive time slices for the horizon function* $(1 - 2M(r,t)/r)$ versus *radius* are *shown, with the top curve corresponding to* $t = 0$ and *lower curves to successively later times. The initial density profile is of the form* $\rho = \rho_0/(1 + (r/r_0)^2)^2$, *and the initial velocity is everywhere zero. There is no horizon initially, but a trapped region quickly forms, since in regions where* $1 - 2M/r < 0$, *photons can only move inwards.*

If pressure is included the pressure gradient causes the internal clock carried by the fluid to run at a different rate from t. The overall picture is similar, however. The horizon forms in a finite external coordinate time and matter has no difficulty crossing the horizon and falling onto the central singularity. The resulting solution is 'maximal' and the end result is a solution of the type discussed in the preceding Section. Again, the 'Newtonian' gauge makes the physics so simple that it surely deserves a prominent place in the description of black hole physics.

6.4 Electromagnetism in a Black-Hole Background

Maxwell's equations in a gravitational background can be written in a number of different forms [9]. From the vector potential A we define the field strength $F \equiv \nabla \wedge A$. The covariant field strength is obtained from F by defining

$$\mathcal{F} \equiv \overline{h}(\nabla \wedge A). \tag{6.35}$$

In terms of $\mathcal{F}$, Maxwell's equations (in the absence of torsion) become [9]

$$\mathcal{D}\mathcal{F} = \mathcal{J}. \tag{6.36}$$

This first-order form of Maxwell's equations is very useful for problems related to the propagation of electromagnetic waves, for example. An equivalent form of the equations is obtained by defining

$$G \equiv \underline{h}\overline{h}(F) \det \underline{h}^{-1} \tag{6.37}$$

and

$$J = \underline{h}(\mathcal{J}) \det \underline{h}, \tag{6.38}$$

so that equation (6.36) decomposes into the separate equations

$$\nabla \wedge F = 0, \quad \nabla \cdot G = J. \tag{6.39}$$

Finally, performing a spacetime split with respect to the γ_0 direction, as at equation (2.15), and defining

$$F = \boldsymbol{E} + i\boldsymbol{B} \tag{6.40}$$
$$G = \boldsymbol{D} + i\boldsymbol{H} \tag{6.41}$$
$$J\gamma_0 = \rho + \boldsymbol{J}, \tag{6.42}$$

equations (6.39) take the conventional form

$$\begin{aligned} \boldsymbol{\nabla} \cdot \boldsymbol{B} &= 0 & \boldsymbol{\nabla} \cdot \boldsymbol{D} &= \rho \\ \boldsymbol{\nabla} \times \boldsymbol{E} &= -\frac{\partial \boldsymbol{B}}{\partial t} & \boldsymbol{\nabla} \times \boldsymbol{H} &= \boldsymbol{J} + \frac{\partial \boldsymbol{D}}{\partial t} \end{aligned} \tag{6.43}$$

where, for spatial vectors, $\boldsymbol{a} \times \boldsymbol{b} \equiv -i\boldsymbol{a} \wedge \boldsymbol{b}$. The form (6.43) offers some insight into how the gravitational field interacts with the electromagnetic field. The tensor $\det \underline{h}^{-1} \underline{h}\overline{h}$ is a generalized permittivity/permeability tensor and defines the properties of the space through which the electromagnetic field propagates. For example, the bending of light by the sun can be easily understood in terms of the properties of the dielectric defined by the $\overline{h}$-field exterior to it.

The problem of interest here is to find the fields around a point source at rest outside the horizon of a radially-symmetric black hole. The $\bar{h}$-function in this case can be taken as that of equation (6.18). The solution to this problem can be found by adapting the work of Copson [26] and Linet [27] to the present gauge choices. Assuming that the charge is placed at a distance $a > 2M$ along the z-axis, the vector potential can be written in terms of a single scalar potential $V(r,\theta)$ as

$$A = V(r,\theta)(e_t + \frac{\sqrt{2Mr}}{r-2M}e_r), \tag{6.44}$$

so that

$$\mathcal{F} = -\frac{\partial V}{\partial r}e_r e_t - \frac{1}{r-2M}\frac{\partial V}{\partial \theta}\hat{\theta}(e_t + \sqrt{2M/r}e_r) \tag{6.45}$$

and

$$G = \boldsymbol{D} = -\frac{\partial V}{\partial r}e_r e_t - \frac{1}{r-2M}\frac{\partial V}{\partial \theta}\hat{\theta}e_t. \tag{6.46}$$

The Maxwell equations now reduce to the single partial differential equation

$$\frac{1}{r^2}\frac{\partial}{\partial r}\left(r^2\frac{\partial V}{\partial r}\right) + \frac{1}{r(r-2M)}\frac{1}{\sin\theta}\frac{\partial}{\partial\theta}\left(\sin\theta\frac{\partial V}{\partial\theta}\right) = -\rho \tag{6.47}$$

where $\rho = q\delta(\boldsymbol{x} - \boldsymbol{a})$ is a δ-function at $z = a$. This was the problem originally tackled by Copson [26] who obtained a solution that was valid locally in the vicinity of the charge, but contained an additional pole at the origin. Linet [27] modified Copson's solution by removing the singularity at the origin to produce a potential $V(r,\theta)$ whose only pole is on the z-axis at $z = a$. Linet's solution is

$$V(r,\theta) = \frac{q}{ar}\frac{(r-M)(a-M) - M^2\cos^2\theta}{D} + \frac{qM}{ar}, \tag{6.48}$$

where

$$D = [r(r-2M) + (a-M)^2 - 2(r-M)(a-M)\cos\theta + M^2\cos^2\theta]^{1/2}. \tag{6.49}$$

The novel feature we wish to stress here is that once (6.48) is inserted back into (6.45) the resultant $\mathcal{F}$ is both finite and continuous at the horizon. Working in the Newtonian gauge enables us to discuss the *global* properties of the solution, rather than having to resort to the 'membrane paradigm' [28] to exclude the region $r < 2M$. One simple way to illustrate the global properties of $\mathcal{F}$ and G is to plot the streamlines of $\boldsymbol{D}$, which are conserved by equation (6.43). The streamlines should therefore spread out from the charge and cover all space. Furthermore, since the distance scale r was chosen to agree with the gravitationally-defined distance, the streamlines of $\boldsymbol{D}$ convey genuine intrinsic information. Hence the plots are completely unaffected by our choice for the g_1 or g_2 functions, or indeed our choice of t-coordinate. Figure 3 shows streamline plots for charges held at different distances above the horizon. Similar

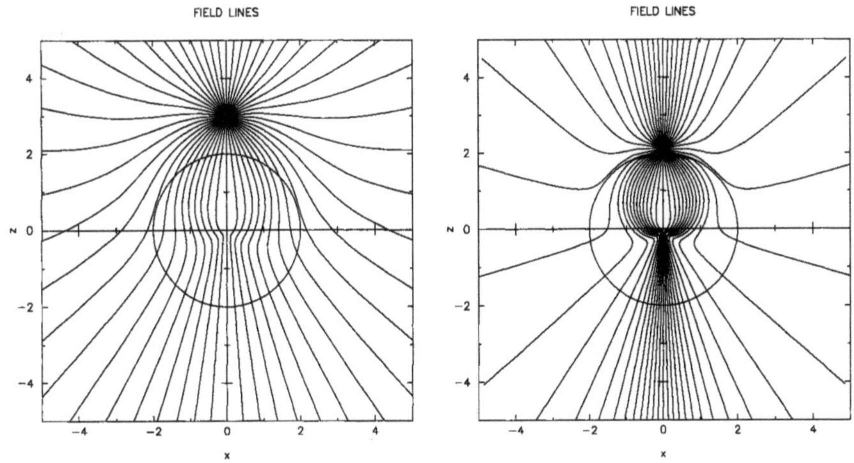

Figure 3: *Streamlines of the* $\boldsymbol{D}$ *field. The horizon is at* $r = 2$ *and the charge is placed on the* z*-axis. In the left-hand diagram the charge is held at* $z = 3$*, and in right-hand diagram it is at* $z = 2.1$. *Note how the streamlines are attracted towards the origin but never actually meet it. Had Linet's additional term not been included some of the streamlines would have terminated on the origin, which is not permitted.*

plots were first obtained by Hanni & Ruffini [29], though they were unable to extend their plots through the horizon. The physical interpretation of these plots differs somewhat from that advanced in the 'Membrane Paradigm' [28]. These differences are dealt with in detail elsewhere [9].

An interesting feature of the above solution is the existence of a repulsive 'polarisation' force [30], the effect of which is that a smaller force is needed to keep a charged particle at rest outside a black hole than an uncharged one. In their derivation of this force, Smith & Will [30] employed a complicated energy argument which involved renormalising various divergent integrals. In fact, as we show elsewhere [9], this repulsion is due entirely to the term that Linet added to Copson's formula. This is the second term in equation (6.48), and produces an outward-directed force on the charge of magnitude q^2M/a^3 — the same magnitude as found in [30]. This is an example of the importance of finding global solutions. The polarisation force is felt outside the horizon, yet the correction term that led to it was motivated by the properties of the field at the origin.

6.5 Cosmology

The equations of Table 1 are sufficiently general to deal with cosmology as well as astrophysics. In recent years, however, it has again become fashionable to include a cosmological constant in the field equations. The derivation of Section 4 is largely unaffected by the inclusion of the cosmological term, and only a few modifications to Table 1 are required. The full set of equations with a cosmological constant incorporated are summarised in Table 2.

In cosmology we are interested in homogeneous solutions to the equations of Table 2. Such solutions are found by setting ρ and p to functions of t only, so it follows immediately from the $L_r p$ equation that

$$G = 0 \quad \Rightarrow f_1 = 1. \tag{6.50}$$

For homogeneous fields the Weyl component of the Riemann tensor must vanish since this contains directional information through the e_r vector. The vanishing of this term requires that

$$M(r,t) = \frac{4}{3}\pi r^3 \rho, \tag{6.51}$$

which is consistent with the $L_r M$ equation. The $L_t M$ and $L_t \rho$ equations now reduce to

$$F = g_2/r \tag{6.52}$$

and

$$\dot{\rho} = -3g_2(\rho + p)/r. \tag{6.53}$$

But we know that $L_r g_2 = F g_1$, which can only be consistent with (6.52) if

$$F = c(t), \qquad g_2(r,t) = rc(t). \tag{6.54}$$

The $L_t g_2$ equation now reduces to a simple equation for $\dot{c}$,

$$\dot{c} + c^2 - \Lambda/3 = -\frac{4\pi}{3}(\rho + 3p). \tag{6.55}$$

Finally, we are left with the following pair of equations for g_1:

$$L_t g_1 = 0 \tag{6.56}$$

$$L_r g_1 = ({g_1}^2 - 1)/r. \tag{6.57}$$

The latter equation yields ${g_1}^2 = 1 + r^2\phi(t)$ and the former reduces to

$$\dot{\phi} = -2c(t)\phi. \tag{6.58}$$

Hence g_1 is given by

$${g_1}^2 = 1 - kr^2 \exp\{-2\int^t c(t')\,dt'\}, \tag{6.59}$$

The $\overline{h}$-function	The ω-function
$\overline{h}(e^t) = f_1 e^t$ $\overline{h}(e^r) = g_1 e^r + g_2 e^t$ $\overline{h}(e^\theta) = e^\theta$ $\overline{h}(e^\phi) = e^\phi$	$\omega(e_t) = G e_r e_t$ $\omega(e_r) = F e_r e_t$ $\omega(\hat{\theta}) = g_2/r\, \hat{\theta} e_t + (g_1 - 1)/r\, e_r \hat{\theta}$ $\omega(\hat{\phi}) = g_2/r\, \hat{\phi} e_t + (g_1 - 1)/r\, e_r \hat{\phi}$

Directional derivatives	$L_t = f_1 \partial_t + g_2 \partial_r$ $L_r = g_1 \partial_r$
Equations relating the $\overline{h}$- and ω-functions	$L_t g_1 = G g_2$ $L_r g_2 = F g_1$ $f_1 = \exp\{\int^r -G/g_1\, dr\}$
Definition of M	$M \equiv \frac{1}{2} r({g_2}^2 - {g_1}^2 + 1 - \Lambda r^2/3)$
Remaining derivatives	$L_t g_2 = G g_1 - M/r^2 + r\Lambda/3 - 4\pi r p$ $L_r g_1 = F g_2 + M/r^2 - r\Lambda/3 - 4\pi r \rho$
Matter derivatives	$L_t M = -4\pi g_2 r^2 p \quad L_t \rho = -(2g_2/r + F)(\rho + p)$ $L_r M = 4\pi g_1 r^2 \rho \quad L_r p = -G(\rho + p)$
Riemann tensor	$\mathcal{R}(B) = 4\pi(\rho + p) B \cdot e_t e_t - \frac{1}{3}(8\pi\rho + \Lambda) B$ $-\frac{1}{2}(M/r^3 - 4\pi\rho/3)(B + 3 e_r e_t B e_r e_t)$
Fluid stress-energy tensor	$\mathcal{T}(a) = (\rho + p) a \cdot e_t e_t - pa$

Table 2: *Equations governing a radially-symmetric perfect fluid — case with a non-zero cosmological constant* Λ. *The shaded equations differ from those of Table 1.*

where k is an arbitrary constant of integration. It is straightforward to check that (6.59) is consistent with the equations for $\dot{c}$ and $\dot{\rho}$. The full set of equations describing a homogeneous perfect fluid are summarised in Table 3.

At first sight, the equations of Table 3 do not resemble the usual Friedmann equations. The Friedmann equations are recovered straight-forwardly, however, by

The $\overline{h}$-function	$\overline{h}(a) = a + a\cdot e_r[(g_1 - 1)e^r + c(t)re^t]$ ${g_1}^2 = 1 - kr^2 \exp\{-2\int^t c(t')\,dt'\}$
The ω-function	$\omega(a) = c(t)a\wedge e_t - (g_1 - 1)/r\, a\wedge(e_r e_t)e_t$
The density	$\frac{8\pi}{3}\rho = c(t)^2 - \Lambda/3 + k\exp\{-2\int^t c(t')\,dt'\}$
Dynamical equations	$\dot{c} + c^2 - \Lambda/3 = -4\pi/3\,(\rho + 3p)$ $\dot{\rho} = -3c(t)(\rho + p)$

Table 3: *Equations governing a homogeneous perfect fluid.*

setting

$$c(t) = \frac{\dot{S}(t)}{S(t)}. \tag{6.60}$$

With this substitution we find that

$${g_1}^2 = 1 - kr^2/S^2 \tag{6.61}$$

and the $\dot{c}$ and density equations become

$$\frac{\ddot{S}}{S} - \frac{\Lambda}{3} = -\frac{4\pi}{3}(\rho + 3p) \quad \text{and} \quad \frac{\dot{S}^2 + k}{S^2} - \frac{\Lambda}{3} = \frac{8\pi}{3}\rho, \tag{6.62}$$

recovering the Friedmann equations in their standard form [31]. The intrinsic treatment has therefore led us to work directly with the 'Hubble velocity' $c(t)$, rather than the 'distance' scale $S(t)$. There is a good reason for this. Once the Weyl tensor is set to zero, the Riemann tensor reduces to

$$\mathcal{R}(B) = 4\pi(\rho + p)B\cdot e_t e_t - \frac{1}{3}(8\pi\rho + \Lambda)B, \tag{6.63}$$

and we have now lost contact with an intrinsically-defined distance scale. We can therefore rescale the radius variable r with an arbitrary function of t (or r) without altering the Riemann tensor. The Hubble velocity, on the other hand, is intrinsic and it is therefore unsurprising that our treatment has led directly to equations for this.

Amongst the class of radial rescalings a particularly useful one is to rescale r to $r' = S(t)r$. This is achieved with the transformation

$$f(x) = x \cdot e_t e_t + S x \wedge e_t e_t \tag{6.64}$$

so that, on applying equation (3.5), the transformed $\overline{h}$-function is

$$\overline{h}'(a) = a \cdot e_t e_t + \frac{1}{S}[(1 - kr^2)^{1/2} a \cdot e_r e^r + a \wedge (e_r e_t) e_t e_t]. \tag{6.65}$$

The function (6.65) reproduces the standard line element used in cosmology. We can therefore use the transformation (6.64) to move between the 'Newtonian' gauge developed here and the traditional 'static' gauge. The differences between these gauges can be understood by considering geodesic motion. A particle at rest with respect to the cosmological frame (defined by the cosmic microwave background) has $v = e_t$. In the standard 'static' gauge such a particle is not moving in the flat-space background (the distance variable r is equated with the comoving coordinate of GR). For this reason we refer to this gauge as being static, even though the associated line element is usually thought of as defining an expanding spacetime [19]. In the Newtonian gauge, on the other hand, comoving particles are moving outwards radially at a velocity $\dot{r} = c(t)r$, though this expansion centre is not an intrinsic feature. Of course, attempting to distinguish these pictures is a pointless exercise, since all observables must be gauge invariant. All that is of physical relevance is that, if two particles are at rest with respect to the cosmological frame (defined by the cosmic microwave background), then the light-travel time between these particles is an increasing function of time and light is redshifted as it travels between them. The value of this redshift is independent of the gauge in which it is calculated [19], and attempting to assign the redshift to an expansion of spacetime, or a change in the speed of light, or any other property of the background space is a gauge-dependent description. The fact that there are alternative gauge choices is often ignored in modern treatments of cosmology, which invariably employ the 'expanding universe' description of redshifts.

Dust Models and Horizons

As an illustration of the utility of the Newtonian gauge in cosmology, we consider horizons in a dust model ($p = 0$). Setting p to zero implies that

$$c(t) = -\frac{\dot{\rho}}{3\rho} \tag{6.66}$$

so

$${g_1}^2 = 1 - kr^2 \rho^{2/3} \tag{6.67}$$

and

$$c(t) = \left(\frac{8\pi}{3}\rho - k\rho^{2/3} + \frac{\Lambda}{3}\right)^{1/2}. \tag{6.68}$$

We are therefore left with a single first-order differential equation for ρ. Explicit solutions of this equations are rarely needed, as we can always parameterise time by the density $\rho(t)$.

Now consider radial null geodesics with a trajectory

$$x(\tau) = t(\tau)e_t + r(\tau)e_r. \tag{6.69}$$

For these trajectories

$$v = \underline{h}^{-1}(\dot{x}) = \dot{t}(e_t - \frac{rc(t)}{g_1}e_r) + \frac{\dot{r}}{g_1}e_r \tag{6.70}$$

and the condition that $v^2 = 0$ reduces to

$$\frac{\dot{r}}{g_1} = \dot{t}(\frac{rc(t)}{g_1} \pm 1) \tag{6.71}$$

$$\Rightarrow \frac{dr}{dt} = rc(t) \pm g_1. \tag{6.72}$$

But, as $c(t)$ is positive for a cooling universe, there is a critical radius beyond which $rc(t) - g_1$ is positive and all photons from this point onwards must follow an outward trajectory. So, if we are at the origin, we cannot receive signals from beyond this critical radius and a particle horizon exists for us at

$$r_c c(t) = g_1 \tag{6.73}$$

$$\Rightarrow r_c = \left(\frac{8\pi\rho}{3} + \frac{\Lambda}{3}\right)^{-1/2}. \tag{6.74}$$

As the density drops this radius increases, though the inclusion of the cosmological term means that r_c never exceeds $\sqrt{3/\Lambda}$.

6.6 The Dirac Equation in a Cosmological Background

As a final application of the theory outlined here we consider the Dirac equation in a cosmological background. At this point it is slightly easier to work with the $\overline{h}$-function in the more conventional diagonal form of (6.65). But clearly there is a problem here if k is positive, since the square root in (6.65) is ill-defined for $r > k^{-1/2}$. This can be viewed as an artefact of the means by which this solution maps a 3-sphere onto Euclidean space [19]. We can remove the difficulty by using a stereographic projection to achieve a globally-defined $\overline{h}$-function, though this is not necessary for the application discussed here. (The stereographic projection is useful in other applications, however, for example it shows up that in a $k > 0$ universe every charge must have an image charge. It is somewhat surprising that this bizarre feature of a $k > 0$ universe is considered acceptable.)

An advantage of the STA formalism is that there is no need to introduce matrices or column vectors to represent spinors. Instead, a spinor can be represented as an element of the 8-dimensional even subalgebra [11]. The even subalgebra is closed under left multiplication by rotors, so can be used to represent a spin-space. In the STA the action of the $\hat{\gamma}^\mu$ matrices on a column spinor $|\psi>$ is replaced by the operation $\psi \mapsto \gamma^\mu \psi \gamma_0$, and multiplication by the unit imaginary is replaced $\psi \mapsto \psi i\sigma_3$. The free-field Dirac equation can then be written

$$\nabla \psi i\sigma_3 = m\psi\gamma_0. \tag{6.75}$$

This equation just employs the vector derivative ∇, so the same gauging arguments as used in Section 3 can be employed to couple in the gravitational field. The only difference is that spinors transform single-sidedly under rotations, so the ω-function only enters on the left-hand side of ψ. The resulting equation is

$$\partial_a[L_a + \tfrac{1}{2}\omega(a)]\psi i\sigma_3 = m\psi\gamma_0, \tag{6.76}$$

which can also be derived from an action integral [9]. It is easily verified that equation (6.76) is covariant under local translations and rotations.

Observables are formed by double-sided application of the spinor ψ on one of the fixed $\{\gamma_\mu\}$ frame vectors. In particular, the charge current is $\mathcal{J} = \psi\gamma_0\tilde{\psi}$ and the spin current is $\mathcal{S} = \psi\gamma_3\tilde{\psi}$. Using equation (6.76) it is not hard to show that $\mathcal{D}\cdot\mathcal{J} = 0$, as is required by charge conservation.

Our purpose here is to find solutions to (6.76) in the background of a dust model for cosmology. With the $\overline{h}$-function in the form of (6.65) the relevant equation is

$$\begin{aligned}\left(e^t\partial_r + \frac{1}{S}[(1-kr^2)^{1/2}e^r\partial_r + e^\theta\partial_\theta + e^\phi\partial_\phi]\right)\psi i\sigma_3 \\ +\tfrac{1}{2}(3c(t)e_t - 2[(1-kr^2)^{1/2} - 1]e_r)\psi i\sigma_3 = m\psi\gamma_0.\end{aligned} \tag{6.77}$$

The question we wish to address is this: can we find solutions to (6.77) such that the observables are homogeneous? There is clearly no difficulty if $k = 0$, since equation (6.77) is solved by

$$\psi = \rho^{1/2}e^{-i\sigma_3 mt} \tag{6.78}$$

and the observables are fixed vectors which just scale as $\rho(t)$ in magnitude. But what happens when $k \neq 0$? It turns out that the solution (6.78) must now be modified to

$$\psi = \frac{\rho^{1/2}}{1+\sqrt{1-kr^2}}e^{-i\sigma_3 mt}. \tag{6.79}$$

But this is no longer homogeneous, and the observables obtained form (6.79) all contain additional r-dependence as well as scaling as $\rho(t)$. In principle, therefore, one could determine the origin of this space with measurements of the current density.

This clearly violates the principle of homogeneity, though it is not necessarily inconsistent with current experiments. The consequences of this fact for self-consistent solutions of the Einstein-Dirac equations are discussed in [19] (see also [32]).

So how can it be that classical phenomena do not see this 'preferred' direction in $k \neq 0$ models, but the quantum spinor ψ does? The answer lies in the gauge structure of the theory. The 'minimal-coupling' procedure couples Dirac spinors directly to the ω-function [9], which in this case is not homogeneous. Dirac spinors can therefore probe the theory directly at the level of the ω-field, whereas classical quantities only interact at the level of the covariant quantities obtained from the gravitational fields (which are homogeneous). Whilst the above result is not conclusive, it does strongly suggest that cosmological models with $k \neq 0$ are inconsistent with the assumption of homogeneity. This does not rule out spatially-flat universes with a non-zero cosmological constant.

7 Conclusions

We have seen how a theory of gravity can be developed as a gauge theory within the framework of a flat background spacetime. The resulting theory is conceptually simple. It employs neither differential geometry nor manifold theory, and the role of the gauge fields can be clearly understood. The theory is also easier to calculate with than GR, and the 'intrinsic' technique outlined here offers the hope of real progress with problems such as finding the gravitational fields inside a rotating star [9]. By studying a radially-symmetric perfect fluid we have found a new gauge which dramatically simplifies the physics, reducing it essentially to a set of classical equations. It has also emerged that, where the global nature of the solutions is important, our theory departs from GR. This is seen most clearly in the description of the horizon outside a black hole. The existence of a global solution enables us to produce fieldline plots for a charge held outside a horizon which continue smoothly across the horizon. The structure of the fieldlines inside the horizon shows up the 'dielectric' nature of the interaction between gravity and electromagnetism.

The application to cosmology has revealed the existence of an alternative gauge choice in which a number of calculations can be performed more efficiently. This new gauge suggests a different interpretation for redshifts, which reinforces the importance of only discussing intrinsic relations between fields, and not the properties of matter relative to the spacetime background. One can only do the latter if it has already been arranged that properties of the background spacetime match intrinsic features of the gravitational or matter fields. A final surprise is obtained when studying the Dirac equation in a cosmological background. The spinor field probes the gauge structure at a deeper level than classical fields and reveals that, for Dirac particles, the only cosmological models consistent with homogeneity are spatially flat.

A fuller presentation of this work is in preparation [9]. This includes the derivation of all the field equations (matter and gravitational) from action integrals. In addition, we treat quantum mechanics in a black-hole background and give an intrinsic treatment of a rotating perfect fluid. It is also argued in [9] that splitting gravitational effects into the $\overline{h}$- and ω-fields suggests a novel route to a multiparticle theory of gravitationally-interacting particles.

References

[1] R. Utiyama. Invariant theoretical interpetation of interaction. *Phys. Rev.*, 101(5):1597, 1956.

[2] T.W.B. Kibble. Lorentz invariance and the gravitational field. *J. Math. Phys.*, 2(3):212, 1961.

[3] D. Ivanenko and G. Sardanashvily. The gauge treatment of gravity. *Phys. Rep.*, 94(1):1, 1983.

[4] M. Gockeler and T. Schucker. *Differential Geometry, Gauge Theories, and Gravity*. Cambridge University Press, 1987.

[5] F.W. Hehl, P. von der Heyde, G.D. Kerlick, and J.M. Nestev. General relativity with spin and torsion: Foundations and prospects. *Rev. Mod. Phys.*, 48:393, 1976.

[6] D. Hestenes and G. Sobczyk. *Clifford Algebra to Geometric Calculus*. Reidel, Dordrecht, 1984.

[7] D. Hestenes. *Space-Time Algebra*. Gordon and Breach, 1966.

[8] S.F. Gull, A.N. Lasenby, and C.J.L. Doran. Electron paths, tunnelling and diffraction in the spacetime algebra. *Found. Phys.*, 23(10):1329, 1993.

[9] A.N. Lasenby, C.J.L. Doran, and S.F. Gull. Gravity, gauge theories and geometric algebra. In preparation, 1995.

[10] S.F. Gull, A.N. Lasenby, and C.J.L. Doran. Imaginary numbers are not real — the geometric algebra of spacetime. *Found. Phys.*, 23(9):1175, 1993.

[11] C.J.L. Doran, A.N. Lasenby, and S.F. Gull. States and operators in the spacetime algebra. *Found. Phys.*, 23(9):1239, 1993.

[12] A.N. Lasenby, C.J.L. Doran, and S.F. Gull. A multivector derivative approach to Lagrangian field theory. *Found. Phys.*, 23(10):1295, 1993.

[13] D. Hestenes. *New Foundations for Classical Mechanics.* Reidel, Dordrecht, 1985.

[14] D. Hestenes. A unified language for mathematics and physics. In J.S.R. Chisholm and A.K. Common, editors, *Clifford Algebras and their Applications in Mathematical Physics (1985)*, page 1. Reidel, Dordrecht, 1986.

[15] D. Hestenes. The design of linear algebra and geometry. *Acta Appl. Math.*, 23:65, 1991.

[16] T.G. Vold. An introduction to geometric algebra with an application to rigid body mechanics. *Am. J. Phys.*, 61(6):491, 1993.

[17] T.G. Vold. An introduction to geometric calculus and its application to electrodynamics. *Am. J. Phys.*, 61(6):505, 1993.

[18] C.J.L. Doran, A.N. Lasenby, and S.F. Gull. Gravity as a gauge theory in the spacetime algebra. In F. Brackx, R. Delanghe, and H. Serras, editors, *Clifford Algebras and their Applications in Mathematical Physics (1993)*, page 375. Kluwer Academic, Dordrecht, 1993.

[19] A.N. Lasenby, C.J.L. Doran, and S.F. Gull. Cosmological consequences of a flat-space theory of gravity. In F. Brackx, R. Delanghe, and H. Serras, editors, *Clifford Algebras and their Applications in Mathematical Physics (1993)*, page 387. Kluwer Academic, Dordrecht, 1993.

[20] D. Hestenes. Proper particle mechanics. *J. Math. Phys.*, 15(10):1768, 1974.

[21] C.J.L. Doran, D. Hestenes, F. Sommen, and N. van Acker. Lie groups as spin groups. *J. Math. Phys.*, 34(8):3642, 1993.

[22] D. Hestenes. Observables, operators, and complex numbers in the Dirac theory. *J. Math. Phys.*, 16(3):556, 1975.

[23] S.W. Hawking and G.F.R. Ellis. *The Large Scale Structure of Space-Time.* Cambridge University Press, 1973.

[24] C.W. Misner, K.S. Thorne, and J.A. Wheeler. *Gravitation.* W.H. Freeman and Company, San Francisco, 1973.

[25] I.D. Novikov and V.P. Frolov. *Physics of Black Holes.* Kluwer Academic, Dordrecht, 1989.

[26] E.T. Copson. On electrostatics in a gravitational field. *Proc. R. Soc. A*, 118:184, 1928.

[27] B. Linet. Electrostatics and magnetostatics in the Schwarzschild metric. *J. Phys. A*, 9(7):1081, 1976.

[28] K.S. Thorne, R.H. Price, and D.A. Macdonald. *Black Holes: The Membrane Paradigm.* Yale Uninversity Press, 1986.

[29] R. S. Hanni and R. Ruffini. Lines of force of a point charge near a Schwarzschild black hole. *Phys. Rev. D.*, 8(10):3259, 1973.

[30] A.G. Smith and C.M.Will. Force on a static charge outside a Schwarzschild black hole. *Phys. Rev. D*, 22(6):1276, 1980.

[31] J.V. Narlikar. *Introduction to Cosmology.* Cambridge University Press, 1993.

[32] C.J. Isham and J.E. Nelson. Quantization of a coupled Fermi field and Robertson-Walker metric. *Phys. Rev. D*, 10(10):3226, 1974.

SEARCH FOR COSMIC STRINGS AT MITO

MARCO DE PETRIS
Department of Physics, University La Sapienza
Rome, Italy, Pz.le A. Moro 2

and

BIANCA MELCHIORRI
Istituto di Fisica dell'Atmosfera CNR
Rome, Italy, Pz.le L. Sturzo 31

and

FRANCESCO MELCHIORRI
Department of Physics, University La Sapienza
Rome, Italy, Pz.le A. Moro 2

and

MONIQUE SIGNORE
Ecole Normale Superieure, Paris
Paris, France, rue Lhomond 24

ABSTRACT

MITO (Millimetric and Infrared Testagrigia Observatory) is a 3 meter telescope dedicated to the study of Cosmic Background Anisotropies. It is located at 3500 m above sea level at Plateau Rosa', Cervinia. We describe the Observatory and discuss two on going research programs: a cleaning procedure in order to remove galactic dust signals from cosmological data and a search for cosmic strings toward double lensed quasars. Both the programs will be carried out by means of an He-3 photometer operating at the frequencies of 5 , 10, 12, 30 cm^{-1}

1. Introduction

MITO is the result of a collaboration among various Italian Institution: Istituto di Cosmogeofisica of Torino, Physics Department of University of Rome, Istituto di Fisica dell'Atmosfera of CNR of Rome, ENEA-TIB of Frascati, the INFN section of Rome. The Observatory is located at 3480 m above sea level over the mountain named Testa Grigia, on the border of Plateau Rosa', Cervinia. The coordinates are $45^0 56'03"$ North of latitude and $07^0 42'28"$ East of longitude.

The metereological properties of the Observatory are collected in Table I

TABLE I: Metereological Properties of Testa Grigia

	Jan	Feb	Mar	Apr	May	Jun	Jul	Aug	Sep	Oct	Nov	Dec
Clean Sky day/months	13	10	11	8	8	7	12	10	11	13	11	13
Minimum Temperature	-30	-34	-35	-19	-22	-16	-9	-13	-16	-2 0	-27	-28
Maximum Temperature	-9	-10	-8	-6	-2	2	4	3	2	-2	-6	-8
Mean Atmospheric Pressure	491	490	491	493	496	499	501	500	500	497	493	493
Mean Relative Humidity	63	68	64	70	68	70	65	6 7	61	60	63	61

A sketch of MITO dome is shown Figure 1 , while Figure 2 is a photography of the telescope in situ. The telescope was assembled inside the dome, togheter with a photometer in June 94 and we expect to be fully operative in November of the same year.

2. The Telescope

The optical configuration of MITO is that of a Cassegrain system. Since the telescope will be dedicated to far infrared observations, a secondary mirror modulation is needed. This requirement posed severe constraints on the optical design, in order to minimize aberrations when the secondary mirror is tilted. We used the program ACCOS V. In Table II the optical parameters of the telescope are quoted in two different configurations, Flux Collector CF and Array AR.

The CF configuration refers to a wide angle system, having a field of view in the sky of 17 arcminutes FWHM and a modulation amplitude up to 1 degree in the sky.

The AR configuration is for an array of 3x3 bolometers with 1 arcmin per pixel and a maximum modulation amplitude of 31 arcminutes. The primary mirror is in any case underilluminated in order to avoid the spurious signals arising from the diffraction at the border of the mirror.

The secondary mirror is tilted around the focus of the primary: this type of modulation is called *neutral point modulation* and it is able to cancell all the aberrations arising from the inclination of the mirror. The remaining abberations are astigmatism and field curvature: in Figure 3 we have shown some results of ACCOS V program for the image on the focal plane. It is clear from these figures that the system has been optimized for a tilted secondary.

By rotating the secondary mirror around a point outside the mirror vertex it is possible to minimize the aberrations and, therefore, the spot size, as shown in Figure 4

TABLE 2

Optical parameters of the 2600 mm telescope

primary mirror diameter	2600 mm
primary focal length	1247.2 mm
primary mirror f/number	0.48

	Flux Collector configuration	**Array configuration**
effective primary diameter	2000 mm	2400 mm
effective primary f/number	0.62	0.52
effective focal length	8151 mm	9273 mm
photometer f/number	4.07	3.86
effective magnification	6.56	7.42
subreflector diameter	410 mm	360 mm
subreflector curvature radius	539.8 mm	434.0 mm
distance subreflector-primary mirror	1018.7 mm	1059.5 mm
distance subreflector-rotation axis	228 mm	181.5 mm
distance primary mirror-Cass. focus	461 mm	334 mm
detector aperture	40 mm	3 mm
field of view (FWHM)	16.6 arcmin	1.6 arcmin
maximum sky field separation	63 arcmin	31.5 arcmin
telescope throughput	1.52 cm^2sr	$3.4 \cdot 10^{-3} cm^2sr$

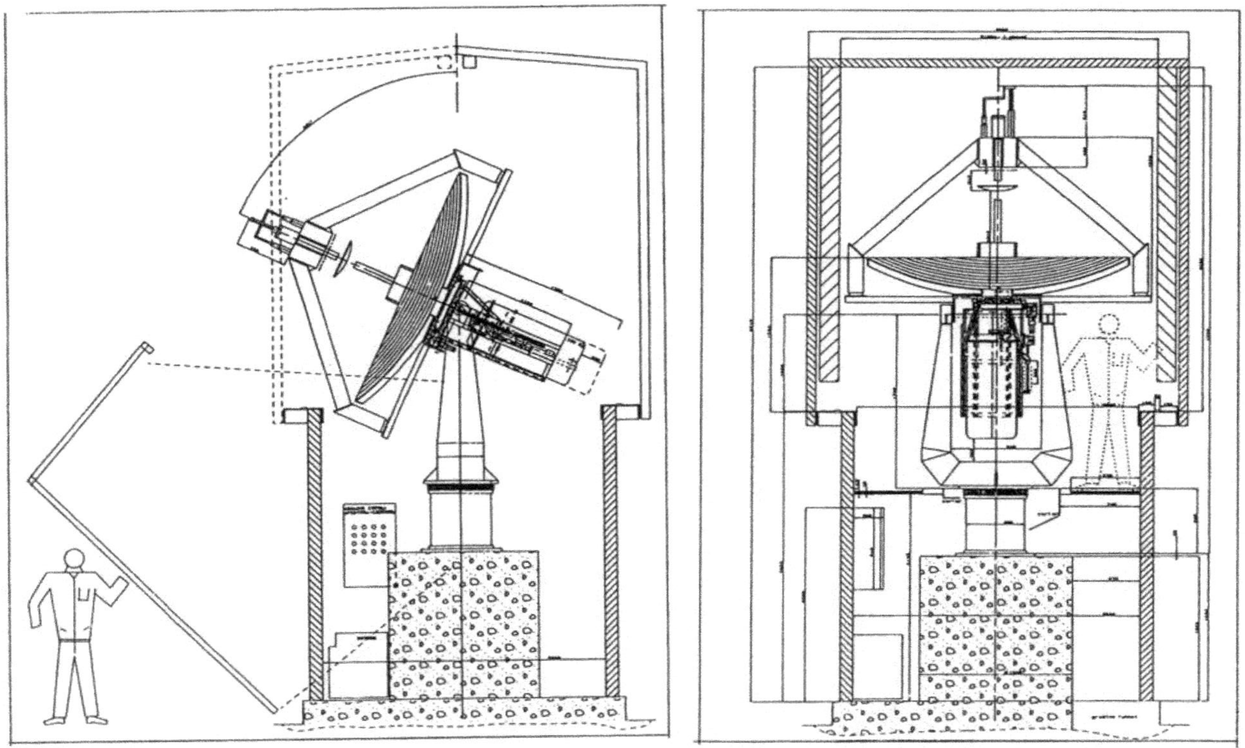

Figure 1: diagrams of MITO observatory: the dome can be completely open by means of an hydraulic system. The telescope has an altazimuthal mounting designed for a pointing accuracy better than one arcsecond. An electromechanical system allows the position and three axis motion of the photometer at the focal plane of the telescope.

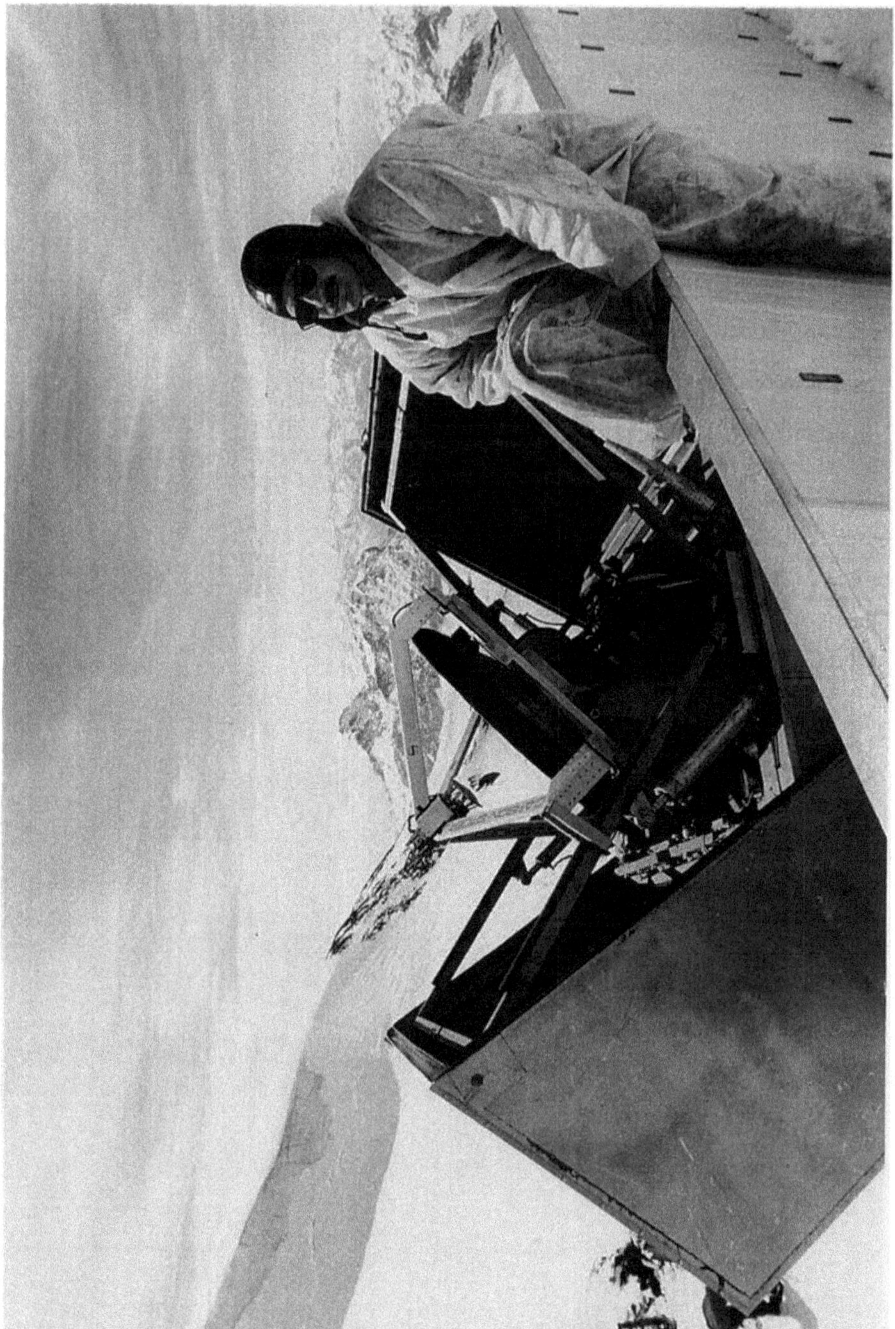

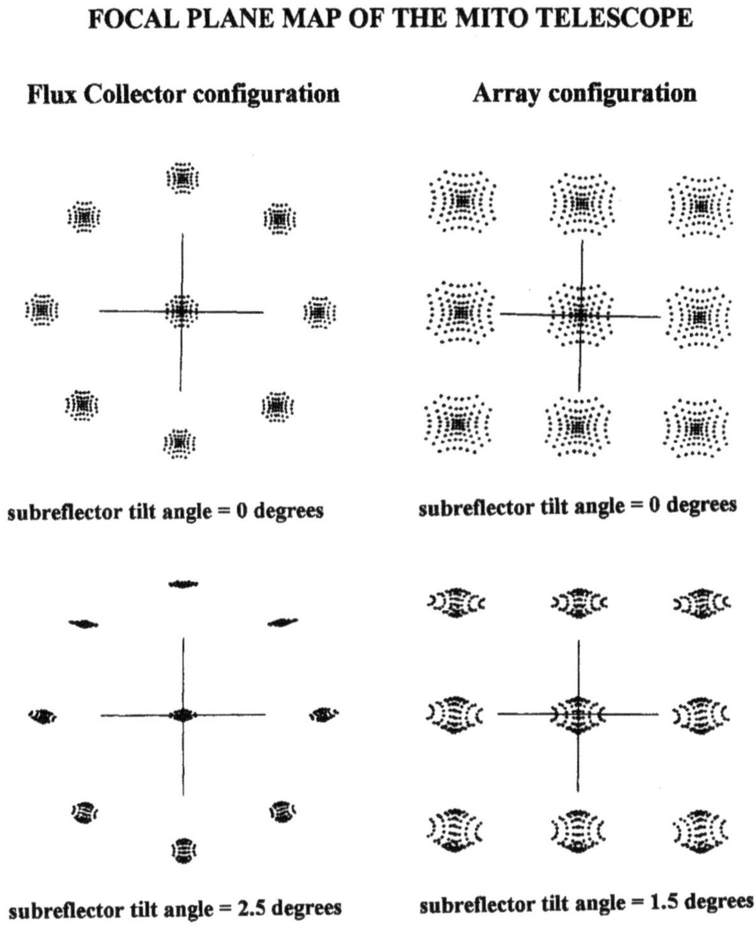

Figure 3: Simulation of focal plane image for a point-like source in the case of Flux Collector and Array arrangement respectively. Note as the aberrations are smaller in the case of a tilted secondary. This choice has been determined by the consideration that MITO spend the largest amount of observing time in the tilted configuration.

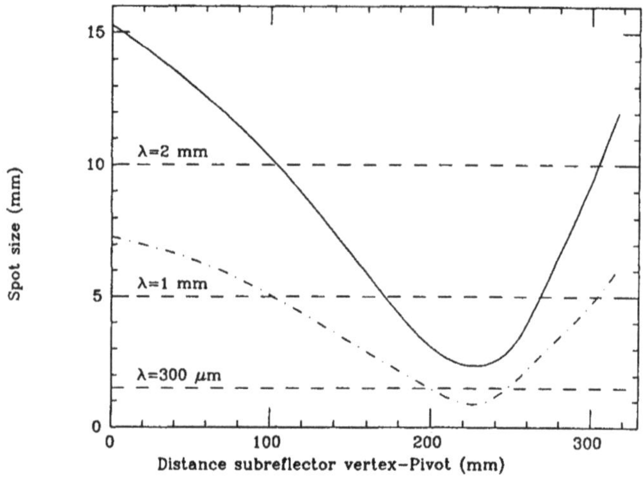

Figure 4:*dimensions of the spot for a point-like source as function of the distance between the secondary mirror vertex and the rotating vertex (pivot point). Dashed lines indicate the spot size due to diffraction (first Airy ring): the solid line is the spot radius, while the dot-dash line represents the tangential dimension of the spot along the modulation direction, the dashed lines are the dimensions of the spot due to diffraction at the relative wavelength of operation*

The primary mirror has been realized in Al (UNI 7257-73 Avio) from Officine Ottico Meccaniche Marcon. The mechanical project has been developed by means of the MSC-NASTRAN program in order to minimize the weight while preserving the structural rigidity and the final weight is only 110 Kg. The accuracy of the surface is close to 0.1 micron r.m.s. while the profile is such that the system is diffraction limited at 100 microns of wavelength. The spider for the secondary mirror is in Carbon fiber, realized by Superelectric.

3. The Modulation System

The modulation system is one of the crucial parts of MITO. Basically it consists of two push-pull electromagnets which drive the secondary mirror with a maximum amplitude of $\pm 1^0$ in the sky and a minimum rise time of 10 ms at a maximum frequency of 20 Hz.

In Figure 5 is a sketch of the system, while a photo is in Figure 6.

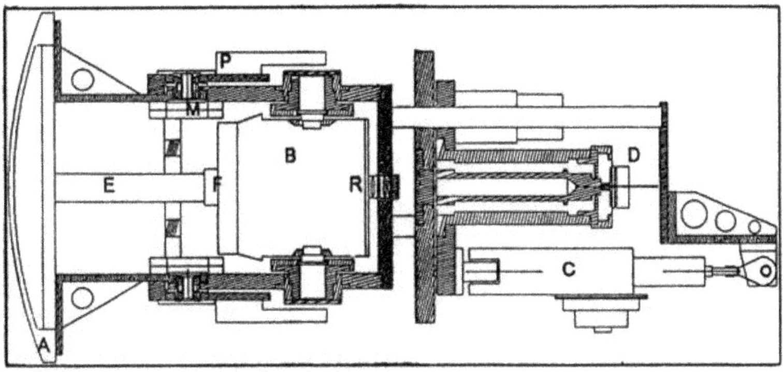

Figure 5: scheme of the wobbling system. A- secondary mirror; E,F- flexible joints driven by the electromagnet B; M- mirror pivot; D-mechanism for rotating the direction of modulation with respect to the horizon; C- linear motor for focusing

The two electromagnets are free of rotating around an axis and two springs take them in the due position. The same is done for the mirror: a precise selection of the two elastic constants minimize the vibrations transmitted to the structure. A linear transducer monitors continously the amplitude of modulation and transmits it to a dedicated hardware. The power to the electromagnets is thus determined by the difference between the position of the mirror and the signal provided by the waveform generator. One can immediately modify the type of modulation (2,3 or 4 fields), the waveform (synusoidal, square-wave), the amplitude (between 0 and 1 degree in the sky) and the frequency (0-30 Hz).

In Figure 7 is a plot of the waveforms obtained with a secondary mirror in Al. A new secondary mirror in carbon fiber is under preparation: it should allow a significant reduction in weight and a better wavefront in the case of square-wave modulation.

The system is provided by a linear actuator which moves the mirror along its optical axis, thereby allowing a rapid focusing. A second rotating motor allows to tilt the direction of modulation with respect to the horizon.

The modulation amplitude has been made stable better than 1 part over 10^3 in order to keep the atmospheric signals below the threshold of $\frac{\Delta T}{T} \leq 10^{-5}$ at $\lambda \simeq 1mm$

4. The Telescope Control System

The control system of MITO consists of three dedicated 386-33 MHz computers for respectively, data acquisition, CCD control, pointing control.

The computer dedicated to pointing is connected to the motors electronics and altoazimuthal encoders: the coordinates of the observed field and the time (given by an ESAT clock sinchronized to the atomic clock of Galileo Ferraris Institute) are transfered to the data acquisition computer.

The CCD computer is connected to a CCD camera Lynx PC+ by Spectra Source Instruments with an electronic intensifier Astrolight Viewer by Electrophysics. It compares the coordinates of a given star with those of a catalogue in memory and records the position in the data acquisition computer.

N-SPIDER

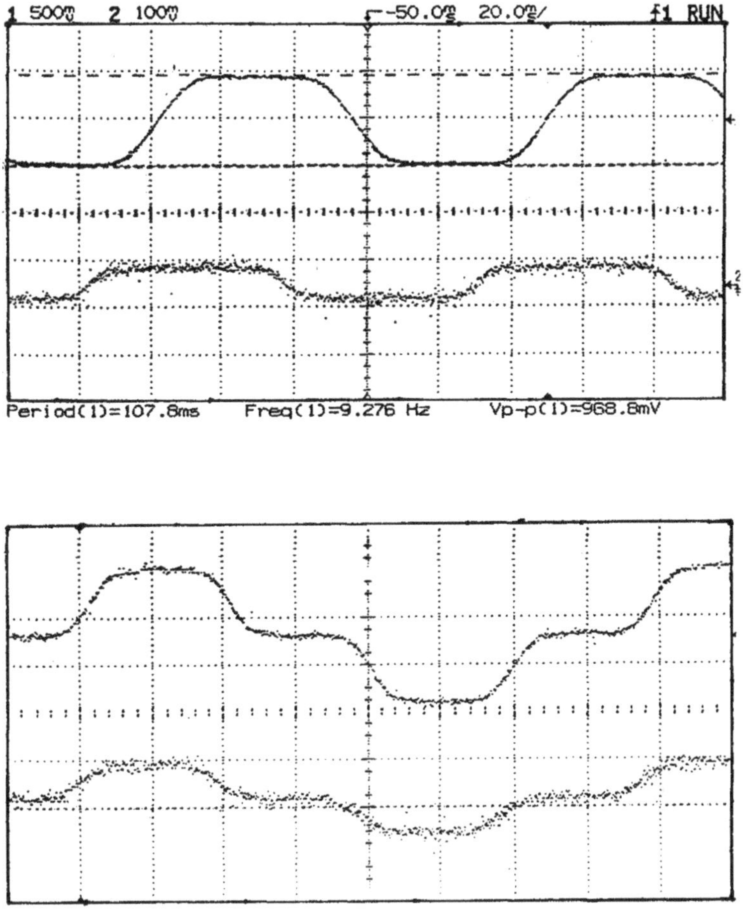

Figure 7: Output from the wobbling amplitude transducer and from the wavefunction generator in the case of two-fields modulation at 9.3 Hz (upper panel) and in the case of three-fields modulation at 4.0 Hz.

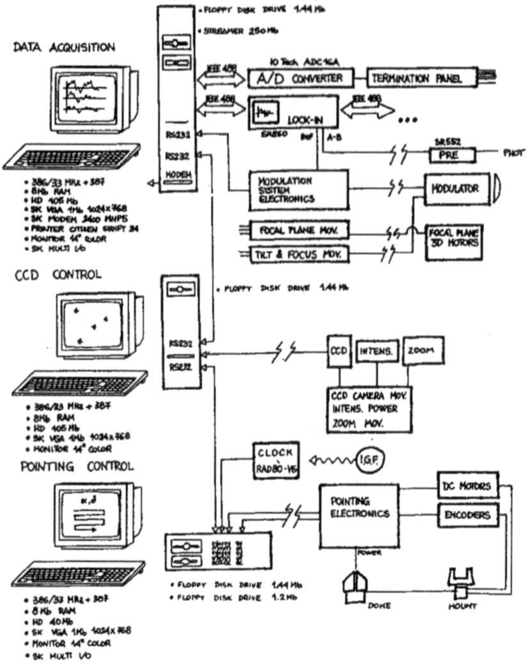

Figure 8: diagram of electronic control system of MITO. Three computers allow to collect the data, the image from CCD and to dirve the telescope.

The data acquisition computer is connected through a IEEE-488 to a 16 bit AD-converter and 4 lock-in digital amplifiers (Stanford Res. SR850). The recorded data are: the time of observation, the coordinates, the four channel signals, the lock-in status, the modulator set-up, the secondary position, the photometer position, the various temperatures on the primary and secondary mirrors, the housekeeping of the photometer.

The acquisition software and data quick-look is provided by a multipurpose program Viewdac from Keithley. The data are recorded on harddisk and on magnetic type cassette of 250 Mb. A modem 2400 MNP5 allows the transmission of short files.

5. Scientific Programs at MITO

5.1. A Cleaning Procedure for CBA Measurements

Although claims of having observed Cosmic Background Anisotropies (CBA) date back as early as 1981 (Melchiorri et al.[5]), it is a widely diffused opinion that "local" disturbances, like galactic patchy emission by dust and ionized gas have largely affected all the experiments. This opinion is partly supported by the spread in $\frac{\Delta T}{T}$ amplitudes quoted by various authors. For instance, in Figure 9 we have collected many of the supposed "detections" and represented them against some theoretical expectations. In this figure we have followed the tradition of representing CBA in terms of spherical harmonics coefficients C_k. To be more precise the anisotropies $(\frac{\Delta T}{T})^2$ are averaged over the sky: the uniformity of the Universe allows us to substitute $2k+1$ to the sum over the second index , so that

$$\left\langle \left(\frac{\Delta T}{T}\right)^2 \right\rangle = \sum_k \frac{(2k+1)C_k}{4\pi} \qquad [1]$$

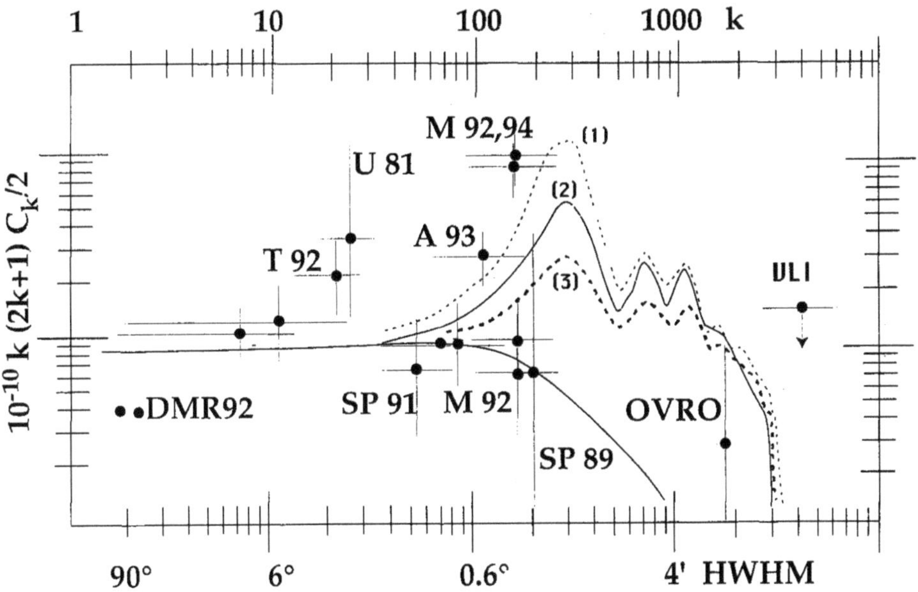

Figure 9: Power spectrum of CBA . The various curves refer to different theoretical models (see Guarini et al. [6] 1994)

TABLE III

Logo/PI	Frequency (GHz)	Bandwidth FWHM (GHz)	Beamwidth FWHM	Modulation FWHM	NET (mK/Hz$^{-1/2}$)	Sky region	$\Delta T/T$ 10^{-5}
IRAM Kreysa	250	42	12''	21''(s.2b)	60	10'x10'	18 UL
BAM Halpern	60-300[(1)]	30	1^0	4^0(sq.2b)	0.3	undefined	0.3 lab
MAX91 Richards	180,240,420	81,84,105	0.75^0	1.4^0(s.2b)	0.5 - 0.7	GUM[(2)]	$4.5^{+5.7}_{-2.6}$
MAX92 Richards	180,240,420	81,84,105	0.75^0	1.4^0(s.2b)	0.5 - 0.7	GUM μPeg	$4.2^{+1.7}_{-1.1}$ $1.5^{+1.1}_{-0.7}$
MAX93 Richards	add 105[(3)]	62	0.55^0	1.4^0(s.2b)	0.3	GUM Iota Drac Sigma Her	$3.7^{+1.9}_{-1.1}$ $3.3^{+1.1}_{-1.1}$ $3.1^{+1.7}_{-1.3}$
Tucker	90	27	12'	23.6' circular		(4)	2.3 UL
Saskatoon Wilkinson	26 - 36	2	1.44^0	2.45^0(s.2b)		5^0 NCP	$1.5^{+0.6}_{-0.5}$
UCSB Lubin	25 - 35	2.5	1.5^0	2.1^0(s.2b)	0.7	$2\text{x}14^0$	1.6 det 0.8 UL
MSAM Meyer	168,261 474,675	10%	28'	0.7^0(sq.2b)	0.7	30% sky	1.1-3.1
ULISSE Melchiorri	300,600	150,300	5^0	6^0(s.2b)	0.5	40% sky	2.0 UL $3.0^{+1.0}_{-1.3}$
ARGO deBernardis	150,240, 360,600	10%	52'	1.8^0(s.2b)	0.3	$3\text{x}30^0$	1.4 - 3
Tenerife Davies	10.45,14.9 33		5^0	8^0(s.3b)		$5\text{x}80^0$	1.5-2.5

(1) Michelson Interferometer; (2) GUM=Gamma Ursa Maior; (3)the same channels as previous MAX flights plus the new one at 105GHz; (4) circles of about 15' of radius in the South Pole sky.

Various experimental configurations are sensitive to different regions of k, the high wavenumber cutoff being determined by the finite beam size σ of the instrument (for a Gaussian shaped angular response σ is just one standard deviation), while the low frequency cutoff is determined by the modulation amplitude Θ. The quoted experimental results have a bar length in k corresponding to $1/e$ of the peak and determined by assuming a sine-wave modulation.

It is customary to plot $k(2k+1)C_k$: it measures roughly the power content $(\frac{\Delta T}{T})^2$ entering an instrument of beam-size $\sigma = 1/k$ and withouth modulation. This quantity is constant at low k for a scale invariant primordial spectrum, as required by Inflation. The few theoretical models have been more or less arbitrarily normalized to a value of $\simeq 10^{-10}$ for small k, close to the available quadrupole estimates.

It is clear from Figure 9 and Table III that a large spread exists in the data, some being consistent with the Doppler peak, due to the motion of barionic matter within the horizon at decoupling , other being more appropriately described by a reionized Universe. This fact justifies the opinion that primary anisotropies are still to be separated by secondary backgrounds. As new satellite experiments begin to be planned it appears mandatory to investigate the role of spurious anisotropies in order to derive the best operating conditions in terms of wavelength, beam size and modulation.

One of the main goals of MITO is just that of exploring the regions where CBA have been claimed by various authors with higher spatial resolution. In order to understand the advantages offered by MITO it is useful to explore for some extent the effects of dust contamination on CBA. (Guarini et al. [2])

Galactic dust emission outside the galactic plane has been studied for the first time by de Bernardis et al.[3] (1984) by means of a double channel far infrared photometer operating at $250 \pm 100\mu$ and $800 \pm 300\mu$ on board of a balloon experiment which covered about 40% of the Northern Hemisphere. From these observations and those of Lubin[4] (1983) the authors derived a spectral index of $n = 2.0 \pm 0.2$ and a Temperature of 20 ± 2 K, values confirmed by IRAS and COBE. de Bernardis et al[3]. (1984) have also measured a large-scale gradient in dust emission and fitted it with a cosec (b) law (where b is the galactic latitude). Once this gradient is removed, significant fluctuations remain which have been interpreted as due to dust clouds.

IRAS satellite has confirmed these founds at the shorter wavelength of 100 microns and has provided the first all-sky map of galactic dust emission. IRAS data too may be fitted by a cosec (b) law of the type

$$S_{IRAS}(100\mu) = A + \frac{B}{\sin|b|} + C \times S_{IRAS}(12\mu) \qquad [2]$$

Where $A = 0.0 \pm 0.8 MJ/sr$; $B = 3.4 \pm 0.9 MJ/sr$: residual contamination by Zodiacal dust is taken into account by estimating it with the 12μ IRAS channel: this explains the presence of C term in the fit. $C = 0.16 \pm 0.04\ MJ/sr$. Removing C from the fit decreases slightly the correlation coefficient from 82.8% down to 82.4%. The second IRAS channel at 60 microns provides data more affected by zodiacal dust: they can be again fitted by a secant law with $A' = -1.3 \pm 0.2 MJ/sr$; $B' = 0.62 \pm 0.02 MJ/sr$ and $C' = 0.58 \pm 0.01$ but in this case removing C' produces a change in the correlation coefficient from 92.1% down to 50%. All these fits refer to data at $b \geq 10^0$ and $\beta \geq 10^0$ where β is the ecliptic latitude.

The existence of a large scale gradient in dust emission thereby suggests to employ modulations parallel to galactic plane while measuring CBR anisotropies. This goal is not easy to achieve in balloon-borne experiments, because another stronger constraint must be fulfilled, namely the modulation has to remain parallel to the terrestrial horizon, due to the much stronger atmospheric gradient. The two requirements can be satisfied for a restricted time of observation, as in the case of ULISSE 1978 balloon flight [3] .

In this analysis we have assumed that a single dust component is present: a colder dust component ($T \leq 10K$) could well have escaped IRAS measurements at 100 microns and several authors have proposed its existence. A systematic study of high galactic latitude clouds has been carried out by ULISSE balloon program (B. Melchiorri, F. Melchiorri[5] 1993) . We do not have evidence for cold clouds at the given level of sensitivity, i.e. their contribute cannot exceede 30 microkelvin at 2 mm of wavelength.

If we assume that dust emission is due to a single component only, then the signals from IRAS may be easily converted into a spurious thermodinamic anisotropy $(\frac{\Delta T}{T})_\lambda$ by the relation

$$\left(\frac{\Delta T}{T}\right)_\lambda = \frac{\Delta I_{IRAS}(100\mu)}{I_{IRAS}(100\mu)} \times \frac{I_{IRAS}(\lambda)}{I_{CBR}(\lambda)} \left(\frac{xe^x}{e^x - 1}\right)^{-1} \qquad [3]$$

where ΔI_{IRAS} are the fluctuations observed in IRAS 100 micron channel, $x = \frac{h\nu}{kT}$.

In Figure 10 we have plotted the relation [3] for $\Delta I_{IRAS} = 1MJsr^{-1}$: we have limited the plot to wavelengths shorter than 3 mm, because at longer wavelengths synchrotron and free-free emission become dominant. We stress that Figure 10 represents the spectrum of the fluctuations and, at least in principle, it could be different from that of the bulk of dust emission, the last being measured by COBE-FIRAS. Therefore, we have indicated in Figure 10 only the measurements relative to the dust fluctuations: they are very few and essentially due to the work of Berkeley and Rome groups.

Statistical properties of "galactic cirrus" have been studied by various authors: for a recent review see Low and Cutri[6] (1994). The main conclusion is that the power spectrum P(k) has a power law dependence on k of the type $P(k) \simeq k^{-\alpha}$, α being a positive exponent ranging from 2.5 to 3.5. (Gautier et al.[7] 1992).

Two questions remained unsolved, however:

a- does the power spectrum index α depend on the averaged total flux, i.e. on the amount of dust present along the line of sight?

b- how the results change if we consider a real experiment in which the direction of modulation is constrained to be parallel to the galactic plane?

As far as the second point is concerned we expect a decrease in α as we remove the long wave effect of the secant law. As matter of facts, if the modulation is parallel to the galactic latitude b we get a signal

$$\Delta I \simeq I(\lambda, T) \cos b \tan b \Delta\Theta \qquad [4]$$

Since practical observations are carried out for $b \geq 30°$, we expect ΔI ranging from $10^{-2}I$ down to negligeable values, as b approaches $90°$. Therefore, the secant law effect is relevant for regions where dust fluctuations are $\leq 0.1I$.

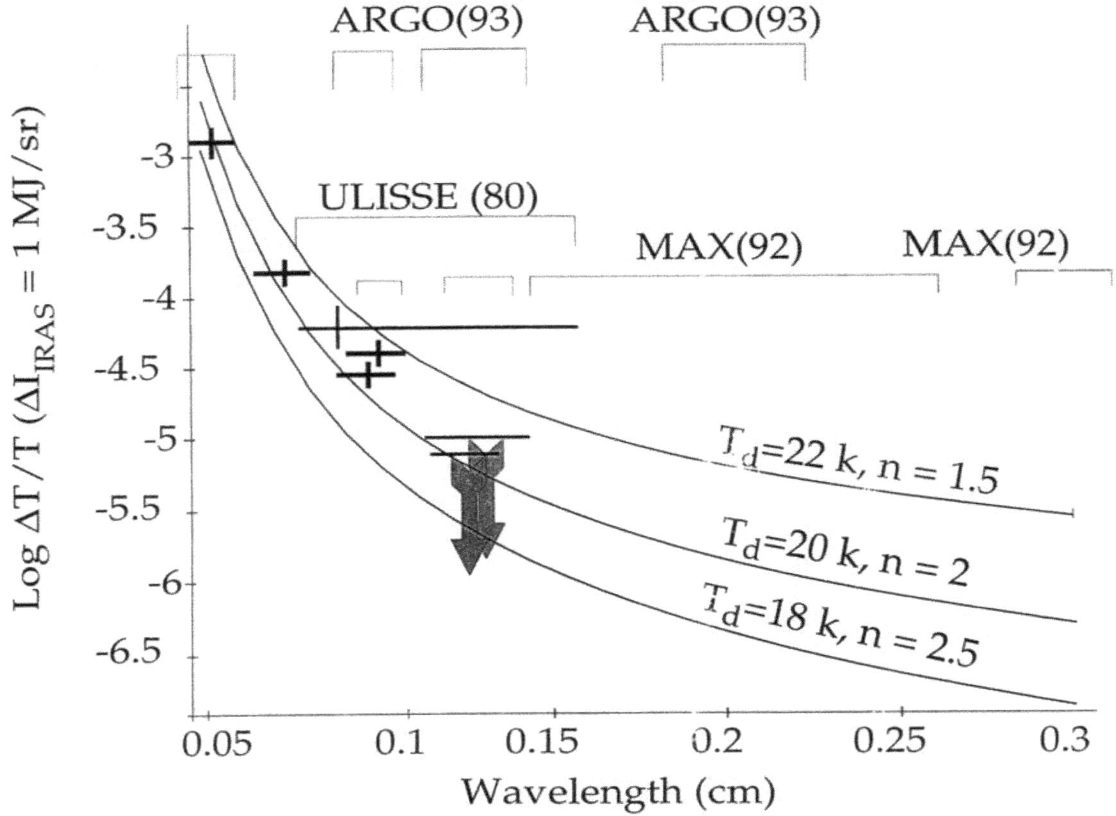

Figure 10: fluctuations in the far infrared produced by the irregular distribution of galactic dust at high galactic latitudes. The few available data are fitted with spectra with different temperature and spectral index.

In order to explore the points (a) and (b) we have arranged our analysis as follows. We have selected three sky regions about 20x20 degrees wide. The first one is toward the North Celestial Pole and is centered around the direction given by Low and Cutri[6] (1994) . The second one is centered on the cleanest zone of the Northern Hemisphere , sometime referred to as "the Lockman Hole" (Lockman et al[8]. 1986). The last region is centered on one of the cleanest Southern regions, selected by BOOMERANG experiment (de Bernardis et al[9]. 1994).

Each region has been scanned on IRAS 100 micron maps by means of a "simulated instrument" having a beamsize ranging from 3 arcminutes up to several degrees and a modulation amplitude ranging from the beamsize up to several degrees. We simulated 10 different scans, all of them parallel to the galactic plane.

The results are presented in Figure 11.

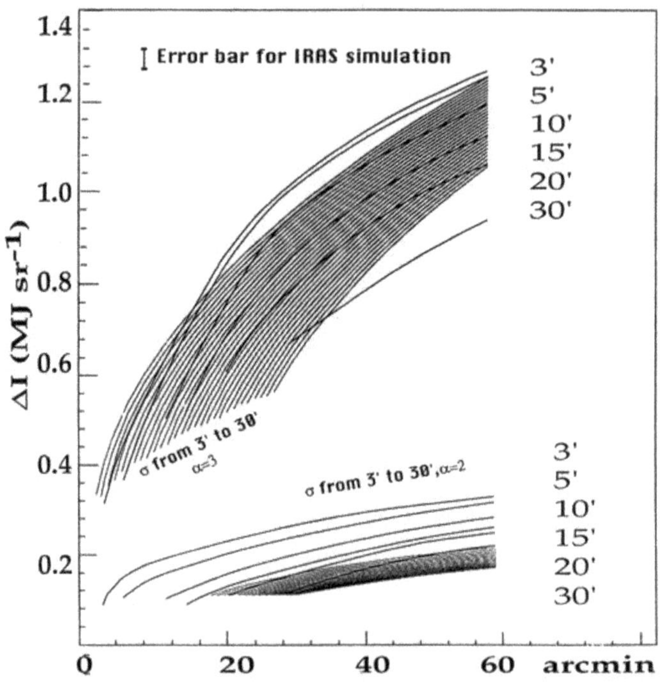

Figure 11: signals produced by a simulated experiment over IRAS sky. The amplitude of modulation is in abscissa, while the various curves refer to different beamsizes, as indicated. The heavy black curves are obtained from IRAS, while the other curves are from the theory, as discussed in the text.

Before to derive any conclusion from these data, let us compare the analysis of Low and Cutri[6] (1994) and our analysis on the same sky region.They derived a power spectrum $P(k) \propto k^{-3}$, i.e. with a spectral index $\alpha \simeq 3$. The index k cannot be immediately identified with the same index in Eq.[1]. In the case of Low and Cutri [6] it refers to a bidimensional (flat) Fourier analysis, while Eq. [1] refers to spherical harmonics. Since we are interested in relatively high values of k the two indexes are not too different, however (just because a region of a few degrees wide on a sphere does not differ substantially from the corresponding figure on a plane). A second correction should be considered, in the sense that we are performing unidimensional analysis on IRAS maps (paths parallel to the galactic plane) and this would correspond to drop a unity in the power index of the bi-dimensional power spectrum of Fourier analysis; since the same would occur in the case of spherical harmonics the effect is irrelevant for a comparison between the two. Taking into account these caveats we can write

$$\left\langle \left(\frac{\Delta T}{T}\right)^2 \right\rangle = \sum_k \frac{(2k+1)C_k}{4\pi} \propto$$

$$\propto \int_{\Delta\Theta_{eff}}^{1/\sigma_{eff}} P(k)k^2 dk \propto \sigma[w-1] \qquad [5]$$

Where $\Delta\Theta_{eff}$ and σ_{eff} are an appropriate modulation amplitude and beam size which take into account the fact that the instrumental transfer function has not been included in the integral. Therefore, we expect that the output of the instrument (proportional to the temperature anistropies) will increase like the square root of σ and $w-1$, within the approximation we have done.

In Figure 11 upper panel we have plotted a serie of curves which corresponds to the above equation and compared them with the IRAS simulation. The agreement is satisfactory, expecially if one takes into account the approximations done. In particular one should note that in this case the output is an increasing function of σ for a given w.

On the other hand the data of Figure 11 lower panel do not agree with the above equation and are better described by a power index close to 2, so that the square of fluctuations is proportional to $Log\frac{\Delta\Theta}{\sigma}$.

Finally in Figure 12 we have quoted the simulation carried out in a region of the Southern Hemisphere where fluctuations are allways $\leq 0.1 I_{min}$, where I_{min} is IRAS minimum flux [1]. In this case the data agree with a power index even less than 2.

---

1 *In the original IRAS data the minimum flux is of the order of* $4-6MJ/sr$. *If one takes into account the results of COBE-DIRBE, this flux has to be corrected for an instrumental offset and becomes 1 - 2* MJ/sr.

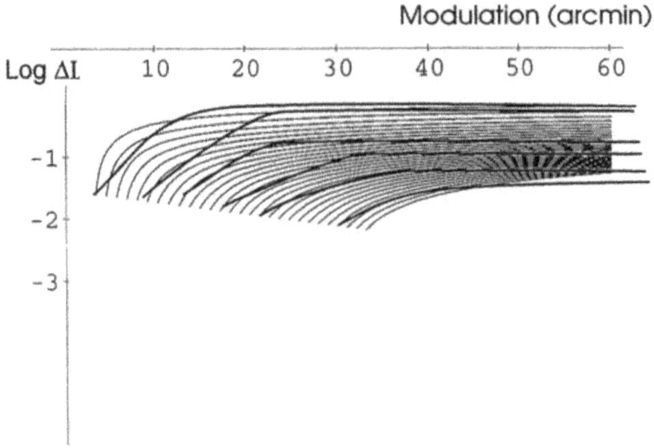

Figure 12: as in the previous figure, but for the cleanest regions explored by ULISSE 78 balloon flight

So far we have limited our analysis to simulations with beamsize and modulation $\leq 60'$. As $\Delta\Theta$ increases beyond 10-20 degrees, the behaviour of ΔI versus $\Delta\Theta$ changes due to the fact that one of the beams start to pick up galactic noise from regions outside the selected clean zone. For instance, in Figure 13 we have extended the analysis of Figure 12 to large values of σ and $\Delta\Theta$. We note that the amplitude of fluctuations depends now on the selected path in the sky. Therefore, in order to take full advantage of the existence of clean regions one has to limit both beam size and modulation amplitude to less than $10-20$ degrees.

The first result of our analysis is the presence of a trend in the power index α: it appears to decrease from 3 down to 2 or less as we explore cleaner regions. The physical reason for this effect is unknown: a Poisson distribution would have $\alpha = 2$, therefore we can even suspect that in the cleanest regions a different (extragalactic?) background is present. This point could be clarified by measuring the spectrum of fluctuations, but IRAS sensitivity is not adequate to this purpose: we think that this could well be a goal for ISO.

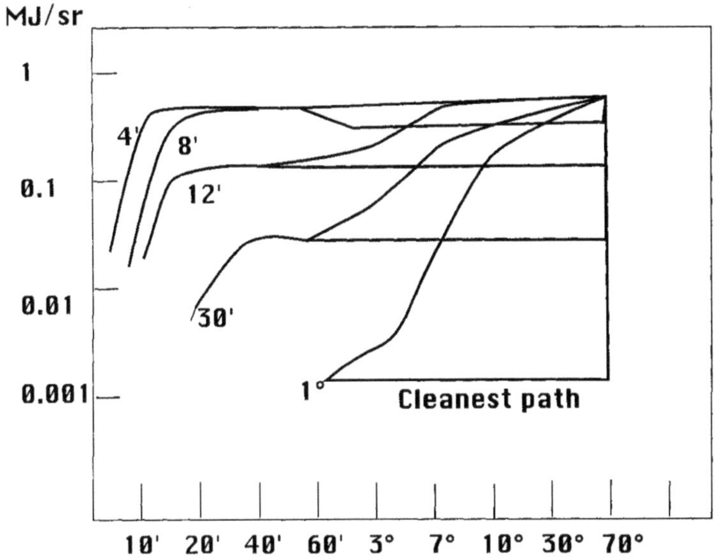

Figure 13: as in figure 12 but for larger amplitudes of modulation. The signals become dependent on the path in the sky and rapidly rise to large values as one of the beams goes out of the clean region.

Results from Figure 11,12, 13 are converted in estimates of spurious $\frac{\Delta T}{T}$ anisotropies at longer wavelengths by means of the spectrum of Figure 10 . We expect rather different situations, depending on the explored sky region. Some of them are in Figure 14a,b. We want to stress that the uncertainties in the dust spectrum of Figure 10 make all our estimates rather qualitative.

First of all, let us restrict our analysis to the few "clean" sky regions : they can be defined as those having IRAS fluctuations smaller than 0.1 MJ/sr at 100 microns for any beamsize and modulation smaller than 60 arcminutes. In these clean regions the power spectrum $\alpha \leq 2$. The relative situations are plotted in Figure 14 a,b: CBR anisotropies are allways bigger than dust anisotropies for wavelengths greater than 1 mm, and stay bigger even for $\lambda = 0.7mm$ in the cleanest regions for $10° \geq \sigma \geq 1°$. This analysis indicates in the case of very low dust background the existence of an '"angular window" around 5 degrees where CBR anisotropies appear to be larger than dust fluctuations even for $\lambda \simeq 700\mu$. We have to stress that this situation holds inside the cleanest regions only. These considerations have important consequences for ULISSE 1978 balloon flight where measurements at $\sigma = 3^0$ and $\Theta = 6^0$ have been carried out just inside the cleanest region in the Northern Hemisphere. Fluctuations detected in this experiment have been attributed to CBR (Melchiorri et al[1]. 1981)

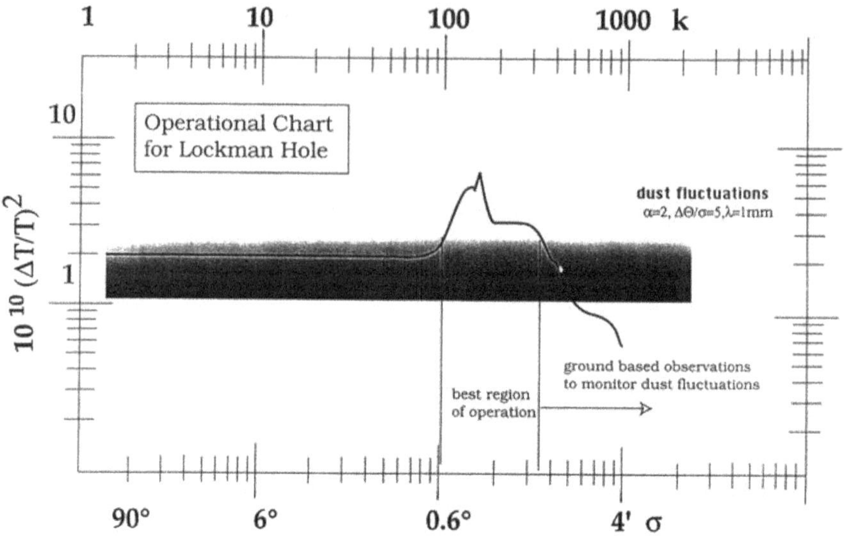

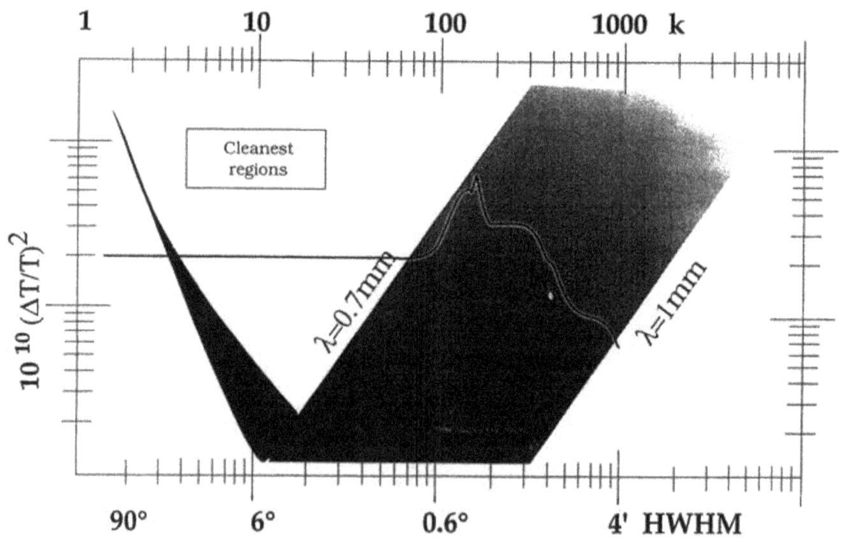

Figure 14: power spectrum of galactic dust spurious anisotropies toward two sky regions where dust contribution is small (Lockman Hole and regions observed during ULISSE flights.

but the possibility of significant dust contamination remained. The main reason of doubts for galactic contamination arose from an MIT experiment carried out at similar angular scale, where huge dust signals have been reported. The present analysis clarifies the situation: in the regions observed by ULISSE 1978 balloon experiment CBR anisotropies are larger than dust fluctuations, if we limit σ and $\Delta\Theta$ to less than 10 degrees. MIT experiment employed a much larger modulation amplitude and correctly reported a much higher dust contamination. Therefore, we can now attribute ULISSE 78 results to CBA and it will be of great interest to compare these observations with those of COBE-DMR at the same angular scale but quite different wavelength.

The second important consequence of our analysis is that the power spectrum of dust is very sensitive to the selected sky region. In the case of a satellite experiment is conceivable that several observations will be carried out in moderately dusty regions, like the North Celestial Pole zone. Dust spurious anisotropies are comparable with CBR anisotropies for $\sigma \simeq 30'$ and $\lambda \leq 1mm$. It follows that the possibility of measuring CBA at small angular scales ($\sigma \leq 0.5^0$) relies on the use of small sky regions and/or wavelengths greater than 1mm. For instance, for $\lambda = 2mm$ $\frac{\Delta T}{T} \leq 10^{-5}$ at $\sigma = 30'$ at North Celestial Pole.

Let us briefly discuss the problem of correcting data for dust contamination. In the cases (a) and (b) of Figure 14 there are two alternatives: first of all, one can operate at different frequencies. Wavelengths shorter than 0.7-1 mm are suitable for observing dust emission. This technique has been already applied in many balloon experiments: its major drawback is due to the possibility that warmer dust could be present at shorter wavelengths, thereby contaminating the final results of data analysis. Figure 14 a,b suggest another possibility, namely that of re-observing the same sky region with much higher angular resolution in order to stay beyond the Doppler peak, where dust anisotropies become again larger than CBA . This offers the advantage of using the same wavelength of operation. For instance, ground based observations around 1 and 2 mm with an angular resolution better than 4 arcminutes seem adequate to explore dust contribution in the cleanest regions.

The main conclusion is that spurious anisotropies due to dust produce different effects depending on the observed sky region. If we limit the observations to the relatively small regions in the Northern and Southern Hemispheres were IRAS fluctuations are ≤ 0.1 MJ/sr, then multifrequency observations togheter with a millimetric map produced by a ground based telescope of about 3 meters in diameter (3 arcminutes resolution) would be able to remove dust contamination at the level of 1-3 μK. This procedure seems particularly suitable for balloon borne experiments, where a limited sky region can be covered. On the other hand satellite experiments are expected to explore much wider sky zones: in such a case correction for dust contamination should be obtained by multifrequency observations.

In the framework of this situation MITO offers a unique possibility: dust emission can be monitored at 3 arcminutes of resolution in the millimetric and submillimetric bands. Therefore, one of the programs of MITO will consist in a systematic survey of the sky regions already explored by Tenerife, ARGO and ULISSE experiments, as well as COBE data, when the combination of the two years map will be available.

The main unsolved question in modern cosmology is that of the origin of the primordial fluctuations responsible for galaxy formation. Two theories appear at present in competition: the first one consider the primordial quantum fluctuations of a linearly coupled scalar field during the inflationary era (Hawking[10] 1982; Starobinsky[11] 1982; Guth and Pi[12] 1982, Bardeen,Steinhardt and Turner[13] 1983).

The second theory explains galaxy formation around cosmic seeds of trapped energy density produced during symmetry breaking phase transitions. (Kibble[14] 1976; Vilenkin[15] 1981; Vilenkin[16] 1985; Turok[17] 1989; Brandenbergen[18], 1992).It should be stressed tha the two pictures of structure formation are conceptually quite different: in the standard, CDM theory the density field of the Universe is taken to be Gaussian random variable; the process in which topological defects act as gravity seeds is distinctly non-Gaussian. It results that the effects of the two types of perturbations on CBR are expected also to be quite diffenret. In particular the main crucial difference between the two types of primordial skies at the decoupling is that of spatial distribution: an almost Gaussian pattern is expected in the case of Inflation, while a strongly non-Gaussian pattern is anticipated for defects like strings and textures. Various authors have analysed microwave background perturbations induced by cosmic strings and compared their spectrum with observations: this has been done numerically by Bennet et al.[19] (1992) and using an analytical model by Perivolaropoulous[20] (1992-1994).

Let us review very briefly the basic properties of cosmic strings before discussing their possible role in the structure formation in the universe and their peculiar observational signatures in CBR.

In a class of gauge theories there arise solutions which are static, have a finite energy and linear structure. The simplest model is the Higgs model. The Lagrangian describing a gauge field $A_\mu(x)$ and a complex scalar field $\Phi(x)$, a Higgs field, is such that

$$L = \frac{1}{2}D_\mu D^\mu \Phi - V(\Phi) - \frac{1}{4}F_{\mu\ni}F^{\mu\ni} \qquad [6]$$

where

$$D_\mu \Phi = \partial_\mu \phi - icA_\mu$$

$$F_{\mu\ni} = \partial_\mu A_\ni - \partial_\ni A_\mu$$

and $V(\Phi) = \lambda(|\Phi|^2 - \eta^2)^2$, is the *sombrero* potential. λ is the self coupling of the Higgs field Φ; η is the value giving the energy scale of the simmetry braking.

The model exhibits a phase transition at a critical temperature T_c (of order η). Above the transition, there are large fluctuations in Φ with zero mean value. Below it, the average value of Φ is close to the minimum of the potential. There exists a "string-like solution" of the field equation. In the case of a string along the z-axis the solution has the form:

$$\Phi = e^{i\Theta} \qquad [7]$$

For this solution the energy density is $\mu \simeq \eta^2$ (mass per unit length) and the width is $w \simeq \lambda^{1/2}\eta^{-1}$. One can show that the theory of the static equilibrium configurations in the Higgs model is completely equivalent to the Ginzburg - Landau theory for superconductivity where strings - topological defects are equivalent to vortex tubes inside a superconducting metal (see for instance Signore and De Luca[2]1, 1991) in the Grand Unified Models, $\eta \simeq 10^{16} GeV$. Hence the width w of the string is quite small and we can think of the string as linear concentration of energy. It is characterized by a single dimensionless number $G\mu \simeq 10^{-6}(\frac{\eta}{10^{16}GeV})^2$. The stress energy tensor for the string can be worked out from the above string solution; it has the form

$$T_{\alpha\beta} = \mu\delta_\alpha(x)\delta_\beta(y) \; diag(1,0,0,-1)$$

for a cosmic string in z-direction.
The metric string produced by such a string is

$$ds^2 = dt^2 - d\rho^2 - dz^2 - (1 - 4G\mu)\rho^2 d\Theta^2 \qquad [8]$$

The space is locally flat but not globally flat: it is cone-shaped. Such a deficit angle produces two main effects: since the conical wedges are identified, the observer will see two images of any object beyond the string, with an angular separation

$$\Delta\Theta \simeq 8\pi G\mu \frac{\ell}{\ell + d} \qquad [9]$$

where ℓ is the distance between the string and the object, while d is the distance between the string and the observer. As a second effect, let us replace "the object" with the last scattering surface. If the string is stationary the effect will be washed out. But if the string is moving at a speed v, a Doppler shift will be generated on the two opposite sides of the string. It results in a line discontinuity in the temperature of CBR given by

$$\frac{\Delta T}{T} = 8\pi G\mu v\gamma; \gamma = (1 - v^2/c^2)^{-1/2} \qquad [10]$$

Perivolaropoulos (1994) has estimated the power spectrum of such an effect for a string network. It is almost flat in the range of wavenumbers 1-200. By normalizing it to COBE DMR results we derive a value of $G\mu \simeq 1.6 \times 10^{-6}$, which corresponds to a value of $\frac{\Delta T}{T} \simeq 4 \times 10^{-5}$ for a single string.

It is therefore possible to observe a linear discontinuity by making a map of CBR around a doubled quasar (if the doubled image is due to the presence of a string). In the map of Figure 15 are indicated the positions of a few doubled quasars having no obvious gravitational lens in front of them and clode to the zenith of MITO observatory. This research program will start in January 1995.

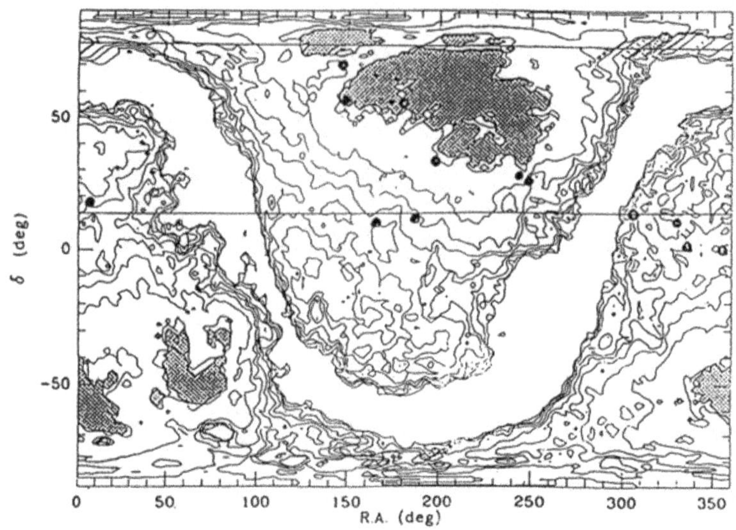

Figure 15: *IRAS map of the infrared sky: dots indicate the position of double lensed quasars: the two parallel lines indicate the region observable from MITO within 10 degrees from the zenith.*

References

1- Melchiorri, F., Melchiorri, B., Ceccarelli, C., Pietranera, L.: Ap. J. , **250**, L1 (1981).

2- Guarini G., Melchiorri B. , Melchiorri F., : in press in Ap.J. (1994).

3- de Bernardis, P., Masi, S., Melchiorri, F., Moreno, G.: Ap. J., **278**, 150, (1984).

4- Lubin, P. M.: in Gamow Cosmology, Varenna School N.86 (1983), F. Melchiorri, R. Ruffini Eds., North Holland 1986, pg 284.

5- Melchiorri, B., Melchiorri , F.: in Current Topics in Astrofundamental Physics, 2nd Course: Eds. N. Sanchez, A. Zichichi, World Scientific (1992), pg. 148.

6- Low, F.J., Cutri, R. M.: Infrared Phys. Technol., **35**, 291, (1994).

7- Gautier III, T.N., Boulanger F., Perault, M., Puget, J.L.: The Astron. J. , **103**, 1313, (1992).

8- Lockman F.J., Jahoda K., Mc Cammon, D; Ap.J. **302**, 432, (1986)

9- de Bernardis et al. in "Present and Future of the Cosmic Microwave Background"; Proceedings of the Workshop held in Santander, Spain, 1993. J.L. Sanz, E. Martinez-Gonzalez, L. Cayon Eds. Lecture Notes in Physics 429, Springer Verlag pg. 188, (1994).

10- Hawking S.: Phys. Lett. **115 B**, 295, (1982).

11- Starobinsky A., Phys. Lett. **117B**, 175, (1982).

12- Guth A., Pi S., Y., Phys. Rev. Lett. **49**, 110, (1982).

13- Bardeen J., Steinhardt P., Turner M., Phys. Rev. **D 28**, 679, (1983).

14- Kibble T.W.B., J. Phys. **A9**, 1387, (1976)

15- Vilenkin A., Phys. Rev., **D23**, 852, (1981).

16- Vilenkin A., Phys. Rep. **121**, 263, (1985).

17- Turok N., Brandenberger R., Phys. Rev. **D33**, 2175, (1986).

18- Brandenberger R., "Topological Defect Models of Strcture Formation", Brown Preprint (1992).

19- Bennet C.L., Preprint ASTROPH-9401012, (1994).

20- Perivolaropoulos L., Center of Astrophysics Preprint N. 3796 , (1994).

21- Signore M., De Luca, A.: in " Infrared and Submillimetric Astronomy from Space", P. Encrenaz, F. Melchiorri Eds., Reiss Romoli (1994).

ECHOGRAPHY OF BABY-GALAXIES

Valerio FERUZZI, Roberto MAOLI
Department of Physics, University La Sapienza
Rome, Italy, Pz.le A. Moro 2

Bianca MELCHIORRI
Istituto di Fisica dell'Atmosfera CNR
Rome, Italy, Pz.le L. Sturzo 31

Francesco MELCHIORRI
Department of Physics, University La Sapienza
Rome, Italy, Pz.le A. Moro 2

Monique SIGNORE
Ecole Normale Superieure, Paris
Paris, France, rue Lhomond 24

Danilo TOSTI
Department of Physics, University La Sapienza
Rome, Italy, Pz.le A. Moro 2

ABSTRACT

The cosmic Middle-Age, extending from Z=1000 down to Z=10, may be explored thanks to resonant scattering of primordial molecules with CBR photons. Moving clouds (correspondent to future galaxies) induce a Dipole Effect at the resonant frequencies. The possibility of detecting such a peculiar signature is discussed

1–Introduction

Up to today it does not exist any theory about cosmic structures formation that is self-explanatory or that fits the available experimental data. The most diffused theory is the "gravitational instability hypothesis", i.e. very small inhomogeneities in the uniform distribution of cosmic matter have grown toward the observed and very large inhomogeneous structures by the infall of the enveloping matter into overdense regions. The available experimental data are: 1) the Cosmic Background Radiation (CBR) Anisotropies, giving information about the distribution of inhomogeneities at $z \simeq 1000$ (i.e. when the universe was about 1000 times smaller and hotter than today), the epoch of the beginning of formation; 2) the statistical distribution of formed structures, the observations range from today up to $z = 4 - 5$. We have no information about the initial phases, when the small perturbation expands less than the universe, neither about the first phases of the real collapse when the infall velocity compensates the cosmic expansion. In this paper we want to illustrate a new method to investigate these phases, related to the role of primordial molecules in perturbation evolution.

2–The phases of perturbation evolution

The problem of galaxy formation in the gravitational instability hypothesis has been studied by many authors (See, for example, the recent book of Peebles, 1993). It is possible to schematize three stages of the evolution of perturbations:

1) linear evolution: the initial perturbation $\frac{\delta\rho}{\rho}$ grows algebraically ($\propto \frac{1}{1+z}$) with time, following the expansion of the universe;

2) turn-around: the perturbation has grown up to about $\frac{\delta\rho}{\rho} \sim 1$. The radius of the fluctuation (generically assumed of spherical shape) has reached its maximum value. The collapse compensates the expansion of the universe and perturbation appears non-moving;

3) non-linear collapse: it is the collapsing phase properly said; density, temperature and pressure increase very quickly up to the beginning of thermal instability and the evolution of the perturbation toward the bound structures we observe today.

It has been shown (Maoli et al., 1994) that objects in phase 1) may be experimentally detected. The physical effect is the resonant scattering of the CBR photons by the diatomic molecules in the proto-objects, the intensity of the effect is

$$\frac{\Delta I}{I_{CBR}} = (3 - \alpha_\nu)\frac{v_{pec,r}}{c}(1 - e^{-\tau_\nu})$$

where α is the spectral index of the photons distribution ($\alpha_\nu = \frac{\nu}{I}\frac{dI}{d\nu}$), τ_ν is the optical depth of the cloud calculated at the observational frequency, $v_{pec,r}$ is the radial component of the peculiar velocity of the cloud and c is the light speed. Here the peculiar velocity has the usual meaning: it is the perturbation to the Hubble flux and it is related to the distribution of inhomogeneities. The effect of resonant scattering keeps its validity in the other two cases and it must be remembered that with the usual concentrations (Puy et al., 1992; Maoli et al., 1994) the LiH molecule is the most effective. In the next sections we are going to analyze more in detail the three situations.

3–The peculiar velocity

To have an idea on the expected order of magnitude we just have an idea about the peculiar velocities of this proto-objects. We can utilize a statistical approach. Let's consider the equation of continuity in the wavenumber space:

$$\dot{\delta}_k + i\frac{\vec{k}}{a} \cdot \vec{v}_k = 0$$

δ_k is the k–th component of the expansion of $\frac{\delta\rho}{\rho}$, $\vec{k}$ is the corresponding wavenumber, $\vec{v}_k$ is its velocity and a is the scale factor of the universe.

Since $\delta_k \propto t^{2/3}$ it is easy to obtain:

$$|\vec{v}_k| = ia(t)H(t)\frac{\delta_k(t)}{k}$$

$$v_{rms}^2 = a^2 H^2(t) \int \frac{k^2 dk}{2\pi^2} \frac{P(k,t)}{k^2} \omega_k^2$$

were ω_k is the window function. For the simple spectrum $P(k) \propto k^n$ it is easy to obtain:

$$v_{rms} = \frac{1}{\sqrt{1+z}} H_0 R_0 \sqrt{\frac{3+n}{n+1}} \left(\frac{R}{R_0}\right)^{-\frac{n+1}{2}}$$

Where R is the size of the cloud and R_0 is the normalization dimension, i.e. for $R = R_0$ we have $\langle\frac{\delta\rho^2}{\rho}\rangle^{1/2} = 1$ today.

This is just to give an order of magnitude of the peculiar velocity. For sake of completeness we have to say we used the cold dark matter (CDM) power spectrum for the real calculation:

$$P(k) = \frac{Ak^{-3+6\alpha}}{(1 + \beta k + \omega k^{3/2} + \gamma k^2)^2}$$

with $\beta = 1.7(\Omega_0 h^2)^{-1}\ Mpc$, $\omega = 9(\Omega_0 h^2)^{-3/2}\ Mpc^{3/2}$, $\gamma = (\Omega_0 h^2)^{-2}\ Mpc^2$, $\alpha = 2/3$ from inflation and $A = [24h^{-1}]^4$ from $COBE$ normalization (Subramaniam and Padmanabhan, 1993) but the integration is very similar to

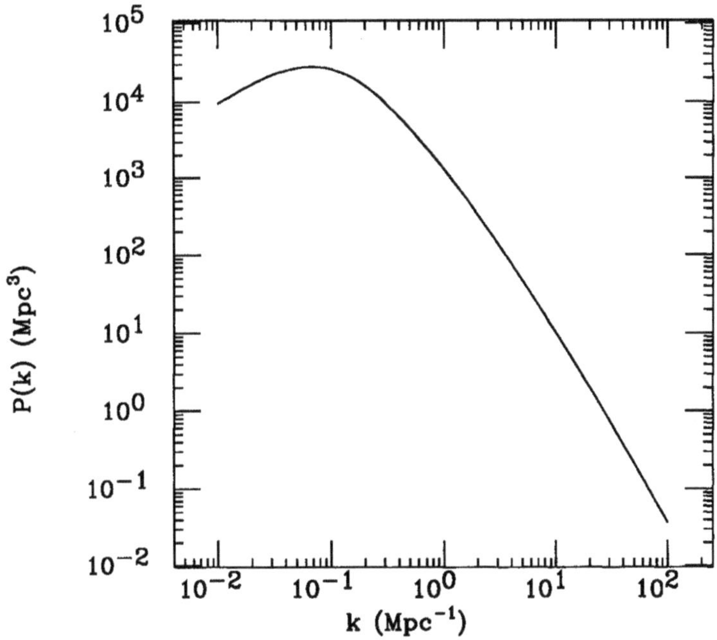

fig. 1a: Cold Dark Matter (CDM) Power Spectrum.

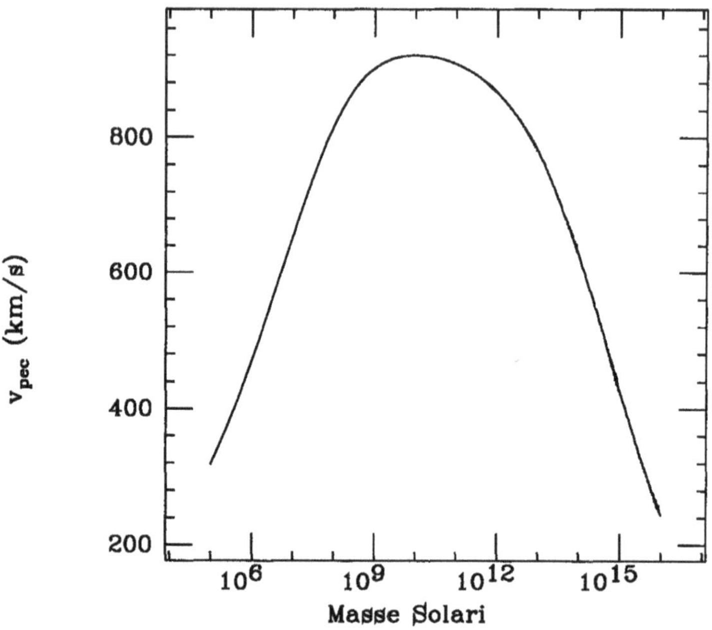

fig. 1b: r.m.s. values of the peculiar velocity, obtained considering the CDM power spectrum plotted in fig. 1a.

the power–law case. In fig 1a and 1b we plotted the CDM spectrum and the corresponding v_{rms}.

4–The experimental consequences of resonant scattering

The numbers in this section refer to the phases as in section 2).

1) the cloud expands with universe and at the same time, due to its (initially small) overdensity it expands less than the universe. The effect is detectable only if it has a peculiar velocity component in the direction of the line of sight. Since the cloud has a finite size D it will exist on a finite range of redshift Δz and the signal will be spread on a corresponding range of frequency $\frac{\Delta\nu}{\nu} = \frac{\Delta z}{z}$.

2) The collapsing velocity compensates the difference between the expansion velocity of the two opposite sides of the cloud. The proto-cloud appears non moving. Due to this all the molecules in every point of the cloud will contribute at the same frequency $\nu = \nu_0/(1 + z_{center})$. If we consider the cloud as formed by many strips orthogonal to the line of sight the total intensity is:

$$\frac{\Delta I}{I_{CBR}} = \sum_{strip} |\vec{v}_{c,strip} + \vec{v}_{pec}|_r \tau_{strip}$$

where $v_{c,strip}$ is the comoving collapsing velocity and τ_{strip} is the optical depth of the considered strip (for sake of simplicity we considered $\tau \ll 1$ but the situation is easily generalizable). Since the collapsing velocity is a function of the cloud radius we can write:

$$\frac{\Delta I}{I_{CBR}} = \sum_{strip} v_{c,strip}\tau_{strip} + v_{pec}\sum_{strip} \tau_{strip} = \tau_{cloud} v_{pec}$$

The line width is the natural one (essentially due to the thermal Doppler effect) and being the temperature of the cloud very low, the line will be very thin.

3) In a more advanced collapsing phase the infall velocity becomes larger than the expansion velocity. The signal is given by

$$\frac{\Delta I}{I_{CBR}} = (\vec{v}_c + \vec{v}_{pec})_r \tau_{strip}$$

Due to the collapsing velocity added to the peculiar one, the signal generally will be larger than in the other two cases and the angular size of these objects will be significantly smaller than in the other phases due to the reduction of cloud dimension in the collapse.

It must be pointed out that the presence of a new kind of velocity (v_c) allows us to have a signal even if the peculiar velocity toward the observer is

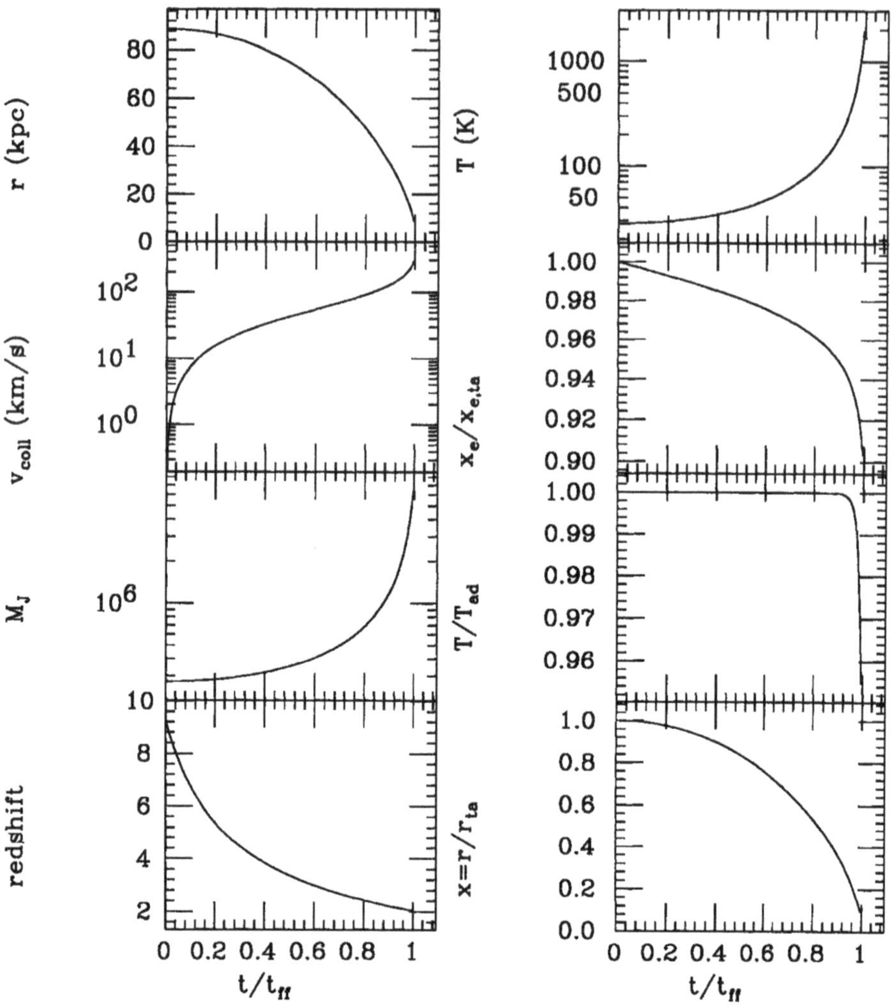

fig. 2: several quantities related to a spherical cloud of mass $M \doteq 10^{12}M_\odot$ during the collapsing phase, calculated following the thermodynamical approach as in Lahav (1986) and in De Araujo and Opher (1988).

zero; the signal will have absorption or emission shape if the collapse velocity of the strip is toward the observer or not.

5–The collapsing phase

To complete the scenario we have to consider how to deal in detail the collapsing phase. We shall summarize our assumptions. We adopted the standard cosmological model with zero cosmological constant, we set $\Omega = 1$, $\Omega_b = 0.1$, $h_{100} = 0.75$; we have used the thermodynamical approach as in Lahav (1986) and de Araujo and Opher (1988). Before all we adopted the CDM power spectrum, maybe this is not the best choice but other assumptions do not modify our results and so it can be considered just as work–hypotheses. The derivation of the evolution laws of of the cloud collapse starts from the first law of thermodynamics

$$dU = -PdV + dE$$

where $U = 3/2Nk_bT$ is the internal energy, $P = Nk_bT/V$ is the pressure, $dQ = -\Lambda V dt$ and $V = NX\mu(r/r_{ta})^3/n_{H,ta}$; Λ is the cooling function (Lepp and Shull, 1983), i.e. the energy loss per unit volume and unit time, we set the initial abundance of molecular hydrogen $h_2 = 1.2 \times 10^{-6}$, following Lepp & Shull (1984) and the initial ionization fraction $x_e = 2.7 \times 10^{-5}(h\Omega_0)^{\frac{-1}{2}}$ as in Lahav (1986), r is the radius of the cloud, ta is the turn-around time, X is the primordial hydrogen mass fraction and μ is the mean molecular weight. To get an equation to describe the evolution of the cloud radius let's utilize the conservation of energy

$$\frac{dE}{dt} = -\Lambda V$$

where the energy E is the sum of three terms: potential energy ($E_p = -3/5GM^2/r$), kinetic energy ($E_K = 3/10M(dr/dt)^2$) and internal energy ($U = 3/2NkT$).

We have considered very large masses of the order of magnitude of galaxies and galaxy clusters masses ($\sim 10^{10} \div 10^{13}\ M_\odot$). An example of the obtained quantities relative to the collapsing phases is in figure 2.

6–Examples and Conclusions

In the figures 3, 4 and 5 we have plotted three examples of the expected signal in the three considered cases.

In the case of collapse, since the two velocities are added vectorially we can have positive or negative values depending on the relative ratio of v_c and v_{pec} and in fig. 6 we plotted the expected signal for different values of v_{pec}.

As it is possible to see in the figures the resonant scattering gives the possibility to detect a cosmic structure in any stade of its life. The molecular character of the effect gives to the expected signal the peculiarity to

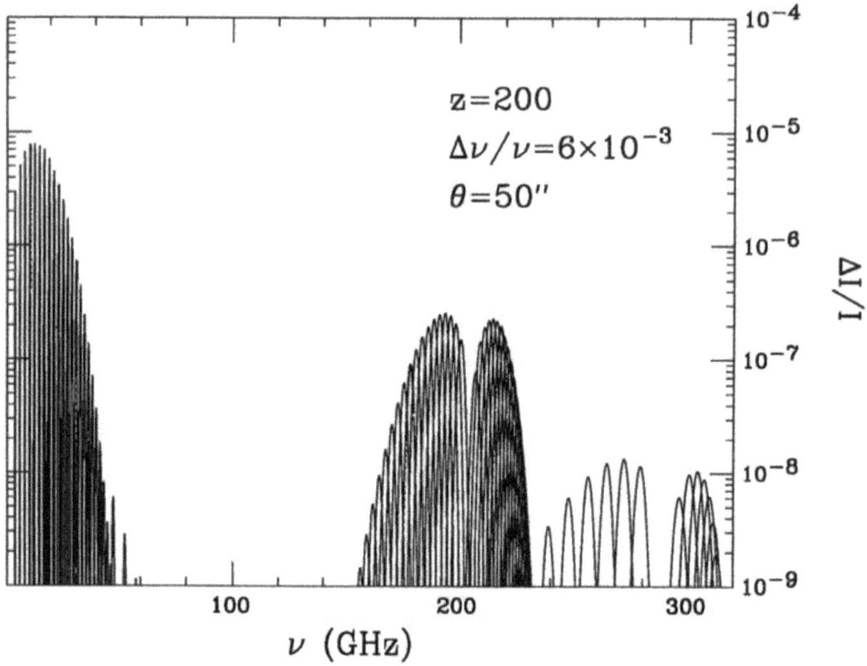

fig. 3: spectrum of a cloud of mass $M = 10^{12} M_{\odot}$ during the linear phase of evolution; the cloud is supposed to be at $z = 200$.

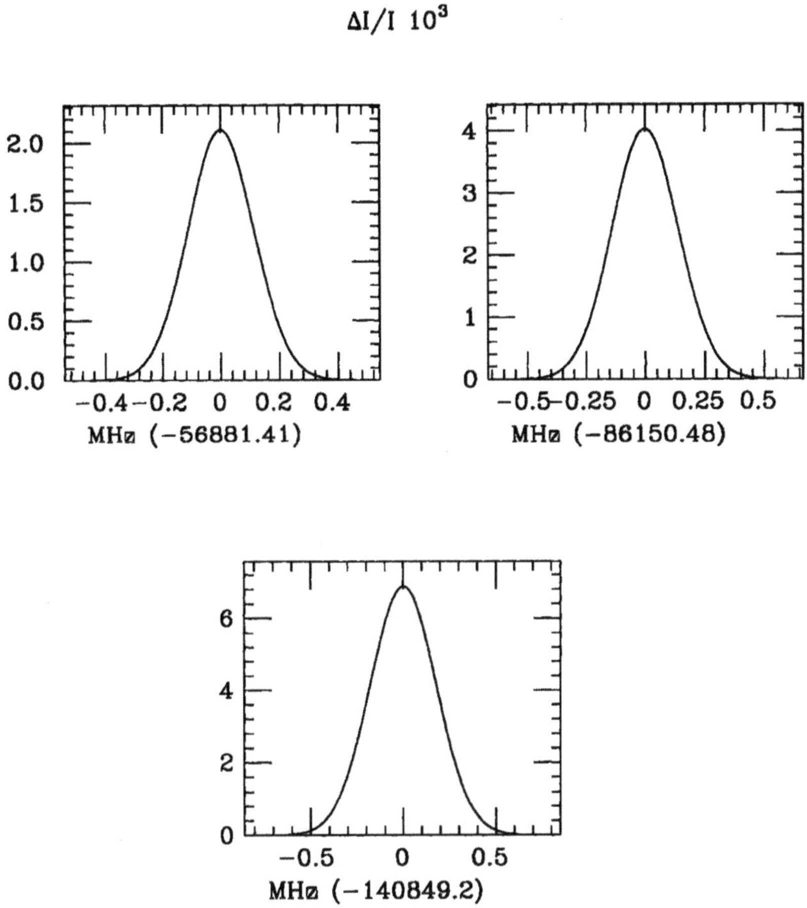

fig. 4: signals from a cloud at turn–around of mass $M = 10^{10}$, 10^{11}, $10^{12} M_\odot$ and $z_{ta} = 14.6, 9.3, 5.3$. The angular dimensions are $\theta = 29''$, $68''$, $168''$. The plotted line is the $j = 1 \rightarrow j = 2$ transition of LiH.

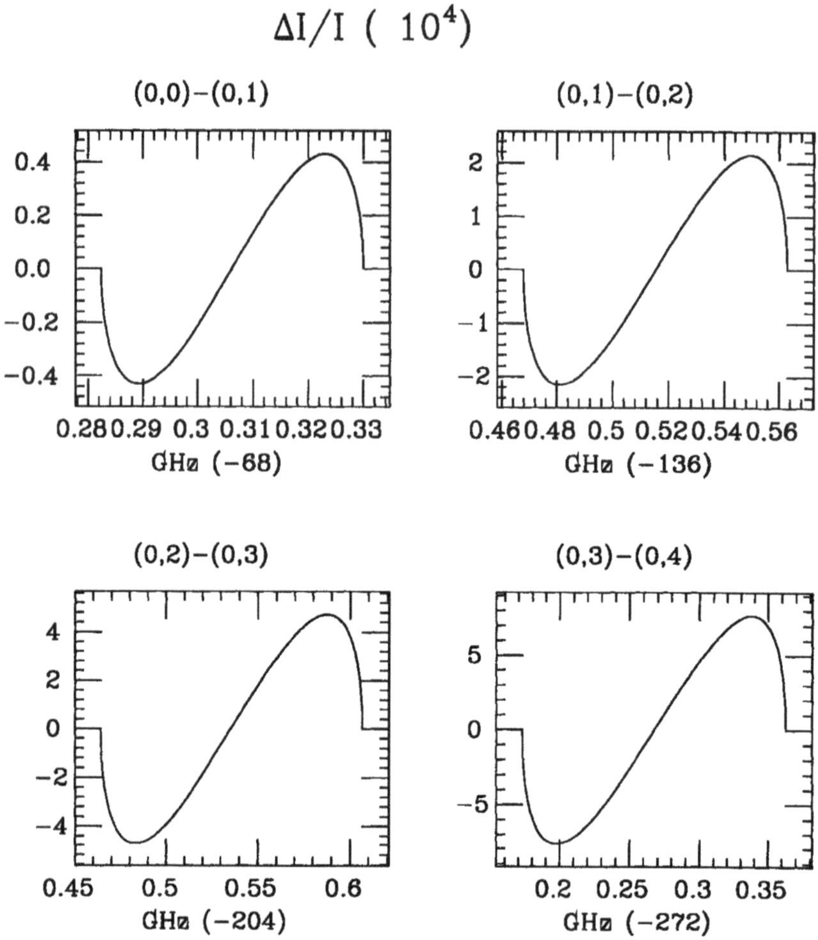

fig. 5: anisotropies due to a cloud of mass $10^{11} M_{\odot}$, $z_{ta} = 20$ when the radius of the cloud is 0.3 times the radius of the cloud at the turn–around. We plotted the first four rotational transitions of LiH. The cloud has an angular dimension of $\theta = 7.6''$.

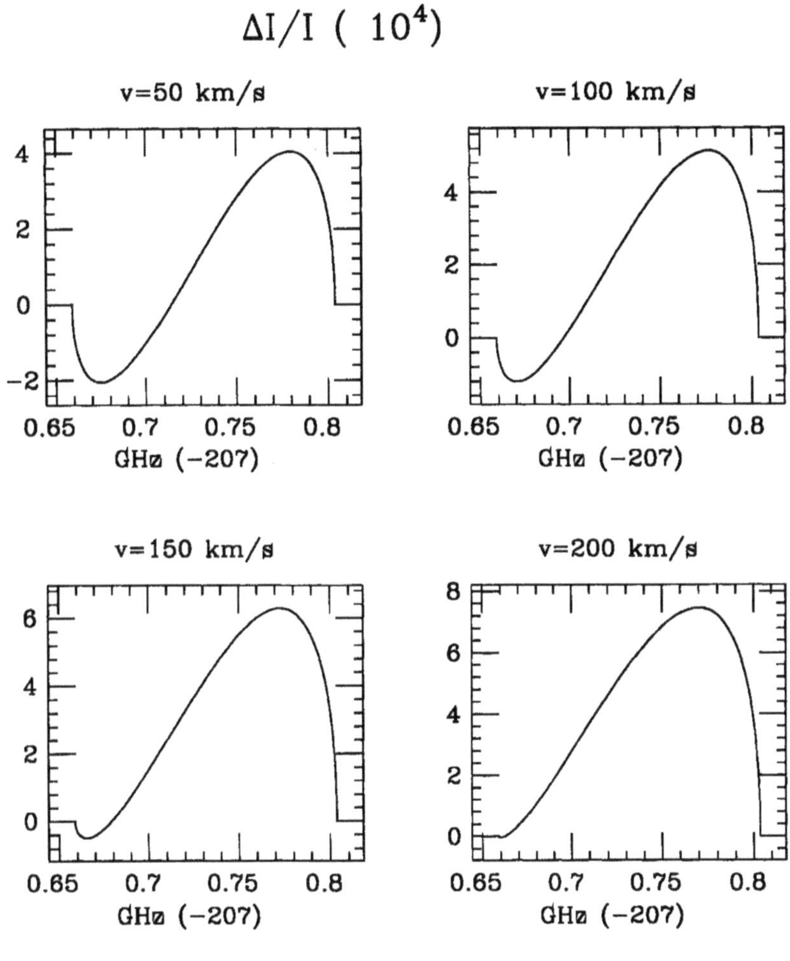

fig. 6: signals from a collapsing cloud with several peculiar velocities. The plotted transition is the $j = 1 \rightarrow j = 2$ of LiH. The parameters of the cloud are: $M = 10^{11} M_{\odot}$, $r = 0.3 r_{ta}$, $z = 5.4$, $z_{ta} = 20$, $v_{coll} = 209 km/s$, $\theta = 7.6''$.

have spectral feature and a detection of two or more lines gives us the possibility to be sure of the detection. Conversely it gives us the possibility to use correlation procedures or interferometric technic to detect the molecular spectrum. Our purposes are now directed toward the experimental search of this effect without considering closed the fascinating aspect of the theoretical study of the formation of cosmic structures.

References

De Araujo J. C. N., Opher R., 1988, MNRAS 231,923
Lahav O., 1986, MNRAS 220, 259
Lepp S. & Shull M. J., 1983, Ap. J., 270, 578
Lepp S. & Shull M. J., 1984, Ap. J., 280, 465
Maoli R., Melchiorri F., Tosti D., 1994, Ap. J.
Peebles, 1993,
Subramanian K., Padmanabhan, T., 1993, MNRAS, 265, 101

PHOTOGRAPHS OF THE CONFERENCE

(Photos: LAZZARI, C. MENDIS, ONGARO)

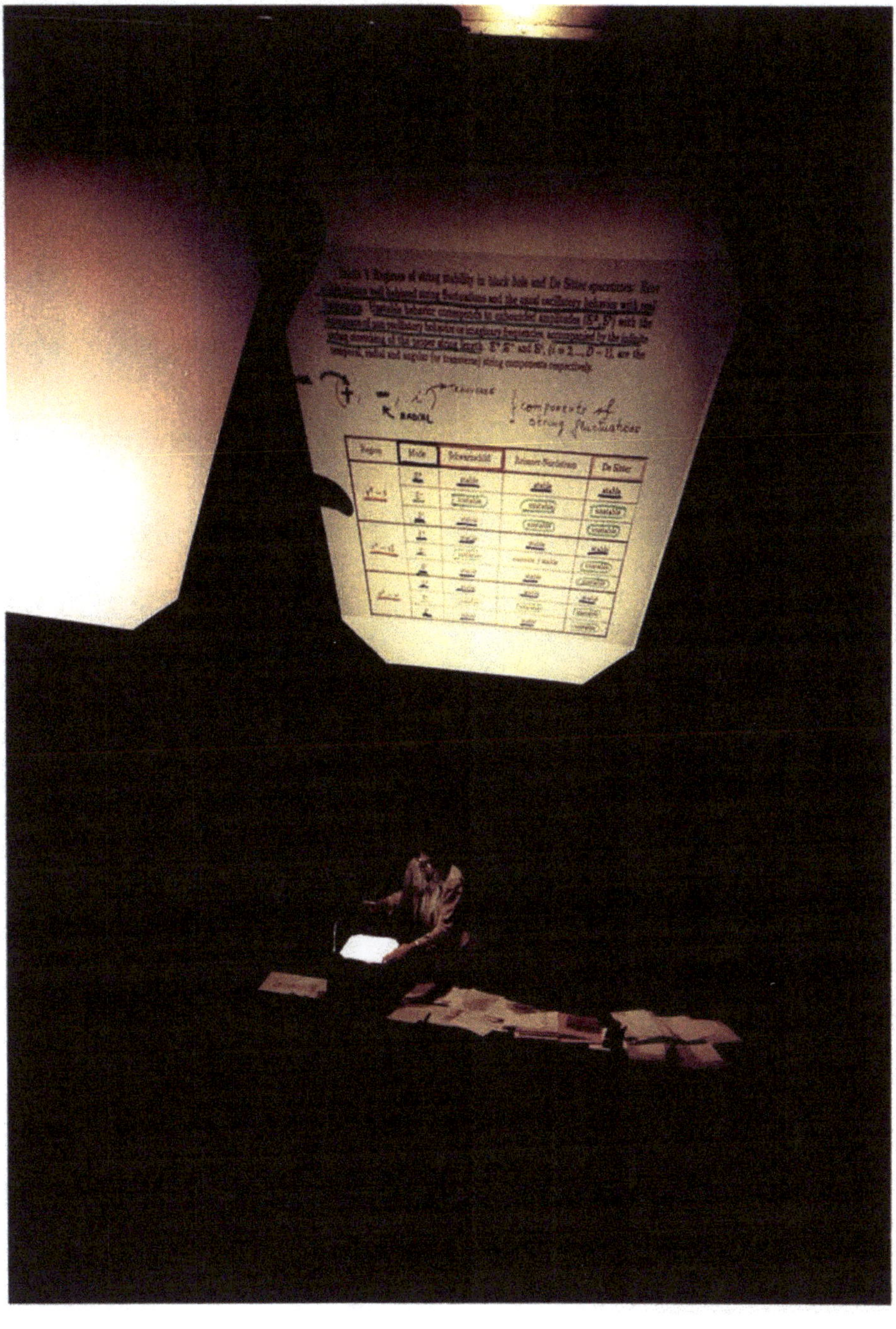

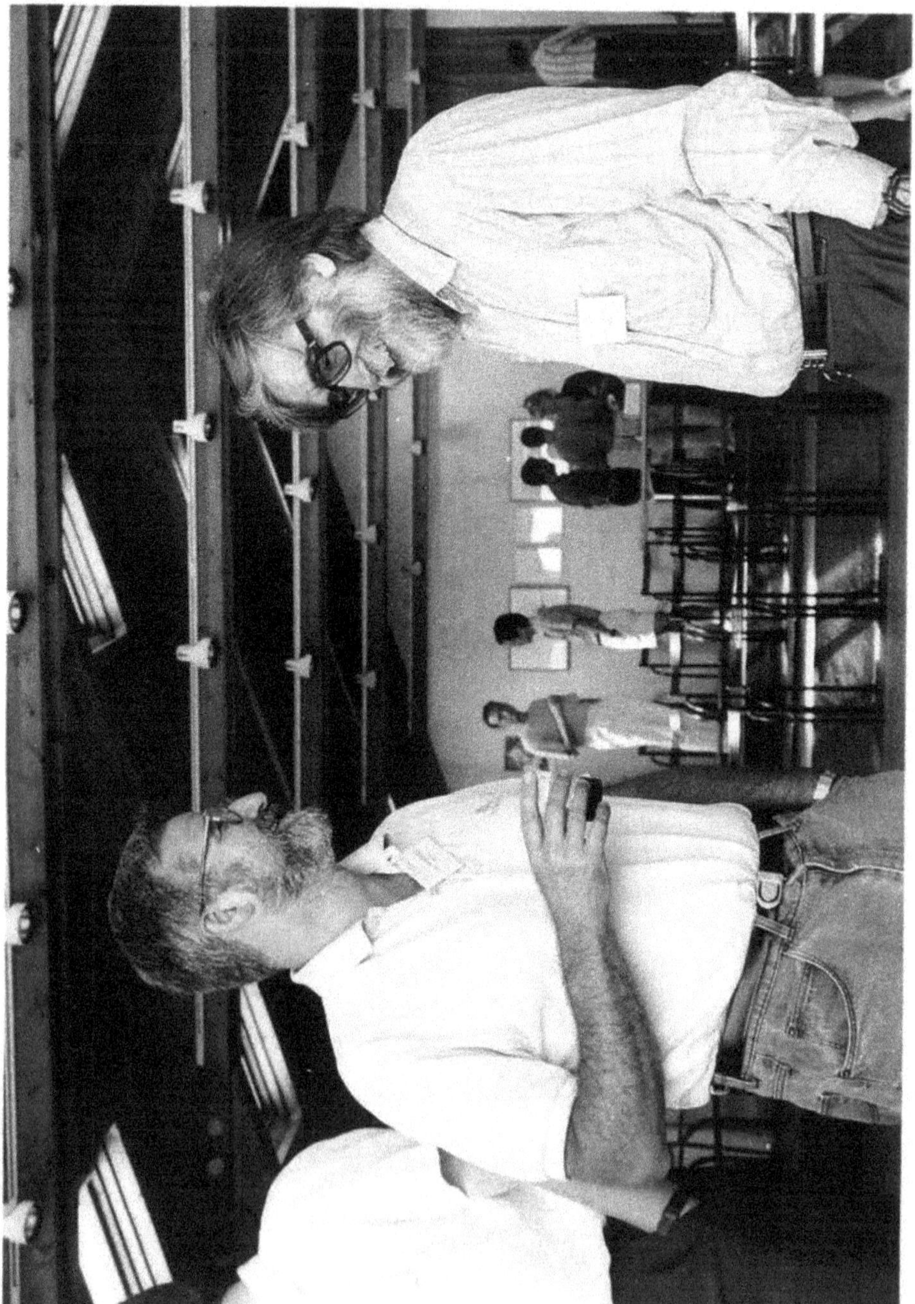

ERICE CITTÀ
DELLA SCIENZA

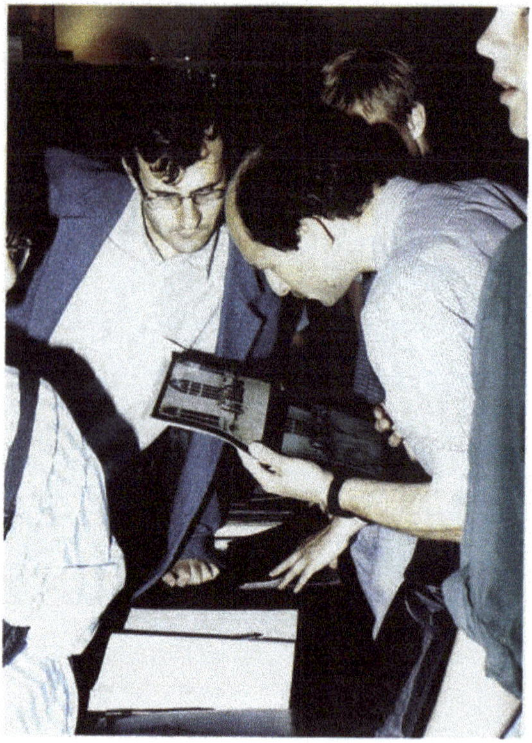

www.ingramcontent.com/pod-product-compliance
Lightning Source LLC
LaVergne TN
LVHW051925120926
847902LV00009B/77